DESIGN IN BRITISH INDUSTRY

DESIGN IN BRITISH INDUSTRY

A MID-CENTURY SURVEY

BY

MICHAEL FARR

WITH A
FOREWORD AND POSTSCRIPT BY
NIKOLAUS PEVSNER

CAMBRIDGE
AT THE UNIVERSITY PRESS
1955

CAMBRIDGE UNIVERSITY PRESS
Cambridge, New York, Melbourne, Madrid, Cape Town,
Singapore, São Paulo, Delhi, Tokyo, Mexico City

Cambridge University Press
The Edinburgh Building, Cambridge CB2 8RU, UK

Published in the United States of America by Cambridge University Press, New York

www.cambridge.org
Information on this title: www.cambridge.org/9780521175951

First published 1955
First paperback edition 2011

A catalogue record for this publication is available from the British Library

ISBN 978-0-521-17595-1 Paperback

TO MY MOTHER AND FATHER

CONTENTS

ILLUSTRATIONS . . . . *page* xi
ACKNOWLEDGEMENTS xxv
FOREWORD BY NIKOLAUS PEVSNER xxvii
INTRODUCTION xxix
DEFINITION OF TERMS xxxvii

PART I: DATA

FURNITURE 3
CARPETS 17
LINOLEUM 26
LIGHT METALWORK 27
SILVER AND ELECTRO-PLATE 33
ALUMINIUM HOLLOW-WARE 42
JEWELLERY 46
DOMESTIC APPLIANCES 51
ELECTRIC LIGHT FITTINGS 62
RADIO AND TELEVISION CABINETS 69
FURNISHING TEXTILES 78
WALLPAPERS 87
POTTERY 92
GLASS 105
LEATHER AND ALLIED GOODS 115
PLASTIC MOULDINGS 122
MOTOR-CARS 129
THE PLACE OF THE CRAFTSMAN IN INDUSTRY 142
  Experiment with New Processes 143
  Experiment with Better Design Standards 146
CONSULTANT DESIGNERS 148

The Retail Trade . . . . page 151

Art Schools 162

Leicester College of Art, *p.* 163
Bradford College of Art, *p.* 165
Birmingham College of Art and Crafts, *p.* 166
London County Council Central School of Arts and Crafts, *p.* 167
Glasgow School of Art, *p.* 169
Edinburgh College of Art, *p.* 170
Manchester Regional College of Art, *p.* 170
Stoke-on-Trent Schools of Art, *p.* 171
Stourbridge Art School, *p.* 172
Kidderminster School of Science and Art, *p.* 173
School for Jewellers and Silversmiths, Birmingham, *p.* 173
Walsall School of Arts and Crafts, *p.* 174
L.C.C. Technical College for the Furnishing Trades, Shoreditch, *p.* 175
High Wycombe School of Art, *p.* 175
Beckenham School of Art and Crafts, *p.* 176

Evaluation 176

The Royal College of Art 180

Comparison: The Bauhaus and The Institute of Design . . . 182

PART II: DESIGN ORGANIZATIONS: PUBLICITY AND PROPAGANDA

The Royal Society of Arts 189

The Faculty of Royal Designers for Industry . . . 191

The Design and Industries Association 192

The British Institute of Industrial Art 198

The Industrial Art Committee of the Federation of British Industries 199

The Committee on Art and Industry 201

The Council for Art and Industry 202

The National Register of Industrial Art Designers . . 205

The Central Institute of Art and Design . . . 207

The Council of Industrial Design 208

Design Centres 222

The Society of Industrial Artists 229

CONTENTS

THE MIDLAND INDUSTRIAL DESIGNERS' ASSOCIATION . *page* 231

SELECTIVE EXHIBITIONS OF INDUSTRIAL ART 232

The Royal Academy Exhibition of British Art in Industry, *p.* 233

The Britain Can Make It Exhibition, *p.* 235

Exhibitions for the Festival of Britain, *p.* 237

PERIODICALS—BOOKS—THE B.B.C. 240

Periodicals, *p.* 240 Books, *p.* 241 The B.B.C., *p.* 247

PART III: CONCLUSIONS

EXPLANATORY NOTE 251

SEQUENCE OF INDUSTRIES ACCORDING TO DESIGN STANDARDS, 1953 251

SEQUENCE OF INDUSTRIES ACCORDING TO DESIGN STANDARDS, 1935 252

DESIGN STANDARDS: Reasons for Differences between Industries, Reasons for Changes since 1935 253

OLD AND NEW INDUSTRIES: The Meaning of Tradition . . . 254

SIZE AND ORGANIZATION OF FACTORIES 256

PRODUCTION METHODS 258

MASS-PRODUCTION AND DESIGN STANDARDS 259

THE CREATOR OF DESIGNS 262

THE MIDDLEMAN 269

THE TASTE OF THE PUBLIC 271

MARKET RESEARCH 273

DESIGN STANDARDS AND PRICE 275

SNOB APPEAL 278

PUBLIC DEPARTMENTS AND PUBLIC TASTE 279

CIVIC DESIGN STANDARDS 281

MARKETS FOR BUYERS AND SELLERS 283

UTILITY SCHEMES—PURCHASE-TAX—EXPORT MARKETS . . 285

ENFORCING DESIGN STANDARDS 287

GOOD DESIGN AND GOOD BUSINESS 288

PART IV: SUGGESTIONS FOR IMPROVEMENTS

University Recognition . . . . . *page* 293
A Museum of Modern Industrial Art 294
Industrial Art Training: A New School for Designers . . 298
The Retailer 303
The Manufacturer 304
The Designer 305
The Aim 306

Addenda 307
Postscript by Nikolaus Pevsner 310
Index 321

ILLUSTRATIONS

BETWEEN PP. 8 AND 9

FURNITURE

I 1. Cautionary dining-room suite.
2. Cautionary bedroom suite.

II 1. Utility furniture, 1943. *Board of Trade.*
2. Sideboard. *Designed by Booth and Ledeboer for Gordon Russell Ltd.*
3. Utility furniture, 1945. *Board of Trade.*

III 1. Dining-room suite. *Designed by R. D. Russell for Gordon Russell Ltd.*
2. Dining-room suite. *Designed by Robin Day for S. Hille & Co. Ltd.*

IV 1. Moulded plywood chair. *Designed by Eric Lyons for E. Kahn & Co. Ltd.*
2. Moulded plywood chair. *Designed by Basil Spence for H. Morris & Co. Ltd.*
3. Moulded plywood chair. *Designed by Carl Jacobs for Kandya Ltd.*

V 1. Armchair. *Designed by Howard Keith for H.K. Furniture Ltd.*
2. Three chairs of beech. *Designed by Geoffrey Dunn and made by Goodearl Brothers Ltd.*

VI 1. Settee. *Designed by Howard Keith for H.K. Furniture Ltd.*
2. Sideboard. *Designed by A. J. Milne for Heal & Son Ltd.*

VII 1. Unit settee. *Designed by Robin Day for S. Hille & Co. Ltd.*
2. Shelves and storage units. *Designed by R. C. Heritage for G. W. Evans Ltd.*

VIII 1. Dressing table and stool. Part of a range of unit furniture. *Designed by the Story Design Group for Story & Co. Ltd.*
2. Desk and chair. *Designed by Basil Spence for the Scottish Furniture Manufacturers' Association for the Chairman of the Furniture Development Council.*
3. Bedroom furniture. *Designed by Paul Bridson for Kandya Ltd.*

BETWEEN PP. 24 AND 25

CARPETS

IX 1. Wilton carpet. *Designed by Ronald Grierson for S. J. Stockwell & Co. (Carpets) Ltd.*
2 and 3. Abstract and naturalistic carpet patterns. *Made by Blackwood Morton & Sons Ltd.*

CARPETS (*cont.*)

X 1. Hand-woven rug. *Designed by Marion Pepler for S. J. Rybczyk.*
2. Carpet showing heraldic motif. *Designed by John Palmer for John Crossley & Sons Ltd.*

XI 1. Axminster carpet. *Made by James Templeton & Co. Ltd.*
2. Two Wilton carpets. *Made by John Crossley & Sons Ltd.*

XII Axminster rug. *Made by Quayle & Tranter Ltd.*

BETWEEN PP. 32 AND 33

LIGHT METALWORK

XIII 1. Window for domestic housing. *Made by Henry Hope & Sons Ltd.*
2. Door closer. *Made by William Newman Ltd.*
3. Bathroom fittings. *Designed by Scott-Ashford Associates Ltd. for W. C. Youngman Ltd.*

XIV 1. Door handle. *Made by William Newman Ltd.*
2. Wood and metal door handle. *Designed by L.C.C. architects Robert H. Matthew and Leslie Martin for the Royal Festival Hall.*
3. Companion stand. *Made by Marris's Ltd.*
4. Dispensing scale. *Designed by John Barnes of Allen-Bowden Ltd. for W. & T. Avery Ltd.*

BETWEEN PP. 38 AND 39

SILVER AND ELECTRO-PLATE

XV 1. Cigarette box with engine-turned decoration. *Designed by S. W. Robinson for J. C. Lowe.*
2. Cigarette case. *Made by K. Weiss Ltd.*

XVI 1. Danish vegetable dish. *Designed by Sigvard Bernadotte for Georg Jensen (Silversmiths) Ltd.*
2. Ogee dish in silver. *Made by Wakely & Wheeler Ltd.*

BETWEEN PP. 42 AND 43

XVII 1. Stainless steel tea-set. *Designed by A. L. Wiggin for J. & J. Wiggin Ltd.*
2. E.P.N.S. stacking tea-set. *Designed by Kenneth Holmes and N. R. G. Poynton for Gladwin Ltd.*

ALUMINIUM HOLLOW-WARE

XVIII 1. Three utensils with a basic handle. *Made by N. C. Joseph Ltd.*
2. Set of saucepans and porringer. *Made by Easipower Ltd.*
3. Milk saucepans. *Made by A. G. Burton Ltd.*

BETWEEN PP. 48 AND 49

JEWELLERY

XIX 1. Gold, sapphire and diamond brooch. *Made by H. G. Mautner.*
2. German matching chain and bracelet set in gilding metal. *Distributed by Walter Graham (Jewellery) Ltd.*

3. Watch-case. *Made by Dennison Watch Case Co. Ltd.*
4. Wrist-watch. *Made by Smiths English Clocks Ltd.*

XX 1. Bracelets. *Designed by Henning Koppel for Georg Jensen (Silversmiths) Ltd.*
2. Glass jewellery. *Designed by F. Lampl for Orplid Glass Ltd.*

BETWEEN PP. 56 AND 57

DOMESTIC APPLIANCES

XXI 1. 'Kabineat' gas cooker, 1935. *Made by Sidney Flavel & Co. Ltd.*
2. 'Renown' gas cooker, 1936. *Made by Parkinson Stove Co. Ltd.*
3. 'New World 84' gas cooker, 1952. *Made by Radiation Ltd.*
4. 'Renown Five' cooker, 1953. *Made by Parkinson Stove Co. Ltd.*

XXII 1. Portable gas fire. *Made by Bratt Colbran Ltd.*
2. Hearth gas fire. *Made by Radiation Ltd.*
3. Electric refrigerator. *Made by Frigidaire Ltd.*
4. Anthracite boiler. *Designed by G. A. Ridgley for Janitor Boilers Ltd.*

XXIII 1. Reflector fire, 1931. *Made by Ferranti Ltd.*
2. Reflector fire, 1939. *Made by H.M.V. Household Appliances Ltd.*
3. Reflector fire, 1953. *Made by Ferranti Ltd.*
4. Electric radiant screen heater. *Designed by John Barnes of Allen-Bowden Ltd. in collaboration with Premier Electric Heaters for Hotpoint Electric Appliance Co. Ltd.*

XXIV 1. Electric kettle. *Made by H.M.V. Household Appliances Ltd.*
2. Electric iron, 1936. *Designed by Christian Barman for H.M.V. Household Appliances Ltd.*
3. Electric iron, 1951. *Designed by D. W. Morphy for Morphy-Richards.*
4. Clock. *Designed by Jack Howe for Gent & Co.*
5. Wall clock. *Designed by Jack Howe for Gent & Co.*
6. Clock. *Designed by W. H. Russell for Thomas Mercer Ltd.*

BETWEEN PP. 64 AND 65

ELECTRIC LIGHT FITTINGS

XXV 1 and 2. Lipless opal glass sphere, 1929. *Made by Merchant Adventurers Ltd.* Sphere fitting surmounted by a metal reflector. *Designed by A. B. Read for Troughton & Young (Lighting) Ltd.*
3. Pendant fitting for diffused light. *Made by Best & Lloyd Ltd.*
4. Pendant light fitting composed of standardized fittings. *Designed by A. B. Read for Troughton & Young (Lighting) Ltd.*
5. Pendant light fitting with pleated buckram shade. *Made by The General Electric Co. Ltd.*

ELECTRIC LIGHT FITTINGS (*cont.*)

XXVI 1. Pendant light fittings with 'Chrysaline' shades. *Designed by Beverley Pick Associates for The General Electric Co. Ltd.*

2. Pendant light fitting. *Designed by Beverley Pick Associates for The General Electric Co. Ltd.*

3 and 4. Two best-selling pendant fittings. *Made by The General Electric Co. Ltd.*

XXVII 1. Convertible light fitting. *Designed by R. C. Hiscock for Hiscock, Appleby & Co. Ltd.*

2. Three-light table lamp. *Designed by R. C. Hiscock for H. C. Hiscock Ltd.*

3. Table lamp with adjustable height. *Designed by Peter Bell for Allom Brothers Ltd.*

4. Counterweighted desk lamp. *Designed by Paul Boissevain for Merchant Adventurers Ltd.*

XXVIII 1–5. Progressive range of fluorescent light fittings. *Made by Thorn Electrical Industries Ltd. (four models designed by Peter Bell).*

6. Louvred fluorescent light fitting. *Designed by A. B. Read for Troughton & Young (Lighting) Ltd.*

BETWEEN PP. 72 AND 73

RADIO AND TELEVISION CABINETS

XXIX 1. Burndept IV receiver, 1923. *From an exhibit in the Science Museum, South Kensington. (Crown copyright.)*

2. Radio, 1934. *Designed by Wells Coates for E. K. Cole Ltd.*

3. Radiogramophone, 1936. *Designed by R. D. Russell for Murphy Radio Ltd.*

XXX 1. Baffle type table receiver, 1945. *Designed by A. F. Thwaites for Murphy Radio Ltd.*

2. Console model with baffle, 1948. *Designed by R. D. Russell for Murphy Radio Ltd.*

XXXI 1. Radio with loudspeaker concealed by plastic shutter. *Made by Pye Ltd.*

2. Table radio showing contrasting textures. *Designed by A. F. Thwaites for Murphy Radio Ltd.*

3. Plastic portable radio. *Designed by Wells Coates for E. K. Cole Ltd.*

4. 'Backless' portable radio. *Designed by A. F. Thwaites for Murphy Radio Ltd.*

XXXII 1. Table television with proscenium surround. *Made by Ferguson Division of Thorn Electrical Industries Ltd.*

2. Table television set. *Designed by Victor Taylor for Ace Radio Ltd.*

3. Console television set. *Designed by R. D. Russell for Murphy Radio Ltd.*

4. Combined television and radio receiver. *Designed by Wells Coates for E. K. Cole Ltd.*

BETWEEN PP. 80 AND 81

FURNISHING TEXTILES

XXXIII 1. Dobby weave. *Designed by Marianne Straub for Warner & Sons Ltd.*

2. Dobby weave. *Designed by Marianne Straub for Warner & Sons Ltd.*

XXXIV 1. Linear pattern in a Jacquard fabric. *Designed by Marion Mahler for Edinburgh Weavers.*

2. Jacquard fabric showing linear pattern. *Designed by Enid Marx for the Utility Furnishing Scheme.*

XXXV 1. Cotton damask. *Designed by Karin Williger for Edinburgh Weavers.*

2. Jacquard woven abstract pattern. *Designed by Marianne Straub for Helios Ltd.*

XXXVI 1. Jacquard woven curtain fabric. *Designed by Marianne Straub for Helios Ltd.*

2. Jacquard weaves with fancy yarns. *Designed by Tibor Reich for Tibor Ltd.*

XXXVII A group of Jacquard weaves. *Designed by Tibor Reich for Tibor Ltd.*

XXXVIII 1. Four examples of moquettes for railway carriage seats. *Designed by the late Ian Colquhoun for the Railway Executive.*

2. Moquettes for dining-car seats. *Designed by the late Ian Colquhoun for the Railway Executive.*

XXXIX Silk-screen-print. *Designed by Anne Loosely and John Drummond for Story (Fabrics) Ltd.*

XL 1. Screen-print on crepe cotton. *Designed by Humphrey Spender for Edinburgh Weavers.*

2. Three-colour screen-print on a cotton satin. *Designed by Sylvia Priestley for Helios Ltd.*

XLI 1. Screen-print on crepe cotton. *Designed by Lucienne Day for Edinburgh Weavers.*

2. Three-colour screen-print on fine cotton. *Designed by John Barker for Edinburgh Weavers.*

FURNISHING TEXTILES (*cont.*)

XLII 1. Early nineteenth-century roller-print, still being produced. *Made by G. P. & J. Baker Ltd.*
2. Modern screen-printed pheasant pattern. *Designed by Sylvia Priestley for Edinburgh Weavers.*

XLIII 1. Screen-print. *Designed by Karin Williger for Edinburgh Weavers.*
2. 'Calyx' screen-print. Judged the best design of the year (1952) by the American Institute of Decorators. *Designed by Lucienne Day for Heal's Wholesale & Export Ltd.*

XLIV 1. Screen-print. *Designed by Graham Sutherland for Cresta Silks.*
2. Roller-print. *Designed by Jacqueline Groag for David Whitehead Ltd.*

BETWEEN PP. 88 AND 89

WALLPAPERS

XLV 1. Block-print. *Designed by Edward Bawden for Cole & Son (Wallpapers) Ltd.*
2. Block-print. *Designed by John Minton for John Line & Sons Ltd.*
3 and 4. Two roller-printed floral wallpapers. *Made by The Wall Paper Manufacturers Ltd.*

XLVI 1. Block-print. *Designed by Graham Sutherland for Cole & Son (Wallpapers) Ltd.*
2. Block-print. *Designed by Armfield-Passano for John Line & Sons Ltd.*
3. Roller-print. *Designed by Roger Nicholson for The Wall Paper Manufacturers Ltd.*

BETWEEN PP. 96 AND 97

POTTERY

XLVII 1. Eighteenth-century earthenware shapes still in use. *Made by Josiah Wedgwood & Sons Ltd.*
2. Earthenware shapes, 1935. *Designed by Susie Cooper for Susie Cooper Pottery Ltd.*
3. Earthenware beer or cider mugs and jug. *Designed by Keith Murray for Josiah Wedgwood & Sons Ltd.*

XLVIII 1. Fine bone china coffee-set. *Designed by J. F. Goodwin, decoration by Victor Skellern, for Josiah Wedgwood & Sons Ltd.*
2. Bone china coffee-set. *Designed by Susie Cooper for Susie Cooper China Ltd.*
3. Bone china teapot, plate, cup and saucer. *Designed by Susie Cooper for Susie Cooper China Ltd.*

XLIX 1. Earthenware tea-service. *Designed by John Adams for Carter, Stabler & Adams Ltd.*

2. Earthenware tea-service. *Designed by Keith Murray for Josiah Wedgwood & Sons Ltd.*

3. Earthenware tea-set. *Designed by William Moorcroft for Moorcroft Ltd.*

L 1. Earthenware jug and beaker using traditional shapes. *Decoration by Richard Guyatt for Josiah Wedgwood & Sons Ltd. for Liberty's.*

2. Jugs and beakers showing well-known 'Garden Implements' design. Late eighteenth-century shape. *Decoration by Eric Ravilious for Josiah Wedgwood & Sons Ltd.*

BETWEEN PP. 104 AND 105

LI 1. Earthenware dinner plate with engraved print and enamel decoration and china dinner plate with engraved print decoration in gold. *Designed by Eric Ravilious for Josiah Wedgwood & Sons Ltd.*

2. Earthenware dinner plate with silk-screen decoration. *Designed by A. Sayer Smith for Johnson, Matthey & Co. Ltd. for Alfred Meakin Ltd.*

LII Bone china coupe shapes. *Made by R. H. & S. L. Plant Ltd.*

LIII 1. Coupe shapes with silk-screen decorations. *Designed by Colin Haxby for Johnson, Matthey & Co. Ltd.* (One design in the range of T. G. Green & Co. Ltd.)

2. Silk-screen transfer. *Designed by Colin Haxby for Johnson, Matthey & Co. Ltd. for Simpsons (Potters) Ltd.*

LIV 1 and 2. Vase, decanter and glass in lead crystal glass. *Designed by W. J. Wilson for James Powell & Sons (Whitefriars) Ltd.* GLASS

3. Decanter and glass in lead crystal glass. *Designed by David Hammond for Thomas Webb & Sons.*

4. Lead crystal glass ice-jug. *Made by The Stourbridge Glass Co. Ltd.*

BETWEEN PP. 112 AND 113

LV 1. Vase in lead crystal glass. *Designed by Clyne Farquharson for John Walsh Walsh Ltd.*

2. Bowl in lead crystal glass. *Designed by Irene Stevens for Webb Corbett Ltd.*

3. Bowl, jug and tumbler in lead crystal glass. *Designed by Irene Stevens for Webb Corbett Ltd.*

GLASS (*cont.*)

LVI 1. Pressed glass. Machine-made imitation of hand-cut glass popular since the late nineteenth century.

2. Fruit bowls in machine-pressed glass. *Designed by R. Y. Goodden for Chance Brothers Ltd.*

3 and 4. Machine-pressed bowl and glasses, mass-produced and inexpensive. *Designed by A. H. Williamson for The United Glass Bottle Manufacturers Ltd.*

LVII 1. Heat-resisting glass casserole, 1953. *Designed by Milner Gray and Kenneth Lamble of Design Research Unit for James A. Jobling & Co. Ltd.*

2. German 'Pyrex' dishes, 1935. *Designed by Wilhelm Wagenfeld for Jenaer Glaswerk Schott & Gen.*

3. Heat-resisting glass dishes, 1953. *Designed by Milner Gray and Kenneth Lamble of Design Research Unit for James A. Jobling & Co. Ltd.*

4. Tea-set in 'Pyrex' glass, 1932. *Designed by Wilhelm Wagenfeld for Jenaer Glaswerk Schott & Gen.*

LVIII 1 and 2. Two fireproof coffee percolators. *The first designed (1934) by Wilhelm Wagenfeld for Jenaer Glaswerk Schott & Gen; the second designed (1949) by Abram Games for Cona Ltd.*

3. Hand-blown vases. *Designed by Wilhelm Wagenfeld for Württembergische Metallwarenfabrik.*

4. Pressed glass vase. *Designed by Wilhelm Wagenfeld for Vereinigte Lausitzer Glaswerk.*

BETWEEN PP. 120 AND 121

LEATHER AND ALLIED GOODS

LIX 1. Valise with zip-fastener. *Made by Papworth Industries.*

2. Suitcase with moulded plastic shell. *Designed by John W. Waterer for S. Clarke & Co. Ltd.*

3. Matching suitcases with special locks and carrying handles. *Designed by Scott-Ashford Associates Ltd., for Papworth Industries.*

LX 1. Matching suitcases with soft sides to save weight. *Made by Parker, Wakeling & Co. Ltd.*

2. Fitted suitcase. *Designed by John W. Waterer (lining by Enid Marx) for S. Clarke & Co. Ltd.*

LXI 1. Suitcase handle in light metal alloy. *Designed by John W. Waterer for S. Clarke & Co. Ltd.*

2. Travelling beauty case with special plastic containers. *Designed by T. C. Maylor for W. A. Maylor Ltd.*

3. Typewriter case with zip-fastener. *Designed by John W. Waterer for S. Clarke & Co. Ltd.*

LXII 1. Pigskin handbag. *Made by Bagcraft Ltd.*

2. Pigskin handbag with zip-fastener. *Designed by F. M. Jennings for J. Jones & Sons (Birmingham) Ltd.*

3. Pigskin wallet. *Made by Mark Cross Ltd.*

4. Pigskin purse, typical of Walsall work. *Made by George Sheldon (Walsall) Ltd.*

BETWEEN PP. 128 AND 129

PLASTIC MOULDINGS

LXIII 1. Two condiment sets of 1932 and 1945. *Designed by A. H. Woodfull for The Streetly Manufacturing Co. Ltd.*

2. Condiment set, 1952. *Designed by W. Bruce Brown for Halex Ltd.*

3. Compression-moulded plastic colander, and other articles. *Designed by Gaby Schreiber for Runcolite Ltd.*

LXIV 1. Tableware in melamine. *Made by The Streetly Manufacturing Co. Ltd.*

2. Three-course meal tray. *Designed by Gaby Schreiber for Runcolite Ltd.*

3. Handleless milk jug. *Designed by R. E. Brookes for Brookes & Adams Ltd.*

LXV 1 and 2. Modernistic and modern door furniture. *Made by Evered & Co. Ltd.*

3. Door furniture. *Designed by Rodney Hooper for Lacrinoid Products Ltd.*

4. Case for electric shaver in leather and plastic. *Designed by A. H. Woodfull for The Streetly Manufacturing Co. Ltd.*

MOTOR-CARS

LXVI 1. Studebaker, 1946. *Designed by Raymond Loewy for the Studebaker Corporation. (Photograph: Alexandre Georges. From the Museum of Modern Art, New York,* 1951 *exhibition, 8 Automobiles.)*

2. Cisitalia, 1946. *Designed by Pinin Farina for Automobile Cisitalia. (Photograph: Alexandre Georges. From the Museum of Modern Art, New York,* 1951 *exhibition, 8 Automobiles.)*

BETWEEN PP. 136 AND 137

LXVII 1. Aston Martin DB2–4 sports saloon. *Designed by Frank Feeley for Aston Martin Ltd.*

2. Allard sports coupé. *Made by Allard Motor Co. Ltd.*

MOTOR-CARS (*cont.*)

LXVIII 1. Austin A.40 sports coupé. *Made by Austin Motor Co. Ltd.*
2. Bentley Mark VI saloon. *Made by Bentley Motors (1931) Ltd.; coachwork by H. J. Mulliner & Co. Ltd.*

LXIX 1. Ford 'Consul' saloon. *Made by Ford Motor Co. Ltd.*
2. Humber 'Super Snipe' saloon. *Made by Rootes Ltd.*

LXX 1. Austin-Healey 'Hundred' sports car. *Made by Austin Motor Co. Ltd.*
2. Rover 75 saloon. *Made by Rover Co. Ltd.*

BETWEEN PP. 140 AND 141

LXXI Jowett R4 'Jupiter' sports car with laminated plastic bodywork. *Made by Jowett Cars Ltd.*

LXXII 1. 'Car of the Future', at the Festival of Britain, 1951. *Designed by E. G. M. Wilkes, Hugo G. Poole and Peter J. Ashmore for the Council of Industrial Design.*
2. 'Tropfenauto': the first streamlined car, 1921. *Designed by Dr H. Rumpler.*

BETWEEN PP. 144 AND 145

THE PLACE OF THE CRAFTSMAN IN INDUSTRY

LXXIII 1. Experimental weaves with fancy yarns. *Designed by Margaret Leischner for R. Greg & Co. Ltd.*
2. Cast aluminium chairs and table. *Designed by Ernest Race for Ernest Race Ltd.*
3. Sample fabric to exploit fancy yarns. *Designed by Margaret Leischner for R. Greg & Co. Ltd.*
4. Bent plywood and steel rod furniture: the joint first prize winner in the International Competition for Low Cost Furniture Design, held by the Museum of Modern Art, New York, 1948. *Designed by Robin Day and Clive Latimer.*

LXXIV 1 and 2. The original tubular steel chairs. *Designed at the Bauhaus by Marcel Breuer (1925), and Ludwig Miës van der Rohe (1927).*
3. Experimental moulded plastic chair. *Designed by Dennis Young.*
4. Experimental chair to exploit Latex Foam. *Designed by Dennis Young for the British Rubber Development Board.*

LXXV 1. Laminated wood chair. *Designed by Peter Hvidt and O. Mølgaard Nielsen for Fritz Hansens Eftfl, Denmark.* (*Photograph by courtesy of Heal & Son Ltd.*)

2. Chair using light alloy die-castings. *Designed by J. W. Leonard for Educational Supply Association Ltd.*

3. Chair of plastic reinforced with glass fibres. *Designed by Charles Eames for Herman Miller Furniture Co. U.S.A.*

4. Chair of steel rods with plastic supports. *Designed by Ernest Race for Ernest Race Ltd.*

LXXVI 1. Silk-screen ceramic transfers. *Designed by Colin Haxby and Jacqueline Groag for Johnson, Matthey & Co. Ltd.*

2. Lead crystal glass with sandblast decoration. *Designed by Irene Stevens for Webb Corbett Ltd.*

3. Deck chair of laminated wood; first used by the Orient Line. *Designed by Ernest Race for Ernest Race Ltd.*

BETWEEN PP. 162 AND 163

LXXVII 1. Ceramic lithograph decoration. *Designed by Eric Elliott at Burslem School of Art.* ART SCHOOLS

2 and 3. A cruet set for B.O.A.C. and flatware for Mitchells and Butlers Ltd. *Designed by Kenneth Holmes and N. R. G. Poynton for Gladwin Ltd. at Leicester College of Art.*

LXXVIII 1. Cut glass decanter. *Stourbridge Art School.*

2. Silver tea-service. *Designed by Alexander Wilkie at Glasgow School of Art.*

3. Stainless steel tea-service. *Designed by Kenneth Holmes and N. R. G. Poynton (assisted by students) for Andrew Brothers (Bristol) Ltd. at Leicester College of Art.*

BETWEEN PP. 166 AND 167

LXXIX 1. Prototype door handle. *Birmingham College of Art and Crafts.*

2. Prototype table lamp. *Birmingham College of Art and Crafts.*

3. Desk. *Designed by J. Padiak at L.C.C. Central School of Arts and Crafts.*

LXXX Hospital lighting fitting. Designed and made under 'factory' conditions. *Designed by D. M. Pearson, W. Schejbaal, P. J. Farrell, J. W. Cooper and C. R. Cheetham at School of Industrial Design, L.C.C. Central School of Arts and Crafts.*

BETWEEN PP. 170 AND 171

ART SCHOOLS (*cont.*)

LXXXI 1. Wallpaper pattern which won Royal Society of Arts Bursary Competition, 1953. *Designed by Margaret M. Stewart at Glasgow School of Art.*

2. Woven fabrics. *Designed by Margaret Broadbent and Jean Falding at Bradford College of Art.*

3. Moquettes and a tapestry winning Royal Society of Arts Bursary Competition, 1953. *Designed by Constance Mary Holt at Leeds College of Art.*

LXXXII 1. Occasional chair. *Designed by L. Dandy at High Wycombe School of Art.*

2. Dressing table using laminated wood for construction and decoration. *Designed by D. Wellstead at Beckenham School of Art and Crafts.*

3. Occasional chair. *Designed by students at L.C.C. Technical College for the Furnishing Trades, Shoreditch.*

4. Laminated table with decorated surface. *Designed by students at Beckenham School of Art and Crafts.*

BETWEEN PP. 174 AND 175

THE ROYAL COLLEGE OF ART

LXXXIII 1. Dobby weaves. *Designed by Ann Jennifer Bell and W. A. Brook at School of Textile Design, Royal College of Art.*

2. Jacquard weaves. *Designed by Ann Jennifer Bell, Margaret Foster, F. Hoswell, Sheila Hankinson at School of Textile Design, Royal College of Art.*

3. Layout for sample Dobby weave. *Designed by Edna Penniall at School of Textile Design, Royal College of Art.*

4. Stacking chairs. *Designed by V. A. Hindley at School of Wood, Metals and Plastics, Royal College of Art.*

LXXXIV 1. Teapot. *Designed by James Rushton at School of Ceramics, Royal College of Art.*

2. Glass dishes. *Designed by G. P. Baxter at School of Industrial Glass, Royal College of Art.*

3. Silver coffee-set. *Designed by David Mellor at Sheffield College of Art.*

4. Silver coffee-set. *Designed by David Mellor at School of Silversmithing and Jewellery, Royal College of Art.*

5 and 6. Flatware prototypes in silver; production models in stainless steel and E.P.N.S. *Designed by David Mellor at School of Silversmithing and Jewellery, Royal College of Art.*

BETWEEN PP. 192 AND 193

LXXXV The Design and Industries Association: exhibition rooms (modern; and best-selling designs, according to the trade). THE DESIGN AND INDUSTRIES ASSOCIATION

LXXXVI 1. The Britain Can Make It exhibition: the 'Sport and Leisure' section.

2. The Britain Can Make It exhibition: kitchen and dining-room for a modern house.

BETWEEN PP. 208 AND 209

LXXXVII Exhibits in the Homes and Gardens pavilion, Festival of Britain. THE COUNCIL OF INDUSTRIAL DESIGN

LXXXVIII 1. The Council of Industrial Design: stand at the Canadian National Exhibition, Toronto.

2. The Council of Industrial Design: portable showcase to explain pottery design.

ACKNOWLEDGEMENTS

It is impossible for me adequately to convey my sincere gratitude to Professor Pevsner. From the first he has taken great active interest in my work. He has advised, corrected and suggested at all stages. It is to the pattern of his book, *An Enquiry into Industrial Art in England*, that any clarity which may be found in these pages must be attributed. It is to him alone that the chief of any merits in my text belong.

This book is the result of the kind and interested help of many people. But above all I must thank my wife for her unceasing encouragement and for her willing transcription of the text.

I was prompted to begin the work by Ian Colquhoun, a very close friend. It is with great sorrow that I record my gratitude to this designer and writer whose tragic death in 1954 cut short a most promising career.

I should like to thank Mr W. L. Cuttle and Dr F. R. Leavis, both of Downing College, Cambridge, for helpful advice in the early stages. Most welcome but unlooked for patronage came towards the end of my research, and it is with sincere gratitude that I take this opportunity of recording the generosity of Mr H. de C. Hastings and Mr M. A. Regan. My grateful thanks are due to the staff of the University Press, Cambridge, who have taken the book untiringly from manuscript to printed page. I should also like to thank Mr T. Culverson for his help and Miss A. Bartlett who worked under great pressure to produce the index.

I am unable to thank all those manufacturers, designers, retailers and art school principals who devoted time and energy to my inquiries. Many of them would, in any case, wish to remain anonymous. But I am glad of this chance to thank Mr Leslie Hardern and Mrs Harvey of the Design and Industries Association, Mr Gordon Russell, Mr Paul Reilly and Mr Hartland Thomas of the Council of Industrial Design, the late Mr Arnold Selwyn of the Design and Research Centre for Gold, Silver and Jewellery, Mr D. H. Tomlinson of the Cotton Board Colour, Design and Style Centre, Mr Kenneth Luckhurst of the Royal Society of Arts, Sir Charles Tennyson, Mr R. D. Best, Mr John Radcliffe, Mr John Gloag, Sir Ambrose Heal, Miss Enid Marx, The Hon. Margaret Lambert, Mr Victor Skellern, Mr and Mrs Eden Minns, Mr John W. Waterer, The Hon. Lady MacGregor of MacGregor, Mr Hyla G. Elkington, Mr E. M. O'R. Dickey and Mr Walter Keesey.

I am greatly indebted to those directors of firms who kindly allowed me to illustrate their products, and to Mr A. B. R. Fairclough, photographic librarian of the Council of Industrial Design, for making available many of the photographs. M. F.

AUGUST 1954

NOTE

The text was completed in the spring of 1952, but wherever possible I have attempted to keep it up to date. Amendments and additions which could not be included appear as addenda on p. 307 to which readers are referred by an asterisk (*) in the text.

M. F.

August 1954

FOREWORD

BY NIKOLAUS PEVSNER

A few words must be said on the origin of this book. In 1934 Professor Philip Sargant Florence of Birmingham University asked me to undertake an investigation into the conditions of industrial design in this country and chiefly in the Birmingham area. I carried out the work in 1934–5, and it was published by the Cambridge University Press in 1937. In 1948 the Press approached me for a second edition. I had to answer that it would be useless to attempt what could strictly speaking be called a second edition. Too much had happened in the dozen or so years. If the Press wished to see the book reappear up to date, it would have to be rewritten. My own work had shifted into different fields in the meantime and I could not do more than find an author and help him as much as possible.

The author was duly found, Mr Michael Farr, fresh from the war and an English degree at Cambridge, a pupil of Dr Leavis and a man with faith in modern design. I put at his disposal the material collected for the first edition and added what had accumulated in odd envelopes since. Armed with this he began, at first supervised, as it would be called in Cambridge, by me in exactly the way that post-graduate research is supervised. This arrangement was carried on through the earlier parts of the present book. The other parts I never saw until I received the complete typescript. Mr Farr in the course of the work had found his own level and his own standards. The book as it is, must be considered entirely his, although the improbable reader who would wish to waste his time by comparing in detail the 1937 volume and this would find occasional paragraphs and more retained—especially in the semi-historical sections of Part II. Also, the general framework remained the same, and that perhaps more than anything else makes Mr Farr's the direct descendant of my book. That is as it should be. Readers of the survey of 1935 should be helped in preserving the feeling that they are now watching the same survey conducted again.

In looking at the two books in that way, they will notice at once numerous differences, superficial as well as essential. Mr Farr's book is not only up to date, it is also considerably longer and better illustrated. The expansion is specially noticeable in the chapters on plastics, motor-cars, and art schools. Entirely new sections have also been added, the most important being that on the design-consultant—a profession which, when I wrote my book, did not exist in Britain and had only just come into being in the United States. The Cambridge University Press, recognizing the position as I have described it, decided to my great pleasure that I should be excused from a joint authorship which could not have been more than nominal, and that the book should appear as Mr Farr's whole responsibility. Since, however, I had so much to do with its inception and growth, and since there are certain general points which are not made by Mr Farr or which appear in a different light to me and to him, or which had not found adequate attention in my own book of 1937, or which, finally, had changed their significance in my mind since, I ventured to suggest that I might add some comments and conclusions of my own. This the Press accepted, and they will be found as a postscript, beginning on p. 310.

JANUARY 1953

INTRODUCTION

'It is to be noted that when any part of this paper appears dull, there is a design in it.' RICHARD STEELE, *The Tatler*, No. 38

We live in a world created by nature, but partly re-designed by man. The composition of our environment is divided between natural things and artificial things. Most of us, in our hours of work and leisure, find that we are much more intimately concerned with artificial things than with natural things. We live in man-made houses, walk on man-made roads, drive in man-made vehicles, work in man-made offices, and so on.[1] The greater part of our immediate environment has, in all its physical aspects, been conditioned by human thought and feeling. We are responsible for the artificial things in our environment, though each of us has had different and varying shares in their design.

This book is concerned only with those things which industrial factories and workshops have made for domestic use. They constitute the largest and, inevitably, the most important part of our artificial environment. But here one hesitates. Are they inevitably the most important part of our man-made environment? The answer, I assume, would be 'yes', but on looking round for proof of this importance, doubts would soon arise in many minds. The more one looks the more depressing becomes the scene. What is the reason for this? There is no shortage of domestic articles. In terms of quantity their importance is adequately stressed. But as regards quality there are wide and seemingly illogical differences between all groups of products. By describing the quality of an article we are, in fact, referring to its utilitarian and artistic elements, in other words its design. And it is in connexion with design that the importance, which we are attempting to assess, appears to have a confusing number of different interpretations.

In order to begin the discussion I am asking the reader to agree that the articles in his home do not all conform to one standard of design. In the shops also there is no single standard expressed in the design of every article. The point can be made with irresistible force by spending a few hours at the British Industries Fair. To be aware of differences we must be able to judge designs according to our preferences. The chief of these is personal; the others, social, ethical, moral, etc., need not detain us

[1] *The Architectural Forum*, November 1948, quotes a recent study in the United States, which affirms 'that urban man spends 88 per cent of his day in an artificial atmosphere, surburban man 70 per cent, rural man 43 per cent'.

here. When we judge a design we estimate its value to us according to our likes and dislikes. We like it, we partly like it, we do not like it; we judge it to be good, indifferent, bad.

I have written this book because I do not like the designs of between 80 and 90 per cent of British domestic articles. I have also written it because I am assured that my opinion is shared by a large number of people living different lives in all parts of the country. I am convinced that dissatisfaction with the prevailing design standards is widespread. Amongst the majority of people this is not expressed, simply because sufficiently rewarding comparisons between good and bad designs cannot be made. This fact has become my central theme. While no one can tell what each individual member of the public wants, it does seem reasonable to suggest that he should be given an opportunity to make up his mind. This means that every shop should offer a choice, which in turn implies, as I shall show later, that many more good designs must be produced. Indeed, the pressing need for good design will be evident to all those with faith in the life-enhancing value of art, and in fact to everybody who believes in the value of education as a means of deepening, enriching and ennobling life.

To plead for better design standards is not new. There are many people who feel that enough has already been said. They remember the exhibitions, the lectures, the questionnaires and all the propaganda on behalf of the Modern Movement in the 1930's. Now, since the war, they are hearing less about the Movement but even more about contemporary design standards. London has always been the favoured centre for design propaganda and for witnessing its results. It is no exaggeration to say that there are more good designs in London than anywhere else in the country. So it is in London, chiefly, that one gains the impression that things are not nearly so bad as they were. This complacency is often found amongst those who were most devoted to the campaign for better design before the war. In some limited circumstances it can be amusing; but in all other respects it is a serious hindrance to the work of those who can appreciate objectively the situation as it is.

The fundamental cause of so much bad design can be briefly dealt with here.[1] It concerns the already familiar story of the Industrial Revolution. From the latter half of the eighteenth century onwards craft-work was progressively mechanized. The intimate contact between men, materials and the market, which is essential to all creative design, was broken by the intrusion of machines. Standardization of designs led to greater production of cheaper articles. New markets were created where quantity rather than quality was valued for its profitable returns. After the defeat of

[1] For detailed analysis see N. Pevsner, *Pioneers of Modern Design*, The Museum of Modern Art, New York, 1949, chap. 1; S. Giedion, *Mechanization Takes Command*, Harvard, 1950.

the craftsman the artist turned and fled. In his imagination he was above the mean standards and mechanical methods of the manufacturer. But the public followed the advances of the new industries and forgot the artist, who in turn became progressively separated from the living thoughts and feelings of his time. Throughout the nineteenth century and into the twentieth century the situation got steadily worse. The change in attitude which we can now detect had its origin in William Morris who showed, in some of his craftwork, how contemporary vitality could return to manufactured articles by bringing the artist back to the workshop. At a succeeding stage in this development, which took place in Germany between 1905 and 1930, the artist was persuaded to enter the factory and design articles for quantity production. In this country the Modern Movement, as it was called, began to establish itself after 1930. Its success with British manufacturers and British artists was never complete. But its underlying belief that works of art can be produced by machinery for the benefit of thousands has been proved conclusively. Today, although we have this proof, we do not sufficiently realize how pressing is the necessity to put it into practice.

This book offers a survey of the conditions and the standards of design in British industries serving the domestic market. To discover the conditions I had to interview manufacturers, designers, retailers, teachers in art schools and a number of other persons. To discuss design standards involves criticism of present methods and suggestions for practicable improvements and that is the main reason why those interviewed remain anonymous in the text. I began my research in September 1949, and continued it throughout 1950. Since then I have completed my material while attempting to keep it up to date. I have personally visited 214 firms in 27 industries; 56 retail stores; 15 art schools; and 38 designers working outside industry. Approximately 48 per cent of the interviews took place in the Midlands; 7 per cent in Scotland; 24 per cent in London; 21 per cent in the rest of the country. I did not attempt to visit or get in touch with all the firms or even all the important firms in each industry. It proved sufficient to visit a selection of characteristic factories covering as far as possible all types of production and all types of markets. However, I must admit that I was mainly concerned with design in mass-production for the general market.[1]

It should be emphasized that I did not confine my inquiries to articles of exemplary design. It seems to be a flaw in most of the publications on art in industry that by dealing only with good design, regardless of price, they are compelled to leave out between 80 and 90 per cent of the national production. To plead for better design in manufactured articles will remain a futile task unless the commercial aspects of the

[1] In the year 1947, out of 20,400,000 wage-earners almost 18,000,000 earned less than £500 per year. To them must be added the several million wives and families dependent on this income group.

problem, unpleasant as they may sometimes be, are carefully taken into account. Such considerations induced me to listen to the views of all people concerned, particularly those of the manufacturer.

With manufacturers, as a rule, the following questions were discussed: general aspects of the trade, usual conditions of design in the trade, history, growth and size of the firm, number of employees, existence of design department or drawing office, existence of head designer, his qualifications and status, firm's reaction to other designers not on the staff, payment of designers, sequence of events in the production of a new design, participation and responsibility of other persons, such as the manufacturer himself, managers, salesmen, foremen or workers, number of new designs produced each year, sales figures and prices of best-sellers, methods of market research, opinion of the manufacturer on the standard of the designs produced, his opinion on public taste, methods of distribution, influence of export markets on design standards, relation to art schools, influence of organizations for design propaganda.

In department stores, chain stores and privately owned shops I discussed general aspects of the trade, methods of presentation, available choice between good and bad designs of comparable price, market served, average rate of stock turnover, methods of payment and training for buyers and salesmen, value of special exhibitions of good designs, opinions on the design standard of stocks, opinions on public taste.

I have not carried out a similar programme of research amongst representative groups of the public. The task would be complex and difficult, and however painstakingly performed would, I believe, be of very little value. The results of interviews in peoples' homes and the answers to questionnaires filled in at exhibitions of industrial art are liable to gross inaccuracies. When confronted with a tempting selection of designs people are apt to forget what they can afford, what they need and what their families need. On such occasions one often finds that people express a gratifying predilection for good designs, but when they are faced with paying cash they seldom feel able to confirm their first choice.

In my interviews with principals and teachers in art schools, I was concerned with these questions: the extent of their co-operation with industry, amount of machinery and equipment, connexion between design, craftwork and fine art, full-time and part-time courses for design students and apprentices, teaching methods, extent of participation in civic affairs, the demand for trained designers, other possible sources of employment. With architects, consultants, free-lance and commercial designers, working outside industry, I was mainly concerned with the design standard of their work. Points discussed were: training, experience of industrial production methods, results of work undertaken for manufacturers, fees charged.

I have compared my material with the evidence presented in *An Enquiry into Industrial Art in England* by Nikolaus Pevsner which relates, in general, to the year 1935.[1] I have also studied the reports compiled by institutions with aims similar to my own. Most of them were concerned with the years before the war: of chief importance are the Board of Trade Committee on Art and Industry, 1931–2 (Gorell Committee), and the Board of Trade Council for Art and Industry, 1934–9 (Pick Council). Some useful evidence is contained in the working party reports on the future of certain industries, which were published between 1945 and 1948 by the Board of Trade. In addition I have paid close attention to the work of the Design and Industries Association, the Industrial Art Committee of the Federation of British Industries, and the Council of Industrial Design.

On the whole I found that manufacturers and others were interested in my research and anxious to help. In no less than 80 per cent of the firms visited I had a private interview with the managing director. This rarely took less than one hour, and frequently lasted all the morning. With very few exceptions I was shown over each factory, retail store and art school I visited. Only 4 per cent of the firms to which I wrote refused to see me. Nevertheless, it may well be that my results in some industries are one-sided and only partly true. I was for instance seldom able to check what I was told. Manufacturers, designers, retailers and teachers may have made wrong statements, they may not have told me the exact truth or the whole truth. I need not say that in cases where I had the feeling that something of this sort was happening, I preferred not to use the evidence at all. The other shortcomings in my report are evident. But I still hope that the facts which the following chapters contain may in the end be of some use. In all cases I have attempted not to assert too much and to prove too little. So often we hear that popular taste is abysmally low, or that designers in firms are not sufficiently well paid, or that artists are inaccessible to commercial propositions, or that manufacturers are not willing to risk experiments, etc., but hardly anybody has seriously attempted to examine what objective truth there is in such contentions.

In Part I of this book I have presented the facts concerning materials, techniques, design, methods of production and distribution and training for designers. I have included historical data where relevant. Part II describes the organizations and institutions concerned with design research and propaganda. It also includes designers' associations, exhibitions and publications. Part III is devoted to conclusions which represent an evaluation of the foregoing evidence. In Part IV I have made some suggestions for improvements, based on the conclusions.

[1] On the relation of that book to mine, see the Foreword, p. xxvii.

It should be made clear that this book is not concerned with industrial art as it is used and misused by the public. Interior design and house planning are matters outside my immediate scope. Nevertheless, one point should be stressed here. Each object of industrial art should be judged according to its intrinsic merits, and allowed to exist in its own right, in harmony with, but not necessarily similar in appearance to, other objects.

The broad aim of this book is to help towards the re-creation of an harmonious understanding between maker and user—with the assistance of the retailer—so that the limitations and desires of each can be appreciated. I have attempted to stimulate interest, to allay apathy and to encourage a greater sense of responsibility on all sides, and in fact, to suggest a way of recovering that sense of collaboration, implying the mutual acceptance of a common standard, which was lost during the first half of the Industrial Revolution and not regained in the second half. If such an objective can be achieved, industrial art will form the basis of a new culture in which taste will become coherent, critical judgements will be everywhere implied, and a popular standard will be acknowledged as a satisfactory level of taste below which nothing should fall.

If I were dealing with music, literature or fine art, the only relevant observations I could make would be directed at the art, not the artist. Expression in these arts is the work of one man and he assumes entire responsibility for it. But in industrial art this is not so. In the evolution of a new design there is a long chain of persons (designer, manufacturer, sales manager, traveller, retail buyer, retail salesman) and circumstances (material, technique, production, cost, market) resulting in complications which must be resolved before the design itself can be criticized relevantly. This book is concerned, therefore, with the material, economic, social and, in general, extra-artistic aspects of the present state of industrial art. I have attempted to provide the reader with sufficient knowledge to reach, through his own efforts of comparison and analysis, a relevant appreciation of industrial art. We must know what can reasonably be expected from industry before insisting that its art should reach a certain standard.

I realize that throughout the handling of the evidence in the following pages my personal standard of judgement is implied. I cannot define this standard satisfactorily, neither would I expect any one else to define theirs. While my object is not to persuade others to adopt my point of view, I have attempted to provoke a critical response. This should of course have begun already, but in order to give the reader some indication of my critical standpoint I offer here the main considerations which I think important.

Let us assume we are looking at a teapot. My first questions would run like this:

Does it make satisfactory tea and can it be operated efficiently? Now, regarding its artistic quality, what degree of intensity of thought and feeling does its form express? Does the decoration enhance the form? Could the identical artistic effect be achieved in another way? If so, there would be no point in continuing the analysis. To have any artistic significance the teapot must represent a creative revaluation of the possibilities of the materials and techniques at the designer's disposal. A useful test is this: Does it make all former teapots slightly less relevant to contemporary artistic needs? Is it something of which we have been intuitively aware, but yet never managed to express? The teapot represents a binding together of materials and techniques, conditioned by the function, to achieve a certain form. Can we say that this form has a vitality which surprises and pleases us? This pleasure should not be obviously presented. It should be achieved by the effort we have made through critical analysis to appreciate the teapot. The final and decisive effect, which we may call the pleasure, eludes explanation. As far as I am aware this final experience can only be described by similes.

Distinct from the teapot's intrinsic values are its evidential values. The contemporary vitality of the teapot depends on them. We recognize vitality, and therefore the evidential values, according to the degree to which we are aware of the *Zeitgeist*, or spirit of our age. We recognize that the matter is complex and in doing so we perceive that today we have no accepted popular basis upon which each of us can erect his personal standard of artistic excellence. We do not enjoy this popular working basis for agreement because we have made insufficient effort to develop our faculty for discrimination. A beginning can be made by accumulating an experience of many works of industrial art in which a similar degree of vitality and intensity of expression is apparent. What they look like is important only as an aid to identification, cataloguing, etc. It is what they do to us and what we do with them that matters. We call the resulting effect pleasure, on account of the effort we have made to possess a new experience. This constitutes the inherent limitation in discussing visual art. An artistic standard cannot be expressed by us in visual terms—except as the most superficial pointer to what may be a rewarding experience for others. It is in this spirit that I have chosen the illustrations for the following pages.

One final point should be made here. I have already made an implicit reference to the life-enhancing value of art. This is generally acknowledged where fine art, literature and music are concerned. In connexion with industrial art it is scarcely considered. This is very strange. Whereas the picture, the poem and the symphony can affect us only at chosen moments when we seek them, works of industrial art surround us and affect us at all times. We use different forms of art for different purposes and none of them need offer us an experience which is more inherently limited than that which is offered by the others. It is once again a matter of personal taste whether one can achieve more pleasure from the appreciation of a teapot than from the appreciation of a picture.

Industrial art suffers from neglect and abuse. The first is the responsibility of the public, as I have already indicated. Responsibility for the second rests with all those concerned with production and distribution. One cannot approve of thoughtless and insensitive designs. Neither can one approve of dishonest designs, such as a pressed glass bowl trying to look like cut crystal glass, a plastic-covered handbag made to resemble snakeskin, an aluminium teapot masquerading as hand-beaten pewter. In the same way one cannot approve of imitations of period designs. As we shall see, the arbitrary invocation of antique styles is a disease from which few industries are free. Such false and meretricious designs attempt to provide a substitute for the needed splendour which all aspects of our environment should be made to concede. The pleasure which most people take in an entertainment so vicarious as the cinema, as well as the pleasure in vulgar and boastful design, is largely accounted for by the universal longing to escape. Looked at from this point of view, the question of industrial art is a social question, it is an integral part of *the* social question of our time. To fight against the shoddy design of those goods by which most of our fellow-men are surrounded becomes a duty.

DEFINITION OF TERMS

I have attempted to use a limited number of terms to describe industrial designs. *Design* has two qualities: *artistic* and *functional*. I use the term *artistic* because it has fewer confusing connotations than the term *aesthetic*. *Functional* indicates the utilitarian, durable and economic properties in a design. The term *modern* is related exclusively to creative work in a contemporary spirit and of positive value. *Exemplary* indicates an outstanding design which may, or may not, be modern. *Good* design and *satisfactory* design are self-explanatory, and also may relate to any period. *Fitness for purpose* and *fitness for use* are two maxims which I have used frequently to describe functional qualities only.[1] *Modernism*, and its adjective, *modernistic*, together with *jazz*, denote fashionable clichés characteristic of many designs during the 1930's. As far as this country is concerned, the style is derived from the exhibits at the Paris Exhibition, 1925. Before that it may be traced through the *Cubist* painters and the German *Expressionist* painters, active from about 1907 onwards. Typical elements in the fashion are zigzag lines, jags and angles. *Borax* is used several times to denote unnecessary blubbery inflations in a design, often accompanied by parallel stripes and bands. In its slickness it is a characteristic mannerism of contemporary design.[2] *Styling* means superficial designing. Its effect is limited to the outward appearance of a design. In the United States *stylist* is the common term for a fully qualified industrial designer, to whom my definition of styling does not necessarily apply.

I have classified all British industrial designers under three headings: *staff designer*, employed full-time in a factory; *free-lance designer*, independent and concerned mainly with two-dimensional designs (the commercial designer is similar, but runs a studio staff of his own); *consultant designer*, also independent, usually manages his own team of specialists and is concerned mainly with three-dimensional designs. To

[1] The phrase 'fitness for purpose' was used by the architect Owen Jones as early as 1852 (*The True and False in the Decorative Arts*, published in 1863). Very similar is the wording in Pugin's *True Principles of Pointed Architecture*, 1841. See also a lecture given by Frank Lloyd Wright in 1903 ('The Art and Craft of the Machine', reprinted in *Modern Architecture*, Princeton, 1931, p. 15). 'Fitness for use', occurs in the first pamphlet published by the Design and Industries Association, in 1915: 'The first necessity of sound design is fitness for use.'

[2] According to Harold Van Doren 'borax' means cheap and shoddy (*Industrial Design* (*A Practical Guide*), New York, 1940). Raymond Loewy gives it another name, 'schmaltz', and attributes it to vulgar, superfluous, gadgety decorations, stripings and curlicues, accompanied by cheap colour (*Never Leave Well Enough Alone*, New York, 1951). See also E. Kaufmann, 'Borax or the Chromium Plated Calf', *The Architectural Review*, August 1948.

these I must add the *architect* and the *artist* (painter or sculptor), both independent. The term *teamwork* is widely used in the engineering and electrical industries to describe all types of design activity. In all products of complicated construction teamwork is inevitable. My use of the term is limited to the very rare cases where engineers and industrial designers collaborate, throughout the development of a product. *Mass-production* means the production of large quantities of identical articles with the aid of machinery. Throughout the book *the war* refers to the Second World War of 1939–45.

PART I

DATA

FURNITURE

There are over 3000 firms in Great Britain making domestic furniture and they give employment to some 120,000 persons. Of these firms 40 per cent are situated in London and the south-east of England, 6 per cent in High Wycombe, and 6 per cent in Scotland. The remaining 48 per cent represents the rest of England and Wales. In order to give a comprehensive account of this large and fundamentally important trade, I interviewed as many as thirty manufacturers. Each one was selected because his firm either represented a particular manufacturing technique or catered for a specific market, both of which would be relevant to a certain section of firms in the trade. My selection of firms was also guided by the fact that nearly 80 per cent of the total output of furniture by value is controlled by only 17 per cent of the trade. This means that the bulk of the furniture produced is in the hands of the larger firms. In the London area, where the trade is concentrated and the largest—and some of the smallest—firms are to be found, I visited twelve firms. In High Wycombe I visited five, in Scotland seven, and in the rest of the country six. I did not interview any independent cabinet-makers, or firms with less than twenty employees because, though numerous, their sphere of influence is very small. Evidence from retailers[1] will be dealt with separately.

Before discussing the facts obtained from these firms it is necessary to have before us a general picture of the furniture available to the public. Let us take a large provincial town. In the centre of the town, in its main streets, side streets and dotted here and there among its housing estates, are the hire-purchase furniture chain stores. They are the first to attract attention for two reasons. They are more numerous than any other type of furniture shop and they indulge in fancy, ever-changing, window and pavement displays. The standard of furniture design in these stores is deplorable. Next in the town are the large department stores. The furniture which they display in solemn, sombre phalanxes of wardrobes, sideboards and three-piece suites, is also deplorable, although it is always possible to find two or three pieces of modern design. That is the complete picture for well over 90 per cent of the buying public in such a town. There remain three or four shops where furniture of the highest quality is available. This quality is usually embodied in period forms, either genuine or imitation antiques, but there is also a fair and varied selection of modern furniture available.

[1] See p. 151.

If timidity or lack of financial resources prevents most people from entering this class of store, there is little chance that they will realize that modern furniture exists.

If that is the situation today, what was it like before the war? What changes have taken place? The chain stores were much the same, but most department stores had no modern furniture. The better-class stores would undoubtedly have some, but it would be relatively much more expensive than the selection shown today. In the years before the war modern furniture was to be seen chiefly in London and the bulk of it was either exclusive to single shops or even to particular clients. There was not sufficient demand for modern furniture to make mass-production worth while. Here, perhaps, is the most significant change which has taken place. Today there are many firms making modern furniture by mass-production and semi-mass-production processes.

The war has had a considerable effect on the furniture trade. Many firms were obliged to change their production from domestic furniture to war-work. In many cases which I have investigated this change has been entirely beneficial. New machines and techniques were introduced which, when the war ended, could be used to make the production of furniture speedier, less costly and more efficient—three factors which have already helped to make modern furniture available to a wider public. But one important war-time technique has not been exploited by the trade. This is the bending of laminated wood as used for the manufacture of aeroplanes, notably the de Havilland Mosquito. Laminated bends presuppose new, and for this country strange, forms and these the trade has generally rejected in preference to forms derived from the more conventional methods of construction. There are a few exceptions. One Scottish firm has used the technique with startling results, but only for high-price furniture. One of the chairs which was made in this way has been put on permanent exhibition in the Museum of Modern Art, New York. Another firm, in East Anglia, has formed an easy chair entirely out of laminated wood. It is reasonably priced and has proved commercially successful.[1]

The attitude of the trade towards the bending of laminated wood is disappointing, especially as it is merely one facet of the furniture manufacturer's deeply rooted antagonism to any change. Another example of this, also an effect of the war, is connected with the story of utility furniture. In 1942 the timber shortage in this country had become acute. The demand for domestic furniture had grown greater because of damage by bombing, and prices showed a corresponding increase. A system for rationing furniture was required, and the President of the Board of Trade appointed a Committee, under the chairmanship of Sir Charles Tennyson, to

[1] For further discussion of this and other new materials and techniques, see p. 142.

advise him 'on specifications for the production of utility furniture of good sound construction in simple but agreeable designs for sale at reasonable prices, having due regard to the necessity for the maximum economy of raw materials and labour'. As the decision was taken to leave no initiative in the hands of individual firms, the Committee employed two well-known designers from High Wycombe to plan the first series, afterwards called the Cotswold range. Since the best equipped firms had already been transferred to war-work, the designs and specifications had to be suited to hand-craft as well as part-hand and machine production. The introduction of utility furniture meant that the uncounted number of designs which was produced up to that time, would be superseded by some thirty pieces of approved design. The trade and the buying public alike had no choice in the matter.[1]

The wisdom of this dictatorship must be judged by the results which have, in fact, been far-reaching. The furniture belonging to the Cotswold range reached the shops early in 1943. No one can claim that it was received with much enthusiasm. Even the Committee which had introduced it was divided in its opinion. 'Some members felt that what the public bought before the war...was what it should be given. That is, about 90 per cent Jacobean period furniture and 10 per cent modern.'[2] In 1945 a new range, called the Chiltern range, was introduced, but it, too, made very few concessions to the known taste of the buying public. Mr Gordon Russell, who was responsible for the work of the design sub-committee, has stated that the aim was to give the people 'something better than they might be expected to demand'.[3] The proportions, the clean outlines, the sense of delicacy and lightness, especially in the Chiltern range, were all excellent, and they enabled everyone for the first time to afford modern furniture. Undoubtedly it was a step forward in the education of the manufacturer and the consumer. But was all this accomplished with too much haste? The criticisms of the public, for example 'boxy', 'severe', 'harsh lines', etc., seem to imply that the designs were too advanced to be generally acceptable. Moreover, the designs presupposed that the public had the ability, both financial and artistic, to furnish a living-room or a bedroom with objects of a related standard. Under skilful management utility furniture could be the *raison d'être* of a variety of excellent furnishing schemes. But without that it would merely look stark and plain. A great opportunity was, for the most part, lost. If it had been possible to persuade individual manufacturers to create something acceptable to public taste and yet maintain the design standard of the Chiltern range, a feat which has been accomplished by several

[1] For further particulars relating to utility furniture, see an article by Gordon Russell in *The Architectural Review*, December 1946, pp. 182–6.

[2] *Ibid.*

[3] *Ibid.*

firms since the war, the crushing reaction, which came when the regulation controlling the appearance of utility furniture was rescinded in November 1948, might never have taken place.[1] The spring sales in 1949 were abundant evidence of this reaction which, in an exhibition catalogue of that year, was expressed in these terms: 'Choice is now playing its rightful part and individual tastes can once again be satisfied. These new 'Freedom' designs in furniture have been given special prominence on the...stand.' A further reason for this reaction must be given. Utility furniture was designed for service, not for selling: for useful work in people's homes and not for 'eye appeal' in the shop window. This made the retailer antagonistic towards the utility designs. Finally, the utility experiment, far from purging people of their love for trite, meaningless ornaments and forms had, in fact, encouraged the trade to return without hesitation to its pre-war standards. Anyone who now asks a manufacturer of 'Freedom' furniture why he does not make modern furniture is immediately disarmed by the reply: 'If utility furniture is what you call modern furniture, then I have tried it and it doesn't sell.'

Before discussing the bulk of the trade which is typified by this remark, it will be necessary to consider why there are more firms making modern furniture today than there were before the war. I visited eleven firms which devote their entire output to modern furniture. Seven of these firms were making domestic furniture of modern design before the war. Two were, and still are, mainly concerned with cabinet work for ships and two had begun work after the war. All these firms are comparatively small. The smallest has 25 employees, the largest 350. The average number would be about 125. Each manufacturer is personally proud of his firm's products. In six firms the manufacturer is also the designer and in the remaining five he takes an active interest in the work of his designer. The chief characteristic is of course the size of the firms. Larger units would be less manœuvrable. Also in few, if any, cases have shareholders to be considered before any risk is taken. The lapse of time from the inception of a new design to its production is rarely longer than three months.

In the seven firms which were making modern furniture before the war an evolutionary pattern of progress is evident. When I made my investigation two factors conditioned their methods of production: the price limit and the demands of the general specification for the manufacture of tax-free furniture, which ensured that it was of sound construction. The second requirement caused little difficulty to these

[1] Although design became free of control, the price regulation for furniture made with the Board of Trade's general specification remained until December 1952. A year after this date there were 2500 firms making this tax-free furniture.

firms, but the first necessitated an unusually long production run on one design and considerable equipment such as jigs and conveyor-belts. Although certain parts of the furniture were standardized and therefore became interchangeable, a high level of design was maintained. The importance of this will be indicated later when the production methods of the larger firms are considered. The lowest output amongst these firms was twenty-five dining-room suites a week. The highest was seventy suites. If we contrast these figures with the average pre-war weekly output of the same firms, about ten suites, the degree of change will be apparent.

The distribution of their furniture is, however, not wide. Steady orders are received from the leading stores in most provincial towns and little effort has been made to widen their market by including the department stores. The manufacturer is not entirely responsible for this limited demand, because it is customary for the chosen retailer in any given town to request that the design be stocked by him exclusively. In one case I was told that 50 per cent of the firm's production was absorbed by one London store and the remainder was taken by selected retailers in the Provinces. Another manufacturer reported that the small private retailer, again one in each town, was his best client. The close link between the retailer and the manufacturer, is, as we shall see later, peculiar to these firms. There is of course the danger that modern furniture will once again become exclusive, but the arrangement has many good points. When a new design is proposed I found that invariably the retailers' opinions were sought. A mutual interest is created in this way long before the design is in production. The manufacturer benefits because he knows that his first run will be sold and the retailer is able to plan his stock in advance. Moreover, the responsibility, both moral and financial, of launching a new design is shared equally by both parties.

The experiences of the two firms which specialize in ships' furniture are instructive. Both of them entered the domestic furniture market after the war. They pointed out that, unlike other firms, they were unhampered by pre-war policy. They could make a fresh start under their own standards. It is significant that both of them chose modern design standards, although when describing their furniture the manufacturers stressed quality rather than design. The possession of a steady business in ship-work was a natural asset in the early months of these experiments, because of the slow and costly process of building up a new market. This market had to be large enough to absorb such quantities as 1500 dining-room suites a year in the case of the firm which was fully equipped for mass-production. The risk entailed was considerable, but within a year the experiment had justified itself. The work for ships was also useful because the special orders sometimes demanded the use of unfamiliar methods and materials and the cost was not one of the first considerations. This gave the designers variety and the freedom to experiment, with the result that the experience gained has been passed on to benefit the domestic furniture.

The two firms which began work after the war were unable to obtain a sufficient allocation of timber. One firm adopted steel and cast aluminium as its media. Armchairs were framed with mild-steel rods, shaped on formers and then welded. The upholstery was prefabricated and then attached to the steel frame. Dining-room chairs and table-legs were made from cast aluminium sections bolted together. From the outset the designs were conditioned by the available materials, in the hands of a first-rate designer, with the inevitable result that they were also modern. The other firm did not turn to a new material, but preferred to use the small amount of wood which could be obtained. As the wood was not of sufficient quality to allow the conventional joinery methods to be used, the manufacturer made it into laminations. For a chair the wood was cemented into strips and moulded in a simple and inexpensive press. Because there are no joints in the construction the chair is exceptionally strong.

From the brief description of these firms two points emerge. First, the use of new materials and techniques has resulted in new forms. Both chairs are more comfortable than many jointed chairs on the market because a spring-like tension is embodied in their construction. The second point relates to the market. In contrast to the pre-war belief that modern design would sell only in the most exclusive shops, these manufacturers have found that today modern or unorthodox furniture will not find a market unless its price is *low*. Their furniture has not yet reached the chain stores generally, but much of it can be seen in every provincial town.

The next group of firms to be described is composed of those which have produced, or are producing, modern furniture not exclusively, but as part of their range. I saw eleven firms in this group. The largest has over 800 employees and the smallest 100. The average number would be about 300. I found that in only two cases was the manufacturer proud of his whole range. Five of the manufacturers are responsible for designing what they call their 'bread and butter' furniture. When something modern is wanted they seldom attempt to design it themselves. These firms are larger than those we have just considered, so the risk of putting a new and untried design into production is considerably greater. Eight of them serve wide markets, ranging from the better class chain store to the department store. The remaining three firms concentrate on the most expensive shops, the bulk of their output being reproduction antiques.

As these three firms are representative of a specialized market it will be necessary to discuss them first. One manufacturer I met was passionately interested in modern

PLATE I

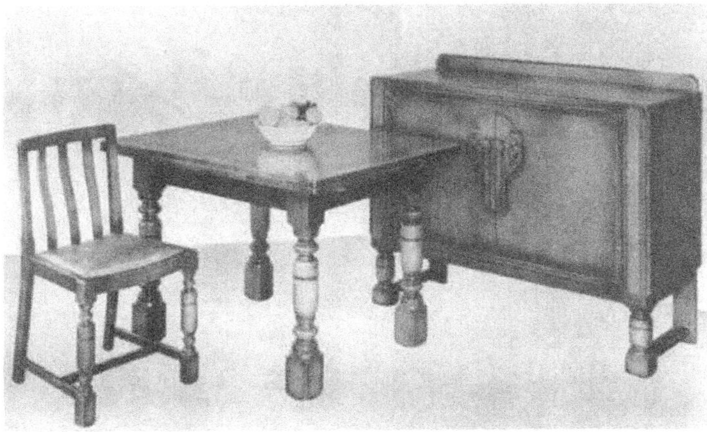

1. How many people realize that there is an alternative to the deplorable furniture designs shown on this page? Yet both dining-room and bedroom suites are representative of the bulk of furniture now put before the public. They are also typical results of the trade's attempt to satisfy public taste immediately after the government restrictions on design for utility furniture were rescinded in 1948.

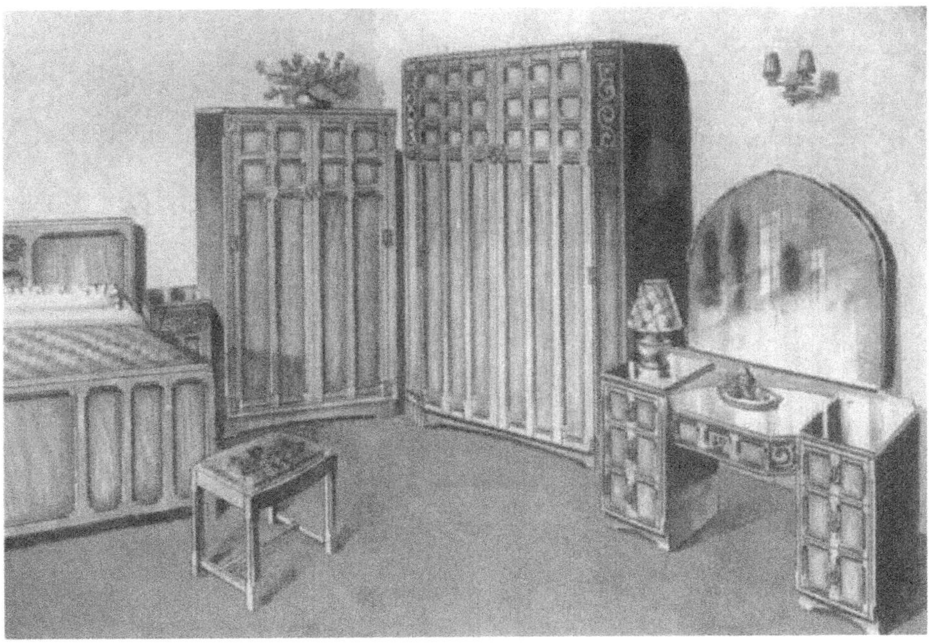

2. A visit to the annual British Furniture Trades Exhibition in London will prove that these two examples of debased design are not exaggerated. In them we can see the nostalgia for past styles, mainly early seventeenth century, contradicting their effort to be useful objects in the contemporary home. On the sideboard modernistic patterns from the 1920's have been added in the form of cheap machine-made ornament. The plates following prove that there is no longer any excuse for this pathetic *pastiche*.

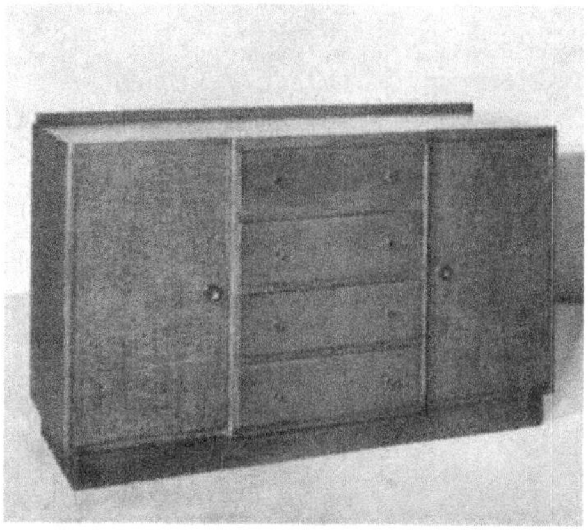

1. Oak sideboard which formed part of the Cotswold range of utility furniture introduced in 1943. These government sponsored designs made by firms in all sections of the trade were too austere to be generally acceptable. But it is to their credit that they were designed for service, not for selling.

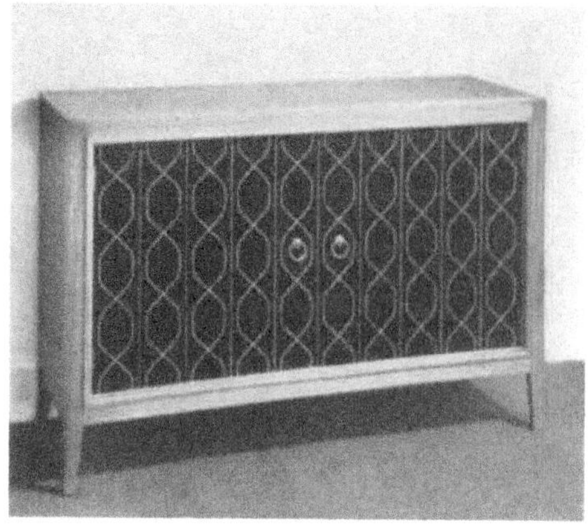

2. People hanker after decoration and furniture manufacturers can satisfy their needs without relying on hand-craftsmanship. In this sideboard a router has cut a pattern in the veneer of Bombay rosewood to reveal the birch below. *Designed by Booth and Ledeboer for Gordon Russell Ltd.*

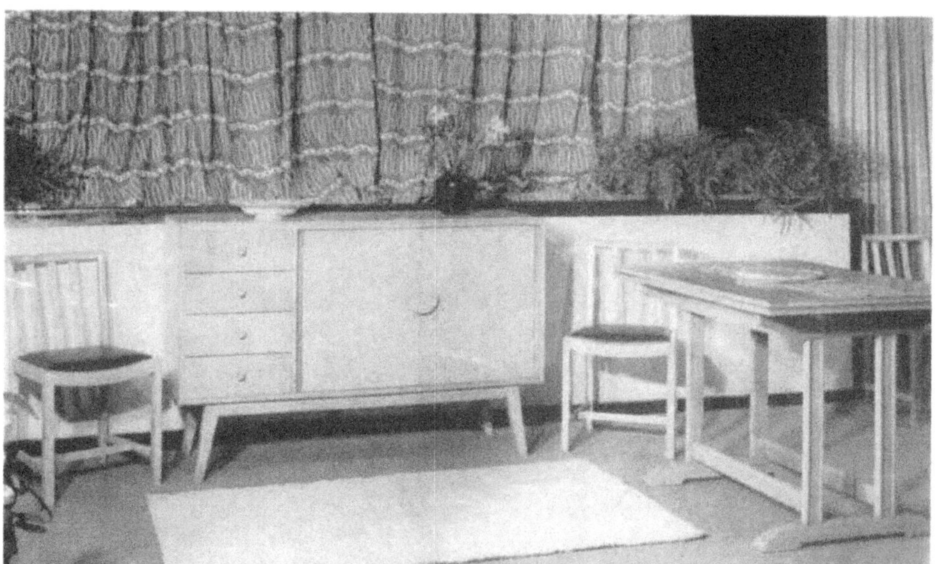

3. A further group of utility furniture produced in 1945. The sideboard belonged to the Chiltern range. Like the Cotswold range it was designed for both hand and machine production to suit all firms. It stood half-way between traditional joinery, as in the table, and the newer Swedish influences, exemplified by the splayed legs of the sideboard. Here was modern furniture that most people could afford, but which everybody had to buy if they wanted new furniture at all.

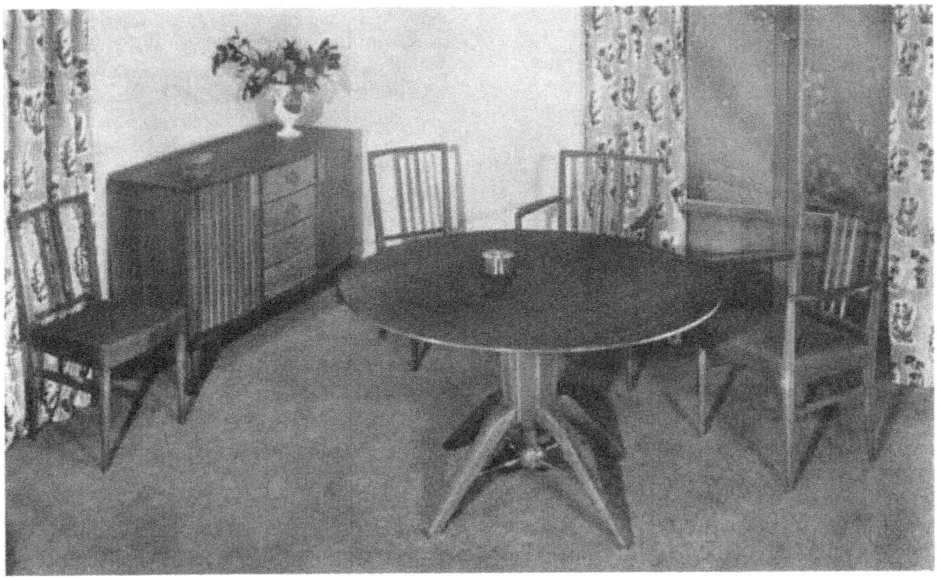

1. Modern dining-room furniture that makes full use of craft-based techniques. It is moderately expensive and fills the purchaser with an immediate sense of lasting quality. The detailing on the sideboard and the table are typical refinements in the work of this and several other leading firms. *Designed by R. D. Russell for Gordon Russell Ltd.*

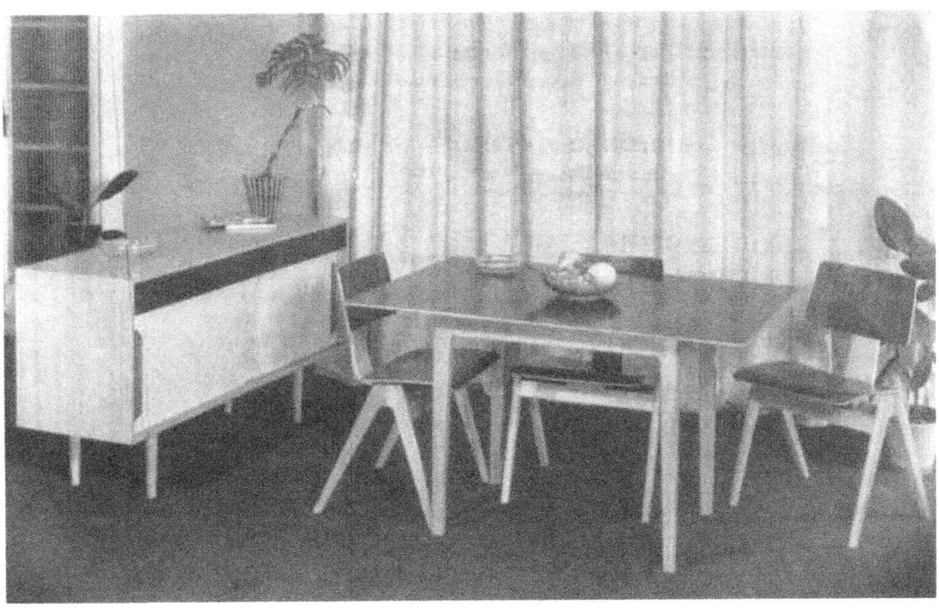

2. Modern furniture of an equally high standard as shown in 1, but relying largely on mechanical techniques for bulk production. Two relatively new elements are involved: the unit principle for the sideboard—called a storage unit—which can support a similarly proportioned bookcase; and the chairs with seats and backs of formed plywood to give economy in production and resilience in use. *Designed by Robin Day for S. Hille & Co. Ltd.*

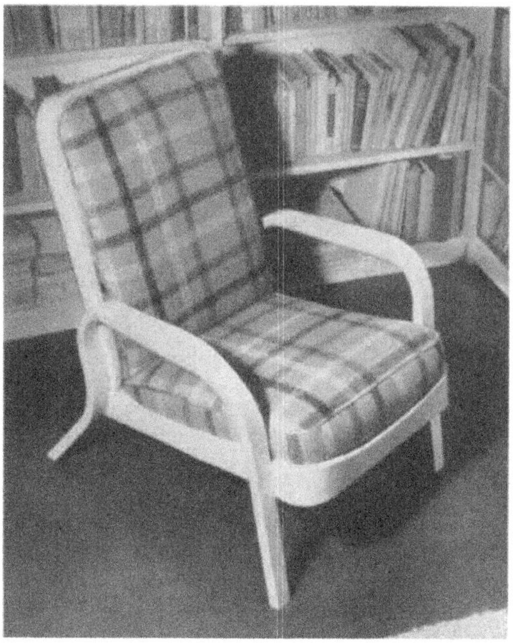

1. Three examples of moulded plywood furniture are shown on this page. This occasional chair has a laminated structure for legs and arms formed in one continuous bend. The laminated seat and back frames are complementary in shape and can be formed in one set of jigs. Added to the attractiveness of the chair's simplicity are a resilience and strength which come from a plywood structure with the minimum of joints. *Designed by Eric Lyons for E. Kahn & Co. Ltd.*

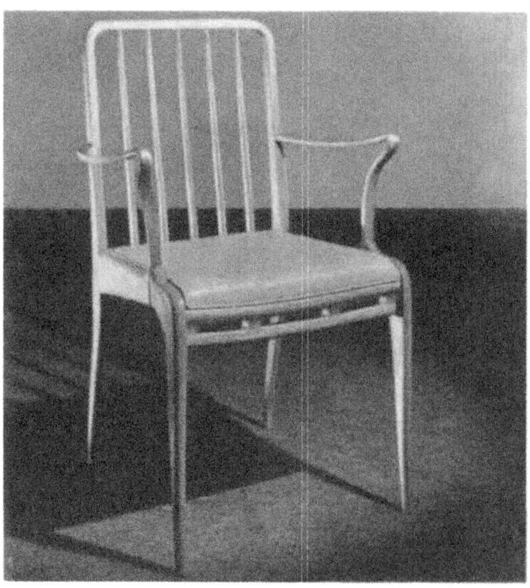

2. A more complex and expensive design which is really a *tour de force* to indicate the possibilities of laminated wood. Alternating light and dark veneers have been cut on the cross, polished and left exposed to demonstrate the visual richness and decorative qualities of the material. *Designed by Basil Spence for H. Morris & Co. Ltd.*

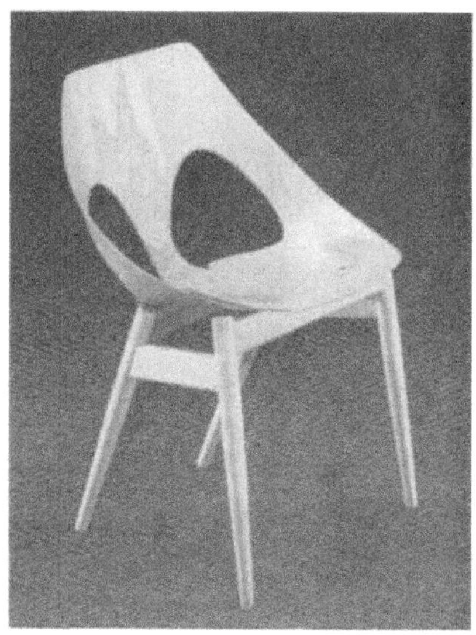

3. The trade generally has been unwilling to make use of plywood in the ways indicated on this page. Tooling and production costs are not high, and the chair shown above, with back and seat in one piece, is surprisingly inexpensive. *Designed by Carl Jacobs for Kandya Ltd.*

PLATE V

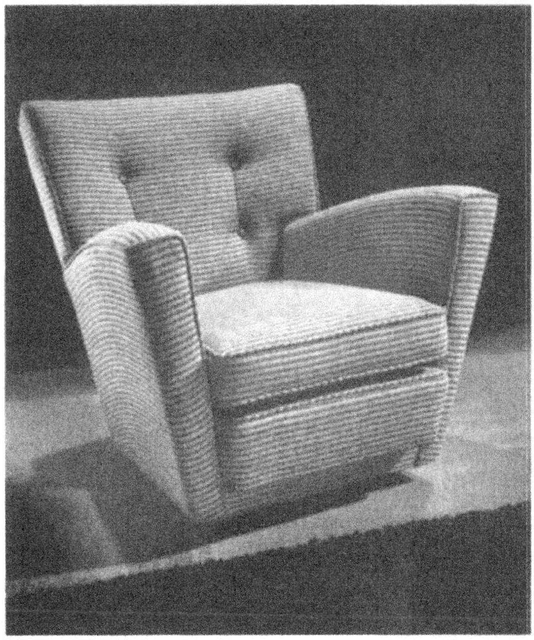

1. Of all the firms sincerely interested in high design standards only a handful make upholstered furniture. The progress from the unrelenting austere shapes of the mid-1930's to a more natural style has been largely due to the energy of Howard Keith. His style is distinctive, although his pieces, each one of which is unusually creative within the limited medium, have been widely imitated. *Designed by Howard Keith for H.K. Furniture Ltd.*

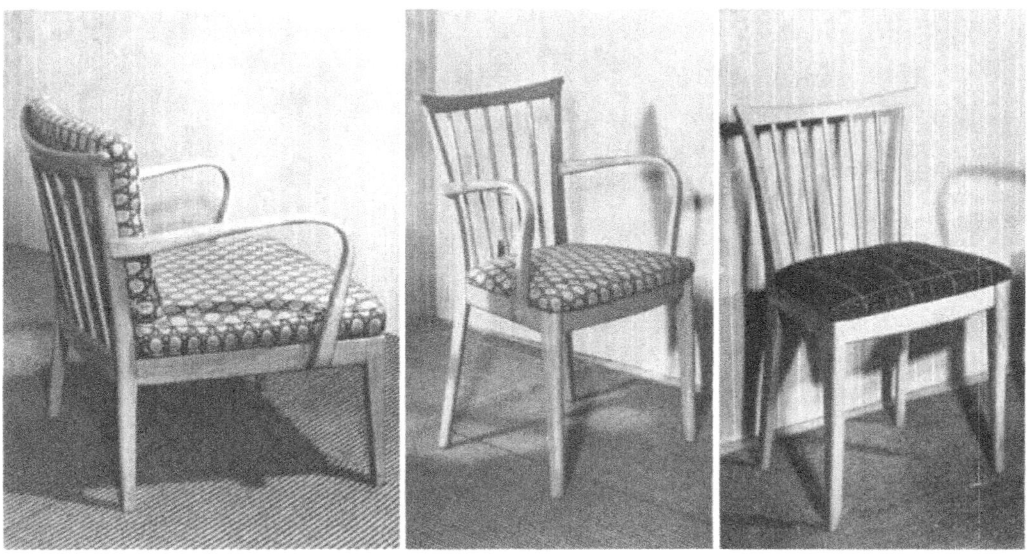

2. Modern design can also flourish in a native craft tradition. These chairs derive their strength from the long-established chair-making industry, particularly associated with High Wycombe. The manipulative capabilities of beechwood, which can be found for example in the many varieties of Windsor chair, are evident here: in Sweden and Denmark this characteristic English trait has been further exploited and has had a wide influence on designers' standards. *Designed by Geoffrey Dunn and made by Goodearl Brothers Ltd.*

PLATE VI

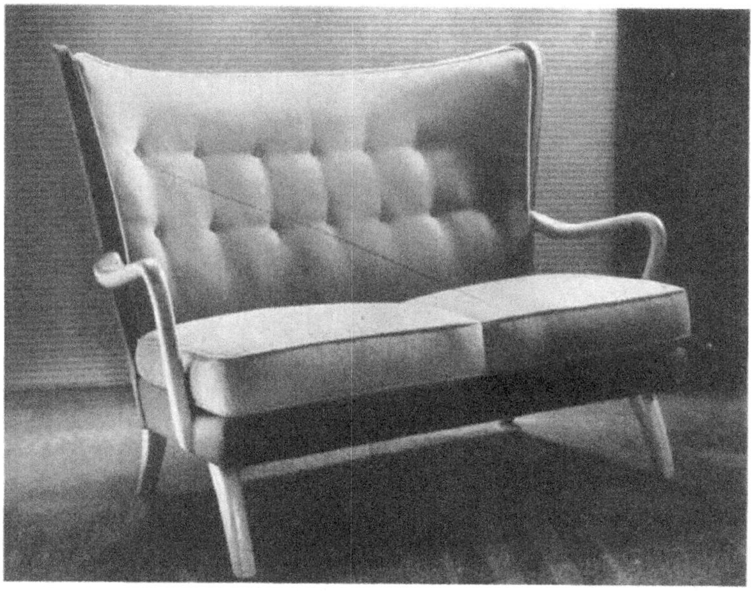

1. Today more people enjoy less space for living than before the war. The need is for small, compact furniture where comfort, durability and appearance must not be sacrificed. A good example is this settee with its slender arms and legs acting as foil to the inevitable bulk of the upholstery. *Designed by Howard Keith for H.K. Furniture Ltd.*

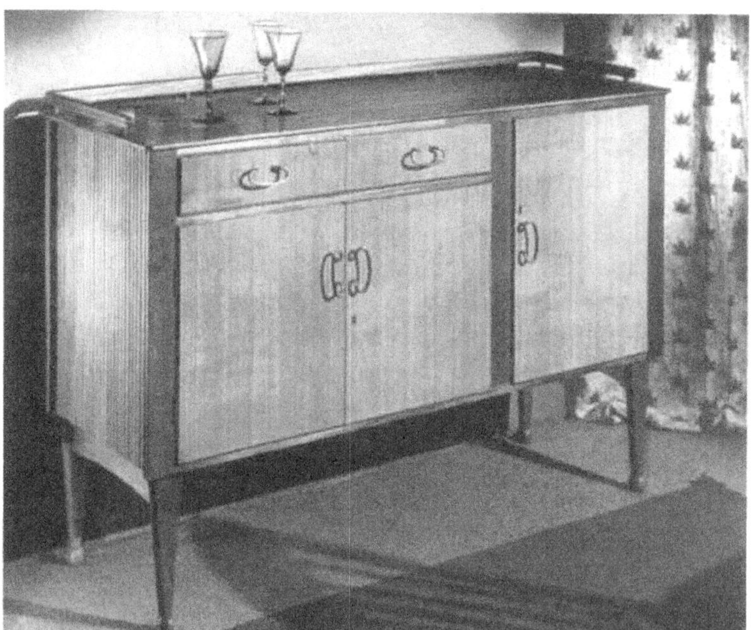

2. The sideboard as a piece of domestic equipment in its own right is now being strongly challenged by unit furniture, where space is often used more economically. The independent quality asserted by this piece comes mainly from care and attention to details and finish. Though no one could deny its usefulness, few would consider this as the governing factor in their choice. *Designed by A. J. Milne for Heal & Son Ltd.*

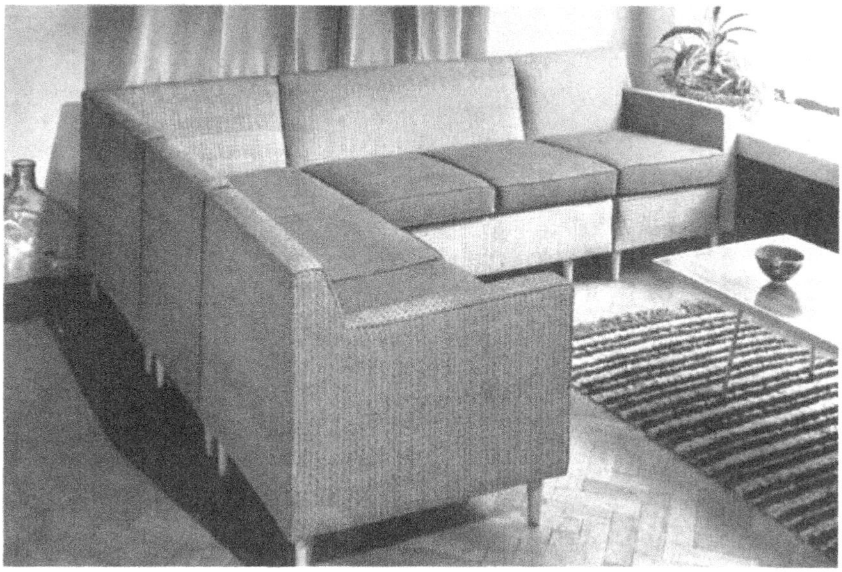

1. The idea of allowing essential furniture to remain simple in line, partly to afford economical manufacture, coupled with the adoption in each set of pieces of a standard dimension, is a growing influence starting from Germany before the war and now coming with more force from the United States. Unit furniture serves to define certain limits in the modern living space; its horizontal simplicity, as in this group, allows the owner to indulge in less expensive forms of decoration, such as wallpaper, indoor plants and studio pottery. *Designed by Robin Day for S. Hille & Co. Ltd.*

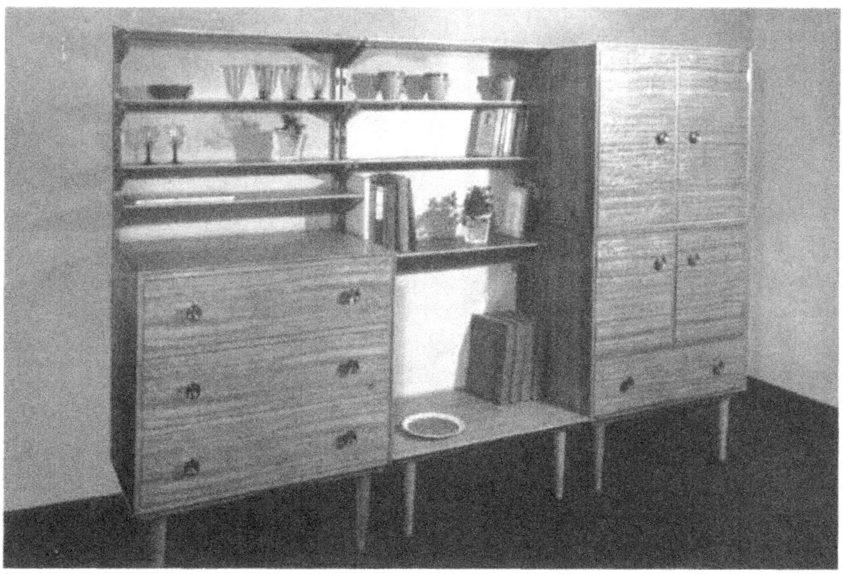

2. The unit furniture in this group can be arranged in different ways and supplemented by other pieces. The designer has begun with a good proportion and increased the effect with a discriminating choice of veneer and fittings. This is a far cry from Plan Furniture of the 1930's, which first stated the case for unit construction in this country, and a great improvement on the bare and impoverished box and shelf units which so many firms have produced since the war. *Designed by R. C. Heritage for G. W. Evans Ltd.*

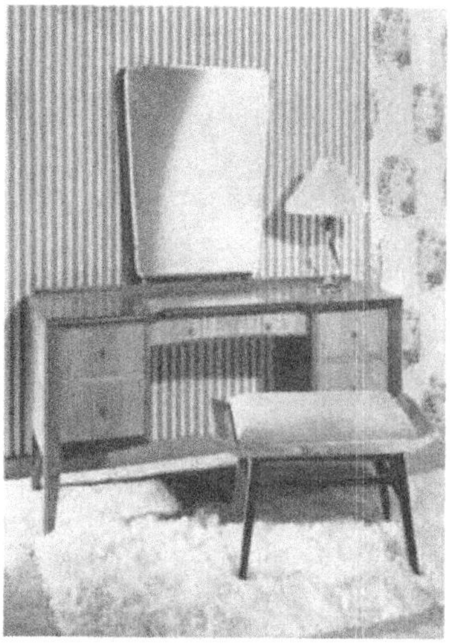

1. Bedroom furniture offers the designer so many opportunities for cunning devices, especially in dressing tables, that good shapes for day-to-day usefulness are frequently ignored. This is one of the few pieces where function and elegance are co-ordinated. *Designed by the Story Design Group for Story & Co. Ltd.*

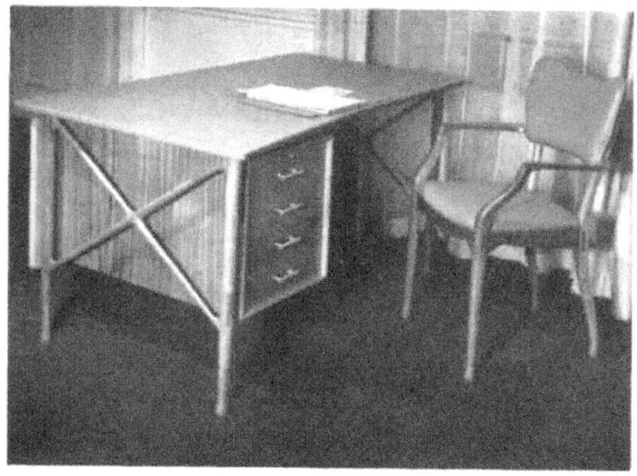

2. Richness and sophistication need not be set aside by modern designers. Though they can be achieved by mass-production, they are more often found in expensive furniture. This desk and chair are indebted more to the English eighteenth century than to post-war Scandinavia. The effect is to make you speak of fine craftsmanship and mean it, feel the surfaces with your fingers and delight in the intricacy of the design. *Designed by Basil Spence for the Scottish Furniture Manufacturers' Association for the Chairman of the Furniture Development Council.*

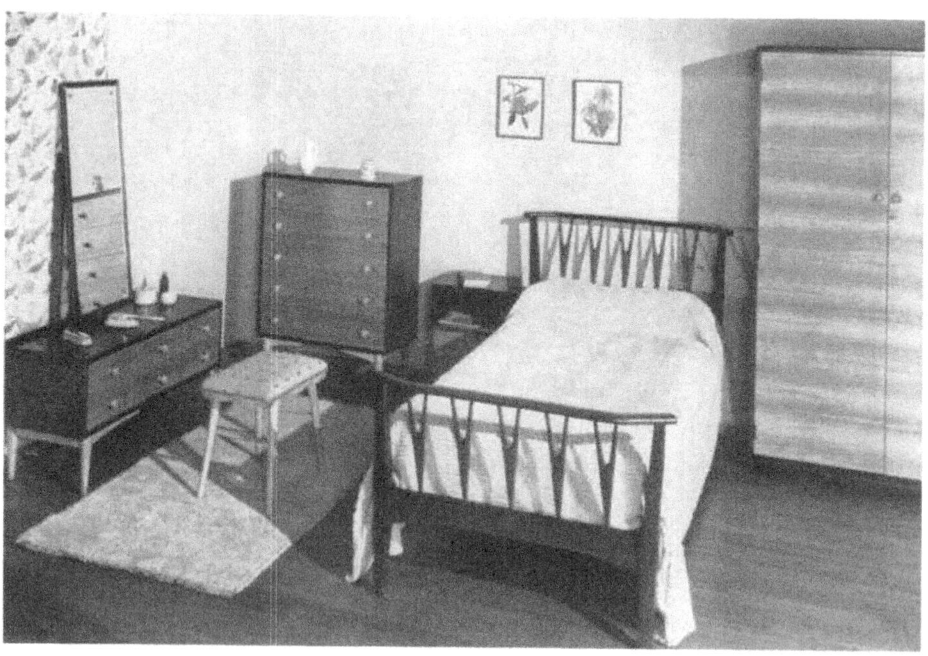

3. Many people tend to put up with the bare necessities in a bedroom and spend their money on the living areas of the house. This is perhaps the reason why makers of the best modern furniture have shown little creative interest in beds, wardrobes and dressing tables. An exception, in a moderately priced range, is shown above where, in the bed at least, a sense of luxury has been restored. *Designed by Paul Bridson for Kandya Ltd.*

design. His method was to procure the most advanced standards by calling in some of the leading consultant designers. In one case a bedroom suite was required, and the consultant was allowed a free hand. There was no collaboration with retailers at this stage, because no specific market was envisaged. When twenty suites were made and put on the market not one of them was sold. The manufacturer then reduced his price so that it came within the tax free range. The increase in sales was not encouraging and the manufacturer suffered a loss. He also quoted several other examples to illustrate his theory that there is no demand for modern furniture. In his experience this generalization seemed to be valid, but in fact it only related to firms which are known by retailers for a certain quality of reproduction work. The point was borne out by the other two manufacturers I met. One of them had been making modern furniture for many years, but he had sold none of it on the domestic market. The furniture, which was designed by consultant designers exclusively, had only been used on exhibition stands. When its short life had ended it was stored in the firm's warehouse. As I was allowed to inspect the shrouded contents of this mortuary, I asked him why he did not produce similar furniture commercially. The prompt reply was that he *knew* it would not sell, and his knowledge was based on the fact that his firm sold its own furniture direct to the public. His customers liked to feel that they were buying exclusive designs at a reasonable price. The type of furniture which fulfilled these specifications was quasi-antique with mouldings, fittings and fancy veneers which could be swapped round or even inverted. I was told that at least thirty different designs could be produced in this way while still retaining a common basic shape. Modern furniture was clearly unsuitable for this treatment. The third manufacturer made reproduction antique furniture. His clients were the successful men of business who required environmental flattery in their own homes. A huge Jacobean carved chair was the best-seller. This, and the fact that only 5 per cent of the firm's output was of modern design, is indicative of the demand for expensive reproduction antique furniture. The inability of modern furniture to compete with reproduction in the high price range is, in the opinion of this firm, a fatal weakness. Such phrases as 'lack of character' and 'lack of durability' plainly indicate the nature of this criticism, as they also expose the quality of mind which formulated them.

Of the remaining eight firms, three have made modern furniture but have given it up. In all cases the retailers and the public were blamed. The first of these firms had a long reputation for good quality furniture which was sold to upper- and middle-class stores. The designs steered a middle course and were inoffensively dull. The firm did not employ a designer, so that any 'new' models were inevitably adaptations of stock patterns. I found traces of the modernistic fashion still lingering in the bulk of the range. Faced with this unruffled façade of mediocrity, it was surprising to learn that this firm had, before the war, produced a range of exemplary modern furniture. This range had a short life, because the firm's particular market was unresponsive and also because the directors did not believe that such furniture was necessary. The same obstacles defeated similar attempts in the other two firms. Both of them supplied chain stores exclusively. Having introduced modern furniture to this market one of the firms found that either the price was too high or the design was too uncompromising. The other firm was large enough to produce as its yearly range twelve different bedroom suites, twelve dining-room suites and eight three-piece suites. Its technique was to mass-produce down to a price and not up to a quality. The introduction of a modern range was entirely due to the zeal of one of the directors, who overrode the apathy of his colleagues. A consultant designer was approached, but instead of considering the firm's production technique and the known taste of its market, he preferred to emulate Swedish furniture. Because of this the experiment was doomed at the outset. The Swedish furniture designs which he 'borrowed' were the product of mechanized craftwork, a method which allows for a high proportion of hand-assembling and finishing and is therefore unsuitable for cheap mass-production. When production began it was found that the designs were uneconomical in manufacture. But having committed itself to considerable preliminary expense the firm decided to continue, and so made very little profit on furniture which was inferior to its Swedish counterpart. Moreover, the furniture proved to be 'very difficult to sell'. The underlying reason for failure was, undoubtedly, the inexperience of the manufacturer. He allowed an outside designer to have full rein and he thrust foreign designs upon a wholly unprepared public. As these three cases illustrate the folly of trying to sell to a wide market furniture too advanced in design, it might be salutary to consider an alternative method of 'educating the public'.

The experience of the remaining five firms in this group illustrates how it is possible to adopt a steady programme which permits one or two modern designs to be included in the yearly range. These firms are vitally important in this context because they are representative of the section of the trade mainly responsible for the steady increase of modern furniture, which we noted earlier. It is not to be expected that the production of these firms should be of the very best modern standard. The point is, that already having a market for their nondescript designs they have been able, by slow degrees, to raise the taste of both retailers and public. It is entirely due to these firms that one can find some modern furniture in the department stores. A responsive and hitherto neglected public is now being served, not only on the basis of quality and price, but on design as well. The manufacturers told me how dissatisfied they were with the bulk of their production. They pointed out that the task of improvement would take many years. To illustrate this, here are some comparisons between the sales of modern and nondescript designs. In one firm the weekly sales of a semi-reproduction fireside chair, i.e. 500, equalled eight month's sales of a modern chair. Another firm had sold 600 chairs of execrable design as against 200 modern chairs. In each case the prices were comparable. In only one firm did the manufacturer blame the retailer for this state of affairs. The rest were satisfied that their figures were an accurate reflexion of public taste.

Why, in view of this, have these firms continued to produce modern furniture? In the first place it would be wrong to suppose that if the production of modern furniture ceased the sales of the other type would rise. The fact that it has sold at all indicates that there is a market for it which could be satisfied in no other way. Moreover, the modern furniture was introduced later, and the sales which resulted have been followed by a corresponding decline in the popularity of the nondescript furniture.

The last group of firms, seven in number, is representative of the largest units in the trade; those which control nearly 80 per cent by value of the trade's output. Their furniture is to be found in all the chain stores and department stores, while a little of it is sold in the high class shops. In these firms I found little enthusiasm for modern design. Although none of them had experimented with it they felt certain that, if it were attempted, the initial turnover would not be sufficiently rewarding.

These firms produce furniture which is heavy, clumsy and lifeless. To alleviate this they apply coarse pressed-wood scroll work, or glue thin strips of wood in jazz formations to various portions of the veneer. They attach fussy handles in gilded filigree brass to the doors and drawers. The ensemble is then coated with a thick, treacle-brown varnish and finished by a bright, glaring polish. Many respected period furniture styles can be traced in these productions. Usually two of them are found together, hence the adjectives 'Jacobethan' or 'Tudorbean'.

I asked each manufacturer in turn why he produced such furniture, and I received the following answers. Without exception they disclaimed responsibility for the designs. The retailer and his buyer were to blame. 'We are entirely in the hands of the retail buyer', said one manufacturer who employed well over 1000 workers, though another manufacturer pointed out that even if the retail buyer were amenable, his own salesmen would have no faith in modern furniture. The retailer constantly demands variety. He does not want something startling, or anything which is in the least unconventional, but just a variation on the last best-seller. This state of affairs has not yet reached the intensity experienced before the war, but in some firms it appears to be very near it.

The long production runs which were possible between the two wars, meant that fewer *basic* designs were available. Firms were forced to supply an 'endless variety of superficially different variants of the same design and [this] resulted in the application of all sorts of decorative features (known colloquially in America as "borax") as a means of creating this variety'.[1] One large manufacturer carried 800 different pieces in his year's range. Trade exhibitions were held more frequently before the war (in 1938 there were eight) and for each one something 'new' was required. 'Design for mass-production', said one manufacturer, 'consists in juggling with a number of standardized parts, so that mouldings and fittings can change round.' Mouldings in any case are essential to some manufacturers, because they can be used to deflect the purchaser's attention from weak points and to cover up bad workmanship. I was told that furniture of modern design demands more accurate craftsmanship than could be afforded for tax-free furniture. For instance, a flush-veneered door would take more time and care to produce than a spindle-moulded door. If the example of utility furniture did not disprove this opinion, then one need only mention the many firms that are making these doors today. Furthermore, if the standardization of parts is essential then good shapes and good proportions certainly do not cost more than bad. The first group of firms we considered has proved this.

To return to the seven large firms. One manufacturer called his furniture 'damned ugly' and said that he made it to please a new market ('the uneducated masses'). He added that the wider his market grew the lower became its standard of design. He admitted that he had made no attempt to discover what the taste of his public was like. All manufacturers stressed that there was a general public reaction against the austerity of utility furniture. In several of the firms it was pointed out that the shareholders would not be willing to sanction any change in policy which might involve financial risk. 'Much of the enterprise which existed in the industry went in finding new ways of making the same thing cheaper, instead of meeting the competition by improving the furniture itself.'[2] As all the firms were large, producing quantities like 1000 bedroom or dining-room suites a week, it was considered unwise to tamper with the established policy for the sake of a better standard of design. Finally, no manufacturer I met felt himself to be morally involved in his business. The favourite phrase was: 'We are not in business for the sake of our health.'

The attitudes, the methods of manufacture, and the design standards in the firms I visited have now been discussed. A number of ancillary factors still have to be dealt with.

[1] *Board of Trade Working Party Report on Furniture*, H.M.S.O., 1946, p. 114.
[2] *Ibid.*

The manufacture of mouldings for furniture forms a specialized section of the trade. While some furniture manufacturers carve their own ornaments, either separately or on the finished article, the bulk of the trade prefers to obtain them from a specialist. There are two main types of moulding in use; pressed wood and machine-carved wood. The first type is made by forcing the pattern of a steel die into the surface of softened wood. The effect achieved is meant to suggest hand-carving and, because of its low cost, is in great demand. Machine-made carvings offer a better imitation of hand-work but are more expensive than the pressed type. In a few cases carving by machine costs even more than hand-carving. This is especially true in connexion with firms at High Wycombe, where hand-carving is comparatively cheap. For the rest of the trade machine-carvings are preferred because they can be supplied in a consistent design for long-production runs.

The manufacturer I interviewed briefly explained the technique he employed. The machine was similar to what is known in Germany as the 'walking chisel', and the tool used is a special kind of router. The machine was capable of following the chisel-cuts of the model, provided that the design was not too intricate. Standardization was inevitable, but the firm's clients were satisfied. In no moulding did I find a conscious attempt to reproduce either the wood-carver's touch or his inconsistencies. In only a few cases was the machine's work finished by hand. This firm supplied the lowest end of the furniture trade, and the standard of its designs was conditioned by the taste of its clients. In addition to a large stock of popular patterns, special designs supplied by clients were reproduced. It was exceptional for a client's designer to trouble to work out his moulding on the machine. The stock patterns available embraced every known furniture style, from 'Medieval and General Antique Styles' through 'Chippendale' to 'Modern'. About 90 per cent of the firm's output consisted of reproduction mouldings. The rest belonged to the jazz era, and was called 'Modern'. I saw only two mouldings which the manufacturer had made for his own pleasure and both of them far surpassed anything available in his stock range. One of them was a simple pattern carved through a walnut veneer so that the wood beneath, which was sycamore, showed through. The possibilities of this 'organic' method of decoration are completely unrealized by the trade in general.[1]

Three types of designer are employed in the trade: the staff designer, the consultant, and the itinerant free-lance or commercial designer. Of the third I found little trace. He sells mainly to the small manufacturer (say, twenty-five to fifty employees) at the rate of a guinea to thirty shillings a sketch. His market does not extend to the mass-

[1] See Pl. LXXX, 2 and 4. For a description of some ways to decorate modern furniture, see C.G. Tomrley, 'Contemporary Decoration for Furniture', *Design*, November 1952, pp. 7–11.

producing firm with a drawing office on the premises, nor to the independent cabinet-maker. In the first group of firms we discussed, six of the manufacturers are also their own designers. They have, however, also used consultant designers with admirable results. The remaining five firms employ a staff of full-time designers. The salary of the head designer would be between £800 and £1000, according to the size of the firm and not the variety or quality of the work required. Consultants are employed occasionally, but the manufacturers I met complained that they knew very little about manufacturing techniques or market requirements. Despite careful briefing, it frequently happened that the consultant had to hand over his ideas to the staff designer so that the production drawings could be made. The overall impression I got from these firms was that the consultant designer is steadily losing ground. Many of the qualities associated with the consultant are now being embodied by the staff designer. For instance, manufacturers are educating their designers by sending them to the Continent more frequently. The design studios are brightly lit and well-equipped. In addition, I noted admirable selections of magazines and other literature concerning architecture and industrial art. These designers may not possess the experience of the consultant who designs for several trades or industries, but they know the full capabilities of a modern furniture factory.

In the next group, the three firms which specialize in reproduction antiques use consultant designers for their modern furniture. In all cases I found that the consultant was given a free hand, and the results, which were always stimulating, tended to be freakish. With no specific public in mind the consultant designer, like any other artist, is inclined to be spasmodically clever and to lack a durable consistency in his art. The three manufacturers who have now ceased to make modern furniture once looked to consultants to supply the designs. Nowadays the manufacturers themselves do all the designing that is necessary. I have been told that this practice is common amongst the majority of High Wycombe firms. The many small firms which are situated there are usually family businesses and the manufacturer's son is also the designer.[1] He may have had some art training at the local Technical Institute, but it is improbable that he would have continued it for more than two years. In High Wycombe I found little creative work. The designs for the lower price range consisted chiefly of copies or adaptations of other Wycombe manufacturers' best-selling lines. The remaining five firms in this group use both consultants and staff designers. Opinions varied as to which type of designer was the more satisfactory. The staff designer usually has some other executive position in the business. In a few cases he is also a director. Salaries vary from £400 to £1000 and more. One manufacturer said

[1] 50 per cent of the firms in High Wycombe employ thirty persons or less.

that he would have little use for a designer who was worth less than £1000 a year. The fees paid to consultants also vary. One manufacturer paid £150 for a dining-room suite and 15 guineas for an easy chair, while another paid a well-known consultant designer £500 a year retaining fee and a royalty on sales. The salaries here are generous and they are a *true* reflexion of the importance which this section of the trade attaches to modern design.

In the last group of firms, where the standard of design is so bad, no consultants were employed. Each firm has an extensive drawing office where designs originate. Generally it is not considered necessary for the designer to travel. One manufacturer said that he required his designers to have a 'decorative flair', and another said that he merely wanted someone to get 'a few ideas from magazines' and adapt them to his production technique. Nevertheless the salaries paid to the head designers are between £750 and £1500, and thus the highest in the trade.

To sum up: The emphasis in this survey of the trade has been placed upon those firms which command a wide market and are prepared to include, year after year, some modern furniture in their range. By doing this, I do not wish to discount the efforts of the first group of manufacturers we discussed. Before and after the war, they have been responsible for formulating the idiom or style which we call modern. Their market has widened significantly in recent years, but their more experimental work, which is fundamentally necessary, will inevitably evoke a limited public response. Their public is drawn from the middle and professional classes, and is concentrated in London.

The section of the trade which this group represents has roots in the tradition established by William Morris, Ernest Gimson, the Barnsleys, Sir Ambrose Heal and Gordon Russell—a native craft tradition which has had, until recently, little to do with the machine and mass-production. This tradition in which the best modern furniture is produced, has been modified by two important factors: the Bauhaus[1] and Scandinavia. The effect of the Bauhaus on furniture design has been entirely wholesome. It emphasized the importance of proportion as a basic consideration in design, and it demonstrated, through its work, the advantages of machine production. In this country the greatest achievement of the Bauhaus type was Plan furniture, which was designed by F. Schuster of Vienna and adapted by Serge Chermayeff. It proved that the standardization of parts for mass-production was not detrimental to the proportions of individual pieces. The fact that this furniture is no longer available is significant because it illustrates our inability to use creatively the accumulated experience of the Bauhaus. The sterility which one notes in the many examples of

[1] For an account of the Bauhaus, see pp. 182–4.

'unit' furniture which are on the market is a further confirmation of this.* The influence which came from Sweden and Denmark, especially since the war, has several aspects. It re-emphasized the pleasing effect which could be obtained from a natural wood finish, it re-introduced to us the proportions which were used by English craftsmen in the eighteenth century, but above all it showed how slender supporting members and 'sculptured' shapes could be used, with economy, to achieve a high standard of design.

These two influences are, in effect, essentially opposed to each other. The one implies the use of the machine; the other the craftsman. To neglect one for the sake of the other would be futile. There is already in this country a tendency to accept uncritically the standards set by Sweden and Denmark and to design furniture in terms of the craftsman's skill. The social limitations of this approach were unwelcome, but occasionally apparent to William Morris. He knew that the work of his firm, all made by hand, was too expensive for the public he would have wished to serve. Today, if we remember the example of the Bauhaus and make use of it, there is no reason why modern furniture should not be within the reach of everyone. If used intelligently the two influences would enable this small but progressive section of the trade, which has suitable machinery (in co-operation with sympathetic retailers and a responsive public) to establish a design standard which is both modern and British.

As an example of what can be done we may consider, briefly, the work of Gordon Russell. Mr Russell started his business by learning the craft of cabinet-making and he has made with his own hands many outstanding pieces of quality. He also designed most of the early work of the firm. Later on he was joined by another member of the family with some architectural training and a progressive taste. By 1937 most of the creative work was in the hands of a staff designer—one of the rare persons in the trade who combines high artistic standards with technical and commercial knowledge. As the work of this firm was decidedly modern and unconventional from the beginning, it proved difficult at first to find an adequate market. Some support seemed advisable, and this was provided by another side of the business, in the hands of the same company. Because of these initial difficulties most of the work turned out consisted of special pieces which were expensive, although as far back as 1929 some medium-price furniture was made, including a table which was only taken out of production in 1949. The firm's ability to produce medium-price furniture in quantity was improved when it started to manufacture cabinets for Murphy Radio, Ltd. A satisfactory balance was attained between mass-production and high quality cabinet-making. In our connexion it is important to draw attention to this instance of the combination of mass-produced articles, radio cabinets, and the making of a comparatively small quantity of articles too advanced and uncompromising in design to be a sufficiently safe proposition by themselves.

Until 1939 the firm's market for furniture was limited to its own retail branch in London. But in that year a new body, the Good Furnishing Group, was set up. It was formed by Gordon Russell as manufacturer and P.E.Gane of Bristol, and Geoffrey Dunn of Bromley, amongst others, as retailers. Its aim was to produce in quantity furniture of good design which it would sell to the middle-class market. A few pieces were made, including a dining-room suite. The war brought the Group's activities to an end but in 1952 it began to operate once more.* The idea behind this project serves to sum up one of the most important points in this survey of the trade: the need for the manufacturer and the retailer to be jointly responsible for putting furniture of a high design standard on the market at a reasonable price.

CARPETS

Carpets were first woven in this country at the end of the seventeenth century. The weavers were French refugees who had settled in the southern part of England, in such places as Wilton and Axminster. During the eighteenth century the demand for carpets increased and the weavers spread northwards, to work in the towns already manufacturing woollen goods. In Kidderminster, where broadcloth was produced, carpet weaving was established by 1763. Carpet looms had been set up in several Yorkshire towns by the middle of the century, and in certain Scottish towns, notably Glasgow, a few years later. Today there are about seventy firms in the industry, although only thirty of them are of any consequence. Of the 27,000 people engaged in carpet manufacture, some 16,000 are employed by the ten largest firms. The industry is situated in three main groups; the West Riding of Yorkshire, Scotland, and Kidderminster, which is the largest group, with twenty-five firms and 10,000 workers.

I visited eleven firms, employing between them more than 14,000 workers. This means that my information represents more than half of the whole industry. Seven of the firms are in Kidderminster, two in Scotland and two in Yorkshire. Five of the firms I saw are amongst the largest in the country (with 3800, 2500, 1800, 1750 and 1700 employees respectively). Two firms I visited are of medium size, with 1200 employees each; while the remainder are much smaller, with 600, 150, 100 and 80 employees. I saw no studio weavers or firms weaving carpets by hand, because the industry, which concerns itself with bulk production by machine, remains uninfluenced by the standards of design in hand-made carpets. In the firms I visited four main types of carpet are made: Wilton, Spool Axminster, Gripper Axminster and Chenille. To these may be added the Brussels weave, though it is now relatively uncommon in the industry. While there is no need to give a full explanation of the various weaving techniques, a few of the limitations which most affect design should be noted.[1] Both Wilton and Gripper Axminster carpets are woven on looms controlled by the Jacquard mechanism. The number of colours which may be used in Wilton is five, and in Gripper Axminster eight. Although in both cases additional colours can be included, by a process known as 'planting', the effect of the carpet when woven may be marred by stripes of colour which tend to break up the design. This is especially true in Wilton carpets, and the most skilful designers in the industry rarely use more than twelve colours, whereas in Gripper Axminster carpets as many as fifty colours may be 'planted' without seriously affecting the design. For Spool Axminster and Chenille there is no such limitation, and sixty or more colours may be used with little trouble.

There are three types of designer in the industry: the staff designer, the consultant and the free-lance or commercial designer. Before the war designers for the factory studio were usually recruited from secondary schools, with little or no art training. For three or more years, depending on their aptitude, they did hack work for the head designer, such as copying out designs on to squared paper (known as 'point paper'). Studios were large, with twenty-five to forty persons according to the size of the firm, and promotion was slow. In most firms the designers were limited in their experience to the factory studio and the routine work in progress, but in some cases they were allowed one afternoon a week to attend the local art school. Visits to exhibitions rarely extended beyond the British Industries Fair. Manufacturers, feeling that there was a lack of brilliance in their studio products, bought sketches from commercial designers as 'food for the studio' (this was the expression used by one works' director). The majority of these free-lance designers were French and German, and they were paid an average of £5 per sketch. The designs were then handed over to the studio and 'mashed up', as one director called it, and not merely redrawn on point paper. A pleasing all-over pattern with, say, some spots and waves in various shades of light grey would, at the end of the uncanny process, have become a centre motif, repeated in the four corners. The colours would be entirely different, the light grey having given way to orange or brown. Consultant designers, when they were

[1] See R.S. Brinton, *Carpets*, third edition revised by J.F.C. Brinton, Pitman, 1947. This extremely useful handbook has been written by the head of a well-known firm in the industry. All aspects of carpet manufacture, including design, are described in detail.

employed, received better treatment. The head designers in small firms earned about £350 a year, in medium firms £450 to £600, and in the few large firms £800 to £1200. The head designer would occupy a relatively unimportant position in the firm and would not have 'a status high enough to enable him to develop a design policy'.[1]

Since the war this situation has improved in many ways. A more enlightened attitude towards design and the designer was noticeable in many of the firms I visited, especially the larger ones. The only exceptions appear to be the two Scottish firms, where I found conditions similar to those just described. In Kidderminster and Yorkshire, young designers are recruited from the junior art department of the local art school, though many still come straight from secondary schools. In a few cases fully trained art school students are employed. As a rule manufacturers feel that, while these students would be of little use to the firm for the first two or three years, they would require to be paid according to age, and not the value of the work turned out. Although this policy appears to be short-sighted, one must bear in mind the fact that the art schools in the main areas of production have in the past failed to gain the manufacturer's confidence in their ability to train carpet designers. Factory studios are smaller now. In the largest firm I visited there were only twenty designers and copyists. For firms of between 1000 and 2000 employees, the studio staffs would number about fifteen, of which one-third would be creative designers. It is generally accepted by manufacturers that their designers should be allowed some time off during working hours. Two firms had formed sketch clubs, and allowed their design staffs to spend one day a week in the country-side. In other firms the student-designers were encouraged to attend the art school for one full day and two evenings a week. In addition, many of the manufacturers I met made special arrangements for their studio staff to visit important exhibitions, chiefly in London. In a few firms the head designer would be expected to accompany the managing director and the sales' manager on visits to foreign markets.

Coincident with these improved conditions for designers, is the fact that very few free-lance designers have been employed in the industry since the war. Manufacturers felt that it was no longer necessary to inspire their staff designers with the sketches of an outsider. Furthermore, they emphasized the difficulties which arise when free-lance designers are employed. For instance, the methods of manufacture, the types and qualities of carpet produced, vary with each firm, and it would be impossible for the free-lance to submit a design which could be sent straight to the weaving sheds

[1] *Board of Trade Working Party Report on Carpets*, H.M.S.O., 1947. The *Report* presents a detailed review of all matters relating to carpet manufacture. The section on 'Design', pp. 18–23, is particularly stimulating.

without alteration. It was also pointed out that the sketches of English free-lance designers were, as a rule, badly presented, and in many cases unsuited to carpets. The free-lance designers were said to have no particular type of carpet in mind when offering their sketches; indeed, these sketches were often designed to appeal to manufacturers of woven textiles and of printed wallpaper, in addition to carpet manufacturers. The salaries of staff designers are also much improved upon the pre-war level. In the small firms the head designers would earn from £700 to £900 a year. The medium-sized and large firms would pay from £1000 to £1500 a year and more, to their head designers, and £550 to £750 to their chief assistants. Junior designers at the age of sixteen would receive from £3 to £4 a week. At the age of twenty-one their salary, which would be about £350 a year, would rise in accordance with their creative ability.

In view of the improvements which have come about since the war, it is surprising to find that the status of the designer remains practically unchanged. With the exception of one firm, where a director is the head of the design studio, I did not meet one designer with an authority equal to that of the sales manager. It appears that in most firms in the industry the designer is attached to the sales side of the business, and therefore responsible to the sales manager. Only in rare cases does he have any contact with the weaving sheds, in fact he may never have done any weaving himself, nor have worked at a power-loom in his art school or in the factory.[1] To appreciate the designer's position it will be necessary to consider, briefly, the stages through which a design must pass before it reaches the weaving sheds. The process begins with an inquiry from the sales manager, often in vague terms; for example, 'such and such colours in an Oriental design'. Perhaps the commonest method is to refer to one of the firm's best-selling patterns and ask for 'something like it, but different'. For each inquiry the designer will produce five or more sketches which will be submitted to a committee, consisting of the managing director, the sales and works managers and the head designer. Although it is the designer's task to cater for the taste of this committee, he can, of course, exert some influence by the way in which he presents his sketches. After this stage the designer's opinion ceases to count. The sketch which has been selected is then drawn out on a point paper and passed to the colourist. The colourist's main job is to be sure that the colours will have the correct weight, or balance, when they appear in the finished carpet. To do this he interprets the designer's colours in terms of his own stock of dyed yarns. Although in some cases he may have to dye

[1] This separation between design and execution persists throughout the whole industry. At least 99 per cent of the carpets and rugs produced in Great Britain are machine-made; in all these the worker has nothing to do with design. And even in hand-knotted carpets it is only in exceptional cases that invention and craftsmanship are in the hands of the same person.

yarn specially, it is left to him to decide which tones and shades are to be used. The creation of a carpet design is, therefore, in the hands of two persons, the designer and the colourist. In two Kidderminster firms which I visited the designer is also the colourist, clearly the ideal situation, but in other firms there is little collaboration between them, and designers often complained that their work had been handled unsympathetically.

There are two main markets for the carpet manufacturer: the contract market (hotels, public buildings, ships, trains, etc.) and the domestic market (chiefly private homes). It is for contract work that the best modern designs are produced. There are three important reasons for this. Many large firms have a small studio in London specially for contract work. Some of the most progressive designers in the industry are to be found in these studios, because their work enables them to experiment more than their colleagues at the factory. When designing a hotel carpet for example, the designer would visit the hotel, meet the manager or his architect, and be able to make suggestions on the spot. This direct contact between the designer and the purchaser has had a salutary effect upon design standards, especially when it is an architect's job to commission the carpet. The absence of this contact is, as we shall see later, partly responsible for the lower design standards in domestic carpets. For all types of carpet a minimum number of yards must be woven before the design becomes economical. A similar length of carpet may be woven for either contract work or the domestic market, but where the latter is concerned the manufacturer must be sure that his choice of design will satisfy an unknown number of people whom he has never met. It follows, therefore, that designs for domestic carpets must be broader in their appeal than designs for contract work, and in the carpet industry this implies a lowering of standards. The point is important, and we shall have to return to it later.

For just over two years during the war no carpets were made in this country. When production began again it was on a very limited scale, and by 1945 the industry's output was only 20 per cent of its pre-war total. Although production has increased steadily in recent years, the demand for domestic carpets exceeds the supply. This shortage in the home market is also intensified by the fact that the industry is exporting nearly 50 per cent of its production (the equivalent figure in 1939 was 17 per cent). From this it appears that carpet manufacturers will be able to enjoy the conditions of a sellers' market for several years. Chiefly because of this situation I found that nowadays fewer new designs are put into production than before the war. 'The total of new designs put into stock ranges in Britain in the one year, 1938–39, must have been well over three thousand.'[1] In one large firm I visited, 300 new designs were

[1] *Working Party Report*, p. 20.

added each year before the war; 100 per year since the war. As the remaining ten firms gave me similar proportions, it would seem that the number of new designs used by the industry is approximately one-third of its pre-war requirements, or about 1000 a year. This means that each design now has a much longer production run. One manufacturer told me that a best-selling design could be expected to stay in the loom for ten years, another said twelve years for his best-seller (before the war five years would have been exceptional), and that the average length of life for all his designs was from two to five years. In one firm before the war a best-selling Axminster would run to 20,000 yards, whereas now the figure would be 80,000 to 100,000 yards. Another firm's output was so large that it measured its best-selling lines in terms of miles. The social consequences of this are considerable, especially where the home market is concerned.[1]

It is openly admitted in all the firms I visited that, when selecting a new range of designs, the needs of foreign buyers are given first consideration. No firm was able to show me a range designed exclusively for the home market.* I was shown the Canadian, the Australian, the Scandinavian and other foreign ranges, but no British range. To answer my question concerning the taste of the home market, manufacturers could only indicate designs which had sold both at home and abroad. 'As for the home market', said one manufacturer, 'it is only too glad to get all the odds and ends we can spare.' In the peculiar circumstances in which the industry finds itself, this remark is entirely justified. My reason for quoting it is not to deprecate the industry's preoccupation with the export market, but to emphasize the fact that no definite conclusions about the taste of the home market, as far as carpets are concerned, can be reached at the present time.

With this in mind let us now consider the design standards of domestic carpets. The first important point is that the modernistic style has almost disappeared. The style was introduced by British carpet manufacturers in the mid-1920's, to counter German competition in the Dutch market.[2] It soon became popular in this country, being the visual correlative to jazz music, and dominated all but the most expensive carpets. The motifs used were derived from a deplorably misunderstood Continental cubism, and consisted of jagged geometrical patterns which were entirely meaningless. The prevailing colours, usually on a fawn background, were a brown, a blatant orange and, towards the end of the 1930's, a grass green no less blatant. Although

[1] For example, if the living-room of a small house requires 10 yards of carpet at the standard width of 27 inches, then 100,000 yards of carpet in one design would be sufficient for 10,000 rooms. From this it will be seen that a considerable responsibility must be attached to all concerned with the choice of a design which may enhance or degrade the lives of 10,000 families.

[2] *Working Party Report*, p. 19.

some of the manufacturers I met made modernistic carpets for the Australian market, the majority assured me that as far as the home market was concerned the style had died in the war. On the other hand, I met one of the largest manufacturers of cheap Axminsters who fully intended to sell his modernistic patterns on the home market for as long as possible. Indeed, for cheap carpets it seems likely that modernistic designs will be made and sold 'until such time as customers show that they no longer favour them'.[1]

There are four main types of design in current production. The first, and the most popular of these, is traditional Oriental design. 'Much ingenuity has been displayed in the copying and evolution of these designs. They meet a wide demand at home and abroad, have unquestionable merit for a great many purposes, and will continue to take an important place among the products of the industry.'[2] If we ask what this 'ingenuity' implies, we shall find that, for the most part, it has been 'displayed' at the expense of the design. If a manufacturer wishes to imitate a Persian carpet on a 'Wilton' loom he will have to contend with certain limitations. It is not possible to weave a Wilton carpet closer than 130 picks to the square inch, and even then fine worsted yarn must be used. But the carpet which he is attempting to copy has been knotted by hand up to 250 or 300 picks to the square inch. In Wilton weave it is impracticable to use more than twelve colours, but in hand-made carpets there is no limit. Again, the dyes used in this country are chemical and very much harsher in tone than the vegetable dyes used in Persian carpets. All three limitations contribute to make the design of the reproduction loose and ill-defined, the colours coarse and unsympathetic. Mr Brinton deplores the attitude of the manufacturer who boasts: 'This is an exact reproduction of the famous carpet of so and so, made in the fourteenth century.'[3] In fact, anyone who compares the two carpets will discover that the reproduction, far from being 'exact', is simply a parody of the original.

It is not possible to classify the second type of design under one heading. It can consist of neo-Classic motifs and be called either 'Adam' or 'Empire'; alternatively it may be based on natural motifs, such as flowers, and be called 'conventionalized floral'. The latter type of design is perhaps less popular today than when William Morris first revived it more than sixty years ago. At that time it provided a refreshing contrast to the raw naturalistic style which had beset the industry. The merits of the style may be gauged by the way in which Morris, Voysey, Crane and others used it. The style was, they said, suitable for all two-dimensional design, especially carpets, because it enables the designer to develop breadth and flatness with simple motifs. Even today these criteria are acknowledged by many carpet designers. Amongst the

[1] *Ibid.* [2] *Ibid.* p. 18. [3] R.S. Brinton, *Carpets*, p. 104.

other rules observed are the following; a carpet design must 'look right' from all directions (as do most Oriental designs), its forms and colours must balance each other, and it must be well-covered in order to make stains, etc., less noticeable.

The third type of design is the chintz. Although chintz designs break all these rules, they are, apart from Orientals, the most popular designs today in all markets. It takes great skill and experience to design a good chintz carpet, especially where the colours are limited. In most firms I found that chintz designs were the special province of the head designer, and invariably his work was claimed to be the best in the industry. The chintz designer works from nature, and it is by nature that his patterns are judged. From our point of view there is little merit in the technical artistry of the chintz design. Little attempt has been made to treat flowers creatively, or to organize them into a pattern which is both serviceable and decorative. Many designers I met spoke about bunches of flowers conceived in a contemporary manner, when they really meant that a certain simplification had taken place.

Fourth and last comes modern design; and by modern I do not mean modernistic. Before the war, if a modern design were required one's choice would be limited to expensive hand- or machine-knotted carpets. The only alternative at a reasonable price was the plain carpet. Today, thanks to the demands of the Scandinavian market, it is possible to obtain a few modern designs of a fairly satisfactory standard in all qualities. I have been told in several firms that when these designs were first displayed the home market buyers were unimpressed. It is only in very recent years that the so-called 'Scandinavian' carpets have been seen in British shops. London appears to have the greatest demand for them although, like modern furniture, they can also be found in a few shops in several provincial towns. But, unlike furniture, it is not possible to find an exemplary modern design in the range of any carpet firm. Most of the designs are unexciting and uninspired, mere copies of an alien idiom. I am in sympathy with the head designer of one firm who described them thus: 'They illustrate the worship of the doodle...mere noughts and crosses...and there is not enough designing in them.'

Nevertheless, the difficulties which face carpet designers should not be overlooked. In hand- and machine-made textiles it is possible to create 'organic' designs by using several different weaves in the same piece of cloth. For woven textiles the designer works in three-dimensional terms, but in machine-made carpets he is limited to a flat surface and no variations of weave.[1] As the carpet designer is unable to obtain subtle

[1] There are two exceptions to this general rule, and both have been introduced to this country from the United States since the war. They are the carved pile carpet, with two or three levels of pile, and carpets with simple weave variations, i.e. one row of cut yarn alternating with another which is looped.

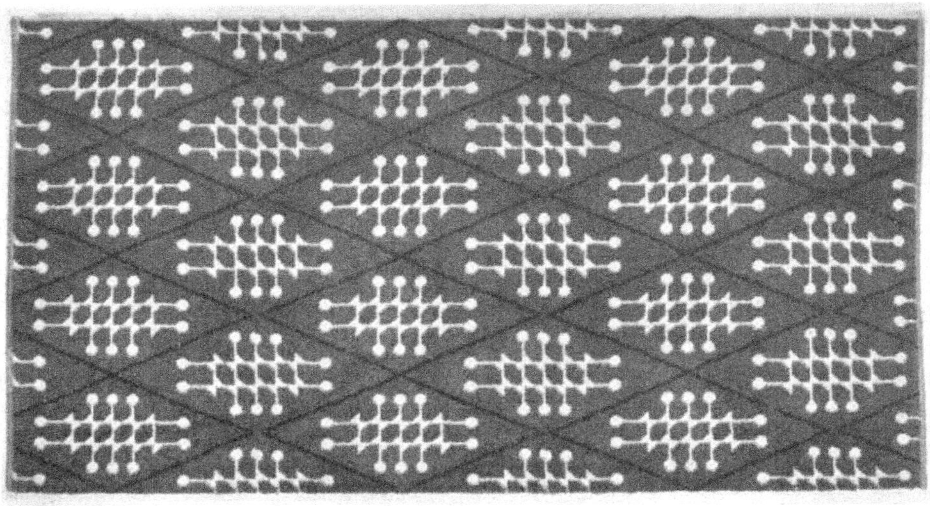

1. Firms dealing direct with large-scale users of carpets, such as architects, are likely to have somewhere in their stocks a small range of modern designs. Where this stimulus is missing, as in the general domestic trade in home and foreign markets, the manufacturer is less inclined to risk untried modern designs. The motif in this Wilton carpet, which may become trying and therefore dated if studied too closely, has been thoughtfully organized. *Designed by Ronald Grierson for S. J. Stockwell & Co. (Carpets) Ltd.*

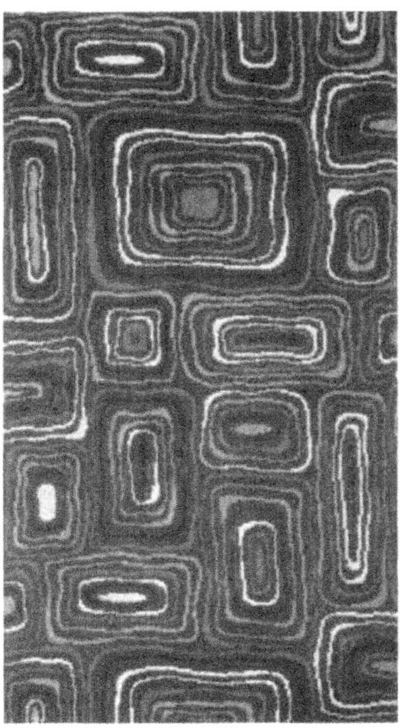

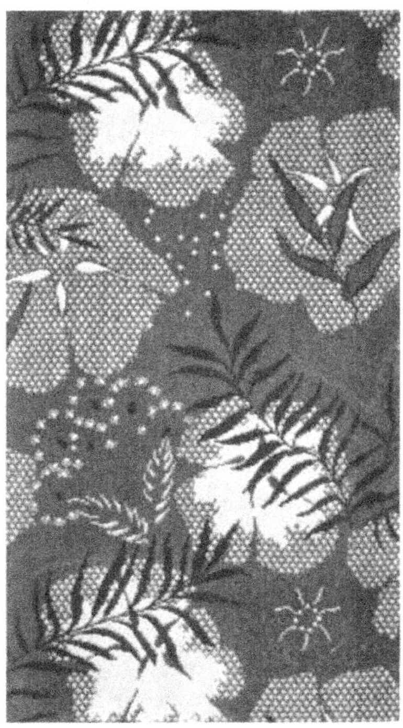

2 and 3. Two contrasting influences in the carpet industry are shown here. On the left an abstract pattern has been constructed with varying, related shapes to make a carpet suitable for all purposes. On the right some perennially popular verdure motifs have a broad treatment over textured shapes so making a distinctive modern pattern. *Made by Blackwood Morton & Sons Ltd.*

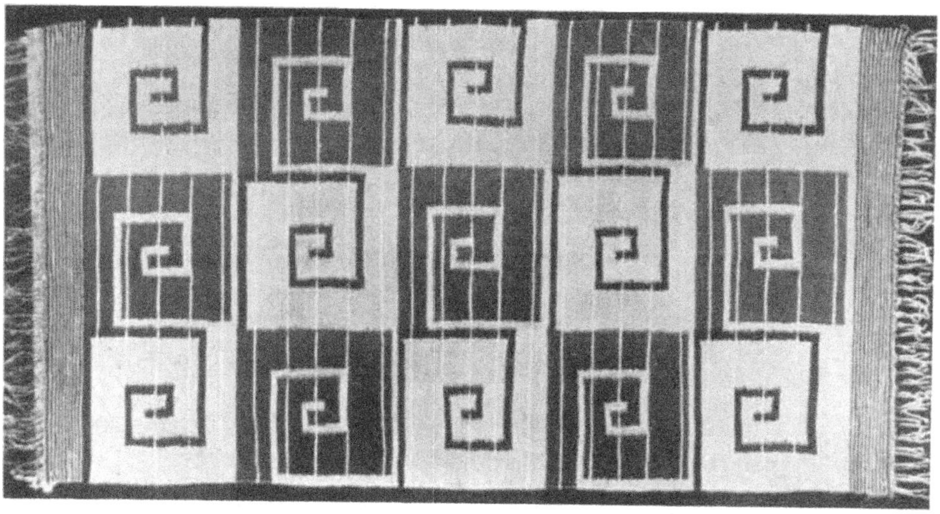

1. The design of hand-woven carpets has little influence on the industry. This is to be regretted, for they are often more adventurous in style and make more use of patterns which could as well be machine-produced in quantity at a lower price. The designer of this rug has been putting forward designs of equal merit for twenty years. *Designed by Marion Pepler for S. J. Rybczyk.*

2. Manufacturers and buyers say that a carpet must 'look right' from all directions. Often this rule is waived for contract work, as in this case the heraldic subject precluded a less formal treatment. This is a *tour de force*, but not essentially a carpet pattern. *Designed by John Palmer for John Crossley & Sons Ltd.*

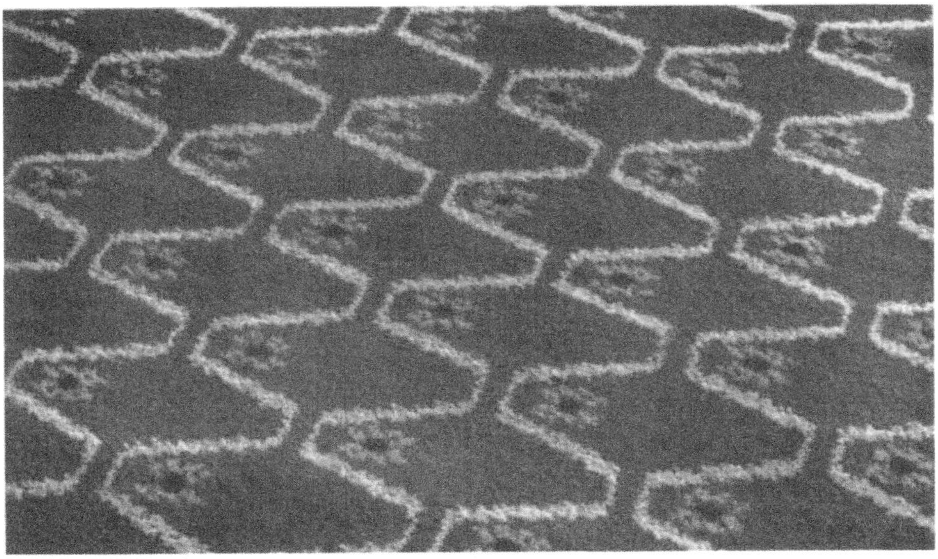

1. Breadth and flatness, essentially two-dimensional characteristics in the handling of simple motifs, are criteria held by carpet designers. For such reasons this example would be approved, although as a good modern design its introduction to the home trade as late as 1952 was almost revolutionary. *Made by James Templeton & Co. Ltd.*

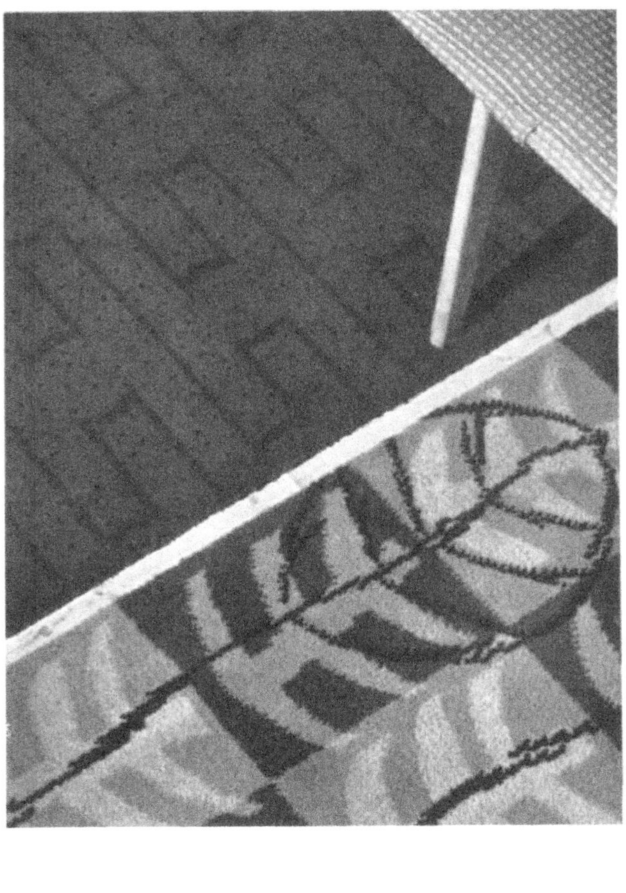

2. In all but the most undisturbed areas it is important for a carpet to be well covered with pattern or strong colour to obscure stains. For contract work this is often the buyer's first consideration. In both examples the 'covering' has been well conceived to meet this point by working to modern standards of taste. *Made by John Crossley & Sons Ltd.*

PLATE XII

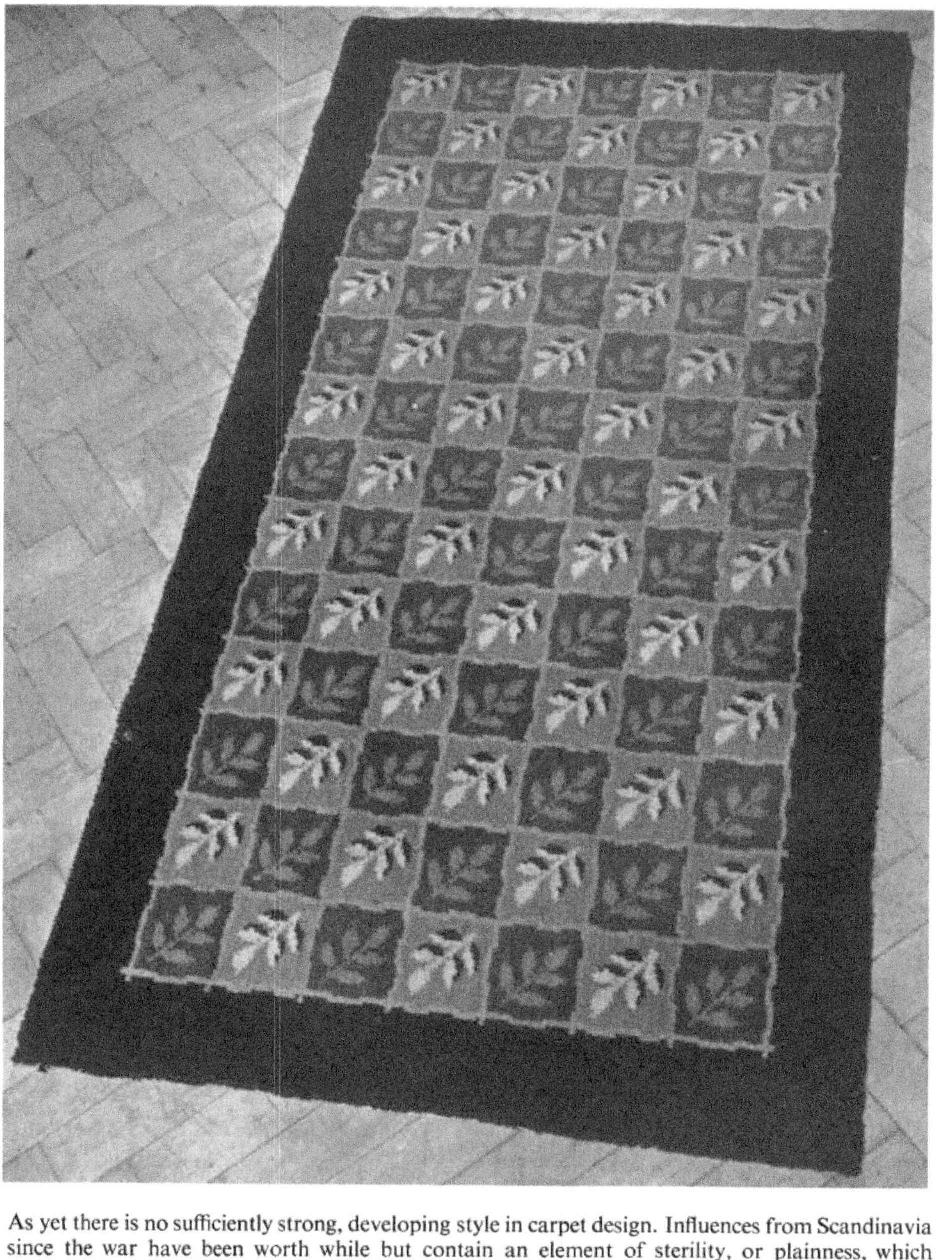

As yet there is no sufficiently strong, developing style in carpet design. Influences from Scandinavia since the war have been worth while but contain an element of sterility, or plainness, which designers have hardly started to overcome. Unlike their Oriental counterparts centuries ago, they find themselves without a living symbolism to inspire their work. Again and again the return is made to Nature, as in this case, where the arrangement has become a pattern and not remained a succession of leaves. *Made by Quayle & Tranter Ltd.*

effects by manipulating his material, he relies almost entirely upon pattern and colour. When deciding upon the patterns to be used the designer invariably turns to nature, chiefly flowers and foliage. Unlike the Oriental craftsman, whose work he copies and adapts, he rarely draws upon any source of living, intellectual symbolism. Several designers have pointed out that there is, at the present time, no contemporary symbolism which could be used to inform their work. They regard Oriental designs as having some meaning, or content, but few would claim that the Tree of Life, the Cone, or the other symbols which are used, have any relevance to contemporary society.

At the present time manufacturers are not encouraging their designers to express themselves freely. Although staff designers now have better amenities, they are still hemmed in by the conventions of their craft, and the opinions of their superiors on public taste. At the same time, I could not help noticing the prejudice which many designers showed when asked to consider the work of certain consultant designers. The presence of the staff designer is, undoubtedly, the first stumbling-block in the path of the consultant. This would not matter if one could feel sure that the staff designer is capable of developing a living idiom for carpet design unaided. But, from the evidence which I have seen in the studios of ten head designers, I am inclined to feel that within the industry there is not the enthusiasm nor the intelligence which the task requires.[1]

Finally, we come to the manufacturer himself, who, as the Working Party points out, '...should comply with, as well as form, the public taste'.[2] When questioned, the manufacturer feels sure that he knows what the public wants. He refers to the sales figures of various designs to prove that his knowledge is correct. The fact that this is no proof at all is irrelevant to the manufacturer, because he knows that his public in the home market will, at the present time, accept almost anything he cares to offer. But, even before the war the manufacturer's attitude was the same, for all firms produced similar designs for similar prices, and the public again had no choice. Very few, if any, manufacturers have taken pains to investigate the taste of the public. In the contract market the manufacturer receives direct stimulus from the purchaser, and produces his designs accordingly, but in the domestic field he has to depend on his intuition and the reports of buyers, both of which are based exclusively on what type of design sold last time. The need, moreover, to make his choice of design approximate to the wishes of several thousand private individuals, means that he must be cautious in his judgement. In a sellers' market the manufacturer has on the one

[1] The degree of artistic bankruptcy which exists in British carpet firms may be judged by reference to *The Carpet Annual*, 1950, pp. 18–32, where designs from this country are shown alongside recent productions from the United States, France and Sweden.

[2] *Working Party Report*, p. 21.

hand the advantage of incurring only small risks, but on the other the great responsibility of deciding what type of design the public is to receive. Although one cannot approve of dictatorship in matters of design, it is at least significant that no manufacturer has taken advantage of the sellers' market to experiment with new types of designs.

In summing up it is necessary to consider one important way in which the industry could react to normal market conditions. To a certain extent this depends upon the attitude towards design of the ten large manufacturers, i.e. those who employ 60 per cent of the total labour force. Their position and responsibility are akin to the large furniture manufacturers which we have already dealt with. If they are prepared to make any notable contributions to industrial art they will, I feel sure, have to begin by employing consultant designers. The present policy of taking on boys at the age of sixteen, and training them within the factory, is clearly unsatisfactory. Alternatively, they can engage graduates from art schools; one leading Yorkshire manufacturer has already done this and, for contract work, has received excellent results.

The only way to escape from the stagnation in which the carpet designer finds himself is for the manufacturer to insist that a small section of his range shall be designed in accordance with the highest standards. As we have seen in the furniture trade, the success of such a policy would be assured, because it is founded upon fact and not on theory, on public taste and not on the private opinions of a few individuals.

LINOLEUM

In many ways the state of affairs in this industry is similar to that which has just been described. The industry is small (in 1938 there were 9200 employees) and 70 per cent of its total output is controlled by four large firms. I visited one of these large firms, and one smaller firm in order to obtain a representative picture of the industry. Three types of floor covering are produced: linoleum, floor-cloth and felt base. Linoleum, which was invented in 1860, is a mixture of pulverized cork and oxidized linseed oil pressed flat on a jute canvas cloth. The design may be either built up of inlaid sections or printed on a prepared surface of linoleum. Floor-cloth has a similar base, and is patterned with several coats of oil colour. Felt base consists of bituminized paper felt, overprinted with a flexible, glossy paint.

The head designers I met did not earn more than £1000 a year, and their status was insignificant. There was little demand for the work of free-lance designers, except as 'food for the studio'. Before the war new designs were required at the rate of 400 a year, in the large firm which made thirty to forty different grades of floor covering. When fully exploited these designs would represent some 1300 colour effects. At the present time far fewer designs are required; in fact the smaller firm was content merely to make changes in the colours available.[1] The sellers' market which accounts for this is also partly responsible for the exceptionally low standard of design in the industry. Unlike carpets, the design of linoleum and the cheaper floor coverings is still dominated by the modernistic craze of the 1930's which exists side by side with 'jaspe' and imitation parquet flooring. One of the best-selling modernistic patterns in the large firm was manufactured at the rate of 15,000 yards a week. The least objectionable designs appeared to be confined to linoleum which had been inlaid by hand and was comparatively expensive, and felt base which no longer suffers from a sellers' market. Finally, the manufacturers I met had made no attempt to evaluate the taste of the home market. It is unnecessary to enlarge upon conclusions from these facts. They are identical with those arrived at in the preceding chapter.

LIGHT METALWORK

Although all the firms which I visited work in metal, they differ widely in methods of manufacture and in standards of design. The heterogeneous range of objects dealt with includes metal windows, ornamental entrance gates, staircase balustrades and door furniture. I confined myself to Birmingham, which is the acknowledged centre for all light metalwork in this country. 'All attempts to give a precise statistical framework to these Birmingham light metal trades must fail. Few firms are content with a single product, and the logic which determines the business man in his choice of a subsidiary product is often hard to find. The shifting histories of the trades, the adaptability of capital equipment, the demands of individual customers, often the whims and fancies of manufacturers, have combined to produce a group of trades integrated by every conceivable tie of cross interest.[2] Out of the nine firms visited, only two employ more than 1000 persons. One of these firms makes metal windows,

[1] 'The average expenditure of the industry on design expressed as a ratio to turnover is under 0·4 per cent.' *Board of Trade Working Party Report on Linoleum and Felt Base*, H.M.S.O., 1947.

[2] *Design and the Designer in the Light Metal Trades*, H.M.S.O., 1944, p. 1. This Report by a committee of the Council for Art and Industry, under the chairmanship of R.D. Best, was completed in 1940. Although it refers to the metal trades as they were before the war, its main arguments are still relevant because few changes have taken place in the meantime. The Report is one of the best factual statements concerning industial art in light metalwork available, and I have drawn on it freely. Its scope is much wider than anything I could attempt.

door furniture and central heating appliances. The other specializes in fencing, hurdles and ornamental ironwork. Three firms, with 800, 600 and 450 employees, are mainly concerned with door furniture and architectural metalwork, but each of them has several sidelines. The remaining firms are small, with 300 to 100 employees and less.

The first product to be discussed is the metal window. A brief account of the recent history of metal windows is necessary, because it illustrates on the one hand an intelligent policy of standardization and on the other the inclusion of architects as a decisive force in the setting of a design standard. The two points are fundamentally important because they contrast with the design methods used in other metal trades. In 1919 the important metal window firms in this country, with the help of several well-known architects, decided to standardize the sizes of their windows. They agreed that 'the proportion of the pane should be the square of the diagonal', and they began from a unit measuring $11\frac{3}{8} \times 9\frac{3}{4}$ inches. From this unit they formulated twenty standard types of window. In order to convince their clients that these windows could be used without loss of freedom in design, the firm I visited commissioned fifteen respected architects to design a range of middle- and working-class homes, incorporating any of the twenty standard windows. Models were made of these designs and photographs of them were issued to the firm's clients. A good many architects, I was told, objected for almost as long as ten years to the use of such standards, as they did not want to be restricted by any agreement between manufacturing firms in designing individually proportioned windows. In the meantime many have realized that this standardization has not only kept prices down, but has also resulted in a considerable improvement in the average standard of window design in builders' houses. The fact that here co-operation between enterprising and enlightened mass-producers and a few technically interested architects has led to such excellent results must be borne in mind.

From time to time additions have been made to the range, which now amounts to nearly 200 types of window, but the basic proportion remains unchanged. In 1945 it became officially recognized in a British Standard Specification. In the period between the wars the most popular pattern in domestic windows showed horizontal lines predominating. During and after the war, it was necessary to use less steel for window frames. The horizontal frames were dispensed with, and the predominating motif became vertical. This, and the fact that a vertical pane of glass is better suited to man, who habitually stands upright, explains why the window on Pl. XIII. 1 is today the firm's best-selling pattern.

The market for ornamental gates and grilles, in fact architectural metalwork in general, has changed considerably in recent years. Domestic users have gradually disappeared and the chief clients are now the directors of large industrial undertakings and department stores; that is, the order for such metalwork no longer comes from an individual with a specific purpose in mind, but from a committee. While the general consequences of this change are beyond the scope of the present survey, it must be recorded as a fact that design in this trade is less adventurous today than it was thirty years ago. In each of the firms I visited I found three ways of initiating a design. In the first case the client (he may be an architect or a business man) selects a design from the firm's catalogue of made-up work, and asks for 'something like it'. The designer then produces two or three alternative sketches and estimates the cost of each. In the second case an architect produces a rough sketch of what he wants and asks the designer to fill in the details, 'without clashing with the general scheme' of his building. The third case is where the architect produces detailed drawings of the ornamental work, and makes himself jointly responsible with the designer for seeing that his design can be made within the limits of manufacture and cost. While this is the least common method, it is also the one which is most likely, as we shall see later, to produce a satisfactory modern design.

In nearly every case the design of ornamental metalwork is conditioned by the architectural style of the building for which it is made. For example, a town hall in a quasi-Palladian style would have metalwork in the same idiom. One would expect, therefore, that if a building were designed in the modern style the metalwork would also be modern. But this is seldom the case. From all the catalogues shown to me I was able to pick out only four examples of modern design. While this unfairly restricts the client's choice, if he depends solely on the catalogue for ideas, it also indicates the inability of modern architects to create a decorative style to suit their buildings. Most modern architects have almost entirely renounced ornament, and have kept aloof from purely decorative work. When they are asked to design a monumental entrance gate, most of them feel at a loss. And, on the other hand, those commissioning the bronze doors of a town hall or a bank usually insist on getting some elaborate design, not a piece of work whose effect lies in the pureness of proportion and the quality of the material alone. Faced with these difficulties it is usual for the architect to hand the designer a rough sketch of what he wants and await results. The designer then as a rule attempts one of two possibilities. He either tries to resurrect and modify an example of eighteenth-century wrought ironwork, or to make an adaptation of recent French work. If the designer wishes to appear original he takes the latter course. I found that the designs of Edgar Brandt were the most popular source, but it was pointed out that the intricacies of Brandt's work were too expensive to reproduce faithfully. The result is the very common contemporary style, based on the exuberant interior ironwork of Brandt, but showing an oversimplification of forms owing to the inclusion of thicker sections to give the necessary strength and durability. In other words, a bad imitation of an inappropriate style.

From the beginning of the century until the end of the First World War the trade was dominated by the style of the Arts and Crafts Movement, followed by Art Nouveau. This was superseded almost immediately by a sparse style composed of long straight lines and sharp angles.[1] One reason for the popularity of this style was that it did not require forge-work and was consequently less expensive than Art Nouveau. A very small proportion of English architectural metalwork of the middle 1920's shows the influence of the jazz fashion. Instead, the austere and economical style just described gradually took on a heavy, lifeless, monumental quality augmented by influences from contemporary French work on the one side and Regency on the other. Although as a style it is entirely devoid of distinction, it has remained popular ever since, chiefly, I was told, because it can be adapted easily to any style of building, except Gothic. It should be emphasized, however, that the main demand has always been for more or less accurate imitations of late seventeenth-century wrought ironwork, notably the designs of Jean Tijou. The reproduction of this work is costly and can only be performed by highly-skilled smiths, and these are, in fact, fast disappearing from the trade. I was told by several manufacturers that before very long wrought ironwork will have to be replaced by welded metal, probably steel. If that is the opinion in the trade, it is surprising to find that no alternative to the wrought scroll and the beaten leaf, the decorative stock-in-trade of the smith, has yet been evolved.

The relationship between the designer and his public, or client, in the remaining trade to be discussed is quite different. The design of door furniture is largely influenced by the taste of the builders' merchant and the ironmonger. The builders' merchant acts on behalf of an architect, a housing authority or an individual builder, and in most cases his task is to fit out small housing estates with door and wall fittings. For this he tends to buy from the cheapest end of the trade, i.e. the small firms with a low standard of workmanship. The people who will eventually live in these houses can exert no influence over his choice of designs, with the result that the standard of builders' brass is on the whole deplorable. For large housing schemes the architect frequently by-passes the merchant and approaches the manufacturer direct. If the architect's order is sufficiently large the manufacturer will produce special designs for him. After a short period these designs will be put into the firm's general range of door furniture. The architect represents a much coveted market for this trade, and most large firms, in order to save the expense and delay in making up special designs, attempt to keep their general range of patterns up to a high standard of

[1] This is perhaps the only trade to anticipate the influence of the Cubists in this country.

design. In this way it has been possible by slow degrees to improve the taste of the builders' merchant.

The design of door and wall fittings is also influenced by the fact that many small firms are located in Birmingham, and very few of them employ designers. 'The presence in the ground of a type of fine sand from which good moulds could be made for casting brass, was one of the factors which played a large part in the location of the early brass industry in Birmingham, and to the present day it is regarded as the centre of the brass casting industry. The long tradition behind it has tended to make the industry conservative in its habits, and firms employing production methods that have seen little change in the past fifty years are still remarkably common.'[1] A typical small brassfounder requires skilled workers for pattern-making—which involves carpentry—casting, finishing and assembling. The pattern is made from a drawing supplied by the designer, if one is employed, or by the manufacturer or his client. It is left to the pattern-maker to decide the exact relationship between one form and another, or the direction and degree of radius of a curve. In all cases which I investigated the pattern-maker's main concern was to form the design in such a way that it was easy to produce. 'As casting is an individualistic process and patternmaking relatively cheap, new designs are fairly frequent and do not involve any disorganization of the factory by their introduction.'[2]

The shortage of skilled labour in the trade has forced many firms, both large and small, to adopt an alternative method of production. This method is to use mechanically operated presses to form sheet-metal into the required shapes. In order to produce an unlimited number of patterns in the same press, interchangeable dies or press-tools are used. The cutting of these dies is much more expensive than making a mould in sand, with the result that longer production runs are necessary to cover the initial cost. For this reason a firm's range of pressed fittings is usually more limited than its range of cast fittings. Hence a design for a pressed fitting should have a wider circulation and a longer life than a cast fitting.[3]

From as early as 1926 until the late 1930's, design in the trade was dominated by modernism and patterns bristled with unexpected jags and angles. 'This craze swept aside the old traditional shapes and introduced far more new designs than could be produced with care.'[4] However, in the middle 1930's there were signs of improvement. A distinguished London architect produced some designs which were remarkable for their simplicity. In particular they showed that a handle or a knob could be com-

[1] *Design and the Designer in the Light Metal Trades*, p. 32. [2] *Ibid.* p. 34.

[3] This economic fact is even more significant in plastic door furniture. See p. 125.

[4] *Design and the Designer in the Light Metal Trades*, p. 34.

fortable to hold and also modern in design. The influence of the Bauhaus was implicit here, as it also was in the simple and well-proportioned designs produced by one of the large firms I visited. These designs had been commissioned by some Dutch architects. In several other firms I was also able to trace the effect of Dutch metal-work design. This was due, no doubt, to the competitive trade in door furniture which this country had with Holland and South Africa before the war. Just prior to the war some of these designs found their way on to the home market. The popularity which they soon achieved forced many of the small firms to produce similar designs.

This tendency towards simplified forms was beneficial to all manufacturers. For castings, these forms required less finishing than the modernistic patterns which had ridged surfaces and jagged edges. For metal pressings, the cost of cutting the dies (i.e. die-sinking) was less because the outlines were either straight or only slightly curved. Yet in spite of these obvious economic advantages the trade in general has, since the war, once again adopted more ornate and fussy patterns. If one tries to analyse them, they appear to be devitalized versions of modernistic designs, with at the same time some influence of American 'borax'. For the purchaser these patterns are not more expensive than the simple modern designs, because they are reproduced in quantities large enough to offset the higher cost of die-sinking. While this necessarily reduces the number of patterns available, the variety is greater than was possible in the simple, modern designs.

Two main reasons account for this latest trend of design in door furniture. The small firms which never employed designers regarded it as necessary not only to copy the simple, modern designs which everybody used but tried, for reasons of competition, to apply a few minor alterations to them. Hence borax, or the conventional three-stripe formation in ridges on the surface of the fitting. The few large firms which attempt to dissociate themselves from this practice will maintain that there is a considerable demand for the simple and unadorned fitting. But it must be remembered that they have always been under the influence of the architect who represents a large proportion of their market. The second reason why borax designs are so prevalent today is because the bulk of the trade caters for the taste of the builders' merchant. Unlike the architect's, his taste is uncreative. His craving is for variety and showiness at the lowest possible cost. One cannot blame the small manufacturer for debasing his standards to suit the whims of his client, when he knows that if he refuses, one of his competitors would eagerly accept the order. The question of the responsibility of the manufacturer for the standard of design of his products, takes on a different aspect in every trade, as will appear more and more clearly from the following chapters of this book. In door furniture the bulk of the trade is in the hands of many

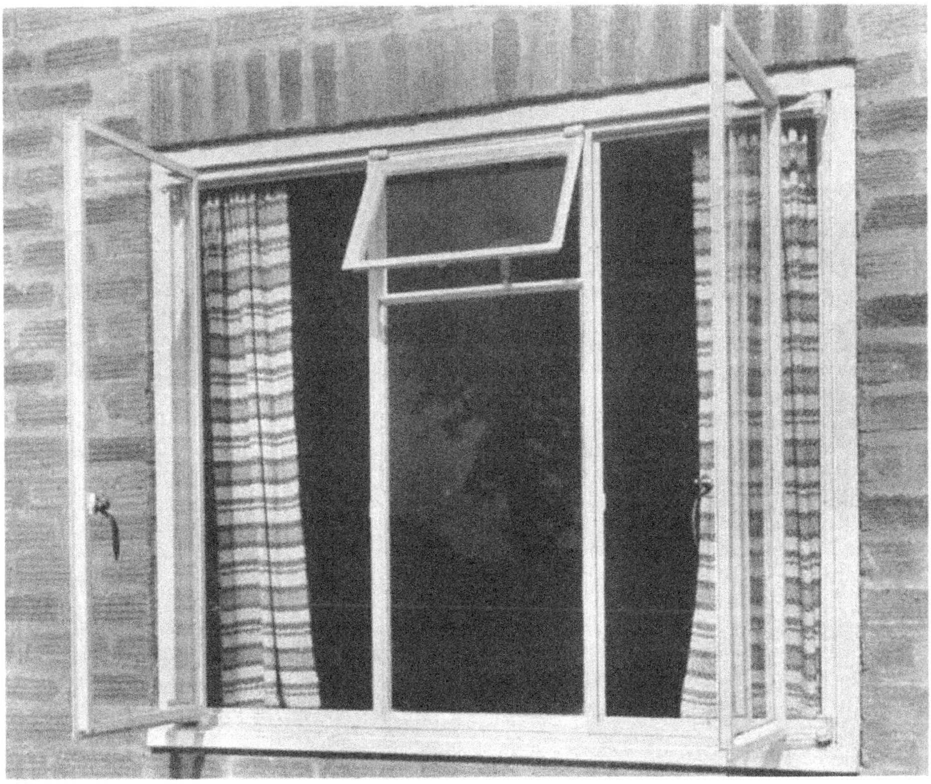

1. The good proportions of standardized metal windows are the result of a collaboration between manufacturers and architects. This window is a best-seller in use for domestic housing. The glazing bars are vertical and suit man, who stands upright, more than the once fashionable horizontal type. *Made by Henry Hope & Sons Ltd.*

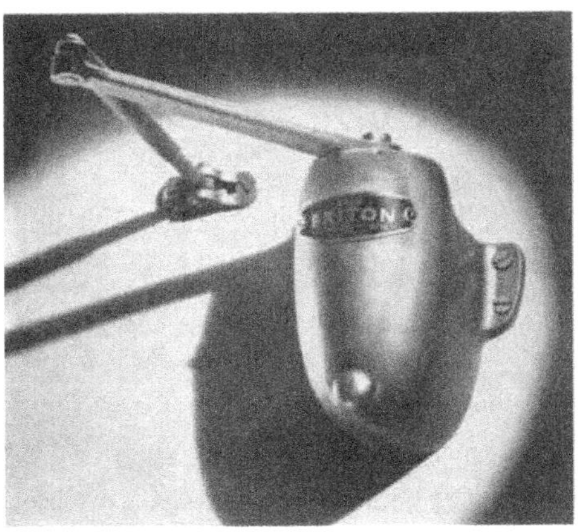

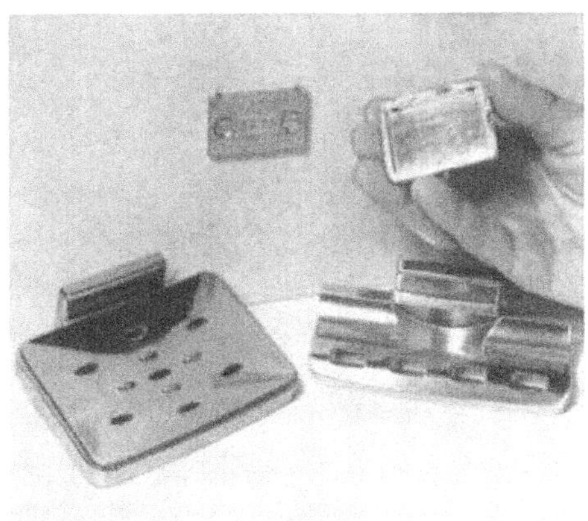

2. The smooth, easily cleaned form is appropriate to this door closer which would otherwise soon look drab with dust. For such purposes complex mechanism often has to be rearranged without losing in efficiency. *Made by William Newman Ltd.*

3. With a growing number of uses die-casting raises a special problem for the designer. These bathroom fittings—in a range with only one standard wall fitting—are the most refined on the market. *Designed by Scott-Ashford Associates Ltd. for W. C. Youngman Ltd.*

PLATE XIV

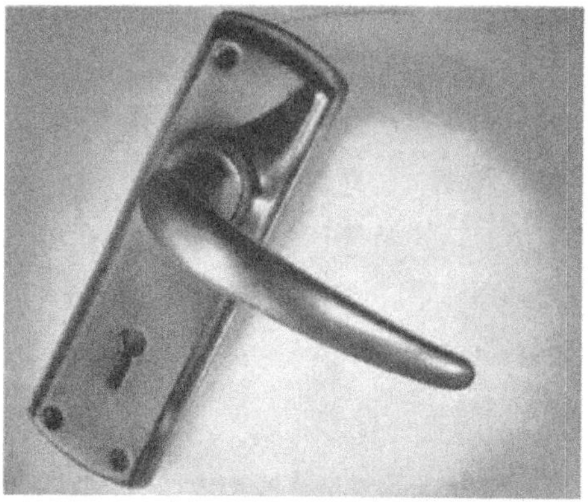

1. Manufacturers of door furniture maintain that jazzy decoration and classical motifs offer more scope for novel designs than the simple shapes now included in their ranges. Achievements in simplicity like this one should now be replaced by modern designs with greater visual interest. *Made by William Newman Ltd.*

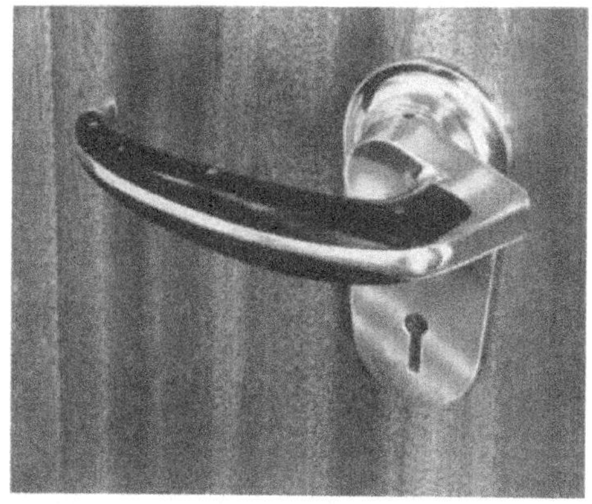

2. It is mainly owing to the demands of architects that the design standard of door furniture has advanced. Here wood and metal are used together for a dual appeal to the eye and the sense of touch. *Designed by L.C.C. architects Robert H. Matthew and Leslie Martin for the Royal Festival Hall.*

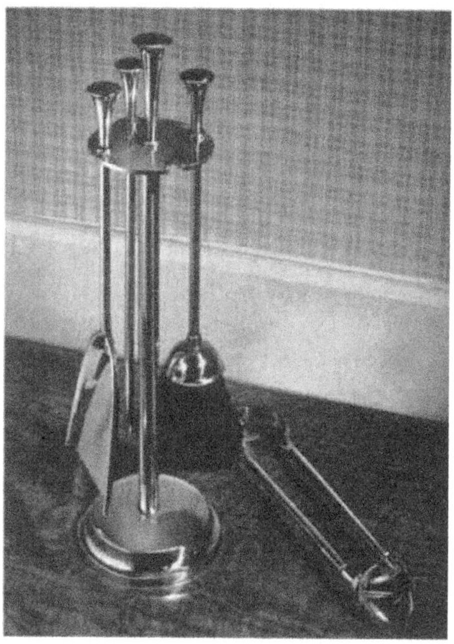

3. Companion stands are in the lists of many firms in the Birmingham brassware trade. For vulgarity and ostentation they are unequalled and it is extremely rare to find a design as good as this one. It represents a refinement on earlier more severe designs beginning in 1937. *Made by Marris's Ltd.*

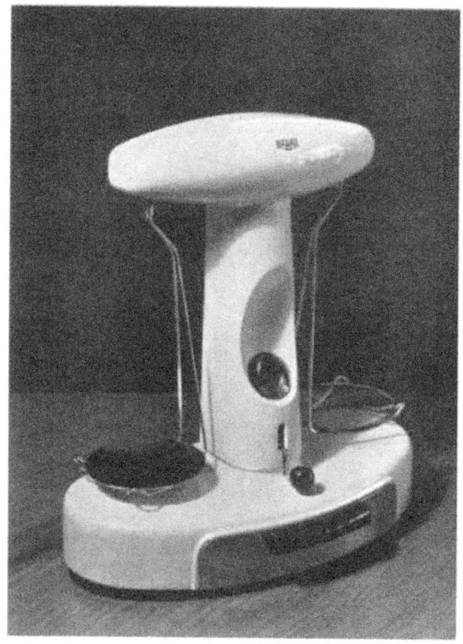

4. Few light metal firms have tried successfully to collaborate with a consultant designer. In all branches the trade needs the stimulus of constructive criticism and a fresh approach to old problems, like the one faced in this dispensing scale. *Designed by John Barnes of Allen-Bowden Ltd. for W. & T. Avery Ltd.*

small firms and very few large ones. If the balance were reversed, as it is in the metal-window trade, it would be justifiable to expect a better standard of design, because one could attach a greater measure of responsibility to individual firms.

The degree of importance attached to the designer in these three trades varies in proportion to the size of the firms. In the firm which made metal windows for architects and the building trade in general, the designer was himself an architect. The firm considered that it was necessary for him to travel abroad and to acquaint himself with new methods and materials relating to the trade. In the firms making architectural metalwork, although the designers were responsible for interpreting the ideas of architects or private clients, no great importance was attached to their creative powers. The salaries of head designers with more than fifteen years' experience were between £500 and £800 a year. For fully competent assistants with three to four years' art school training, £300 to £450 a year was the rule. The best designs which I have seen in door furniture came from a large firm where the designer is also a senior director. Another large firm with high design standards employed a designer at a salary of £600 a year. In the smaller firms very few designers were to be found, although I came across traces of the work of free-lance designers. In these firms 'it could easily be argued that a good designer would study and bring about economies of production, and would attract custom to an extent that would more than support his cost'.[1] However, 'the beliefs of the business man rather than economic facts... govern most industrial production, and it must be remembered that many small and medium-sized firms provide their owners with a fixed net income and a large expansion of their trade would be as embarrassing as a diminution'.[2] 'Except in certain cases where manufacturers are altruistically enlightened, designers will not be employed, unless some deliberate stimulus is given by conditions connected with the market.'[3] How it may be possible to develop that 'deliberate stimulus' in these and in other trades and industries is one of the main questions to be discussed in Part III.

SILVER AND ELECTRO-PLATE

'One of the most striking things to the layman who studies the British silverware industry is the small attention which appears to be paid to design and the small honour in which designers appear to be held.'[4] I begin with this quotation which

[1] *Design and the Designer in the Light Metal Trades*, p. 4. [2] *Ibid.* [3] *Ibid.*

[4] *Board of Trade Working Party Report on Jewellery and Silverware*, H.M.S.O., 1946, p. 37. The Report is of considerable value as a recent statement on conditions of manufacture and design in the two trades.

comes from the Working Party Report, because through my own researches I have reached the same conclusion. As in the light metal trades, I have concentrated on Birmingham, which is the largest centre for the industry. Every category of silversmith's ware is produced in Birmingham, from the least expensive article to that of the highest grade, and a large number of silversmiths combine the production of sterling silver with electro-plate. I saw twelve firms altogether, nine of them in Birmingham. In addition I interviewed two artist-craftsmen and two well-known consultant designers.

The two artist-craftsmen have small workshops with very few assistants and make special pieces, such as trophies and rose-bowls, for private customers. Five of the firms I saw work mainly in sterling silver. One firm, with more than 200 employees, makes expensive tableware in gold, silver and electro-plate. The other firms are smaller, with twenty, twenty-five, forty and sixty employees. They are representative of the very considerable section of the trade in the hands of small firms. Their products include brush-sets, powder-compacts, cigarette-boxes and general tableware. All of these firms produce special items, often commemorative, in the shape of vases, chalices and sports trophies. The remaining seven firms are chiefly concerned with electro-plated articles. These firms vary in size; the largest has 700 employees and specializes in hotel tableware, while the smallest, with fifteen employees, is concerned with cheap butter-dishes, toast-racks, etc. Nevertheless, most of them are larger, more mechanized and have longer production runs than the silversmithing firms.

The general effect of purchase-tax on design standards will be described later. In this trade the tax of 75 per cent on articles made in sterling silver is particularly harmful. In all the silversmithing firms which I visited I was told that the high rate of tax was chiefly responsible for the reactionary and unadventurous designs produced. To relieve this situation, the Design and Research Centre for the trade proposed that articles of sterling silver which were entirely hand-made and of approved design should be exempt from tax.[1] This was agreed to by the Chancellor of the Exchequer in July 1948. To administer the scheme the Centre has formed a panel of judges, consisting of six manufacturers, three designers, two retailers, and two independent members. A design is first submitted on paper and, if approved, it is again judged after it has been made-up. A further panel of judges, representing the Crafts Centre of Great Britain, is then asked to give its opinion on the workmanship of the article. If this panel does not approve the workmanship, the article may be sent to the Director of the Victoria and Albert Museum for final judgement. Once the design

[1] The other activities of the Design and Research Centre for the Gold, Silver and Jewellery Industries, are described on pp. 225–6.

and the workmanship of the article have been approved, it may be sold free of purchase-tax. A further concession, applicable only to articles of domestic use, allows the original and five copies of it to be granted relief from tax. The Design and Research Centre estimates that about 60 per cent of the articles submitted are passed.

The method by which the purchase-tax relief scheme is worked and its effect upon the trade are important to us in this context. Apart from the fact that the scheme is unique, it shows how the sale of a manufacturer's product can be controlled by the Government and a panel of judges. The many far-reaching conclusions which may be drawn from this must be held over for a later section. Since the introduction of the scheme the standard of design in hand-made articles in sterling silver has improved slightly.

The two artist-craftsmen whom I saw work in gold and silver, and each of their designs is unique. Both have been trained at the London Central School of Arts and Crafts, and were awarded scholarships to study on the Continent. One of them professed a special debt to the work and influence of Omar Ramsden, and considered it his duty to carry on, in part, the Ramsden tradition. The artist-craftsman has few overhead costs, practically no machines, a staff of highly skilled assistants and, at present, a ready market. He does his own designing and supervises the production of each piece of work. The chief advantage which the artist-craftsman has over the larger manufacturing silversmith, is his ability to experiment with new materials, methods and designs. Furthermore, he is in a position to receive considerable benefit from the purchase-tax relief scheme. For these reasons it is not surprising to find that his standard of design is the highest in the trade.

The late Frank Pick, in his Report on this trade, showed how the teachings of William Morris could be related to industrial production.[1] Throughout the Report the importance of the artist-craftsman to the trade is continually stressed, the ideal designer being the one who makes as well as designs. There was some evidence of this relationship in the five silversmithing firms which I visited. The staff designers employed are intimately concerned with workshop technique. The articles produced are either hand-made or turned into shape on the spinning lathe. This lathe represents the intermediate stage between hand-beaten and machine-stamped silver. Although spinning requires a craftsman of great skill, it does not demand creative ability, because the silver is spun into shape against a piece of wood, or chuck, representing the inner profile of the article. This, it will be seen, is a technique basically distinct from that of drawing a shape to be made by a craftsman's hand. The craftsman's

[1] *Report on the Jewellery, Silversmithing and Allied Trades, for the Council for Art and Industry*, H.M.S.O., 1937.

touch, while the silver is between his fingers, contributes noticeably to the appearance of the finished article, whereas in spinning little human sensitivity comes in. Apart from the preliminary stamping which is necessary for some articles, the only other mechanical device used in this trade is the engine-turning machine.

Two of the five firms visited, with less than thirty employees each, have no designer. It was pointed out that they were too small to support one on a full-time basis. One of the firms has an agreement with a free-lance designer, who is also on the staff of the Royal College of Art, which enables it to call on him for special work. Apart from originating the design, he is expected to take part in any alterations which might be required during production. The bulk of the firm's output is designed by the manufacturer himself. The other firm claims that it achieves better results by evolving its own designs, for it felt that the artistic benefits which might result from the employment of a free-lance designer, are outweighed by his ignorance of manufacturing methods. The firm turns out some pieces of undistinguished modern design.

The remaining three firms employ their own designers. In two of them the designers are given a fairly free hand by their employers, and they produce some fairly creditable modern designs. But as most of these designs are made to suit the requirements of the purchase-tax relief scheme and are sold to private clients, we must rule them out as a true reflexion of public taste. Another firm specializes in expensive tableware, sports trophies and other commemorative pieces. The designer is content to take his inspiration from anything in silver between the reign of Elizabeth I and the Great Exhibition of 1851. Only the trophies designed for motor racing competitions contain the spirit of the twentieth century, and they are inclined to be modernistic in shape and pattern rather than modern.

The demand for new designs in these silversmithing firms is very small. All manufacturers agreed that the market, both home and abroad, is unwilling to accept modern design. Although only a small quantity of a particular design is made at one time, there is in these firms little desire to change from a reproduction to a contemporary style. In some sections of the trade accurate reproductions of antique styles are being constantly produced, but in the firms I visited the designers' chief task is to make suitable variations on popular patterns. The fact that the few pieces of contemporary, and in some cases modern, design which do exist are made by hand, indicates the general lack of confidence in any design which does not recall the past.[1] As hand-made articles are considerably more expensive than machine-made articles, it is not surprising that manufacturers experienced difficulty in selling them. Even if the hand-

[1] In 1937, 80 per cent of the industry's output was in reproduction silver and electro-plate (see the *Report on the Jewellery, Silversmithing and Allied Trades, for the Council for Art and Industry*).

made articles *were* relieved of purchase-tax, they would still be too costly for a very large section of the public.

With this in mind, let us now consider the product of an allied trade which *does* reach a wide public. Electro-plated nickel-silver articles are less expensive than articles in sterling silver, and the purchase-tax on them is only 50 per cent. The electro-plate trade exists to meet the needs of those people who want silver in their homes and cannot afford the real thing. The technique of plating nickel-silver electrically was originated and patented by a Birmingham manufacturer in 1838. The object was to imitate expensive silver by a process which was relatively cheap. From the design point of view this was an unfortunate beginning for a trade which today accounts for well over 75 per cent of the industry's total output. Then as now design in the trade was dictated by the characteristic forms of silver, and was therefore bound to follow the silversmith and never to exist in its own right. Other more recent industries and trades, such as stainless-steel tableware, have refused to accept a similar handicap.[1] Even if they have at the start copied the forms and the traditional techniques current in similar trades, in order to attract the market, they have soon launched out on their own. They have realized that the materials and the machines used had qualities which required to be expressed in ways peculiarly their own. The electro-plate trade has never realized this; it has steadfastly pretended to produce articles in sterling silver. True to its original intentions the trade has always lagged behind in design (this is especially noticeable at the present time), and has never given its designers an opportunity to use their initiative.

Since the trade has elected to follow the design standards of the silversmith, electro-plated ware must be judged by his standards. The disadvantages of this are at once apparent. Silversmithing is the province of the workshop and the hand-craftsman. Electro-plate, because it serves a much larger and poorer public, belongs to the factory and the machine. Until the trade decides to proclaim honestly that a craftsman's die in a power-press, and not a craftsman's hammer on the bench, is the method of manufacture, it cannot hope to develop a character and style of its own, let alone a standard of modern design.

As we have seen, the majority of today's designs in sterling-silver are reproductions. It is impossible to say which style is most frequently used, but it appears that the Late

[1] I visited a firm of 100 employees making stainless-steel tableware, excluding cutlery. Production began in 1930. The metal was found to have unique properties which demanded a new and unbiased approach from the designer. The work of a free-lance designer proved to be unsatisfactory, and his place was taken by a senior member of the firm. As ornamentation (embossing, chasing, etc.) appeared to be unsuited to the metal, the merits of the designs had to depend entirely upon proportion and surface finish. At the present time no attempt is being made to imitate the characteristic forms of silverware.

Georgian, Regency and Mid-Victorian styles are more common than those of the Early Georgian and pre-Georgian periods. In other words, the excellent proportions and simplicity of the earlier styles are less popular today than those which are more ornate and pretentious. Since the electro-plate trade, as far as design is concerned, follows the silversmith's, the bulk of its output for the domestic market is also designed in these latter-day styles. 'There has been too much copying in the past of beautiful hand-made articles. It has only needed to be suggested that a particular piece could not be reproduced by a machine, for the manufacturer to take this as a challenge of honour and devote all his ingenuity to making a machine produce it.'[1] Apart from the seeming futility of this, it is evident that the machine-toolmaker (die-sinker), can never produce more than a semblance of the hand-made original. In some cases where the pattern is undercut he cannot operate at all. In order to meet the requirements of the power-press, the subtle modelling of the silversmith has to be simplified and often obliterated. A curve that would take a silversmith a week to execute to his satisfaction, has to be reproduced by the machine in a fraction of a second. There is another, and perhaps more persuasive reason why the trade should adopt its own design standards. The cost of producing an article of ornate design is greater than that required for a comparable article of simple design. There are two reasons for this. First, the die for the ornate design is more complicated and therefore takes longer to produce. And secondly, it is more difficult to polish an article which has a recessed pattern.

The die-sinker (who should be a designer and not merely a skilled technician) should be given the opportunity to create new patterns directly suited to the method of manufacture. In the seven electro-plate firms which I visited, I found that this was only possible when tableware for the hotel trade was required. In one large firm the management is concerned with mass-production exclusively, and every article is stamped into shape by the power-press. Instead of a designer, a production planner and a staff of draughtsmen are employed. The method of bringing out a new design is as follows: The works manager and the production planner are asked to evolve about a dozen samples of different tea-sets for the hotel trade. From these samples one will be chosen for full-scale mass-production. The first considerations are low cost and general ease of manufacture. As the samples are to be made out of existing dies, it is necessary to inquire into the sales results of the tea-sets for which the dies have originally been cut. When the best-sellers are selected, the next step is to give each sample some individual features. This is accomplished by exchanging handles, spouts and other movables, until the samples comprise between them the best-selling features

[1] *Working Party Report*, p. 38.

PLATE XV

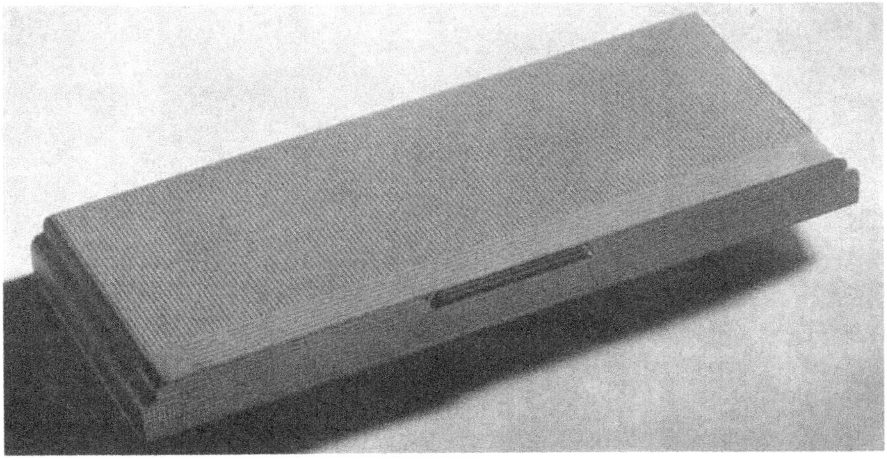

1. Reluctance to accept modern design standards as the basis for fresh experiments is the hallmark of the silverware trade. Expensive metal, production in small quantities and a high purchase-tax are other factors which make this cigarette box with engined-turned decoration rare amongst a host of pseudo-antiques. *Designed by S. W. Robinson for J. C. Lowe.*

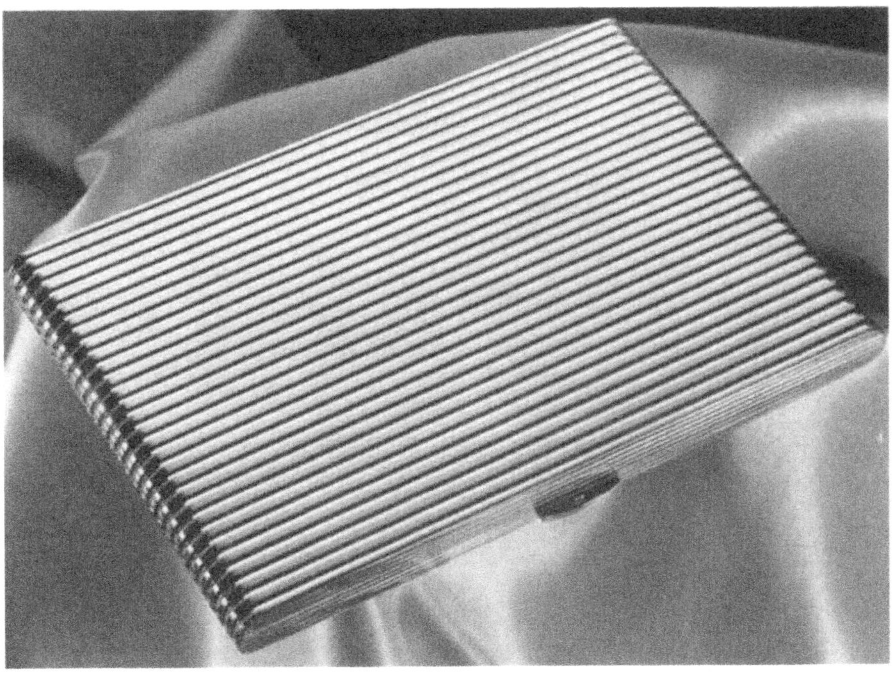

2. The possibilities of engine-turning as a method of decorating gold and silver articles remain largely unexplored. For the general market the demand for rich simplicity, as in this gold case, is said to be outweighed by the popularity of figurative, classical designs which look more obviously expensive. *Made by K. Weiss Ltd.*

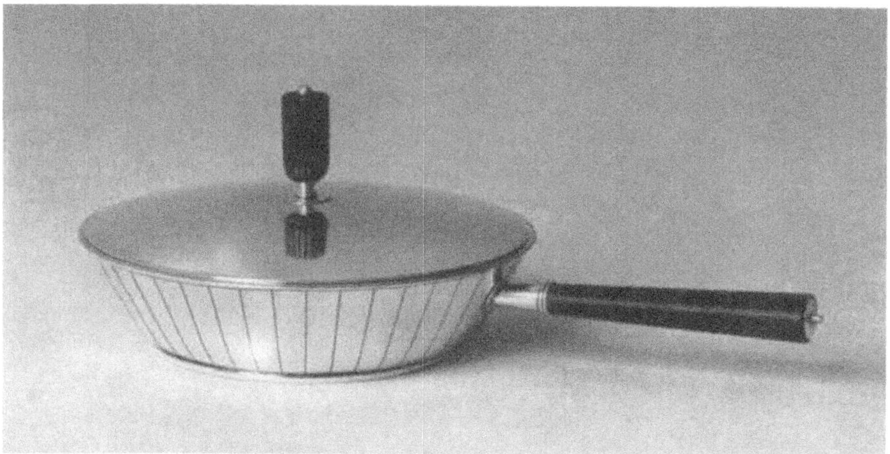

1. Overwhelming proof that British firms are lagging behind modern taste, even for expensive articles, comes from the Danish firm of Georg Jensen where all the work produced for home and export trade is of a high design standard. This vegetable dish is a typical example. *Designed by Sigvard Bernadotte for Georg Jensen (Silversmiths) Ltd.*

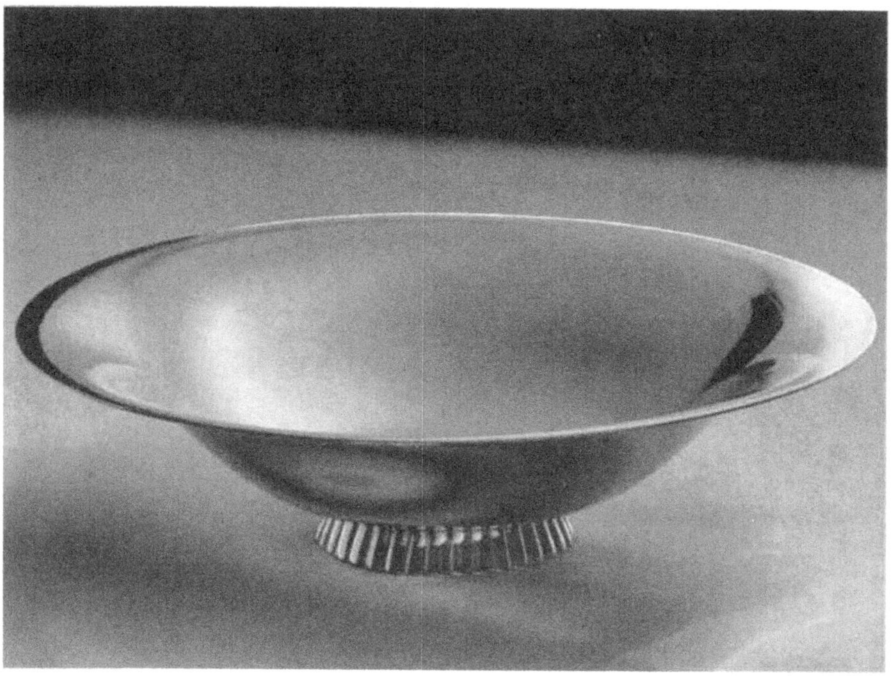

2. As in other trades orders from enlightened customers often persuade silversmithing firms to include good new designs in their ranges. Today this dish with a fluted foot is an example of the trade at its best. *Made by Wakely & Wheeler Ltd.*

of perhaps thirty different tea-sets. Known market preferences for inefficient shapes vie strangely with the clean, modern lines dictated by the power-press. The effort made to compromise has unhappy results in some cases, and I feel that the most satisfactory sample is the one which openly proclaims that it is machine-made. It is, incidentally, a tea-set that would be easy to use, hygienic and altogether fit for its purpose.

This design procedure has been described at length for three reasons. First, the sample chosen was simple in outline and therefore cheap to produce, which proves, if only from an economic point of view, that a machine style has its value. Secondly, here is a serious attempt by an enterprising firm to design for mass-production and satisfy the taste of the market at the same time. Thirdly, the procedure chosen is representative, though on a much wider scale, of the methods employed by five of the remaining six firms. These firms, which are mainly concerned with the domestic market, may be grouped together, for they profess a common standard and a common aim. None of them employs a designer. In each firm a senior member of the management is responsible for producing designs. In many cases his invention is limited to the patterns in stock, for which dies have already been cut. His task is to modify these patterns without involving the expense of cutting new dies.[1]

Undoubtedly, a designer who has the status of a senior partner is in a very strong position for directing the design policy of his firm. But from our point of view, that policy will be of little value unless he has an understanding of, and a training in, design. These men know the technical and commercial aspects of their trade thoroughly, but they do not possess the artistic sensibility of a trained designer. Free-lance designers are not encouraged because they are said to lack experience of machine production.[2] One manufacturer suggested how this could be remedied. The free-lance designer should be encouraged to study the limitations of die-sinking and learn the various techniques used in the assembly shops. But the responsibility does not lie wholly with the designer. If manufacturers would take more trouble to explain exactly what they require from the free-lance designer, much of the prejudice which is manifest on both sides would disappear.

The remaining firm produces civic regalia, sports trophies and motor-car badges. The designer is kept extremely busy inventing slight variations on existing patterns. One of his problems was how to express in modern terms a coat of arms couched in a pretentious Victorian style, without shocking the municipality concerned.

As we have seen, the relatively satisfactory standard of design in hotel tableware is

[1] The cost of cutting a die for a small ash-tray would be about £40.

[2] See *Design and the Designer in the Light Metal Trades*, H.M.S.O., 1944, p. 19.

due on the one hand to the correct use of the machine, and on the other to the specific purpose for which the tableware is made. But, as soon as the manufacturer thinks of his products in terms of the 'home' an entirely different set of standards is raised. 'Fitness for purpose' and the characteristic virtues of the power-press (factors which tend to safeguard designs from artistic excess) are replaced by the styles of the Late Georgian and Victorian periods. All the electro-plate manufacturers I have met assured me that reproductions of antique silver are in great demand, thereby implying that the public gets what it wants. No doubt there is indeed a demand for period imitations. Some people, because they associate an article of value, such as a silver teapot, with the past, are inclined to think of electro-plated wares in the same way.[1] Throughout the eighteenth century, but more especially in the nineteenth, a family's wealth and refinement was often expressed in terms of silverware. Although most families today prefer motor-cars or holidays on the Riviera to be expressive of their wealth, the tradition of 'having some silver' in the house still lingers. Many families, too, have silver which has been handed down through generations. When a replacement is needed, they quite understandably prefer to match the set with an imitation rather than purchase an up to date design. But even where a completely new set is required, period design is given preference. One reason for this is that very few people in the country have ever seen silver and electro-plate of modern design. But more important is the fact that antique silverware recalls a mode of living which was enjoyed in less confused and troubled ages. Thus tableware of antique design helps people to escape, even for the brief space of tea-time, from the alleged austerity of the modern home.

In both the silversmithing and electro-plate trades, the status and salary of the designer are unrelated to the size of the manufacturing unit. Three of the silversmithing firms visited employ designers with salaries between £400 and £600 a year. The one electro-plate firm to employ a designer pays a salary which is also within that range. The remaining eight firms do not employ designers. If these salaries appear to be extremely low when compared with other trades, we must remember that in most silver and in all electro-plate manufacture, there is very little demand for creative designs. In addition, I found that free-lance designers are frequently employed by the smaller silversmithing firms for special work. Their fees vary from £20 to £30 for a brush-set to £60 and more for a tea-service.

At the present time, with the restrictive rate of purchase-tax on silver, there seems to be little chance that any large silversmithing firm will give serious consideration to

[1] It is significant that modernistic designs have never been popular, except at the cheaper end of the electro-plate trade.

modern design. If we remember the high cost of silver and turn to the electro-plate trade, where the tax is less, we find each manufacturer convinced that modern design will not sell on the domestic market. From evidence in other industries, we have seen that the introduction of a few items of modern design has been ultimately rewarding to the manufacturer. There is therefore some reason for supposing that a similar experiment in this industry would also be successful. In the Working Party's view the public does not get what it wants. 'The industry as a whole has been most unenterprising in regard to design, and in general most unready to learn that its public of today is very different from its public of forty or fifty years ago.'[1] As it is impossible to find one large British firm selling modern designs to a wide public, it is I think excusable to consider a foreign example. A brief account of the origin and development of Georg Jensen Ltd., has therefore been included here for comparative purposes.

Georg Jensen, after serving his apprenticeship, became a journeyman goldsmith in 1884. From that time until 1904, when he opened his own workshop in Copenhagen, Jensen gained a wide experience as a sculptor, a potter and a worker in precious metals. Of particular significance is his visit to France and Italy, and especially the museums in those countries, which convinced him that the work of the artist-craftsman is as important as that of the painter and sculptor. From the start his desire was to serve the Danish middle-class with articles it could afford. 'Good, solid family silver was to be the basis of his business.'[2] This policy of keeping the work of his firm within a specific price range is illustrated by Jensen's willingness to make use of machinery. Moulds and dies were not rejected by him, but assimilated in his workshop as other tools had been. Jensen's factory in the Ragnagade, which was opened in 1919, employs 300 workers today. In 1950 the firm had four designers, of whom Harald Nielsen, Jorgen Jensen and Sigvard Bernadotte are particularly well known. They were paid fixed salaries, and received royalties on the sale of their work. The factory makes no reproduction work, but only those designs which have been created by Georg Jensen himself, or by its own designers. While the bulk of the firm's output consists of articles in gold and silver, a new metal, stainless steel, has recently been introduced. It is significant that the designs for cutlery and hollow-ware in this metal are totally unconnected with those for silver.

There is little doubt that the methods of design in this firm make a striking contrast with those practised in British factories. As far as I am aware it is not possible to point to one artist-craftsman in this country who has, in recent years, enlarged his

[1] *Working Party Report*, p. 36.

[2] See Ivan Munk Olsen, *Georg Jensen Silversmith*, København, 1937.

workshop, introduced machines, and so increased his production that he might serve a wider public. A silversmithing firm in this country, equal in size and output to Georg Jensen Ltd., employs one designer only, and more than 90 per cent of its production is in antique styles. In defence of the British industry it may be said that a high proportion of its output is in antique styles because of the demands of foreign markets. But the work of Georg Jensen Ltd., all of which is modern in design, accounts for 87 per cent of the total export of the Danish silver industry. Again, electro-plate in the hands of British firms is merely a medium for imitating sterling silver, whereas stainless steel for Georg Jensen Ltd., demanded a fresh approach, both in manufacture and design. Finally, the maintenance of a high standard of design in this Danish firm is not merely fortuitous. Its standard, which far surpasses that of any silver designs in this country, exists because the firm has adopted an enlightened policy towards its designers and its public.

ALUMINIUM HOLLOW-WARE

Until about 1880 cast and wrought iron were used for the manufacture of domestic pots, pans, kettles, etc. Stamped steel (tin-plate) was then introduced and used extensively by the trade. Like iron used for the same purposes it was enamelled, chiefly in brown and cream colours. The next change came during the First World War, when the production of aluminium was developed. Aluminium is now steadily replacing all other metals for culinary utensils, though enamelled tinware is still in demand for food storage purposes. The chief reasons why aluminium has become such a popular metal are its cheapness, lightness, pleasing surface finish and the ease with which it can be manufactured. Although aluminium discolours when in contact with certain fruits and vegetables during cooking, it has been proved not to be detrimental to health. In recent years chromium plating has added to the popularity of aluminium hollow-ware, for chromium will sustain a much higher polish and will not tarnish. The latest development is the anodizing of aluminium to attain a wide variety of colours.

To obtain information about the designing of domestic hollow-ware in aluminium I visited three firms. One of them, with 350 employees, is entirely concerned with domestic ware of all qualities. The second, with 550 employees, balances its production equally between domestic articles and industrial pressings in a variety of light metals. The third firm is larger, and produces mainly aluminium storage tanks and condensers

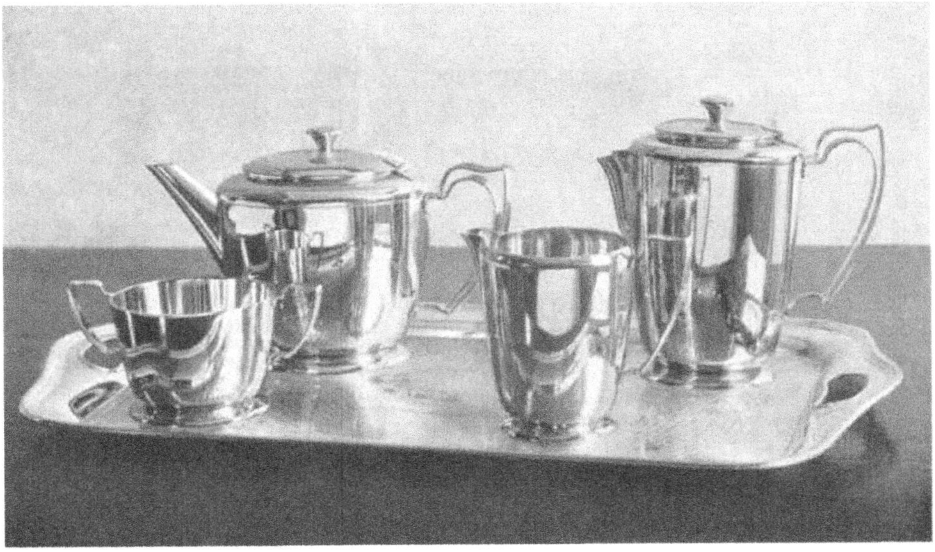

1. Firms making domestic articles from stainless steel find themselves with a relatively intractable material with unique properties to be exploited. Unlike E.P.N.S. there can be no hankering after antique silver shapes, so that results like this one tend to express the material at its best in simple, well-formed shapes. *Designed by A. L. Wiggin for J. & J. Wiggin Ltd.*

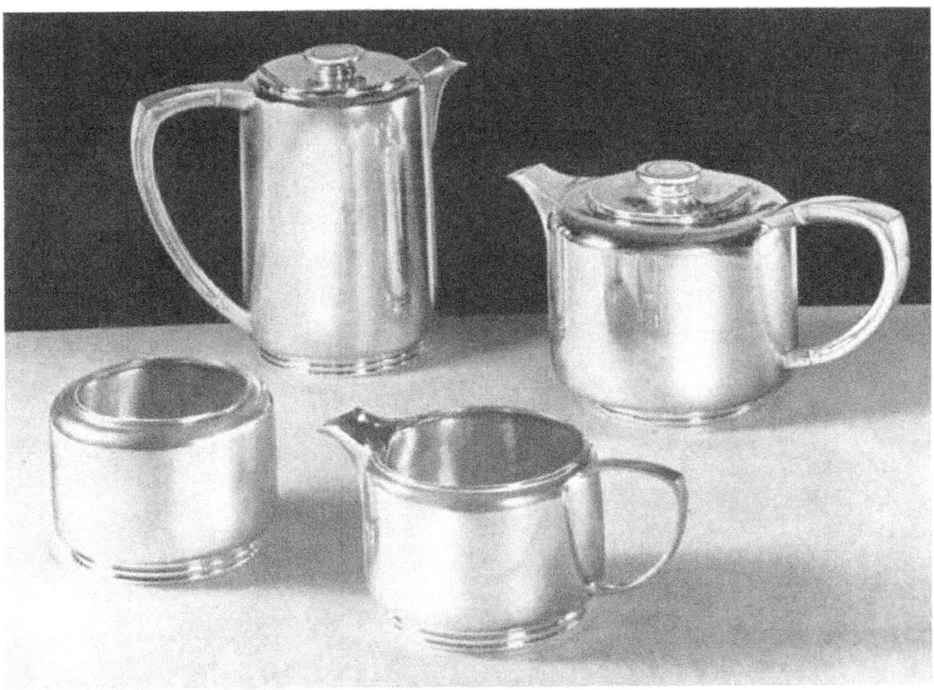

2. The fact that most machine-produced electro-plated ware imitates in a less expensive material the characteristic forms of hand-made silver, means that progress in design is handicapped. Only with large contracts are manufacturers likely to meet a customer who insists on a design as good as this stacking tea-set. *Designed by Kenneth Holmes and N. R. G. Poynton for Gladwin Ltd.*

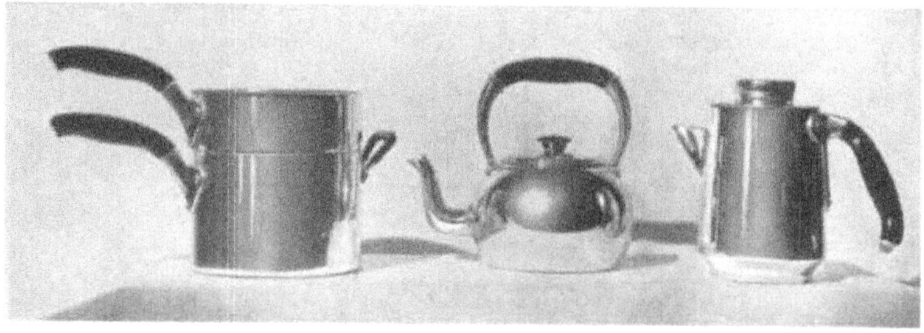

1. The best designs in aluminium hollow-ware are to be found in cooking utensils. Intelligent economy can be seen in this example where the same plastic handle has been designed to suit all three articles. *Made by N. C. Joseph Ltd.*

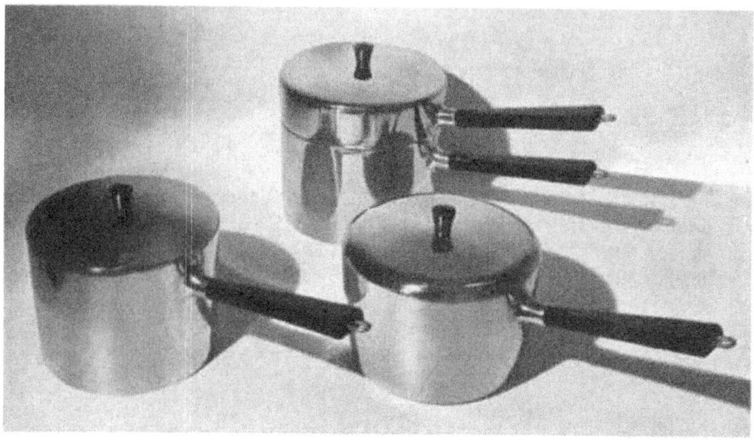

2. These items are fit for a purpose that does not vary in the domestic kitchen. Although all firms have basically similar patterns, few have produced shapes as graceful as these. *Made by Easipower Ltd.*

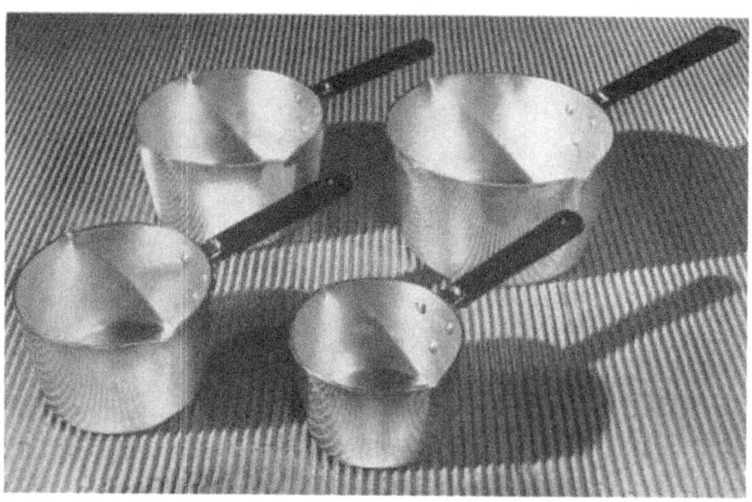

3. Handles are no less important than shapes and the best-looking examples invariably come from firms that insist on special designs for each new range of utensils. The tradition of using good functional designs, as in these articles, is nowhere to be found in aluminium tableware. *Made by A. G. Burton Ltd.*

for breweries and chemical plants. Only 40 per cent of its production consists of domestic ware.

All these firms are run on mass-production lines. The machines used are the power-press and the spinning lathe. While both of these machines have been mentioned in the preceding chapters, it will be necessary briefly to reconsider them here because of their effect upon the design of aluminium articles. The more pressure is applied in shaping an article between two dies the stronger it becomes. In all cases the pressed article is stronger than the spun article, because the spinning process tends to thin the metal and so weaken it. For example, the metal of a spun teapot body will be thicker at the lip than it is lower down the side. The thinnest section has to be so arranged that it coincides with the greatest curvature of the body. In this way sufficient overall strength is retained, although it is usually possible to dent a spun article more easily than one which is pressed. For comparatively short runs spinning is more economical than pressing. One manufacturer said that he would use the spinning lathe for a production run of less than 1000 articles. When longer production runs are needed the power-press is used. In one firm I was told that for a set of three saucepans the dies would cost between £500 and £600, which means that more than 50,000 sets would have to be sold before the initial outlay could be recovered. In another firm, the various dies required for a tea-set cost about £1000.

Full-time designers are not employed by the firms I visited. The reasons are these: The basic shapes of most aluminium kitchen utensils are copies of older designs in iron and tin-plate. They are known in the trade as 'parish goods', and the majority of them are twenty-five to thirty years old. The demand for them has remained steady because the purposes for which they are made remain practically unchanged. At the same time manufacturers said that they were unwilling to introduce new and improved designs because of the high cost of dies. In one firm the die which had been cut for a pudding basin was, with very few modifications, used for stamping the body of a teapot. The result, though hideous, had sold well on account of its low price. In all firms it was stressed that the fullest possible use must be made of existing patterns. To some extent, therefore, the presence of satisfactory designs in aluminium ware is due to chance. New or modified designs are produced when trade is slack, and their arrival is usually timed to coincide with the British Industries Fair. In addition, the manufacturers I met were ready to copy a competitor's design, if it were popular on the market. As an instance of this, the recent introduction by one firm of a so-called streamlined kettle, prompted many other firms in the trade to produce similar designs. 'Copying...is generally accompanied by some greater or lesser alteration from the original model. The original may be fresh and happy in its appeal, but the alterations are often made by unskilled adaptors of undeveloped taste, when the results are likely to be of the nature of a "Rake's Progress".'[1]

As in electro-plated articles, 'fitness for purpose' is a constant safeguard in the design of aluminium ware, especially when cooking utensils are concerned. But when a manufacturer considers his tableware production, teapots, jugs, coffee percolators, etc., he is liable to attach a false meaning to the word design. He thinks of design in terms of decoration only, and not as something which controls the function, shape and appearance of the entire article. Representative of this loose thinking about design is the case of the manufacturer who said, 'a thing always looks well if it works well and has a few fal-lals on it'.[2]

Before the war there was little demand for ornate designs. In one firm the more expensive tea-sets were engraved by hand, while a small proportion of the cheaper ware was stamped with a wickerwork, or 'basket' pattern. In the other firms I found practically no trace of applied decoration in their pre-war ranges. The most popular patterns were plain, and on the whole a satisfactory standard of design was maintained. Today we find the reverse; most of the best-selling patterns are decorated. When I asked manufacturers why this change had come about, they pointed out that very little aluminium for domestic use was produced during the war. As a result of this shortage the public was prepared to accept anything which was available. But more important is the fact that manufacturers believed that the public's reaction against austerity, which we have noted in other trades, has increased the demand for decorated articles. With a sellers' market on the one hand and no staff designers on the other, manufacturers were forced to fall back on the decorative small change which has been current amongst metal workers for centuries. These patterns are embossed, or stamped into the metal. They take the form of hammer-punch marks, oak leaves, Rococo scrolls and the 'Willow-Pattern', and are reproduced with varying degrees of inaccuracy. There is no need to dwell upon these inappropriate and meaningless forms of decoration, except to say that had designers been available in these firms, some new and vital forms might well have been created.

The trade's experience of free-lance designers appears to have been unsatisfactory, for I do not know of any firm which has regularly employed one. The manufacturers I met were convinced that the employment of an outside designer would waste both time and money. Their opinion is echoed by evidence which is given in the Report on this trade. 'Most of the experiments have been with free-lance men, and the committee never heard of one which was an unqualified success.'[3] Several cases of failure

[1] *Design and the Designer in the Light Metal Trades*, p. 51. [2] *Ibid.* p. 11.
[3] *Ibid.* p. 18.

are quoted, including one which concerned 'an eminent industrial designer of international reputation'.[1] The designer knew little about the technique of manufacture and refused to modify his work in order to remove certain production difficulties. But on the other hand a 'large aluminium firm' is reported to have had 'several quite satisfactory' designs from a free-lance designer, although none of them was a best-seller. The Report offers no solution to the problem, but mentions that two firms have successfully combined the duties of designer and advertising manager.

In the firms I visited new designs were evolved, as one manufacturer put it, 'between myself and the foreman'. In the case of a teapot the method was to 'take a bit off here and put a bit on there'. Anybody in the factory may, as another director said, 'have a brain-wave'. The task of making-up a prototype of the new design would be left to the toolmaker. If the design happened to be original, and there was some doubt about its reception on the market, a few gross would be made on the spinning lathe. These would be tried out on the firm's chief customers and, if successful, dies would be cut and from then on the design would be mass-produced by the power-press.

In summing up the state of affairs it is important to stress that in kitchen utensils the reason for the satisfactory standard prevailing is once more the fact that these objects are made for use and not for ornament. In a few articles of tableware the same standard applies, for it is possible for the public to choose between shapes which are simple and straightforward in design, and shapes which are disgracefully decorated. The differences in price are negligible, although they are inclined to favour the decorated shapes. By quoting their sales figures, manufacturers are able to claim that the public chooses decorated shapes in preference to plain shapes. This is understandable because an article with an embossed surface tends to obscure small dents and scratches. It is perhaps this utilitarian quality, rather than any artistic appeal, which prompts people to buy decorated articles. But, as it is impossible to find one decorated article of satisfactory design, we are unable to say that the public is satisfied artistically. In this matter considerable responsibility must be attached to manufacturers for two reasons. First, the sellers' market in aluminium hollow-ware makes the public uncritical of the designs produced. Secondly, the vast quantities which are turned out enable manufacturers to sell their articles more on price than on design. In one firm I was told that 16,000 teapots in one design were made every week. Because these teapots were very cheap to buy, the public was forced to accept the manufacturer's choice of design. It rests with the manufacturer, therefore, to decide what standard of design the public is to accept.

[1] *Ibid.*

JEWELLERY

Throughout the nineteenth century Birmingham was the centre of the British jewellery trade. About 1820 the workshops of the goldsmiths and jewellers were still in the city, in Colmore Row, Great Charles Street and Newhall Street, while most of the workmen lived in the newer quarters to the north. The growth of the trade enabled more and more workers to set up independent firms of their own. The new firms were located in the houses of the former workers in Frederick Street, Vittoria Street, Vyse Street, etc. First, only one room would be converted into a workshop, later the whole house would be adapted and fitted up as offices, while the hand-work was done in a shed at the back, where a little garden had been before. Such is the origin and the appearance of most of the firms constituting the jewellery trade in Birmingham, and many an old manufacturer may well remember having played as a child in the room which is now his showroom or warehouse.

Today there are nearly 400 jewellery manufacturers in Birmingham, although 60 per cent of them employs no more than ten workers. All classes of jewellery work are undertaken by these manufacturers: diamond-setting, work in precious and semi-precious stones and metals, in marcasites and glass 'stones', and cheap novelties. I visited firms in all these classes, and in addition I saw in London one firm of high class precious stone setters, one firm specializing in all-glass jewellery and three costume, or fashion, jewellery firms. I also saw two firms making vanity-cases and one watch-case factory.

As long as all jewellery with precious or semi-precious stones sold in this country is subject to a purchase-tax of 75 per cent conditions must remain critical at the expensive end of the trade. Manufacturers of jewellery consisting of diamonds and other precious stones and metals which, before tax, cost between £300 and £1000 receive little encouragement and no guidance in matters of taste from the home market. In fact, only about 35 per cent of the total production of expensive jewellery reaches the home market. The prevalent taste is still dependent upon the Rue de la Paix. In Bond Street some of the most famous houses have French names. Tradition rules undisputed. The wealthy classes for whom this trade caters are particularly conventional in their taste. Most of the customers, I was told, belonged to the older generation. Younger people of ample means preferred spending their money on something more up-to-date.

One can even go further and state it as a fact that the demand for high-priced jewellery appears to be dying out. To wear glass chains and cheap imitation brooches or clips matching a particular dress and discarded with it has become the common habit in all classes. The close relation between this development and more general social changes is evident. In European society as it is and as it will be, there is no place for a jewellery trade of Victorian splendour. Manufacturers, as far as they survive, already complain of not being able to find suitable workers. Enterprising and skilful boys prefer to start in more promising trades, such as motor-cars, motor-accessories or plastics.

In the firms I visited I found that each article of precious jewellery was unique and many were made with the active collaboration of the customer. Often the stones were provided by the customer, who went to a shop in order to get a piece of jewellery remodelled. The retailer, who usually had no designer on the premises, then asked the manufacturer to work out some suitable designs and submit them to the customer. The design selected was then passed to a craftsman who used nothing but hand-tools. The capabilities, ingenuity and inventiveness of the craftsman were allowed for by the designer, and therefore made a considerable impression on the appearance of the finished article. Although the standards were based on Continental fashions, and magazines from Paris and Vienna were studied for this purpose, in practice the design became anglicized long before it reached completion. An economic factor was partly responsible for this. A one-carat diamond, which before the war cost £14, now costs about £40. Manufacturers have been compelled to limit the number of diamonds by spacing them out with strands of platinum and gold wire. It is noteworthy that there is no suggestion of zigzag jazz shapes; in fact the tendency has been the other way, with precise and graceful curves which have brought the art of diamond and gem setting into greater prominence.

Inexpensive fashion jewellery consists, chiefly, of artificial stones in stamped mounts. It is subject to a purchase-tax of 50 per cent. The design of fashion jewellery in this country originally followed the lead of Pforzheim in Germany and Jablonec in Czechoslovakia. Nowadays the chief influence comes from the United States. For the manufacture of inexpensive jewellery, tool-making and die-sinking are the only necessary forms of skilled labour. And since tools and dies are usually made by specialist firms, the cheap jewellers' factories are run with unskilled operatives only. Their task is to use the tools and dies for producing stampings, into which the paste is cemented. A bracelet or brooch consists of two or more such stampings assembled in a certain way. Most of the variations between different products in firms of this description, are achieved by manipulating the limited number of existing stampings. All the time the manufacturer or his designer has to invent new combinations, and this puzzle never comes to an end, since retailers ask for new samples several times a year.

The production methods of one firm of less than sixty workers are worth describing in detail because, apart from the fact that a full-time designer is employed, they are representative of general practices in the trade. Every year the designer and the sales manager visit Paris, not so much to copy what they see as to enable them to anticipate the styles that will be popular in six months' time. While British jewellery of this class has to satisfy the home market, it also has to compete with that made in France, Sweden and Switzerland for sale in those countries. The designer, whose studio is equipped with a drawing board and a work bench, builds up each new design from ready-made stampings of gilding metal—the metal in which it will be produced in the factory. From the same stock components the designer formulates an average of 300 new designs a year. In some cases, however, when an unprecedented design is required, the firm can resort to a new and relatively cheap method of casting, known as investment casting. A wax replica of the component, or complete article, to be cast is coated with a refractory plaster. The wax is then burnt out and molten metal, such as tin-antimony, is cast in the cavity which has been formed. The cost of each casting is only two or three shillings.[1] Glass 'stones' from Austria and synthetic pearls from Paris in standard sizes complete the article. About 700 to 800 reproductions of the same design are assembled by batch-production methods. Each piece is finished in gilt, chromium or rhodium plate.

Although the standard of design in this firm is far from satisfactory, it is at least higher than that attained by most firms in the cheap jewellery trade. In these firms the practice of imitating whatever is fashionable, regardless of its artistic merits, has ruled out all serious consideration of design. Indeed, one works director put it bluntly by saying that 'design has nothing to do with cheap jewellery'. Another manufacturer said that he obtained his 'slickest' and 'cutest' designs for lapel badges by copying American samples. A few firms make use of the refreshing notions of an occasional free-lance designer. It is generally thought that the work of a staff designer becomes stale after a very few years, while that of the free-lance is always fresh. Accordingly, sketches are bought for sums ranging from £10 to £50, and then applied to the firm's particular method of manufacture. This application of an idea on paper to a fixed manufacturing method necessitates, as a rule, modifications, and the design suffers in consequence. What was once perhaps a lively and original idea becomes an unimaginative piece of jewellery.

The fault lies, of course, with the manufacturer who should allow a designer, a man of proved artistic ability, to oversee the entire operation. There is, after all, no reason

[1] For a description of this new process the reader is referred to *Bulletin No. 6* of the Design and Research Centre for the Gold, Silver and Jewellery Industries.

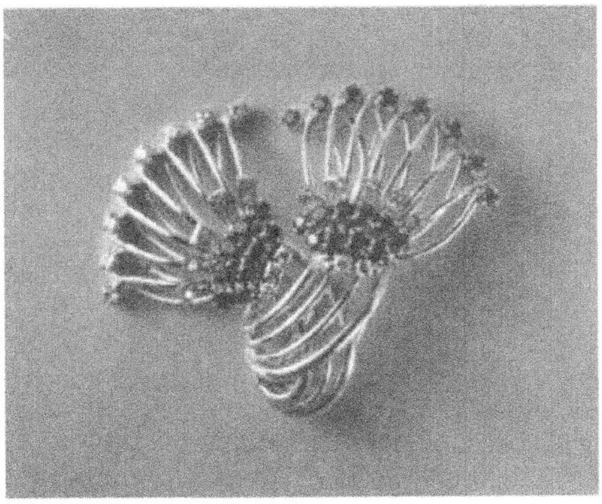

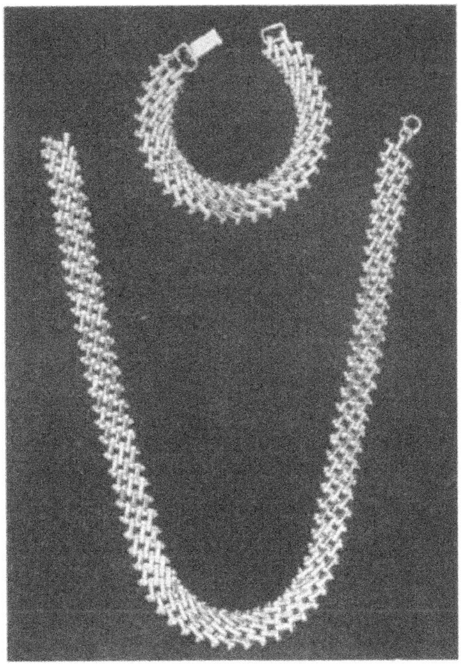

1. *Above*. Precious stones, because of their very high cost, must be used economically. This perhaps explains why the only discernible trend is for openwork patterns, but few are as good as this one. *Made by H. G. Mautner.*

2. *Right*. In British jewellery dependence on past styles is dominant. In this chain and bracelet from Germany there is mechanical precision, typical of today, which immediately attracts. *Distributed by Walter Graham (Jewellery) Ltd.*

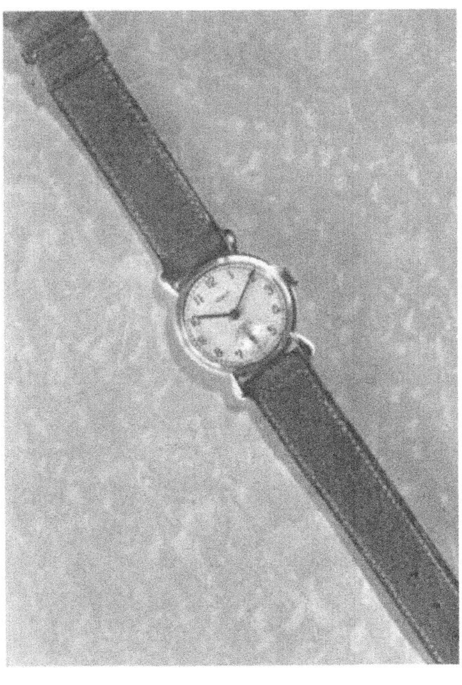

3. The wrist-watch is for use: jewellery is for adornment. Watch-cases have reached a generally good standard with only occasional lapses in the design of figures. *Made by Dennison Watch Case Co. Ltd.*

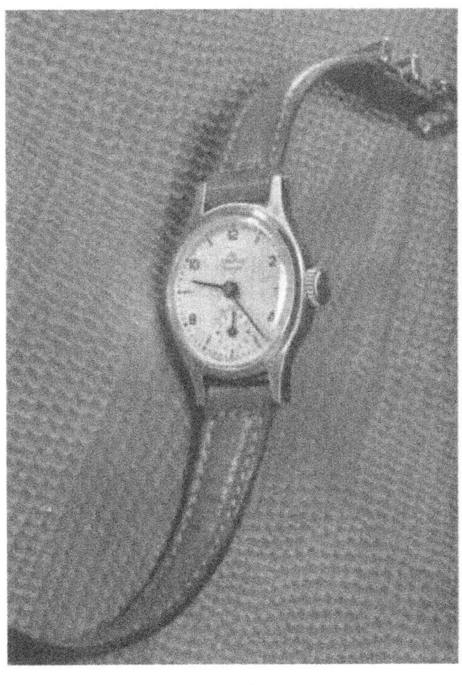

4. In watch-cases there is little desire for sham magnificence. Expression of mechanical accuracy is of first importance and this can best be achieved by precise attention to detail. *Made by Smiths English Clocks Ltd.*

PLATE XX

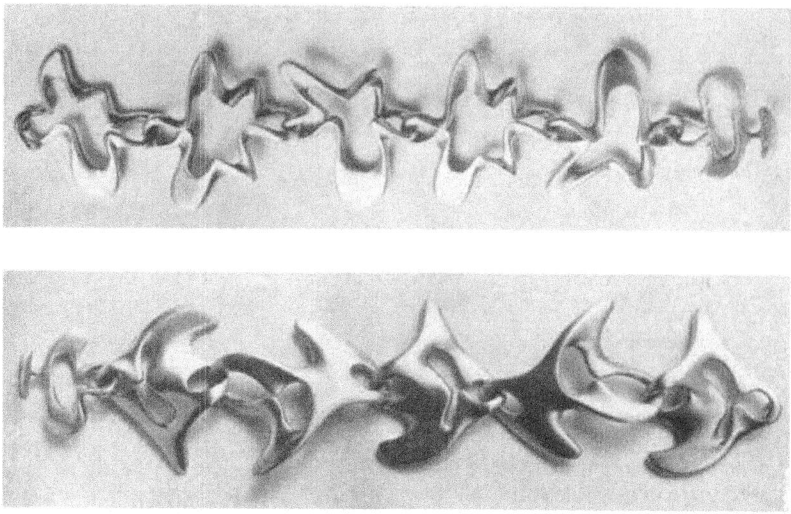

1. These bracelets from Denmark should not be dismissed as yet another stage in the quest for the unconventional in jewellery. Although they have a great vogue today, British firms have not even begun to use metals in an adventurous way, or to explore the possibilities of unadorned, interrelated shapes. *Designed by Henning Koppel for George Jensen (Silversmiths) Ltd.*

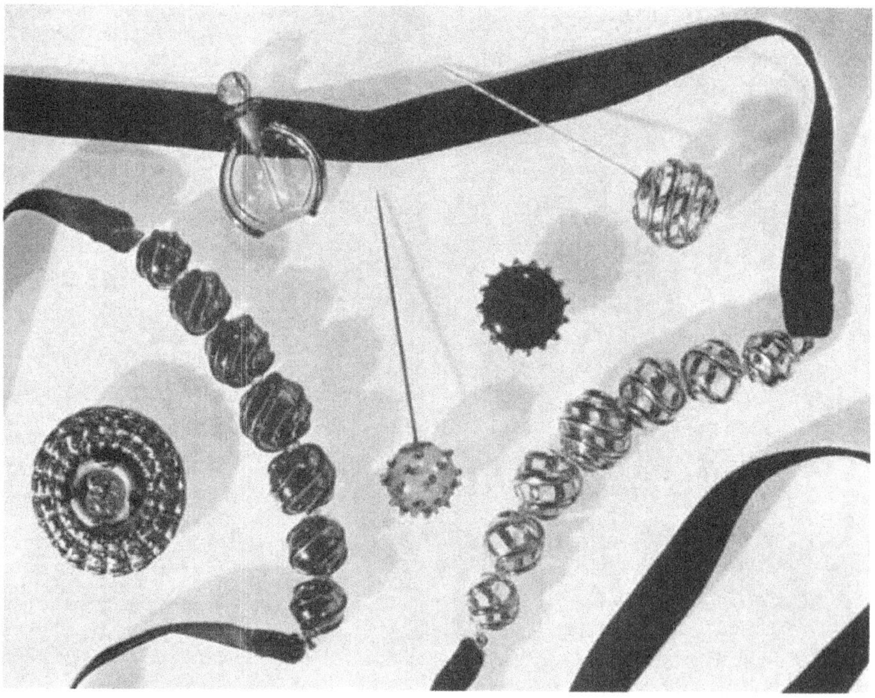

2. Glass jewellery as shown here forms no part of the trade as it is carried on in Birmingham, but helps to suggest that at least some original designs are reaching the market. This is not the only new medium for the creative jewellery designer, though it has by no means been exploited to the full. *Designed by F. Lampl for Orplid Glass Ltd.*

why a staff designer should not be as fresh with his ideas as any free-lance, if he is granted reasonable freedom to travel and to associate himself with the fashion trade. In addition, he would have a sounder knowledge of jewellery manufacture as it affected his own firm, than any free-lance designer could hope to possess. Every year the art schools of London and Birmingham provide a large number of potential designers. One of the main reasons why these graduates are not entering the trade as staff designers is the low level of salaries offered. In nearly every case which I investigated the salary is less than that paid to teachers of craftwork in secondary schools. However, one isolated case proves what can be done in a firm of less than 100 workers. The designer is paid a fixed salary, but receives royalties of 2 per cent on any increase in the firm's turnover. He estimated that he earned between £1300 and £1500 a year.

The firm I saw which specializes in all-glass jewellery employs techniques totally unlike those so far described. The manufacturer, who was trained at the Vienna Art School, does all the designing himself. His standards are his own and not derived from the vanities of Paris. Buttons and brooches are formed in a steel die, and when cold they are gilded. The freer forms, also shown in the illustration on Pl. xx. 2, are either blown or roughly shaped in the flame. The peculiar properties of glass, its ductility and transparency, may be exploited in turn to provide novel and satisfying effects.

The manufacturers making vanity-cases depend entirely upon free-lance designers, who receive about 15 guineas for each design. The vanity-case is usually of stock size and it is for the top or cover that designs are required. This is a difficult medium for the designer, because there is little he can do apart from applying a picture to the flat, shiny metal surface. Most of the designs I saw did not suit the medium, because in composition they were unrelated to the strictly limited area of the case. The cost of the transfer process which applies the design to the case is high, and it is not economical for a manufacturer to produce less than 10,000 of each design. The market is largely British and demands, so I was told, plenty of colour and a certain over-emphatic liveliness, especially in floral and sporting motifs. On the other hand, the best-selling lines are more quiet and subdued but, as far as modern design is concerned, totally uninteresting.

The watch-case trade is very small in this country, and only two or three firms are of any importance. Most of the movements and the majority of the cheap cases come from Continental countries, notably Switzerland. American firms through their British subsidiaries also play an important part. Designing, when necessary; is done in a drawing office on the premises. In the firm I visited about ten new designs a year are needed, but, as most of these would be interchangeable, at least thirty new patterns would eventually go into the range. Watches, which are certainly worn more often than any particular piece of jewellery, are invariably of better design. Whatever shop you go into you will find attractively shaped wrist-watches both for men and women. It appears ludicrous even to think of floral decoration in connexion with a watch. And yet how easy it would have been to start a modernistic or borax fashion. There are several reasons why watch-cases and dials have achieved a high standard of modern design. The wrist-watch is a new type of object altogether. When it was first developed no tradition stood in the way, no basic desire for showiness, no detachment from the standards of twentieth-century life. Moreover, the person who buys a piece of jewellery because it is pretty buys a watch because it works. This attitude in itself is likely to protect people from any desire for sham magnificence which would only detract from the actual value of the watch. The manufacturer of watch-cases looks at the problem of design in the same way. He begins and ends with an instrument of use. In contrast with this the jewellery manufacturer starts off at a disadvantage, for all he can do is produce, say, a brooch, which is merely pretty. A brooch or a lapel badge can have no other *raison d'être*. Unlike a watch it has no movement, no function, save that it conceals a pin. In the showroom of the factory I visited I was confronted with more than 200 different watches. I found it almost impossible to select one example of really bad design. Given an equivalent number of necklaces, brooches and rings, I should not have had the same difficulty.

In this country designers of jewellery have not yet created a style really independent of past standards. Hitherto the Modern Movement in architecture and industrial design has almost completely overlooked jewellery. And no wonder, for the Modern Movement means the superiority of architecture over the fine arts. This wholesome state of affairs, natural to the Middle Ages and still more to the Baroque of the seventeenth and eighteenth centuries, came to an end when the rapid growth of industrial towns overwhelmed the architect's foresight. The new machinery ruined handicraft, and the new superiority of the painter, basking in the light of his message to the world, separated artists and public. Morris was the first to oppose this fatal development. Gropius's Bauhaus was the most advanced expression of the new doctrine which was to guide industrial art in the twentieth century.[1] The ardent desire for cleanness, directness and precision led to the abolition of all that flamboyant decoration which had to the nineteenth century meant architecture and decorative art. Any surviving decoration was rigidly subordinated to the simple lines and surfaces of modern building. That is clearly applicable to furniture, lighting fittings, fires and many other

[1] See N. Pevsner, *Pioneers of Modern Design*, Museum of Modern Art, New York, 1949, chap. 1.

objects in the house. But where could jewellery come in? The wearing of jewels springs from the primitive and natural wish to excel, to attract notice and to conquer. How can these basic functions of jewellery be combined with the canons of modern design? It is typical that chains and chain bracelets, built up from simple, precise and smooth joints, or cigarette cases in gold, shagreen, or even sometimes enamel, are so much more satisfactory than the richer classes of jewellery. In designing a brooch the modern designer obviously feels uneasy. Unless there is a change in this state of affairs (and I do not see how there could be such a change without deep changes in the universal outlook on life), unless some of the best creative designers concentrate their serious study on the reformation of taste in jewellery, and unless, on top of that, the younger generation of consumers insists on buying only what it really likes—unless something of this sort occurs, there is no hope of more than occasional improvements.

DOMESTIC APPLIANCES

This survey of the industries producing domestic appliances is not intended to be exhaustive. The products dealt with are divided into two categories, products operated by gas and by electricity, and have been selected because they illustrate the design methods adopted for a wider range of appliances. For instance, the method of designing gas cookers receives attention because it also relates to fires, water heaters, space heaters, boilers and slow-burning grates. I visited seventeen firms altogether, and from the information obtained I have attempted to indicate the design trends that are being followed. The industries to which these firms belong can be given no specific geographical location. I collected information from firms chiefly in the London area and the Midlands. The gas appliance industry may be said to be concentrated in the Midlands, but electric appliances are made by firms scattered throughout the entire country.

In most firms the design of appliances is in the hands of an engineer, and very few trained industrial designers are employed. In all appliances intended for use in the kitchen, 'fitness for purpose' is the ultimate criterion. Where appliances designed for the living-room and bedroom are concerned, this functional safeguard against artistic excess tends to become obscured. In such cases the engineer, unless guided by a designer, is inclined to make use of period or slick modernistic motifs. Although there is considerable scope for consultant designers in these firms, especially where a small number of new designs is required each year, manufacturers generally are unwilling to employ them. This is at present partly due to the existence of a sellers' market which persists for many types of appliance. As we have already found, a docile and uncritical public is not conducive to a high standard of design.

Until the early 1930's gas cookers had inherited the structure and appearance of solid-fuel appliances. Many of the firms making gas cookers had also made cooking ranges, and cast iron was the chief material for both types of appliance. With all manufacturers the primary consideration was 'fitness for purpose', and gas cookers were designed to cook food and not to appeal to the housewife's artistic sense. All cookers conformed, despite minor variations, to the standard arrangement of a set of boiling rings at the top, a grill immediately below, while the remaining space was taken up by an oven with a heavy cast iron door. The first indication of the new trend occurred in 1933, with the Parkinson 'Minette', No. 1212, which was partially enclosed by a hinged cover over the burners and sheet steel side panels. In the following year Flavel's introduced the 'Kabineat'. The basic structure and arrangement of the cooker remained, but it was boxed-in by sheet steel panels which were finished in heat-resisting enamel. The exterior of this cooker was remarkable for the ease with which it could be cleaned, and for the importance which the manufacturers had attached to its appearance. An improved version of the 'Kabineat' was included in the Exhibition of British Art in Industry at Burlington House in 1935. In the same year Parkinson's, in conjunction with Mrs D'Arcy Braddell, produced a special cooker for showing at the Ideal Home Exhibition. It was designed to be built into the kitchen of a house, a gift to King George V, on show at the Exhibition. Although this cooker was, like the 'Kabineat', constructed of cast iron, the steel sheets used were pressed into shape so that they totally enclosed all structural members, leaving no sharp edges. Two further points should be noted: the drop-door, which provided an additional working surface, and the central placing of the thermostat, which enabled the cooker to be built-in flush with other kitchen equipment. The original model was further developed, and put on the market as the 'Renown' in 1936. Although for a short time its unfamiliar appearance caused what is known as 'sales resistance', the cooker soon achieved popularity and is still being produced by the firm.[1]

The rest of the industry was quick to follow the lead given by these two firms, and by 1939 all cookers had become more or less enclosed by steel panels. Apart from covering the unattractive cast iron frame, these panels, when curved, or given sufficient radii at corners and edges, enabled the high gloss or colour of the enamel to assert

[1] The market's reaction to the 'Renown' is shown by the following figures which are quoted by kind permission of the Parkinson Stove Company Ltd.: 1107 models were sold in 1936, 11,119 in 1937, and 25,052 in 1938. In 1953 Parkinson's introduced the 'Renown Five', a basically similar cooker, but with many refinements and technical advantages. See Jack Stafford, 'Gas Cooker', *Design*, May 1954, pp. 16–19.

itself. As the enamelling of sheet steel presented difficulties, especially when the steel was structural, manufacturers were unwilling to discard cast iron altogether. A further difficulty was that the gas cooker, unlike the refrigerator, required a door which had to fit the frame perfectly. This demanded a rigidity which, they explained, could only be obtained by using cast iron.

Before passing on to consider the design methods of the industry, it is important to note a comparable achievement in the design of gas fires. The earliest gas fires were, like cookers, not designed from the point of view of appearance. The fire was black, grim and as inconspicuous as a gutter. Indeed, black was the only colour available until the early 1930's, when a range of pastel shades was offered as alternatives. When the possibility of applying 'art' was discovered a phase of artistic outrages followed. Even where something new was aimed at, and the style of the Arts and Crafts and later some motifs characteristic of Art Nouveau and then Cubism were adopted, the results seem hybrid now. The decisive changes which eventually came about, were concerned more with the shape of the radiants than the external casing of the fire. Early radiants were hollow clay cylinders which allowed much of the flame to run up to waste. The introduction of the flame-shaped radiant did much to check this waste, and also provided a pleasing, appropriate, and so to speak, symbolic appearance. In 1932 Bratt Colbran introduced the 'Portcullis' fire, in which the radiant formed a grate or gridiron of portcullis shape. This was a functional and artistic improvement of great consequence. The radiant was the result of experiment by the firm's engineers to attain an increase in the heat output. Since they considered this problem seriously and comprehensively, and endeavoured to carry it to a solution as near perfection as was within their power, they arrived at the same time at something which was artistically satisfactory. Nevertheless, it took about a year before the retail buyers and the municipal gas companies could be converted to the new radiants. The outside casings for fires, made of cast iron or pressed steel were, until the beginning of the war, inclined to be modernistic in pattern. It is only since the war that casings have achieved a design standard comparable with contemporary cookers.

To collect information on design methods I visited five firms making cookers, fires and other gas appliances. One firm is representative of a large group of companies employing about 8000 workers. Its products include coal and coke ranges and grates, gas cookers, water heaters and fires. Three other firms visited have 1000, 1100, and 1200 employees respectively, and are mainly concerned with gas cookers and fires. The remaining firm is smaller, with 300 employees, and specializes in gas fires. The following is the typical procedure when developing a new cooker. From the researches of a full-time staff of chemists and engineers, the firm's directors, in collaboration with the sales manager, decide upon the type of new design which is to be put into production. With specific terms of reference the research staff, in the charge of an engineer, will then carry out experiments and tests, and finally build a prototype. At all stages of development the basic considerations are: ease of cleaning, appearance and function, and more than one engineer told me that the housewife, when buying a cooker, pays more attention to the first two of these than to the third. Staff designers as such are not employed. Although I found that in most firms there is a senior draughtsman who has had some art school training, his chief work is to prepare detailed plans for the engineer in charge or, in the early stages of development, an isometric air-brush drawing for the directors and sales manager. On matters of artistic judgement his opinion would probably be sought, but it would never be decisive. There is an important exception to this. In the firm specializing in fires a designer with architectural training is employed, and it was clear that his taste had exerted a strong influence on the firm's products. The salary of a senior draughtsman would be from £700 to £1000 a year, and about £600 for his chief assistant. One firm is in the habit of calling in a consultant designer at a late stage in the development of a new appliance. The managing director explained that this is done to avoid the 'handwriting' which characterizes the work of his development staff. Here again, the designer has little influence, although I was told that all possible alterations would be made to suit his taste before the model went into production. It is significant that in the other firms I visited the directors are particularly proud of the continuity which can be traced in the design of their products. It is felt that the influence of a consultant designer, however limited, would disrupt this.

Before the war it was usual for a firm to add one completely new cooker to its range each year. Modifications to existing models were being carried out continually, and the average number of cookers in the year's range would be between thirty and forty. Nowadays a new cooker is added to the range only once every four or five years. Modifications are still carried out, but the average number of cookers in the year's range has dropped to four, or less. The enamel colourings available have also been reduced from five or more, to two or even one since the war. Firms concentrate on the pre-war best-selling colours: i.e. white and cream for cookers, bronze and black clouded gold for fires. In fires there is a wider range of colours available than in cookers. The quantities of each design produced are consequently greater than before the war. Two firms gave me their production figures. In one, the present-day maximum output of all types is 12,000 to 15,000 cookers a week, and about 400,000 models in one design over a period of five years. In the other firm, 100,000 cookers in one design are produced each year. Comparable figures can be quoted for gas fires.

There are two reasons why the number of designs available is limited. One is the sellers' market which exists because very few gas appliances were made during the war. For the other reason we must turn to methods of distribution. Before the war there were about 1000 gas undertakings in the country. With the exception of small fires, space heaters, etc., all the products of the industry were sold through the gas companies. Many gas undertakings had individual requirements which manufacturers had to take into account when designing a new appliance. 'If all the gas undertakings had been of similar size and buying power, this would have resulted in dear appliances and incredible complications in maintenance and manufacture of spare parts.'[1] As it was, the undertakings varied in size, and by far the largest of them was the London Gas, Light and Coke Company, with headquarters at Watson House. Although the Company handled 20 per cent of the industry's total output, its influence reached much further. In the laboratories at Watson House tests were carried out on the appliances which manufacturers wished to sell through the Company. It was understood that an appliance should not be tested beyond the limit of performance claimed by the manufacturer. Furthermore, many of the Watson House specifications coincided with British Standards which, in any case, had to be observed by all manufacturers. Just prior to the war the specifications which these tests had established became generally recognized by the smaller gas companies, who began to refuse a manufacturer's appliance if it had not been approved at Watson House.

After the war the influence of Watson House steadily increased. A committee, consisting of the controller of research, the commercial manager and the public relations officer, was established to review each new appliance after it had been tested. From our point of view the public relations officer is important, because it is his task to assess the artistic merits of the appliance. On the recommendations of this committee a new appliance may be accepted, rejected or sent back to the manufacturer for modification. Since the nationalization of all gas companies on 1 May 1949, and the establishment of only twelve Gas Boards for the whole country, the influence of the Watson House committee has been extended to include all but a few of the industry's products. When questioned, manufacturers said they favoured this dictatorship for a number of reasons. When a new appliance is accepted by the committee the manufacturer is sure of receiving a sufficient quantity of orders to enable him to commence a long production run. He is also assured that the twelve gas boards will agree to accept the same model without modifications or alternative parts. In addition, manufacturers are satisfied with a policy which has partly affected them for twenty years, and they respect the judgement of the committee as being the

[1] Leslie Hardern, 'Design and the Distributor', *Design*, October 1949, pp. 10–12.

best criterion for the design and performance of their products. That the public, too, seems to be well served by this policy, can be judged from the appliances displayed in all gas showrooms.

The responsibilities of the committee are great. Some 750,000 cookers of approved design were sold by the Gas Boards in 1950, and out of the total number of temporary houses built since the war, 170,000 have been equipped with gas appliances approved by the committee. The Gas Boards have women's home service advisers, whose duties are to demonstrate new appliances, to lecture to women's organizations and to visit homes after an appliance has been installed. No official design propaganda is handled by these advisers. Unless the director of the local Board holds positive views upon the appearance of the appliances in stock, the function of the advisers on design matters is likely to be negative—the mere recording of complaints. Although we may question whether it is advisable to give one small committee the overriding control of a whole industry and its public, we must admit that at the present time, gas appliances have attained a satisfactory and often very high standard of design.

Nevertheless, the chief credit for this high design standard must, of course, go to the manufacturers and their engineers. The fact that cookers are, in appearance, better designed than fires, is proof that the engineer, when left alone to create a useful instrument, naturally produces a design which often cannot be improved. Before the war gas fires were modernistic: today they still have modernistic motifs but, like some types of door furniture, these have been softened by the borax trend. Although many fires are of exemplary design, there is a tendency for the engineer, when thinking of his products in terms of the hearth and not the kitchen, to insist upon superfluous decoration. The point is important, and it will recur later.

For information on electric appliances I visited eleven firms. Two firms are mainly concerned with electrical engineering for industrial purposes (transformers, switch-gear, etc.) and a relatively small number of their employees produces domestic appliances. Two other firms are primarily concerned with radio and television receivers: one of them has a household appliance section and the other a heating and lighting section. I visited two factories which concentrate on refrigerators, and three which make irons, kettles and toasters. I also interviewed one manufacturer specializing in fires and one specializing in clocks. I completed my investigation by seeing three consultant designers. Firms dealing exclusively with electric cookers are omitted for three reasons. First, electric cookers are not so extensively used as gas cookers; the ratio is about 1 : 4. Secondly, the uncertainty of receiving an adequate supply of electricity may have prejudiced the public's choice of power for cooking equipment. Thirdly, the design of electric cookers has undergone an evolution so similar to the

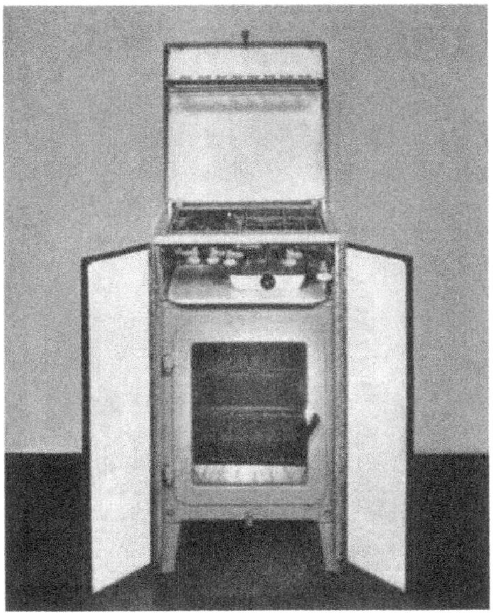

1. 1935 'Kabineat'. The first major attempt to design a modern gas cooker. A cast iron frame was boxed-in by enamelled sheet steel panels. The plate rack folded and the splash-back could be laid flat on the hotplate. *Made by Sidney Flavel & Co. Ltd.*

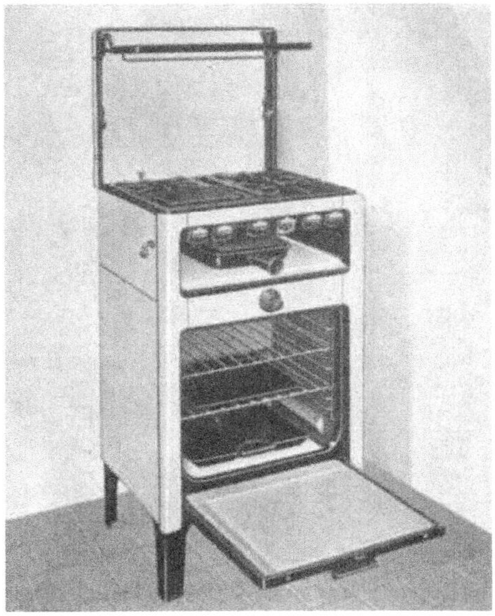

2. 1936 'Renown'. Here sheet steel over the frame forms an integral part of the design. A counter-weighted door drops down to form a working surface and the cooker could be built into a modern kitchen flush with other equipment. *Made by Parkinson Stove Co. Ltd.*

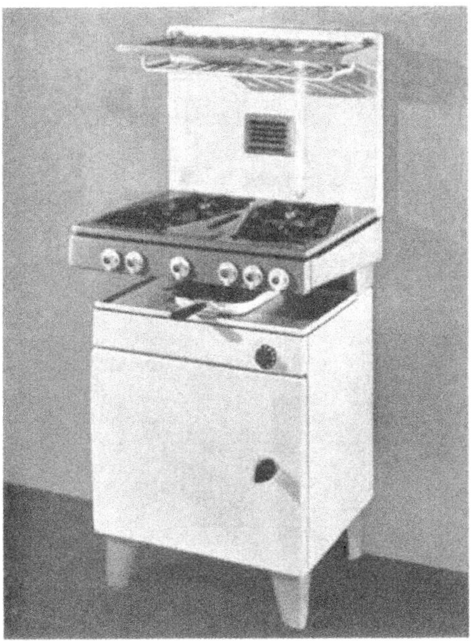

3. 1952 'New World 84'. An example of refined post-war design, characteristic of the industry, but with an original feature. The hot plate is cantilevered from the main steel frame to give an accessible working area beneath. *Made by Radiation Ltd.*

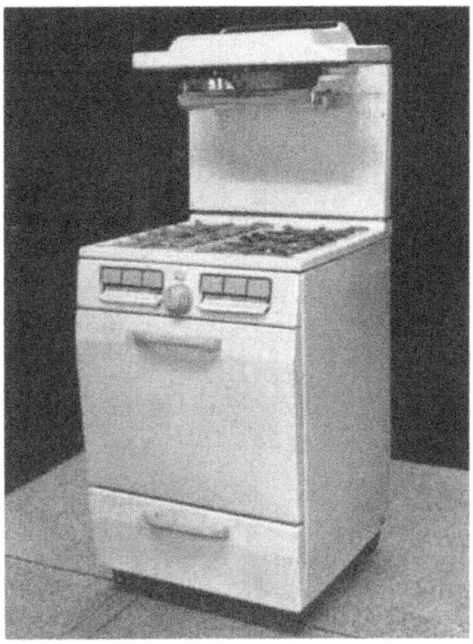

4. 1953 'Renown Five'. The successor to the 'Renown' has evolved logically to meet the more exacting demands of function and appearance. A warming drawer has been included and the grill has been placed at the top of the splash-back. *Made by Parkinson Stove Co. Ltd.*

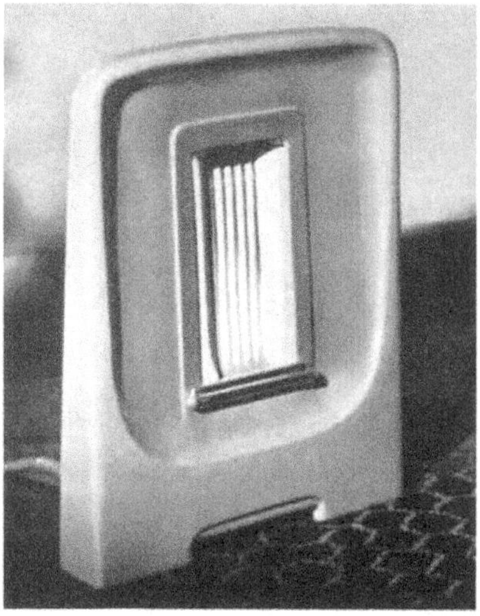

1. The sculptural quality of the casing helps radiation and also does much to enhance visually the purpose of this portable gas fire. *Made by Bratt Colbran Ltd.*

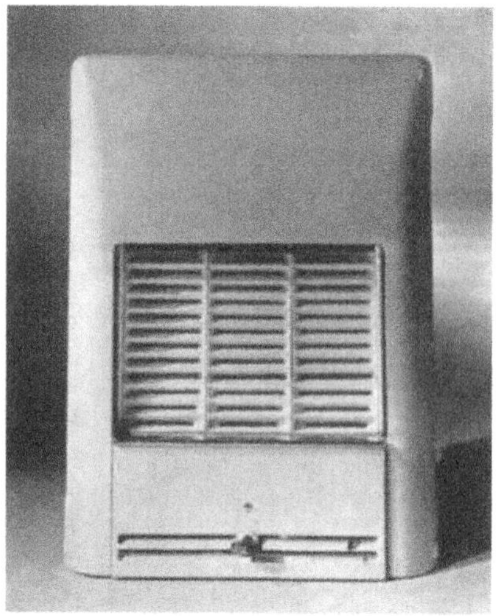

2. A gas fire with triple radiants controlled from the front, made for close fitting in a hearth. One of the most graceful designs on the market. *Made by Radiation Ltd.*

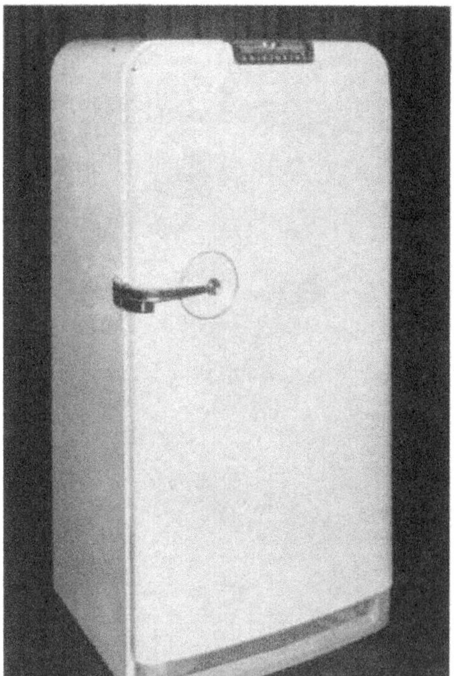

3. Few firms have withstood the temptation to 'enrich' their refrigerators with chromium-plated stylish decoration. This is a particularly restrained and well-balanced model. *Made by Frigidaire Ltd.*

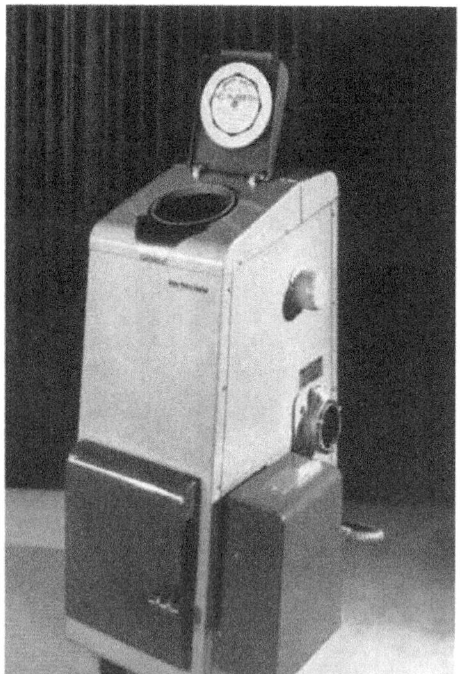

4. Straightforward engineering frequently produces designs of good appearance. This anthracite boiler pretends to be no more than an efficient appliance. *Designed by G. A. Ridgley for Janitor Boilers Ltd.*

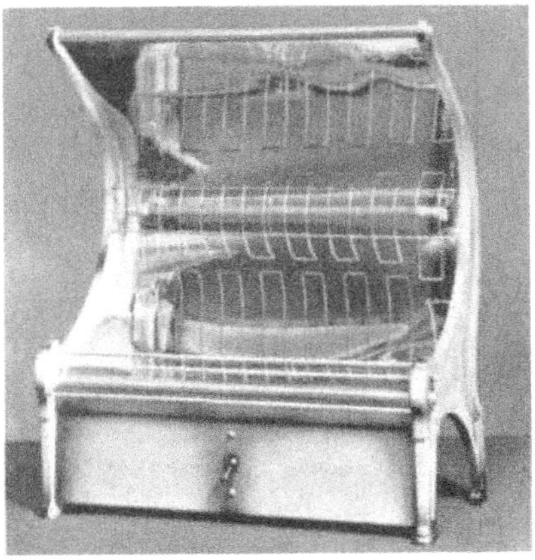

1. 1931. A development of the revolutionary concave reflector fire of 1926 which was designed by the late Dr de Ferranti, an engineer of wide experience and imagination. *Made by Ferranti Ltd.*

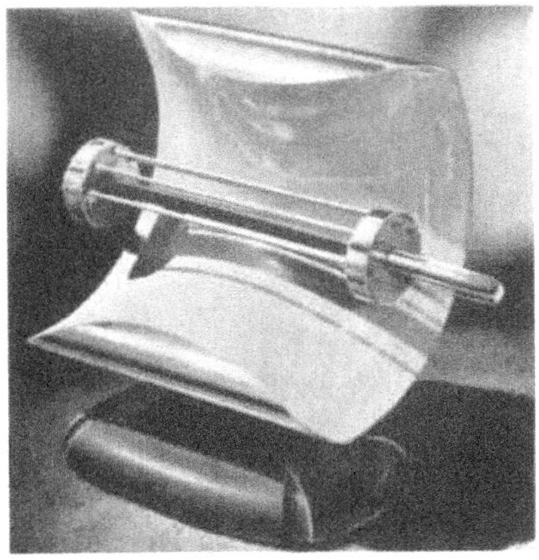

2. 1939. Working to a similar principle as in 1, this fire uses a backward-bent parabolic reflector to radiate heat over a wide area. *Made by H.M.V. Household Appliances Ltd.*

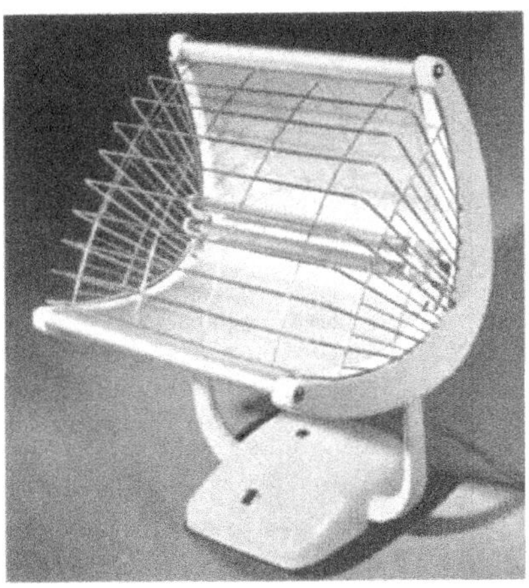

3. *Above.* 1953. The shape of the reflector has been re-designed for greater efficiency and supported on a cast iron stand. The guard meets the new regulations for safety. *Made by Ferranti Ltd.*

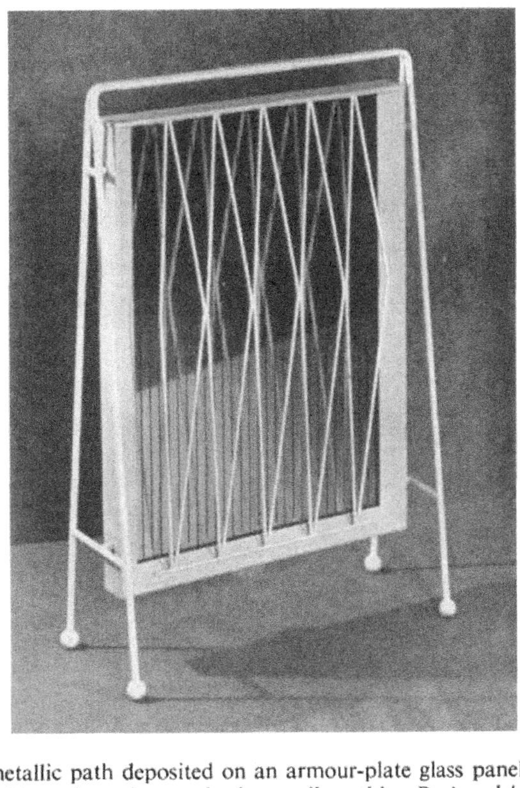

4. *Right.* 1952. The heating element is a metallic path deposited on an armour-plate glass panel. Few ideas to pose entirely new design problems have been solved as well as this. *Designed by John Barnes of Allen-Bowden Ltd. in collaboration with Premier Electric Heaters for Hotpoint Electric Appliance Co. Ltd.*

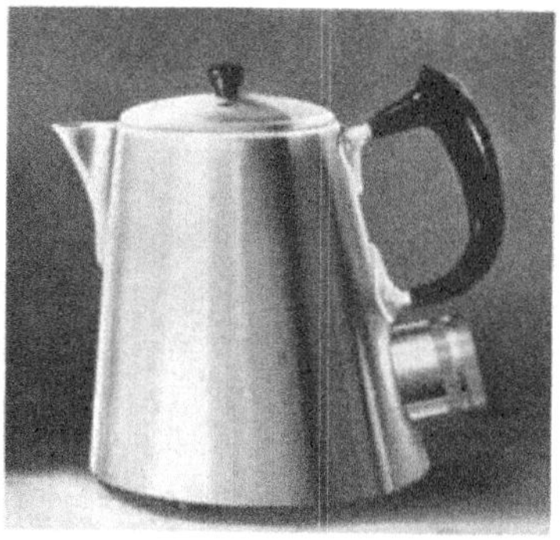

1. Only recently has it been realized that the electric kettle is in no way associated with the stove. The trim design of this model makes it suitable for use anywhere in the house. *Made by H.M.V. Household Appliances Ltd.*

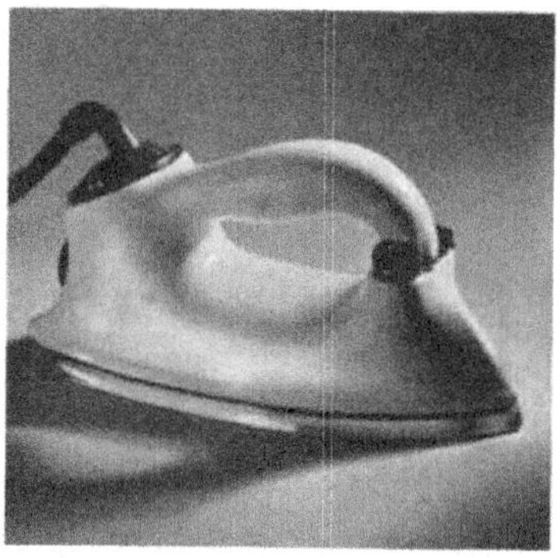

2. 1936. The first British electric iron to dissociate itself from the conventional flat-iron shape. *Designed by Christian Barman for H.M.V. Household Appliances Ltd.*

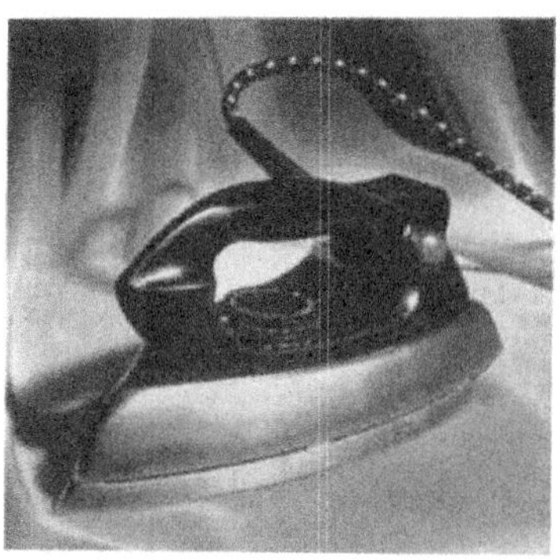

3. 1951. Iron with plastic handle moulded to the grip, with heat regulator for various types of cloth. The flex may be plugged in at either side. *Designed by D. W. Morphy for Morphy-Richards.*

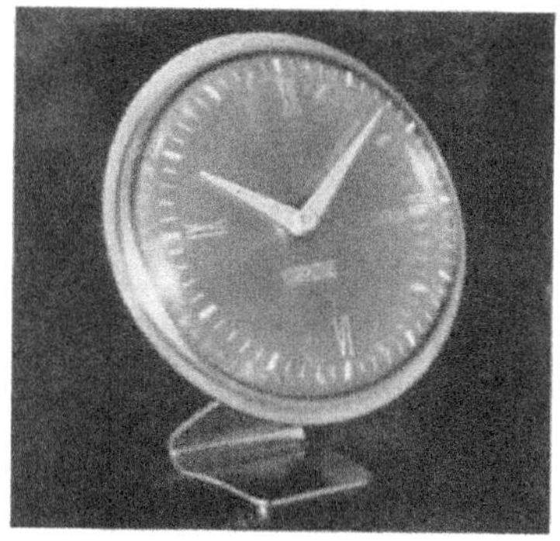

4. A bold attempt to inject fresh thought into an industry hidebound with hackneyed precedents. Manufacturers rarely risk such interesting designs. *Designed by Jack Howe for Gent & Co.*

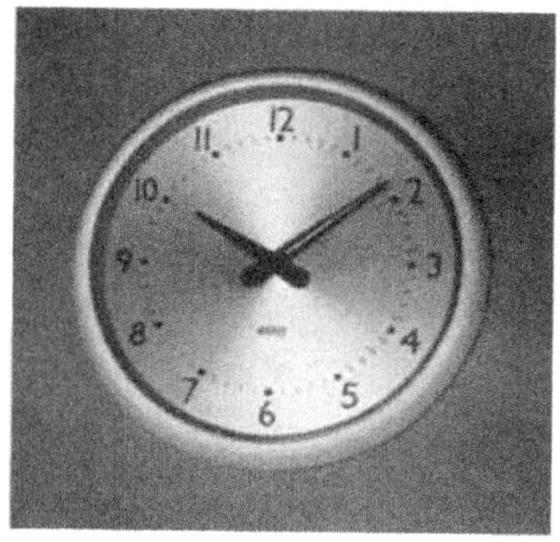

5. Partly owing to demands from architects good wall clocks like this recessed example are fairly common. Note the scale within the numerals. *Designed by Jack Howe for Gent & Co.*

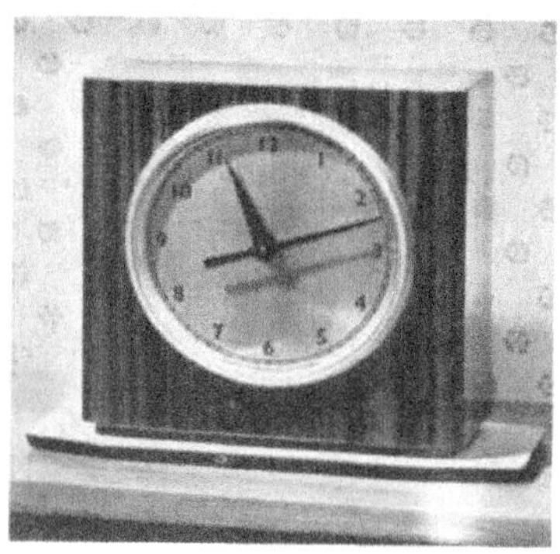

6. To be a respectable ornament on the mantelpiece a clock need not be tricked out with scrolls and brass trimmings, yet this example is still exceptional. *Designed by W. H. Russell for Thomas Mercer Ltd.*

gas cooker that discussion would be repetitive. In general layout and appearance also the two very closely resemble each other.

Before dealing with refrigerators, it should be noted straight away that manufacturers and designers in this country owe a great debt to the pioneering work which has been done in the United States. In the engineering laboratory of one firm I found examples of all the well-known American and British refrigerators. Each one had been stripped, examined and reassembled, so that any new details or devices could be copied and adapted to the firm's own models. This procedure is common to most engineering industries; its practice becomes vicious only when it prevents the designer from creating anything that has not already been tried on the market by a competitor. Refrigerators are perhaps more susceptible to unnecessary stylish embellishments than any other type of kitchen equipment. In recent years the method of manufacturing refrigerators in large firms has been standardized for reasons of economy and efficiency. The two sides and the top are formed out of one piece of sheet steel, bent in a press. This is held rigid by a steel back-plate and the internal cross-members which support the storage chamber and the power unit. The door and lower front panels are also separate steel pressings. Given this standard method of construction, there is little a designer can do to make his model different from a competitor's.[1] Although the proportions of a refrigerator may be excellent, many designers have to satisfy the salesmen's demand for 'eye-appeal'. This prompts them to over-design the nameplates and handles, and to apply chromium-plated strips to an otherwise smooth exterior.

In a large firm mass-producing refrigerators the following procedure was used when bringing out a new model. A market research organization was commissioned to interview a representative number of people to obtain their views. The Electrical Association for Women, which has its own methods of studying the wishes of consumers, was also asked for advice.[2] Given this information on the taste of the market, a committee, consisting of the production director, the commercial manager, two engineers and the sales manager, made suggestions for a new model, with estimates of performance, numbers required, cost, etc. With these preliminary specifications the engineers planned the model with the help of a consultant designer. Although the designer made three or four mock-ups in wood, each of which embodied the specifications demanded, his chief task was to design the fittings, handles, badges, etc. So great was the importance attached to this work, that the designer received an average of £1000 for each refrigerator. As tool costs were very high, between £25,000 and

[1] Some of the factors which condition the design of mass-produced refrigerators have been examined by H. Van Doren in *Design*, October 1949, pp. 2–5.

[2] See *Towards Perfection*, The Electrical Association for Women.

£30,000 for each design, it was essential for a new model to be produced at the rate of 600 per week for the first three years. A very satisfactory model, with a few modifications to the fittings, might be expected to run for at least five years. The bigger the production the more money could be spent on 'improving' the design. This meant in practice that best-sellers were steadily enriched with unnecessary chromium strips.

Early electric fires in this country were straightforward in design, and consisted of a round copper bowl with the element coiled in the centre. Like gas fires, they went through a decorative phase, when Tudor and Adam surrounds flourished, and all the characteristics of the traditional solid-fuel fireplace were introduced into the designs. A development on these lines took place in 1920, when Berry's brought out its 'Magicoal' fire. This type of fire, which has been copied by several firms, incorporated a device which sent a flickering red light across a semi-transparent bank of real or imitation coals. This interest in a cunning device is, as we shall see in subsequent chapters, related to that preoccupation with techniques which is so frequently thought to be an adequate substitute for artistic merit. Since its introduction there have been many variations on the flicker-coal fire, with a range of styles extending from the Middle Ages, chiefly wrought iron work, to Regency, chiefly embossed sheet steel. While there are some inexpensive models, this type of fire makes its strongest appeal to the high-price market.

From our point of view the decisive change took place in 1926, when Ferranti designed the first rectangular concave reflector fire. The appearance of the fire was gradually improved by casting off some unnecessary frills. (Like the 'Portcullis', this fire did not enjoy immediate popularity on the market.) The exemplary design which was achieved for the 1931 model has remained practically unchanged for more than twenty years.[1] The conception and development of this fire was largely the work of Dr de Ferranti, who was an engineer of wide experience and considerable imagination. For both the 1926 and the 1931 models, the firm employed no special artistic designer. Although this type of reflector fire has since been copied by many manufacturers, the idea behind it was further developed by the engineering and design staff of H.M.V. Household Appliances. The H.M.V. fire, which was first produced just before the war, has a conventional rod-type element, but with a backward-bent horizontal parabolic reflector. The heat radiated is evenly distributed over a wide area, and is not concentrated in the front. Here again, the logical solution to a mechanical problem, when not interfered with, results at the same time in an exemplary design.

The method used for designing fires varies in each firm I visited. As it would be impracticable to deal with firms individually, I prefer merely to emphasize certain

[1] In 1953 Ferranti's introduced a new version with a reflector of greatly improved efficiency.

characteristics, especially where they differ from the design methods already discussed. In one firm making antique reproduction fires, some of which incorporate the flickering device, the managing director sketches out 'new' designs and then calls in his chief engineer to make technical suggestions. This approach, which means putting the artistic aspect before the functional, seems to be quite common among firms serving the expensive end of the market. In other firms making plain, functional fires which are also cheap, a satisfactory standard of design is maintained for one of two reasons. Either the managing director says that the low price of the fire prevents him from applying any 'design' (or 'art') to it, or an engineer-designer is employed. In the first case a large increase in the production of one design, as we have already noted when dealing with refrigerators, is likely to be reflected by artistic excesses. In the second case this would probably not occur and the design of all fires produced, including the best-sellers, would maintain a constant and satisfactory standard. It is inevitable that a great deal depends upon the status and the personal taste of the designer-engineer. In all firms I visited his status is lower than that of the sales manager. Usually he is a man with a technical college rather than an art school education and he is, like most of the directors I met, inclined to confuse design with decoration.

The same problems arise in the majority of firms making toasters, irons, kettles and other small kitchen equipment. Only where the functional requirements of an article are such that any unnecessary embellishment will reduce its efficiency, is a satisfactory design achieved. In the firms I visited there seems to be no definite method or policy when developing a new design. Many manufacturers are content to let their engineers imitate a competitor's model. One engineer told me that he collected magazine photographs of American appliances and then, after an interval of perhaps five years, designed something like them for the home market. Another said that his wife thought up new ideas for him, while yet another maintained that the basic shapes for irons, kettles and toasters were established anyway, and that it was his job to streamline them wherever possible. In rare cases consultant designers have been employed for one or two items (the H.M.V. iron by Christian Barman for example), but their influence upon the industry has been too spasmodic, and the interest shown in their work too slight, to have any lasting effect upon design standards.

I found that the situation in the clock-making industry is fundamentally similar. The standard of design seems to depend entirely upon the price of the clock and the position which it will occupy in the home. One of the largest producers in the country gave me the following information. Of the firm's total output by value, 60 per cent is in wooden cases and 40 per cent in metal, plastic and glass cases. Wooden cases belong almost exclusively to medium and high price mantelpiece clocks. They are decorated with appliqué scrolls, beading, etc., and are antique or semi-antique in appearance. The higher the price the more decoration needed. The modernistic style, which lingers in many domestic appliances is still used for clocks, especially the so-called 'cocktail bar' clock. As far as I am aware, the only advance upon pre-war standards is the fact that circular dials are now more popular than other shapes. Some plastic cases have broken away from the modernistic craze, but the firm's best-selling cheap clock (500,000 have been made) is embellished with jazzy lines. It is significant that no expensive movements are put into plastic cases. The manufacturer explained that although this appeared to be illogical, his distributors (mostly jewellers) habitually consider plastic articles as makeshift, and to be associated with the lowest class of customer. As no industrial designer is employed, the manufacturer depends upon his own ideas, together with his ability to modify existing patterns, to produce about forty new designs a year. Although he was responsible for all the designs produced, he was sufficiently conscious of their low artistic standard to refuse my request for a photograph of his worst-looking best-seller. In another large firm the standard of design is no better, even though a staff designer is employed. The only acceptable design in his entire range is a cheap plastic kitchen clock.

In the firms visited it appears that both management and design staff are incapable of producing clock-cases of satisfactory design. As soon as the clock ceases to be a kitchen instrument and becomes a respectable ornament for the mantelpiece, the worst artistic outrages can be expected. That this state of affairs is typical of the whole industry may be proved by visiting the British Industries Fair. I know of only two firms which can be excepted from this general criticism, and the better design standards they have achieved are entirely due to the work of consultant designers. Both firms commissioned mantelpiece clocks, and the results are very satisfactory commercially and artistically.

In this survey the work of the engineer has received a good deal of criticism. The engineer is frequently instructed to make a simple, functional instrument respectable enough to be associated with the furnishings of a living-room. Having accepted the implied criticism (which invariably comes from the sales manager) that his instrument is not presentable outside the kitchen or the laboratory, what approach does the engineer take? The basic shape of his instrument has already been determined in terms of efficiency, and cannot be altered. This leaves the outside case which may or may not be susceptible to changes in proportion and finish. In cases where no dimensional alteration can take place, the engineer must confine himself to adding some decorative features to obscure the simple, unpretentious lines of his instrument. At this point the confusion between design and decoration arises. Although no satisfactory result can be obtained once this process has begun, the kind of decoration used depends upon the engineer's awareness, or ignorance of, contemporary fashion trends. The degree, or amount of decoration applied, depends upon the money available; the better an appliance sells, the more money can be spent on enriching its design.

Who is to blame for this state of affairs? Surely not the engineer. His main task is to design an instrument that functions efficiently and can be produced in a simple and inexpensive manner. When he is permitted to do this untroubled by considerations of respectability, he often produces appliances which are artistically excellent. As we have seen, most domestic gas and electric cookers illustrate this, and for a further example we can refer to the 'Janitor' boiler. The exterior appearance of the boiler has been conditioned entirely by the internal structure and mechanism. It was conceived and executed by G. A. Ridgley, an enterprising and imaginative heating engineer.

Some manufacturers, conscious of the engineer's dilemma when asked to superimpose artistic motifs on his product, attempt to solve the problem by calling in a consultant designer, or stylist as they prefer to call him. As this is invariably done at a late stage in the development of a product, the result is still unsatisfactory; in fact it is probably no better than that which would have been achieved by the engineer alone. Few manufacturers realize that in order to obtain a satisfactory standard of design, they must enlist the services of a designer in the initial stages of a product's development. The designer must be allowed to give his opinion before the final layout and shape of the product is established by the engineer. Only in this way can the artistic quality necessary become an integral part of the product's design.

For an illustration of the practical application of this, we can refer to the design methods of one large and well-known company manufacturing electrical appliances. The company has several factories, each with its own staff of development engineers. In 1947, an independent industrial design section was established in the company's London office. Although the designer is responsible to the sales director, he is able to operate as an internal consultant on design problems in the factories. His authority is not yet sufficient to enable him to develop a design policy, but his influence is clearly beginning to show in the company's latest products.

ELECTRIC LIGHT FITTINGS

The proper balance between form and function, which we call good design, is difficult to achieve. In the previous chapter we considered the argument against applying decoration to a finished product which is purely functional. From the deplorable results of making a product 'presentable', we have seen how vitally important it is to consider the appearance of the exterior at the initial stage of development. The question which concerns us here is essentially the same, only the argument must be reversed. Manufacturers begin with the appearance of a light fitting and afterwards relate it to its function. Although this may appear impossible, one must remember that electric light was preceded by gas, which in turn was preceded by oil and candles. There is an inexhaustible supply of 'forms' which have been associated with artificial illumination, and manufacturers today do not hesitate to draw upon it. By this method of design no balance between form and function can be achieved. In fact, the approach is disastrous and results in a low design standard throughout the industry.

The two main centres where light fittings are produced are London and Birmingham. I visited seven firms. Two of them have more than 4000 employees each, and are among the four largest in the industry. One of these firms specializes in fluorescent fittings, the other manufactures all types of lighting equipment, both industrial and domestic. Two more factories which I saw at Birmingham employ 150 and 80 persons respectively, and produce a wide range of domestic fittings in all price groups. The remaining firms are smaller, with 60 persons and less, and are included because of their high design standard in a limited market.

The industry has only grown up during this century, but many of its firms had previously made gas pendants and brackets. The first electric light fittings retained the characteristic forms of the gas brackets, etc., from which they had been adapted. There was little change in this approach towards design until 1910, when the alabaster bowl was first introduced by a Birmingham firm.[1] The bowl and the new type of fitting which it demanded soon became popular. 'With it came a great increase in the desire to conceal the actual mechanical source of light and to illuminate the room with a more diffused light reflected from the ceiling.'[2] Apart from alabaster bowls, there were hardly any but imitation antique fittings between the First World War and, say, 1925. It was the ominous Paris Exhibition of that year which must be held

[1] See R.D.Best, *Brass Chandelier*, Allen and Unwin, p. 120.

[2] *Design and the Designer in the Light Metal Trades*, p. 27.

responsible for the introduction of modernistic designs.[1] Jazz fittings with pseudo-cubistic, angular sheet-glass panels, clipped together by thin metal supports, became the fashion. Similar spiky forms had had a great vogue in Germany about 1922–3, in connexion with the style which dominated in 'Expressionist' painting, during the most restless years of post-war change. In Germany, however, a reaction against the vulgar commercial caricatures of this type set in almost immediately. The Bauhaus introduced some simple spherical fittings in 1923. Since then, unadorned lighting fittings such as the spherical, tubular and square box types, steadily gained ground in Germany. For a short time they seemed to have almost eclipsed the more elaborate shapes. But the final result was a compromise, and chandeliers with three, four or six arms carrying tubular or bowl-shaped holders for lamps became the favourite kind of ceiling fitting. In Britain, the box fitting was introduced about 1927, the lipless opal glass sphere in 1929, and the tubular fitting a little later. While the merit of having achieved this great improvement is due to a few enterprising firms, it should be noted that, to begin with, all the glass parts for these fittings were imported from Germany. Of all the products of the Bauhaus, the lipless sphere has been the most enduring. It is unique and unalterable, and is today one of the best-selling patterns, particularly for offices and public buildings.

Apart from the spherical fitting, the influence of the Bauhaus, through the few British firms which followed its example, was gradual and hesitant. In the early 1930's the industry was content to rely upon the catalogue of the Paris Exhibition, as a welcome and inexhaustible source of sham splendour. One of the most popular types of domestic ceiling fitting was a modernistic pendant with jagged orange glass panels. Improvements on this were confined to fittings of medium price. In 1934–5, the best-selling pattern in one large firm was no longer the jazz model, but a pendant similar to the German chandeliers with upright tubular holders. In 1935–6, public preference was for bowl-shaped models of German or Swedish character. As many firms were unable to obtain suitable glass in this country, they had to rely on the stock patterns available from Thuringian and Czechoslovakian glass-works. In this way the Bauhaus-type designs became, almost unawares, part of the industry's currency. For cheap fittings the designs remained modernistic, while period styles were preferred for expensive fittings. By 1939 the demand for mock-antique designs had increased, not only in expensive fittings, but for all types.

After the war the vogue for reproduction work appears to have become even more pervasive. In the catalogue of one firm I visited, the styles are vaguely termed French, Flemish, Tudor, Georgian and Regency. The fact that these styles were originally

[1] Exposition Internationale des Arts Décoratifs, Paris 1925.

related to essentially different methods of illumination, appears to be a matter of no importance to the manufacturer. What does concern him is his inability to reproduce by machine that which was originally made by hand. The styles are used for ceiling fittings and wall-brackets made of copper, brass and thin gauge mild steel—so-called 'wrought iron'. The fairy-tale-type electric lanterns are usually detailed to look Gothic. They are made of iron (as if by a rural blacksmith), and given an 'antique armour' finish. One manufacturer explained that he got this effect by painting the iron with black paint in patches, and then polishing it. I was told that in all reproduction work the 'finish' is very important. Amongst the more popular types are antique brass, bronze, old silver colour and satin silver. The shades are made of glass, real or imitation parchment, or translucent plastic, described in one catalogue as 'natural imitation vellum'. There is no need to dwell upon the futility of this neo-antique craze which supports, at one end of the trade, huge Baroque 'electroliers', and at the other that feeble 'light in the shape of an elliptical bulb emerging from an imitation candle'.[1]

The remaining styles in use are numerous and varied. The influence of the Bauhaus, together with modernism, has produced a bastard, but very common style. Forms which were created at the Bauhaus have now become design clichés through years of unimaginative imitation, while the zigzags of modernism have been softened by the borax trend. The resulting style, which is called 'modern', has several characteristics. The glass bowls have loose and indefinite curves. They are decorated with veined patterns to imitate alabaster or painted with bands of colour in the manner of pre-war pottery. The plain opal bowls and globes are frequently surrounded by a flat, horizontal Saturn-like ring. All attempts to mould glass decoratively have been uninspired. Amongst the more eccentric light bowls are the shell and flame formations, and the imitation cut glass diamond patterns. The chromium-plated metal fittings are in stepped formation for both pendant tubes and wall-brackets. In addition wall-brackets are often cross-banded with three parallel lines. As one might expect, the fittings designed exclusively for the bathroom are generally excellent. They are produced by the same firms responsible for the appalling, but 'respectable' fittings we have just dealt with.

There are several conditioning factors which help to explain why the design standard in this industry is so depressing. Very few firms employ designers who have had a full-time art school training. I only know of two cases in recent years where a consultant designer has been able to design a complete range of fittings. Instead, I found that engineering draughtsmen with no special artistic training were used. The head

[1] *Design and the Designer in the Light Metal Trades*, p. 27.

PLATE XXV

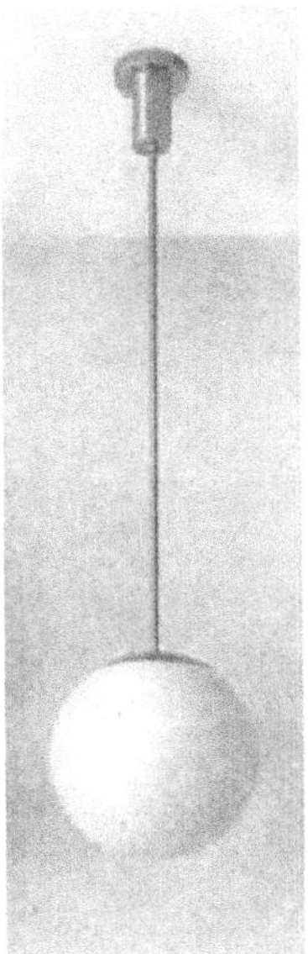

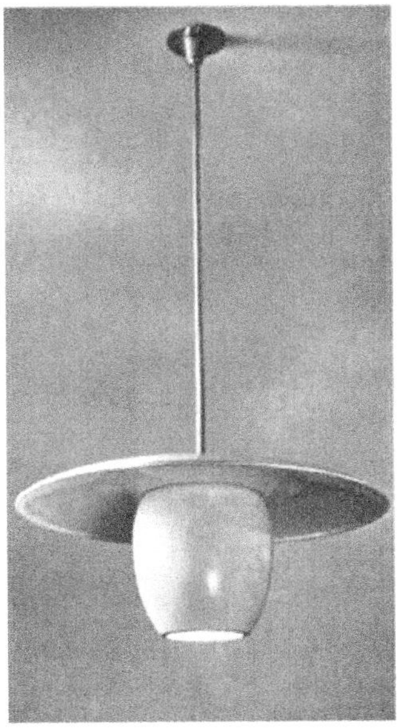

1 and 2. *Left.* The lipless opal glass sphere first introduced in this country in 1929. It is based on an original Bauhaus design. *Right.* One of the best variations on the sphere having a metal reflector. Left, *made by Merchant Adventurers Ltd.* Right, *designed by A. B. Read for Troughton & Young* (*Lighting*) *Ltd.*

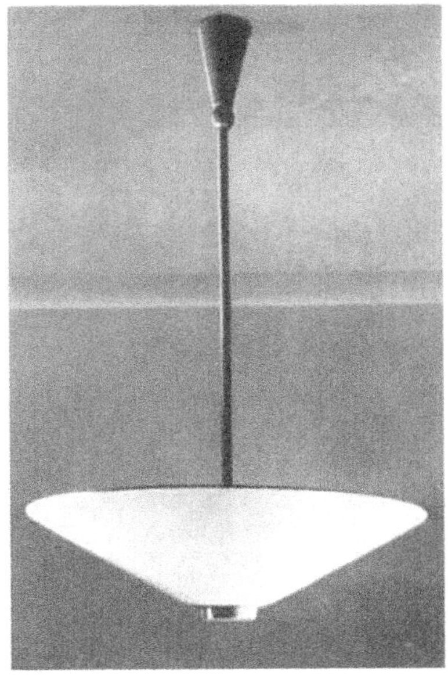

3. All light fittings attempt in different ways to hide the electric bulb. Here a shallow aluminium cone reflects the light on to the ceiling, where it can be diffused over a wide area. *Made by Best & Lloyd Ltd.*

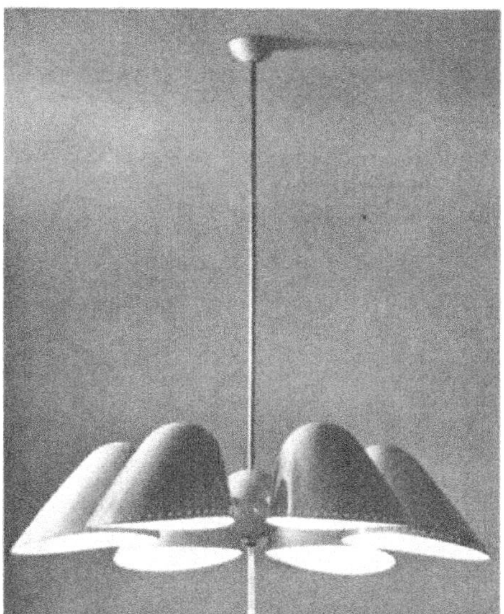

4. For economy in manufacture and variety in the range it is necessary to standardize a limited number of components and use them in different combinations. These fittings can be rearranged in many ways to achieve always a balanced design. *Designed by A. B. Read for Troughton & Young* (*Lighting*) *Ltd.*

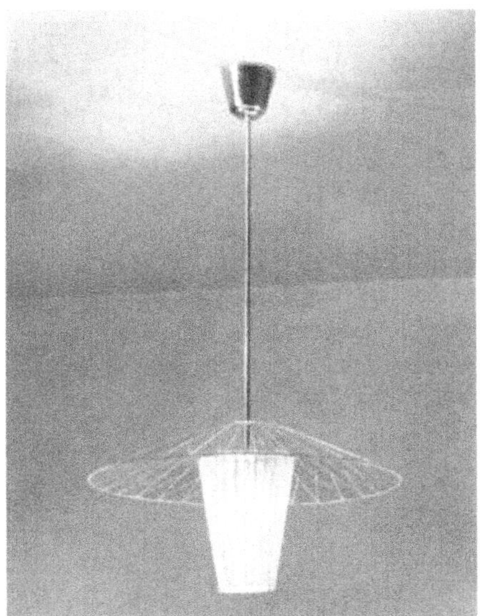

5. Pendant fittings present a special problem for they often have to hang in the same position for decades. Here there is a decorative intricacy in the glass canopy to relieve the austerity of the cone; the shade is pleated buckram. *Made by The General Electric Co. Ltd.*

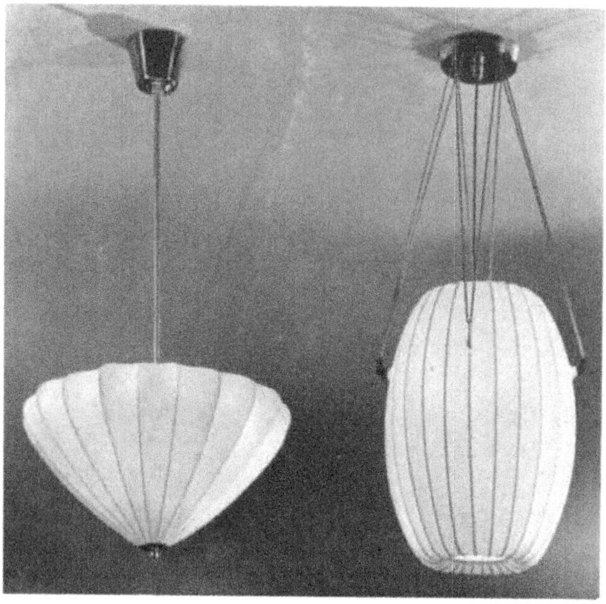

1. Another method of shading light has been developed for these 'Chrysaline' fittings. Liquid plastic is sprayed over the wire frame and allowed to shrink into position to form a series of double curvatures. Here a new material has been designed to produce satisfying new forms. *Designed by Beverley Pick Associates for The General Electric Co. Ltd.*

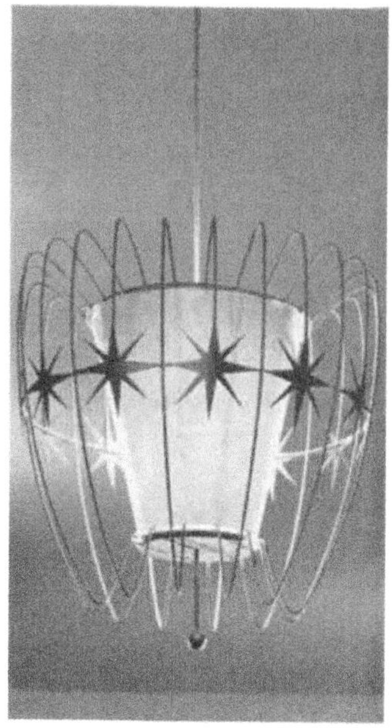

2. The designer's puzzle is how to integrate the fitting with the basic function of the light itself. Here it is done decoratively. *Designed by Beverley Pick Associates for The General Electric Co. Ltd.*

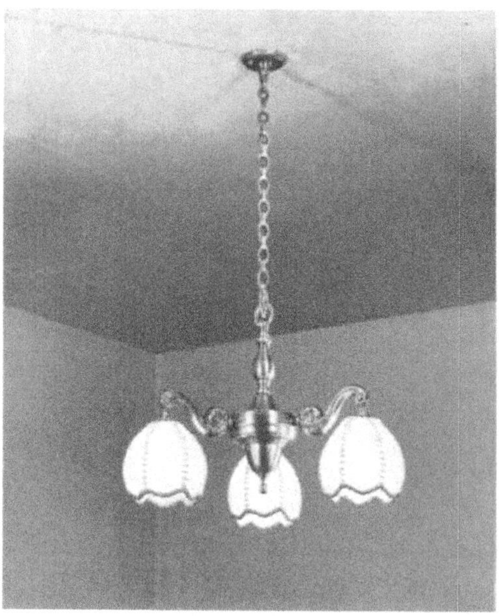

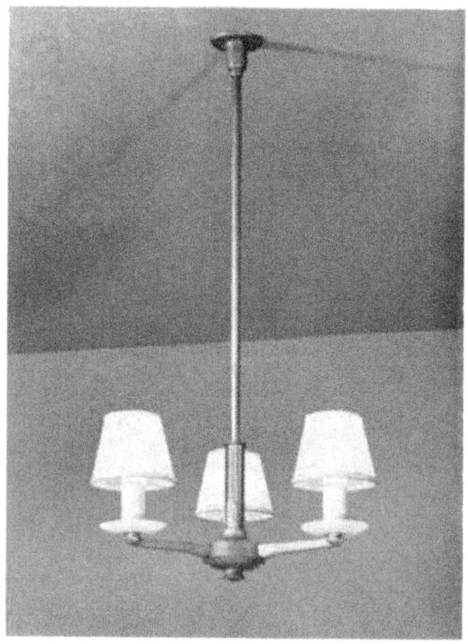

3 and 4. The supply of antique 'forms' to dress up the light fitting is inexhaustible. Here are two best-sellers: the pendant in the left shows how to hang the lamps themselves as mere appendages to a grandiose chassis; the other might be called 'Georgian Modern'. Although fitted with drip trays the plastic 'candles' have no imitation wax dribbles. *Made by The General Electric Co. Ltd.*

PLATE XXVII

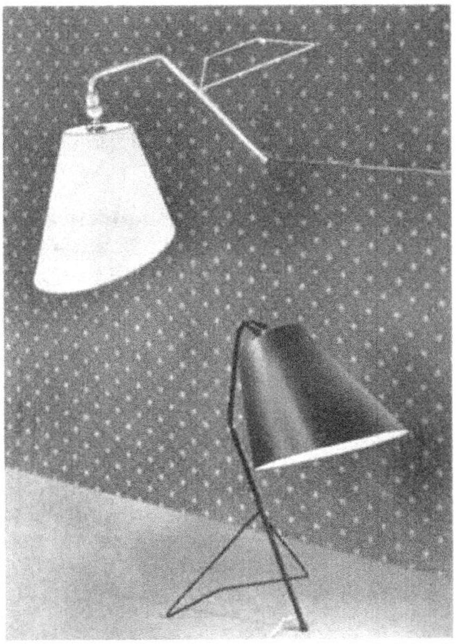

1. Where informality is needed in the domestic interior occasional fittings may be used. This lightweight model of thin steel rod supporting a card shade can be hung from the wall or stood on a desk. *Designed by R. C. Hiscock for Hiscock, Appleby & Co. Ltd.*

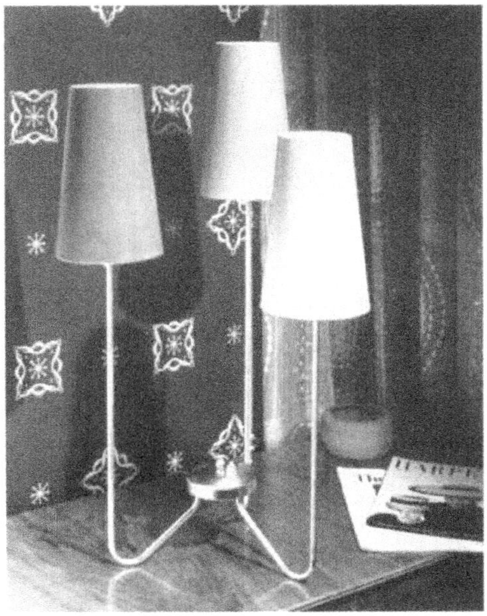

2. Four or five lamps of low power can be used to provide individual lighting in a room. Here three are grouped together with varying heights and different coloured shades making a 'sculptural' feature. *Designed by R. C. Hiscock for H. C. Hiscock Ltd.*

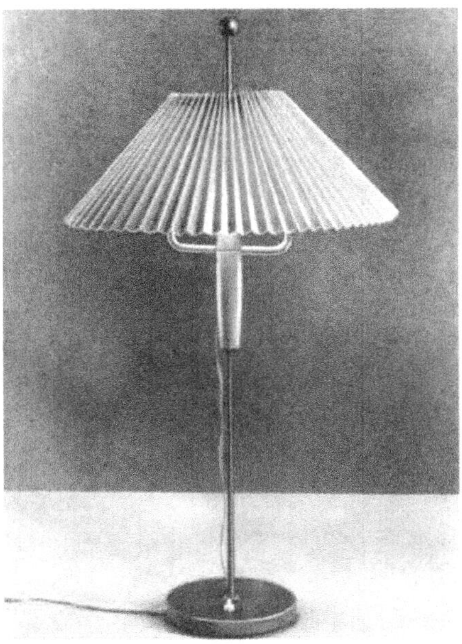

3. The sliding plastic sleeve is a feature common to other models in this maker's range, and the skilful way it has been shaped is enough to justify its protrusion beneath the shade. *Designed by Peter Bell for Allom Brothers Ltd.*

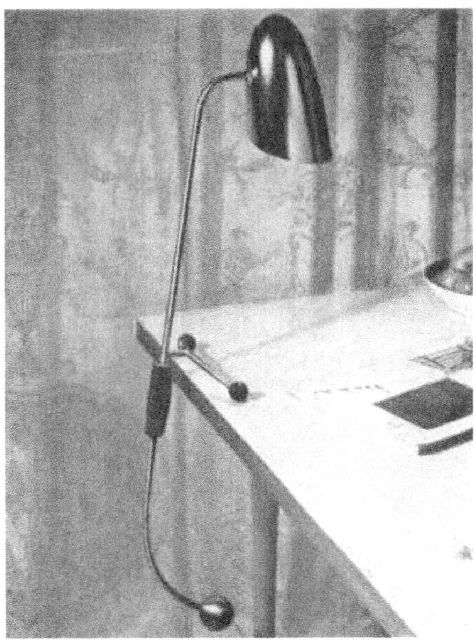

4. The counterweight allows this portable lamp to balance on a flat surface. Few of the mechanical problems of light fittings are worked out with such subtle simplicity. *Designed by Paul Boissevain for Merchant Adventurers Ltd.*

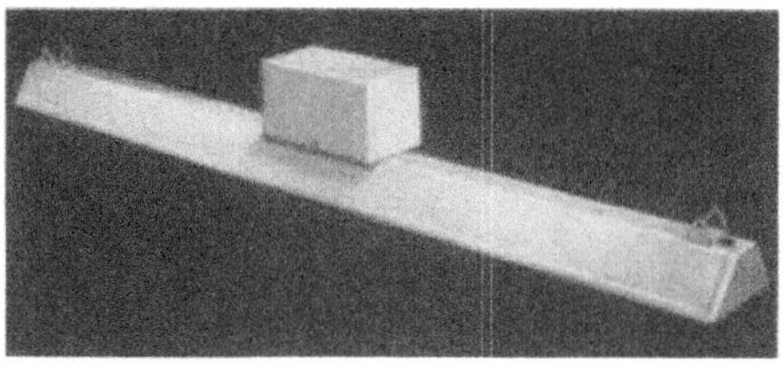

1

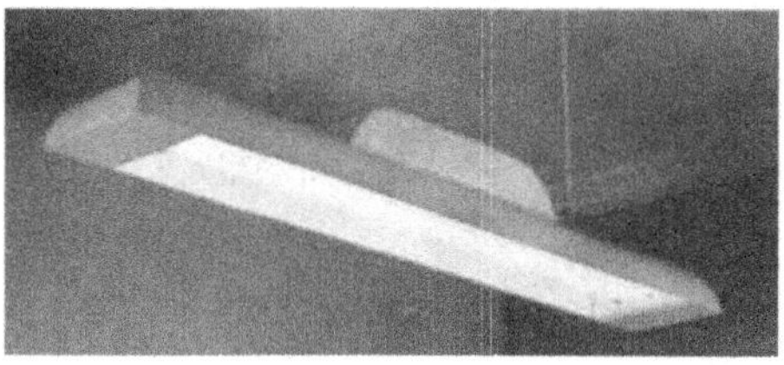

2

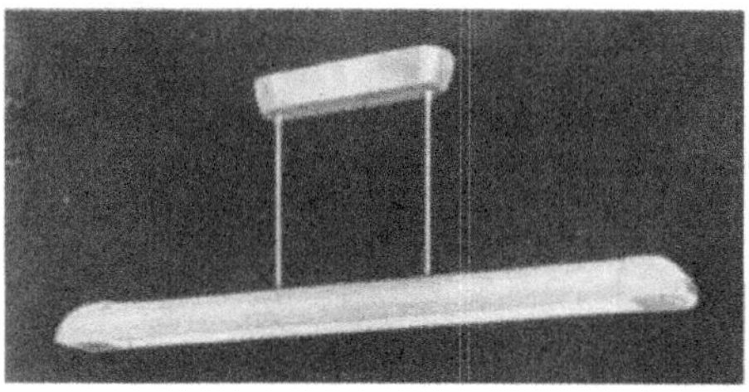

3

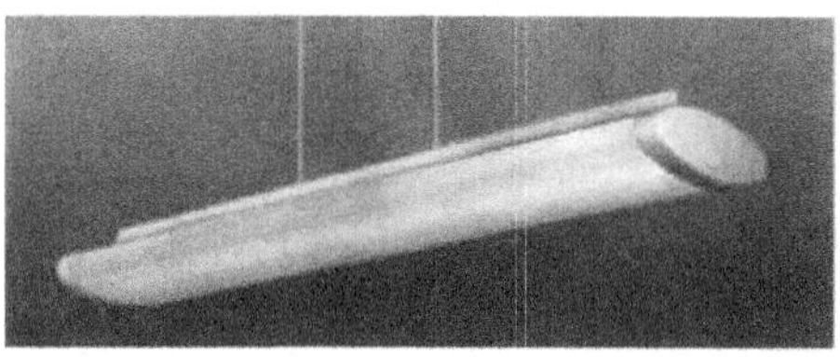

4

1. 1941. The first fluorescent lamp fitting to be introduced by Thorn Electrical Industries. Many designs in their earliest phase come out as crude functional statements. An inverted 'pig trough' holds the lamp: a separate box the control gear.

2. 1946. The appearance of the fitting became more important as fluorescent lamps were used in shops and offices. Metal flaps conceal the ends of the lamps and the control gear, still bulky, is more related to the reflector.

3. 1946. To reduce glare light is diffused upwards through a translucent plastic reflector. A feature is made of the indispensable control gear by mounting it on the ceiling.

4. 1947. The fitting has been designed to incorporate the control gear: the lamps are totally enclosed. A reeded perspex diffuser throws light upwards and on all sides. The fitting can be mass-produced for assembly in varying lengths.

5. 1949. The ethereal nature of artificial light is recaptured here and emphasized by the plastic wrapper and end caps which distribute light evenly over the entire fitting. The louvred version is for direct downward illumination. This range of fittings represents a plain statement of purpose being progressively refined, without the interference of pre-conceived decorative 'forms'. 2–5: *Designed by Peter Bell for Thorn Electrical Industries Ltd.*

6. It is partly the fault of the designer that fluorescent lighting has not become popular in private houses. Although designed for general purposes this fitting has an attractive intricacy which suggests a good approach to the domestic market. *Designed by A. B. Read for Troughton & Young (Lighting) Ltd.*

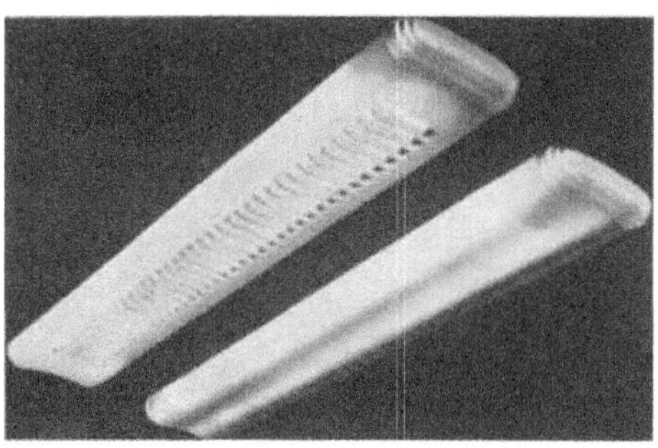

5

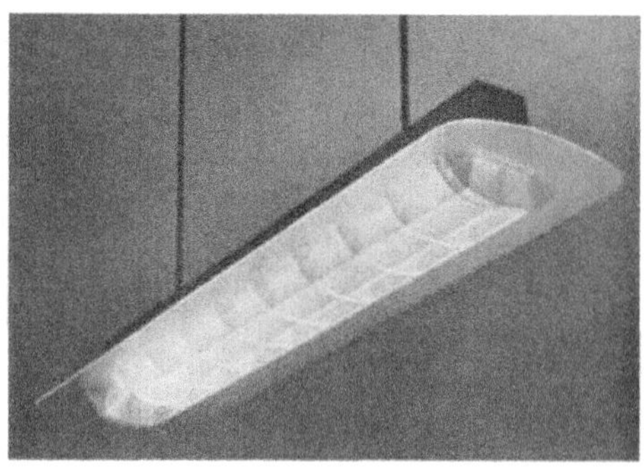

6

draughtsman occupies a relatively inferior position and is responsible to the works manager. His salary would be between £500 and £600 a year. He would not be expected to visit Continental countries or to take an interest in exhibitions in this country, apart from the British Industries Fair. The task of the draughtsman is not enviable. As an average of only one gross of each design is sold, production runs are short. The draughtsman finds that the need to keep manufacturing costs as low as possible prevents him from creating original work. On the one hand the sales manager demands fifty or sixty new patterns each year, and on the other the works manager restricts the number of new components to be used.[1] In order to make each new design economical in production, the draughtsman has to use castings and moulds already in existence. The fact that most firms prefer to buy their glass from British and foreign glass-works instead of making it themselves, presents a further difficulty. As moulds for glass are expensive, e.g. £80 for a small ceiling globe, the draughtsman is rarely allowed to design a completely new shape. The problem of design is therefore similar to that in the cheap jewellery trade, where the designer has to create new patterns by juggling with stock components. I found that most draughtsmen prefer to use antique and modernistic styles, where the various scrolls and angles can be formed into an apparently endless series of 'new' patterns.

While these factors may tempt the manufacturer to adopt a haphazard attitude towards design, they do not, of course, relieve him of the responsibility for setting a reasonable design standard. If the number of components available for new fittings is limited, owing to cost, it does not follow that the design of those fittings should be artistically unsatisfactory. I found abundant proof of this in three of the firms I visited. The firms are similar to each other in many ways. They began work about the same time (the first in 1924), they are small, and they serve the same limited market. From the start the principle of 'fitness for purpose' was adopted for all fittings, and throughout their development period in the 1930's these firms adhered to the design standards set by the Bauhaus. The chief client is the architect, to whom the price of the fitting is not the only governing factor. Nevertheless, it is necessary to standardize the metal and glass components, and this is done by planning an entire range of fittings in which each component is conceived in relation to several others. No matter in what combinations the components are used, the result will be artistically satisfactory.

Only since the war have these firms made a serious attempt to enter the domestic market. Their reluctance to do this is shared by all firms in the industry, who find that

[1] Part of the blame for this state of affairs must be borne by the retailer, who demands a constant flow of new patterns, but only buys one or two of each. The reasons for this must be held over for a later chapter. See p. 155.

contract work for architects enables them to plan long production runs and so economize on tool costs. As soon as they attempt domestic fittings they find themselves in a critical situation. They have to compete with masses of pretentious, yet popular designs, and their own standard products are too uncompromising in appearance to find a ready response amongst retailers and their public.

What would be needed for these firms today is a much more subtle campaign. The revolutionary ideas behind the Bauhaus or the recent utility furniture scheme have not won over the general public which desires fittings giving both light and decoration. Somehow the public must be satisfied without a lowering of standards. The point of departure must be the few existing fittings which sell well and yet are of exemplary design. No doubt progress in this direction will be slow, for it is handicapped by the apparent inability of modern architects and designers to include a decorative quality in their work. The problem will not be solved by importing pleated linen and paper shades from Denmark, and superimposing them on pieces of hand-thrown pottery. This can have no useful influence upon designers who work with metal and glass. Recent examples from Italian workshops may prove to be more inspiring, for they are more allied to the manufacturing methods in the industry. For the industry in general the only possible solution is for the leading firms to engage fully-trained designers who have already proved that they can use the peculiar materials and methods of manufacture creatively.[1]

In order to demonstrate that the use of such designers is a practical solution, I shall have to refer to another product of the industry, the fluorescent fitting. The first electric discharge lamp was of the cold cathode type, introduced into this country in 1933. It was used mainly for restaurants, ball-rooms, liners, etc., and with its wide colour range it was essentially decorative. The tubular lamps, which had to be more than 8½ feet long for efficient operation, could be bent into a variety of shapes. Semi-circles, squares and hexagons were popular, and there was a tendency for modernistic patterns to predominate, especially in the bulky casing for the control gear attached to the lamps. The very high voltage of these lamps made them dangerous to touch, and very few, if any, were used for domestic purposes. The hot cathode fluorescent lamp was first used in 1939. As this type of lamp cannot yet be made into shapes more decorative than the circle it has been, from the start, considered simply as a functional instrument. Short tubes can be used efficiently and the lamps, even when alight, are harmless to the touch.

[1] Although one or two of the leading consultant designers in this country might achieve better results, their fees would be too high for firms requiring fifty to sixty new designs each year. In fact, it is doubtful if more than £10 per design could be afforded, on account of the short production runs.

The successive stages of development may be followed by referring to the illustrations on Pl. XXVIII, from the range of Thorn Electrical Industries. Throughout the war fluorescent lamps were used mainly in factories and hospitals. In their first phase they did not have shades, but in 1941 the so-called 'pig trough' was introduced as an early crude fitting (1). The transition and growing importance of appearance design is shown by (2). The control gear was large and ungainly, and had to be housed separately on top of the fitting. In 1945 a consultant designer was engaged to work with the firm's development team, and the attempt was made to re-design the fitting so that it could be used in shops and public buildings.[1] A diffused upward light was necessary and perspex reflectors were used (3). By 1947 a new fitting was designed to incorporate the control gear. Clear, ribbed perspex was used, and for the first time the entire fitting was enclosed (4). This type of fitting indicates the main line of development which is being followed in this country. As the tool costs were high (about £2000), it was necessary to proportion the design so that it could be assembled in varying lengths. By careful planning for mass-production, it was possible to reduce the selling price to a figure much below those of earlier fittings. In the United States, on the other hand, the metal louvred fitting has become more popular. Under certain conditions it is preferable to the plastic diffuser, because it directs light with greater emphasis, while still masking the source. It was first used in this country in 1948, and most firms have two or three adaptations of it in their ranges. One of the main recent developments (5), shows how the fitting can be designed in a manner appropriate to the ethereal nature of artificial light. This has been achieved by reducing the metal parts to a minimum, and by using a translucent plastic wrapper and end-caps, so that the light is distributed evenly over the entire fitting. An equally significant (and similar) development, was the lipless sphere for incandescent lighting, which we have already dealt with.

Since about 1948 attempts have been made to encourage the use of fluorescent lighting in the home. (For kitchens and bathrooms it has, of course, been in use since the war.) To do this manufacturers have to meet considerable prejudice from both retailers and public. There is a widespread opinion that fluorescent lighting has a harsh, unnatural colour. The so-called 'daylight' lamps used during the war are largely responsible for this. Despite the introduction in recent years of the 'warm white' and the 'peach' coloured tubes, it can still be claimed that fluorescent light has different tonal qualities from incandescent light. Although both types are tonally inexact when compared with daylight, we have after more than fifty years grown accustomed to the

[1] For an account of the teamwork involved see Peter Bell, 'Operation Fluorescent', *Art and Industry*, November 1949.

colour of white incandescent light. We call it normal artificial light and we have, for example, chosen the colours of our furnishings to suit it. Again, it is said that fluorescent light recalls the factory and the office. It is the designer's job to counter this opinion, and there are several fittings on the market of a high standard of design completely suited to domestic interiors.[1]

Nevertheless, the most effective argument for a wider use of fluorescent lighting has nothing to do with design. To begin with, a 40-watt fluorescent lamp has a light output equal to that of a 100-watt incandescent lamp. This means that it requires 60 per cent. less fuel to provide the same amount of light. The endurance of the fluorescent lamp is about 3500 hours, and therefore many times greater than that of an incandescent lamp. Although the initial cost of a medium size fluorescent lamp and fitting may be as much as £12, its long and economical life suggests that, in future, it will be widely used in private homes.

This brief survey of fluorescent fittings is instructive because it raises one very important point. All new products reach a stage when they can no longer remain merely functional instruments, confined to the factory and the workshop. It is realized that they have a wider application to a market where the artistic element in a design is a significant factor. The important question is: How will the manufacturer approach this problem? In earlier chapters we have dealt with this question and found that in most cases the manufacturer is content to enrich the appearance of his product after the engineer has finished with it. For incandescent light fittings the reverse is true, with artistic appearance as the dominating factor, despite the revolutionary Bauhaus designs. In fluorescent fittings, where this crucial stage has just been reached, it seems possible that a similar violent reaction will be unnecessary. Already in at least two leading firms consultant designers work in conjunction with the engineers throughout the evolution of a new model. In this way the balance between form and function can be achieved. If a domestic fitting were required, it would be unthinkable to make use of antique or modernistic styles. Once the leadership of a few firms is established at this stage, it is reasonable to suppose that the bulk of the industry will follow their example.[2] From our point of view the question is fundamental, and it will be dealt with more fully in the following chapter.

[1] Nevertheless, the problem in this country has been tackled coyly by designers who prefer indirect illumination from tubes hidden behind pelmets and bookshelves. The designs of Gino Sarfatti of the Italian firm Arteluce are different, for they make a positive and exciting feature of the fluorescent tube itself.

[2] From 1952 onwards, in the market for occasional domestic fittings using incandescent light, a progressive and expanding group of small manufacturers, producing many original modern designs, has been meeting mild competition from some of the largest firms in the industry, whose products have been made similar and often cheaper.

RADIO AND TELEVISION CABINETS

The design problem in the case of radio and television cabinets is similar to that dealt with in connexion with domestic appliances and lighting. Radio and television *sets* are considered to be, in themselves, unattractive. It is generally recognized in the industry that a cabinet of wood, plastic or metal, is required to cover up the working parts of a set. It is as if the cabinet became the radio set's passport to respectability, for it makes an unsightly collection of wires, valves, coils, etc., 'presentable' and therefore acceptable in a domestic furnishing scheme. This misguided conception lies at the root of the few disgraceful and the many mediocre design standards which the industry maintains. In order to discover why this approach towards cabinet design is today so pervasive, it is necessary to review the main lines of development in this young and virile industry.

Although it is usual to say that the industry is a product of the early 1920's, its problems are really older, as they are very much the same as those of gramophones: and gramophones have been manufactured on a commercial scale since the beginning of the century. Gramophones, like wireless sets, required cabinets, although the first experimental models, such as Edison's 'phonograph' (1877), and Emile Berliner's 'gramophone' (1887), were hand-operated instruments and remained uncovered. In 1897 the first satisfactory clockwork gramophone was produced by The Gramophone Company. This model, which later formed part of the H.M.V. trade mark, had a cabinet which enclosed the motor, but left the turntable and the sound horn exposed. No important changes took place until 1909, when H.M.V. introduced an upright, or console model. The sound horn was reduced in size and was enclosed, together with all working parts, by an oak cabinet in the Sheraton style. This very early example of cabinet work is instructive for two reasons. It shows the natural tendency in all new industries to fall back on traditional forms in order to gain public support for novel and unfamiliar products. It also marks the point at which the gramophone ceased to be a mere instrument, and became a 'respectable' piece of furniture.

Early wireless sets followed a similar pattern of development, but their progress from the instrument stage (which was essentially amateur and non-commercial), to the 'piece of furniture' stage was speedier, chiefly because of the results already achieved by the makers of gramophone cabinets. The commercial production of wireless sets began after the First World War. In 1919, public broadcasting was started by a private company in the United States; in this country, the Marconi Company made its first public broadcast in 1920, and in 1922 the British Broadcasting

Corporation was formed. For several years crystal and valve sets remained the province of enthusiasts, who constructed their own, while sets produced commercially made little pretence to be more than instruments. A typical set in the 1920's had a vertical panel with indicators and knobs in front and the coils, valves, etc., exposed behind. To this set the batteries and earphones (later to be replaced by the loudspeaker) were attached by wires (see Pl. XXIX, 1).

About 1928 the horn loudspeaker was replaced by the cone type, and it is at this point that radio cabinet design began. The early cone loudspeakers were fragile and had to be mounted on wooden boards, which also acted as sound baffles. From this it was a short step to designing a complete case for the speaker (which was still separate from the set), and many varieties of artistic 'overcoats' were to be seen on the market. One firm offered a selection of no less than seven cabinets for one type of loudspeaker, with sound holes that ranged from a plain circular opening to a fretwork Gothic window. It was axiomatic that if loudspeakers were considered to be unsightly instruments, to be heard but not seen, then the wireless itself should also be covered up, and by 1930 several models had appeared with the wireless and speaker together in one cabinet.

To follow the development of radio cabinets through the 1930's it is necessary to distinguish between two approaches to the problem of design. The first and more common approach implies that the cabinet is a piece of furniture, suited to a certain standard of interior decoration. The easiest and the safest way to achieve this is for the designer to make use of period styles. In its earliest phase, therefore, the appearance of English radio cabinets was dependent on the historic styles of furniture. Georgian, Queen Anne, Tudor, in all kinds of adaptations and mixtures, were popular. Then about 1932 the radio set reached the state of wanting its modernity to be expressed in the cabinet. The immediate result was that design became modernistic, again borrowing forms from the furniture trade. This was the period of the flamboyant and cubistic sound holes, and for some time, in the cheaper trade and a large part of the more expensive trade, most of the cabinets were adorned with jags and angles. The craze reached its peak in the middle 1930's, and then steadily declined in popularity. By 1939 most radio cabinets were straightforward and unpretentious, depending for their appeal on the colour and quality of the wood veneer and, to a lesser extent, upon proportion.

In order to account for the industry's remarkable change towards a satisfactory design standard by 1939, we must refer to the work of two firms, Murphy Radio and E. K. Cole. In 1931 Mr Frank Murphy decided that in order to get cabinets of really distinguished (and at the same time distinguishing) character, it would be worth trying one of the most uncompromisingly modern cabinet-making firms in the country. He approached Gordon Russell's, and the resulting designs by R. D. Russell proved to be highly satisfactory in every respect. The sales figures were rising when the experiment was started: they continued to increase steadily after the introduction of the new models. Only the first of these designs was particularly striking; the following designs were of a quiet, dignified style which owed much to the best architecture of the period. The appeal of these cabinets rested upon the subtle effects gained by contrasting surfaces of woods in natural colours, upon simple contours and accentuated horizontal lines. As they were exemplary designs at that period these cabinets went a long way towards establishing the radio as an instrument which could exist in its own right.

The second approach to improved cabinet design was the instrument approach. As we have seen, the earliest sets claimed to be no more than instruments. Although the masquerade which followed them was partially checked by the Murphy example, the low design standards of the majority of furniture firms encouraged radio cabinet manufacturers and designers to adopt an irresponsible attitude towards their work. Clearly a more straightening example had to be made in order to clear away the dross of sentimental (or at the worst meaningless) ornament. It is significant that the designer who began this work of revaluing radio cabinet design was also responsible for the Bauhaus-type Plan furniture, which we have already dealt with. In 1932 E. K. Cole approached Serge Chermayeff and Raymond McGrath, both of them architects, for cabinet designs which could be moulded in plastic. The designs were to be entirely new and regardless of any conventional forms already on the market. Each designer produced six sketches for which he received adequate fees. The design chosen for production was one of Chermayeff's, and his fee was increased by another £50. Nothing comparable with this design existed either in Britain or abroad. It was the result of a careful study of function combined with genuinely artistic imagination. The firm in thus breaking away from all traditions took a great risk, and some opposition within the firm had to be met. If sales did not exceed 10,000 sets in the first year, it would be a serious failure. What happened remains an immense credit to the British public. The new cabinet, so uncompromisingly functional in appearance, became exceedingly popular from the first month. About 100 per cent more sets were sold in the 1933–4 season than in the previous season.

Greatly satisfied with this experience, the firm called in for its 1934 production Chermayeff, McGrath and another architect, Wells Coates. The choice fell on two of Wells Coates's designs, of which the circular one quickly became the most famous set in British radio. The most important part of the set, the speaker, was placed in the centre and the scale was made unusually large and legible. Conceived as a simple plastic moulding and as a straightforward, functional instrument, this is, one is tempted to say, the obvious, logical solution to radio design, wherein the cabinet and the set become integrated. Wells Coates approached the problem as one of designing a piece of modern machinery, whereas in the Murphy cabinets the governing idea was to create good modern furniture. Both conceptions have had a salutary influence upon the rest of the industry.

From this survey two main points emerge. First, the industry is new and it has shown an astonishingly rapid development in less than twenty-five years.[1] Secondly, it has followed two main approaches towards design, both of which have been decisively influenced by consultant designers, most of them architects. To discover how the industry is now tackling the problem of design I visited ten firms. One firm is concerned with electrical engineering in general, with only a proportion of its 14,000 employees occupied with radio and television receivers. Of the remaining nine firms, the largest employs 5000 workers, the smallest 150. The average size is between 1500 and 2000. While I have attempted to cover all price groups, I was able to collect more information from firms catering for the middle range of prices extending from 12 to 120 guineas than from others. A firm of average size produces every year about five new table radios, three television sets, two radiograms and two portables. Probably not more than three of the new models have a plastic cabinet; the remainder are constructed entirely of wood. Very few factories have workshops for cabinet-making, and the usual practice is to give the work out to one or more furniture firms. Most firms have their plastic cabinets made by trade moulders, also outside the industry.

I found that manufacturers, engineers and salesmen all showed a keen interest in the work of the designer, for they believed that about 80 per cent of their customers for radio sets bought by appearance alone. The five-valve superhet receiver is now almost standard for radio sets in the middle price range, and the slight improvements which are made each year are unlikely to affect the sales. In these days when a firm has already established its reputation for reliable radio receivers, according to the various price groups, it is realized that the outward appeal of a new model is of the utmost importance. This is, of course, not true of the television sets made by firms which seek to improve the technical aspects of their models every year.

Before considering the types of designer employed, it is important to understand the process, or sequence of events through which a new design must pass. The following is a typical example. The first move is made by the managing director who, with guidance from his sales manager, decides that a new portable radio set is required.

[1] In 1929 nearly 3,000,000 wireless licences were issued. By 1939 the figure had risen to 9,000,000.

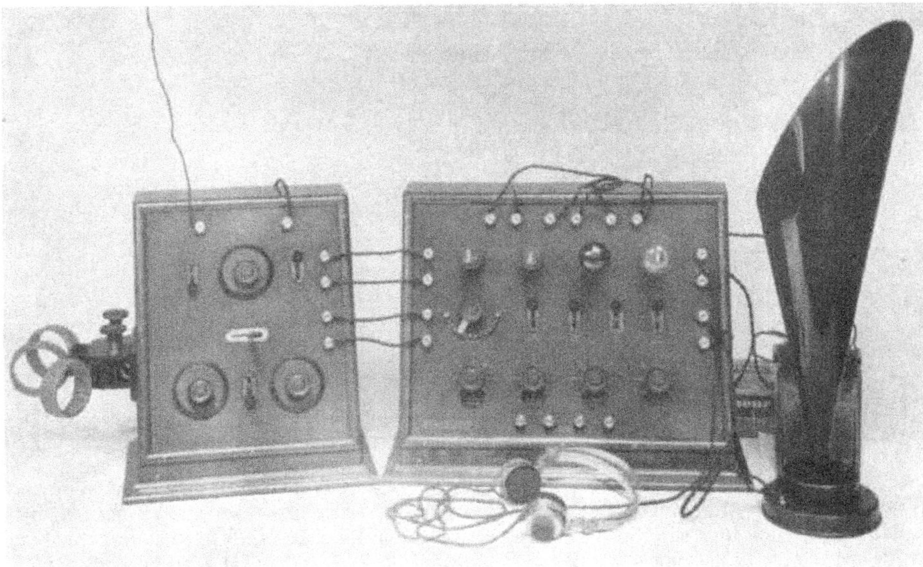

1. 1923. The instrument shows the transitional stage between the wireless set of separate parts and the all-enclosing cabinet. The complicated controls are precisely laid out: although the main chassis is covered the makers did not think it unsightly to leave such items as valves and coils exposed. Burndept IV Receiver, in the Science Museum, London.

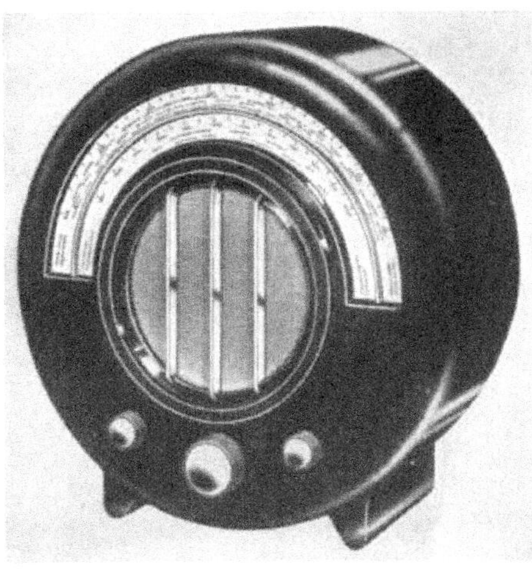

2. *Above.* 1934. The case for the instrument is restated by a designer conscious that his product must be attractive and easy to use. Logically the speaker is placed in the centre of the plastic cabinet and a large calibrated scale curved round it. The tuning knob, for symmetry and simplicity, is placed beneath. *Designed by Wells Coates for E. K. Cole Ltd.*

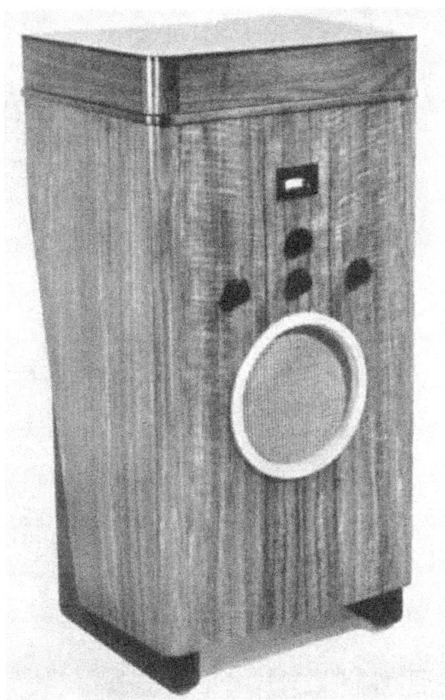

3. *Right.* 1936. A furniture designer very much aware of the Modern Movement in architecture was responsible for this radiogramophone. Although made of wood like furniture, the model asserts its independence as equipment designed for a specific purpose. *Designed by R. D. Russell for Murphy Radio Ltd.*

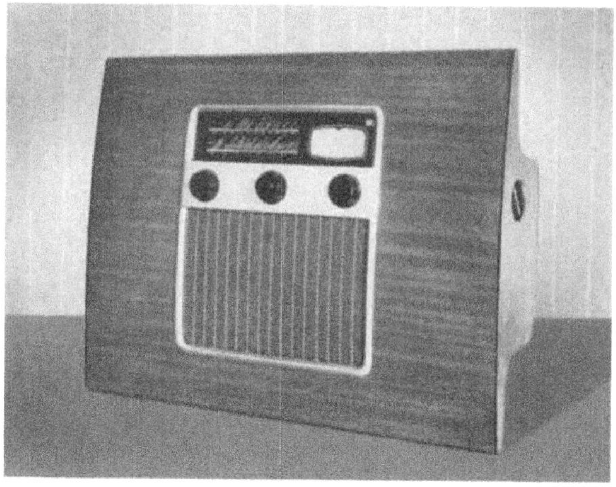

1. 1945. The principle of the baffle-board, once used extensively by amateurs, was taken up and developed for a table radio. The designer has made it the *raison d'être* of an original and distinctive shape: later, four variations in successive years, for the sake of always having something new to show, were not so finely wrought. *Designed by A. F. Thwaites for Murphy Radio Ltd.*

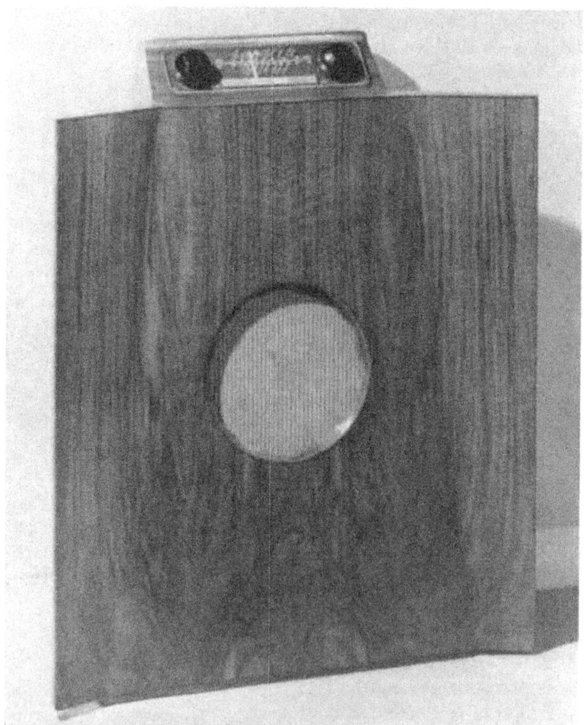

2. 1948. The demand for a console model enabled the same sound principle to be thought out again. Here the baffle is more effective because it is larger, and forms a tapering concave frame to the sound hole. The waist-high controls are an integral feature of the main design. *Designed by R. D. Russell for Murphy Radio Ltd.*

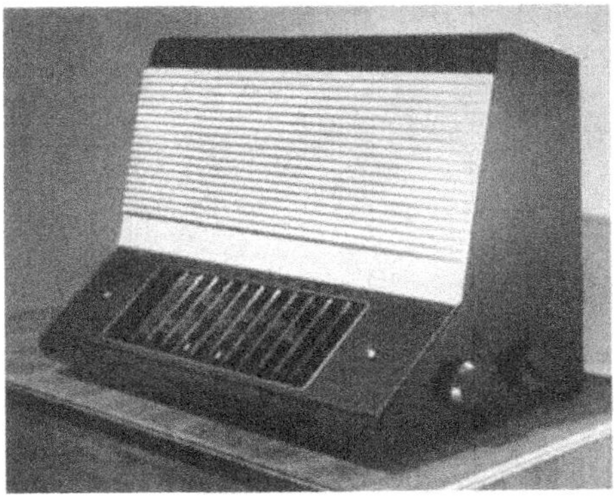

1. *Left.* The most distinctive feature of a radio, the loudspeaker, need not be emphasized. Here it is concealed by a plastic shutter in a design which disposes its main elements where they can be most effective. *Made by Pye Ltd.*

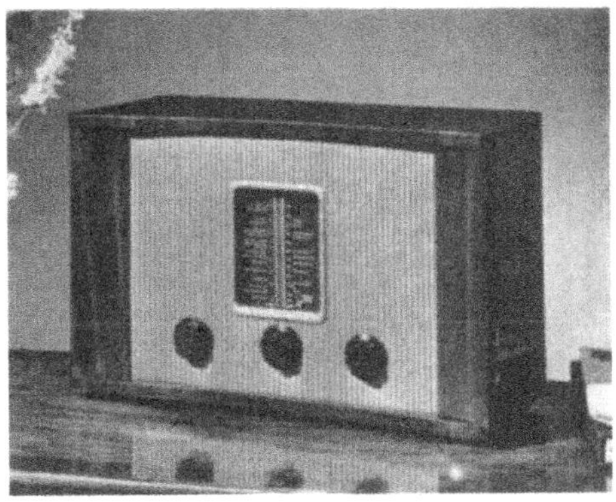

2. *Right.* As a relatively new product the radio is still establishing its own characteristic design standard. The vitality of this model comes from the direct contrast of two different textures. *Designed by A. F. Thwaites for Murphy Radio Ltd.*

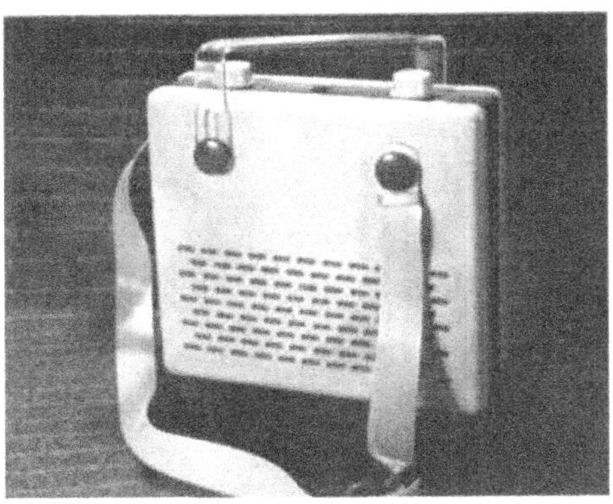

3. *Left.* A plastic portable radio case consisting of a central blue band and two identical cream coloured end caps to accentuate its slimness. The sound hole is exploited as a decorative feature on both sides of the set. *Designed by Wells Coates for E. K. Cole Ltd.*

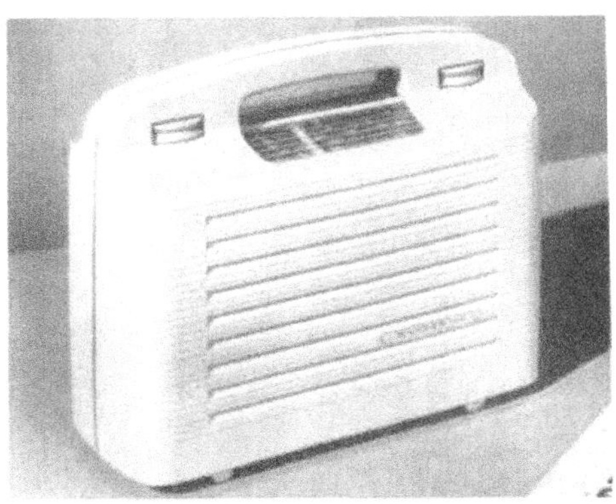

4. *Right.* Portables can be seen from any angle. To produce a design with a neat exterior and economize in production this 'backless' model uses the same moulding on both sides. This homogeneity is completed by the indigenous handle. *Designed by A. F. Thwaites for Murphy Radio Ltd.*

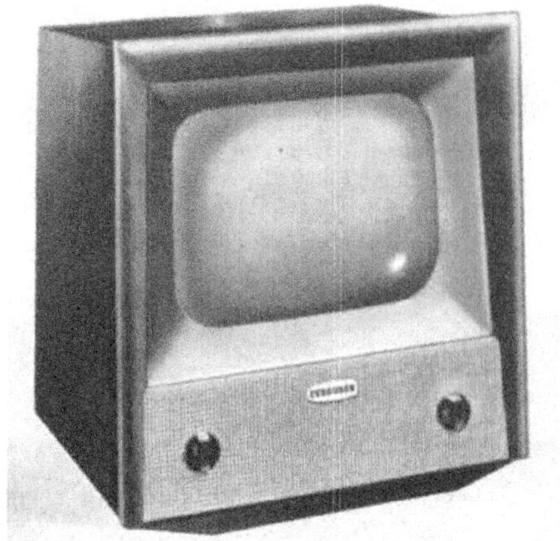

1. The present size of the cathode-ray tube sets a visual problem which few designers can overcome without considerable cost. In this case it has been intentionally featured by superimposing a stage-like proscenium. *Made by Ferguson Division of Thorn Electrical Industries Ltd.*

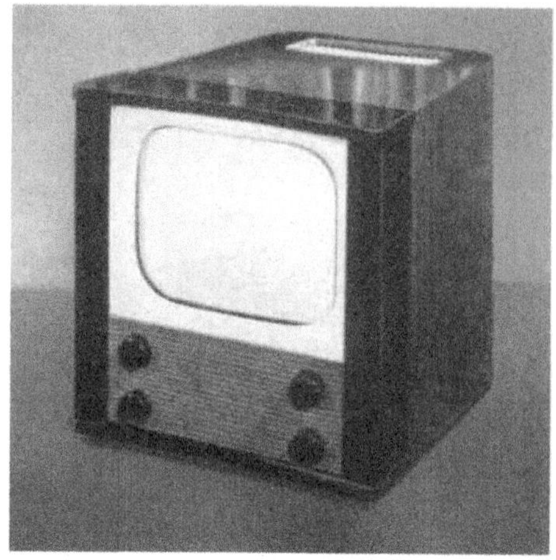

2. When not in use the gaping white eye of the television screen can be balanced, as in this model, by the inclusion of fabric texture and plastic knobs. *Designed by Victor Taylor for Ace Radio Ltd.*

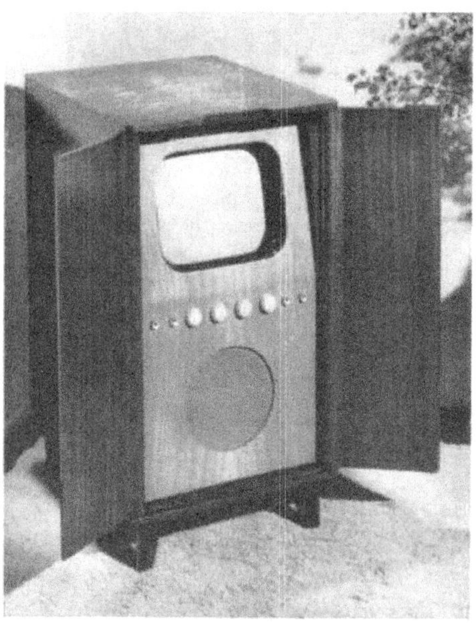

3. Thinking of the television cabinet as a piece of furniture can lead the designer to adopt irrelevant design standards. Although in this model there is no masquerade with typical furniture 'forms', a less discriminating designer might produce disastrous results. *Designed by R. D. Russell for Murphy Radio Ltd.*

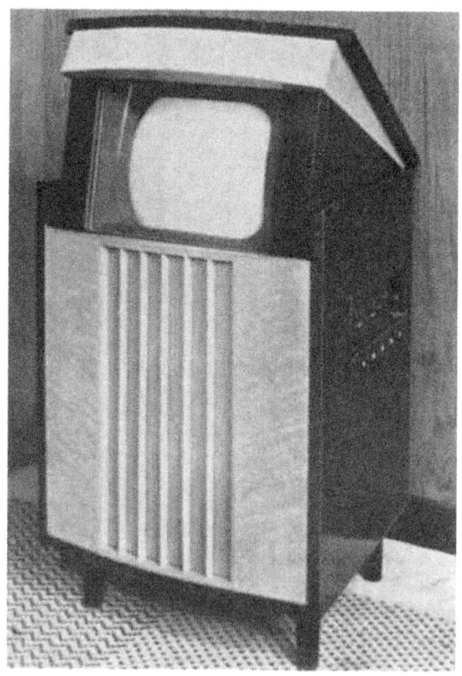

4. In this combined television-radio set, the screen can be lowered out of sight when not in use. The formal simplicity of the front gives prominence to the loud speaker, and the controls, including those for pre-set radio stations, are recessed in the side. *Designed by Wells Coates for E. K. Cole Ltd.*

The approximate selling price of the new model is determined and the sales results of its predecessor are assessed, together with the general trend of public taste in this particular price group. Market research in the industry is invariably carried on at one remove from the buying public. The dealer may give the firm's representative some notes on the reactions of his customers to the designs he has in stock, but the decisive information comes from the sales chart. I know of only one case where a firm gets in touch with a section of the public by sending out photographs of forthcoming models together with a questionnaire. The next step is to call a conference consisting of the managing director, the engineering and electrical chiefs, the works and sales managers and the designer. In formulating the programme to be followed they would, for instance, decide whether an existing chassis should be used or whether the cabinet should be designed first and the chassis afterwards or—and this is preferable—the two should be designed together. If the cabinet is to be of plastic, an assurance would be required from the sales manager that a large number of sets, say 25,000, would be sold to cover the high cost of the moulding tool. The matter is then left in the hands of the designer who, in collaboration with the chief engineer, works out a number of preliminary sketches. When the next conference is called the designer will have an opportunity to elaborate and explain his ideas. In a few rare cases the designer will have an authority second only to that of the managing director. The result of the conference usually shows that each party has had to compromise a little. A mock-up model of the set is now made and considered by the works manager, who points out any snags that may arise when the set goes into production. Matters such as the positioning of the knobs, and the scale with its complicated pointer mechanism, must be worked out, so that by this time the design of the chassis and cabinet has become a team job. When the work is complete a prototype has to be approved by the managing director and the sales manager. If any radical alterations are suggested at this stage, it is usually preferable to scrap what has already been developed and start again. In the case of the more revolutionary type of design, it is quite common for the development period to last for two years.

The most interesting point is that the design of a radio set involves teamwork. Each team must have a leader, and it is significant that in the firms where the best cabinet designs are produced, the designer is also the leader of the team. It would appear that he merits the position, because he is, after all, the person who converts a series of facts and figures into a three-dimensional shape. While it may be necessary for him to compromise on several points, a design that only shows the handwriting of one man is preferable to one that bears the marks of many.

There are three types of designer in the industry: the free-lance, the staff designer and the consultant. It may be stated at once that the free-lance designer has had little success in this industry. I was shown an instance where a free-lance was asked to suggest a few new shapes for a table model. While there seems little doubt that he was inadequately briefed for the job, his ideas proved to be totally unallied to the engineer's requirements. The firm, however, was undismayed by this and, as soon as the designer had departed with his fee, the engineer set about making several drastic alterations to the design. The result was disastrous financially and artistically.

Most firms have a staff designer with two or three assistant draughtsmen. I found that the training and experience of staff designers were very similar to those of the engineers in the domestic appliance industries, but with this difference: the staff designers I met, while being technically competent, were not afraid to express themselves artistically. As we have seen, the designer's presence is continually required at the factory when a new model is being developed. He must be available to give his views concerning the smallest detail. It seems, therefore, that the staff designer, at work in an office adjacent to that of the engineer, is indispensable. When a consultant designer has to be called in, time-lags are bound to occur. From the manufacturer's point of view this is an important factor. On the other hand, the problem has one very different aspect. If a designer has been on the staff of a firm for a long time, he is inclined to lack the freshness and the imagination which characterize the consultant. His job requires him to spend five days a week designing nothing but radio and television cabinets. Unlike the consultant, the staff designer does not work in other industries and his experience of new industrial materials and methods is limited. Although this could be partly remedied, I know of only two firms which periodically send their designers abroad to study market trends and to gain a wider knowledge of industrial design generally. The lack of such stimuli is a serious handicap to a man who is essentially an artist.

In four of the firms I visited, where both staff and consultant designers are used, it is the management's policy to allocate the work in such a way that the consultants get the projects which are likely to create a new demand in the market, and the staff designers are left with the task of supplying a demand which has already been estimated with some accuracy. In fairness to the staff designer it should be made clear that no fixed rule of policy of any particular firm is implied. There are several cases where the staff designer has been given a new project and has worked it out with excellent results. The Murphy baffle-board receiver is an example of this, and a brief account of its development may help to clarify the relationship of the staff designer to the consultant, as it exists in the most enlightened firms. The head staff designer and the chief engineer collaborated to produce a radio which gives the best possible reproduction without the disadvantage of 'boom', or box-resonance. Although the principle of the baffle-board had been employed for many years by amateurs, it had never been applied to a table model by a commercial firm. On the Murphy set the usually large dimensions of the baffle-board have been reduced to the minimum size reconcilable with efficiency. The first model was introduced in 1945, and was followed by four more baffle sets, two of them designed by a consultant. In 1948 the firm decided to include a console model in the range and the design work was given to another consultant, also retained by the firm. The size of the baffle in this set is much greater than in the table model, and consequently its effectiveness is much improved. In each model the designer's task was to transform a mechanical principle into a saleable article. This was not done by making the cabinet suit a particular furnishing scheme, good, bad or indifferent, but by designing in such a way that the radio can exist in its own right. These Murphy sets have so vigorous a spirit of their own that they will give a convincing account of themselves in any room (see Plate XXX).

Two manufacturers I met had employed consultant designers without success. In so far as one can assess it, the fault appears to lie with the manufacturers. In one case the designer was not given sufficiently exact instructions, but a completely free hand to create what he thought was suitable for the firm's public. The models turned out were stimulating and gathered much publicity, but they did not sell. Faced with an entirely new range which was unsaleable, the firm's staff designer attempted to modify the designs—'to tone them down' as he put it. But the success achieved was not sufficient to prevent the withdrawal of the models as soon as a new range was in production. In the other case a specialist in display and packaging design was used. He was unsuited to three-dimensional work, and the models were styled rather than designed. Again, they were withdrawn from the range as soon as possible.

On the other hand, it is possible to find several cases where suitable and sympathetic consultant designers have been successful. For example, in 1945 the Decca Record Company developed an entirely new instrument for reproducing records. A new instrument involved a new design, and an architect was engaged. The design for this very expensive instrument was uncompromising, but not clever or stylish. Although it met with disfavour when first introduced, its undoubted success may be measured by the fact that nearly 4000 had been sold by 1949.

However, the progressive methods used by Murphy, E. K. Cole, Decca, and a few other firms, are exceptional in the industry—a statement which can be verified by any visitor to Radiolympia. In four of the firms I visited there are no staff designers, only engineering draughtsmen. They are entirely in the hands of the chief engineer and the sales manager. Their work is imitative, mediocre, and subject to all manner of fashionable crazes. Furthermore, the table models, consoles and television sets produced by these firms are timid in design, self effacing and characterless. The portable sets are more adventurous in the vulgar manner, resorting to borax or bright, garish colours. Again, the manufacturers are responsible, but in all four cases there is little hope of improvement because they are embarrassingly proud of their products.

The salaries paid to designers in the radio industry vary. A staff designer with ten or more years' experience in a fairly large firm, say 2000 employees, would receive £800 to £1000 a year, his chief assistant £500 to £600. A consultant designer to the same firm working on a new project, might receive £200 to £300 altogether, and this would probably be paid in instalments as the work progressed. In addition, the consultant would probably be paid a retaining fee. One manufacturer who employed a consultant, but no staff designer, told me that he set aside the sum of £1000 each year for design work. It is not common to pay royalties or a commission on sales to the designer, because he is looked upon as a member of a team. To pay him and not the others would be considered unfair. However, in firms where the designer is the leader of the team, I found that royalties were paid.

All the consultants I have met in connexion with this industry have had an architectural training, while the staff designers have, almost without exception, received their design training in a radio firm. As the foregoing pages have attempted to show, the consultant tends to maintain a higher design standard than the staff designer, who has not had the benefit of an art school or architectural training at the beginning of his career. While such a training undoubtedly aids the designer, it should be emphasized that the really vital qualities which distinguish one man's work from another's, are intelligence and the creative ability to integrate the elements of art and mechanics.

But before any work of merit can take shape the manufacturer must be confident in the ability of his designer. Far too many radio and television sets today are the result of superficial styling. Several manufacturers I met had a fixed preconception of the final product, even before the designer was briefed for his job. In the first place, they insist that the radio shall be a 'piece of furniture', and secondly, that it shall fit into the type of furnishing scheme which they admire or think their public admires.

This brings us back to the two conceptions of radio design; either the cabinet is to be a 'piece of furniture', or the radio is to remain an instrument, undisguised. I found that the majority of manufacturers preferred to consider their cabinets as furniture. Here are some extracts from 'hand-outs' collected at Radiolympia, 1949: 'A combined television and powerful all-wave radio receiver in figured walnut, and a really fine piece of furniture.' Or, '...a floor model television receiver, the lower part forming a useful bookcase...provides an attractive piece of furniture'. And again,

'... you will look for a handsome piece of furniture and in —— you will assuredly find it... all models are tastefully designed [to] make impressive additions to the best of furnishing schemes'. What, then, is 'a piece of furniture'? What are the qualities which make radio cabinets suitable for 'the best of furnishing schemes'? According to most radio manufacturers, 'a piece of furniture' is something which is bulky and solid, relieved by contrasting wood veneers (preferably walnut) but toned down again by a thick, dark brown stain, suggestive of treacle. The stain is finished with a high polish. The most commonly used powder for moulded plastic cabinets is dark, mottled brown, similar in appearance to burr-walnut. In the many wooden consoles and radiograms which are monolithic in character, there is a downward heaviness which appears to symbolize the force of gravity. The fact that all this is totally inappropriate to the essential nature of a 'wireless' instrument, is merely incidental to the manufacturer intent on producing cabinets which equate exactly with the bulk of department and chain store furniture. At the more expensive end of the trade, period styles, especially Chippendale, are again fashionable. The higher standards of cabinet design which were noticeable just before the war have generally lapsed, to leave the best models of Ekco and Murphy even more isolated in their pre-eminent position. Little attempt is now being made to follow their example, and most manufacturers appear to have lost the enthusiasm which they displayed during the 1930's.

The very few manufacturers who have followed the furniture approach to design and yet succeeded in maintaining a high standard have, and this is important, used consultant designers frequently. They have acknowledged the standards set by the best modern furniture designers; in some cases they have surpassed these standards. Nevertheless, their success does not alter the fact that this approach is fundamentally wrong. The approach gives too little emphasis to the essential qualities of radio and television. The better, the more honest and straightforward approach is that which has been followed, in general, by E.K.Cole, and several manufacturers of portable sets. Again referring to Radiolympia, I found a new Wells Coates television receiver advertised in this manner: 'The cabinet, in its graceful, elegant simplicity, is worthy of its place among the finest furniture for itself alone....' The change in emphasis is unmistakable. This surely is the approach which manufacturers should make towards radio design. Once it is accepted that a radio can exist *in its own right*, the design standards of the furniture trade, whether good or bad, need not bother us at all.

The objection may be raised that this approach is too insular, too unsympathetic to those of us who wish to make impressive additions to furnishing schemes. In reply one can point out that a high standard of design in a radio set depends, not upon its formal similarity to the well-designed products of another industry, but upon the way in which it embodies on the one hand the essential qualities of radio and on the other the spirit of the age in which it is created. As copying is the designer's deadliest sin, I think it is justifiable to insist that radio and television receivers should erect their own standards of excellence.

FURNISHING TEXTILES

There are about 220 firms producing furnishing fabrics in this country. They form part of the textile industry, which is located in three main areas: Lancashire, Yorkshire and London. Manchester, which forms the focal point of the Lancashire industry, is mainly concerned with printing on cotton and synthetic fibres. The large towns of Yorkshire, such as Bradford, Halifax and Huddersfield, specialize in weaving woollen cloths. In both Lancashire and Yorkshire the biggest firms cater for the middle and lower price groups. In addition to furnishing fabrics many of these firms produce dress, household and other textiles. In London the industry is confined to smaller firms manufacturing higher price printed and woven fabrics.

I visited twenty-four firms in the three areas of production, and four elsewhere. Amongst these firms there are ten which manufacture both printed and woven textiles. Two of the largest firms control several mills and each employs more than 8000 workers. I also saw six firms each employing between 1000 and 3000 workers. The remaining firms are smaller. It would be misleading to give individual employment figures for these firms, because many of them did not carry out all operations concerned in the production of their fabrics. For example, I found that a merchant converter, while being responsible for the designs, kept only a small office staff and engaged other firms to weave and print his fabrics.

Before dealing with the work of these firms it is necessary to emphasize the difference between printed and woven textiles. Briefly, in printed textiles the weave of the cloth is of less interest than the pattern which is applied to it. The cloth can be regarded as flat, and therefore calls for a two-dimensional design. In woven textiles the pattern is in the weave, and all weaving is, consequently, a process in three dimensions. For this reason designing for weaving requires a different approach from designing for printing. Although it is true of all woven textiles that the pattern *is* the weave, there are yet two interpretations open to the designer, and it is necessary to distinguish between them. According to one, the cloth and the pattern are one and the same; according to the other, the pattern, although in fact produced in the same process as the cloth, is in the mind of the designer a separate element applied to the cloth. In order to clarify this I want to start my survey of factories with woven textiles, and begin with three firms using the first method of interpretation.

In these firms the designer begins with the yarn. In some cases he is even responsible for the construction and quality of the yarn, and supervises the spinning of it himself. The many varieties of coloured yarn available constitute the basic material with which the designer works. In order to create the fabric he uses a hand-loom. The warp, which runs longitudinally through the loom, is made up of, say, eight sections. The width of each section would be about 6 inches and it would consist exclusively of yarn in one quality and colour. At right angles to this warp mixture the designer weaves the weft threads. These threads may comprise several strands, each of a different material. For example, a strand of cotton may be included for strength, wool for warmth, rayon for a bright surface quality. The fabric is constructed as the weft interlaces with the warp. Its qualities of texture and colour depend upon the yarns and threads used, and the effect of their juxtaposition. By using a 'mixture warp' the designer can determine the composition of the required fabric more quickly than if an ordinary one-colour warp were used. Once the design has been selected, its warp and weft combination is set up on a power-loom of the Dobby type and full-scale production begins. The design method here may be said to be organic, because the structural and artistic elements in the fabric are interdependent.

From this it will be seen that the texture of the fabric depends equally upon the yarn and the type of weave. It follows that the designer constantly requires a wide variety of suitable yarns. This has been appreciated by several spinning firms, but notably by R. Greg & Co. Since 1900 the firm had made special fancy yarns, but designers in the industry appeared to be unaware of their true potentialities. During the slump period in the early 1930's, the yarns were used to 'liven up' designs which would otherwise have been drab. Except in one or two firms, no attempt was made to use these yarns as organic elements in a design. Consequently, in 1944 Greg's engaged Margaret Leischner, who had studied at the Bauhaus, to demonstrate how new designs could be built with the yarns. This was accomplished by weaving sample lengths of cloth which incorporated the fancy yarns in various combinations. The samples were then carried by the firm's sales staff, not as designs which could be bought or copied, but to show how the yarns could be used successfully in simple machine weaves. In addition Miss Leischner creates for herself some of the yarns which she afterwards works into patterns on the loom.[1] The three firms I visited where this approach to woven design was followed, are representative of a minor but progressive section of

[1] See Pl. LXXIII, 1 and 3.

the industry. All three are small, and in each case a senior director is responsible for design. However, before referring to these firms in detail, we shall have to deal with design methods in the bulk of the weaving industry.

Every one of the firms I saw has a staff designer with three or four assistants and copyists. Very few free-lance designers are employed because they lack the technical knowledge considered to be essential for the creation of all woven fabrics. The majority of staff designers I met have had a full-time art school training which in some cases has been followed by a period of part-time work at the local technical school. As for juniors who enter the design studios at sixteen, most firms are prepared to send them to the art school for one day a week. For the necessary technical experience, they are expected to gather enough merely by watching looms and talking to weavers in a firm's mill. I found that it was rare for the juniors to have worked a power-loom under the supervision of a trained designer. Instead, it seemed that a junior, who later often becomes a designer had, as it were, a list of rules setting out the limitations to which he must work.

This divorce between design and execution is a serious handicap to the designer. Although he has an accurate knowledge of the mechanical limitations of the looms available, he cannot tell how the colour and texture of the yarns will affect the pattern he has designed. Although for each new design a sample length is woven so that he can compare it with his original pattern on paper, the designer takes no active part in weaving the sample. Because of this designers are looked upon as originators of new patterns, and not of new fabrics. It follows that the designer's approach to the problem is a negative one. He adheres to the limitations set down and is not prepared to risk any combination of yarns which has not been tried before.

As a result of this the success or failure of a design is, before the looms are set up, judged entirely from the pattern. The pattern, which may be representational or abstract, is conceived two-dimensionally. Its application from point paper to the fabric is a technical matter. The problem is simply how to weave the pattern and still maintain a consistent cloth structure. And the yarn, far from influencing the designer in his work, is forgotten. Its three-dimensional nature is not considered. Instead, an illusion of depth is often produced deceptively by means of colour-shading.

A large number of furnishing fabrics are woven on looms controlled by the Jacquard mechanism. The Jacquard is necessary for patterns which are complicated—for example, figured and floral motifs. While the Jacquard does not exclude consideration of yarn texture, it tends to preoccupy the designer's mind with linear problems, and the way in which the colours can be worked *into* the fabric. The process is so complicated and so utterly different from any which the designer could experience at an

PLATE XXXIII

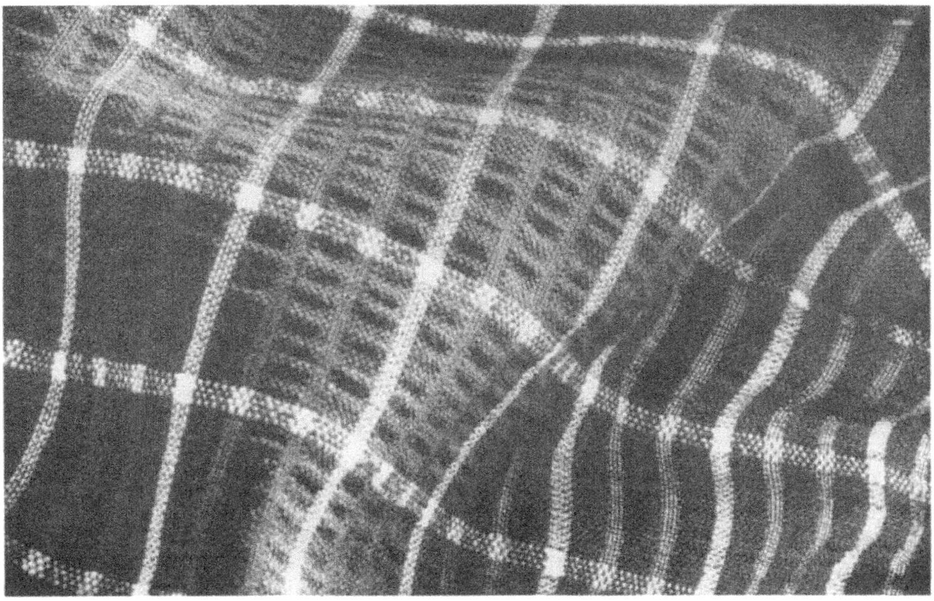

1. In weaving the designer's basic material is the yarn, which can be obtained in an infinite variety of weights and textures. This Dobby weave has been built up the interlacing yarns to become a three-dimensional fabric. *Designed by Marianne Straub for Warner & Sons Ltd.*

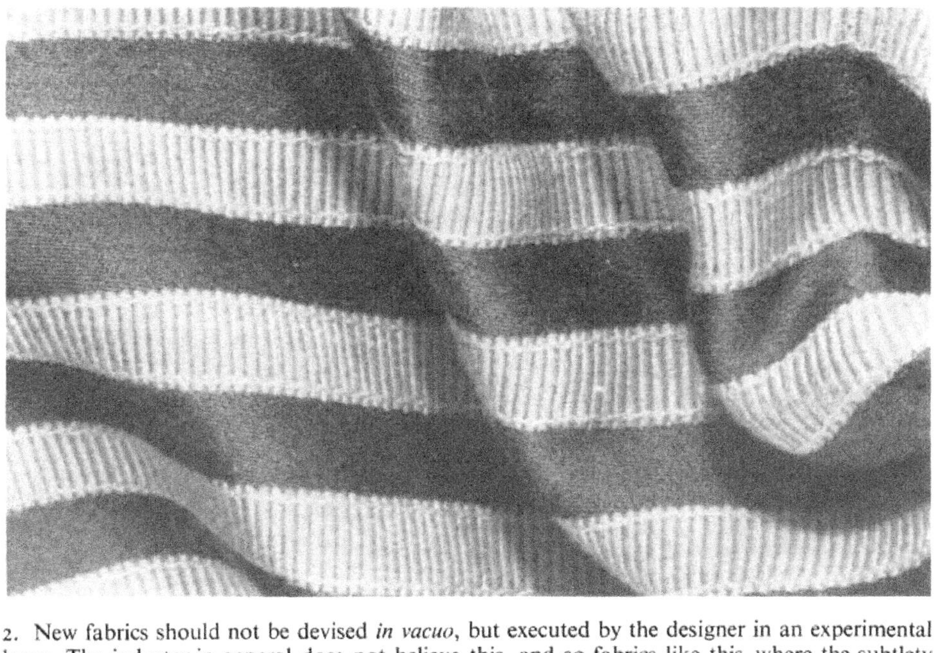

2. New fabrics should not be devised *in vacuo*, but executed by the designer in an experimental loom. The industry in general does not believe this, and so fabrics like this, where the subtlety of contrasting textures is vital, are rare. *Designed by Marianne Straub for Warner & Sons Ltd.*

The fabrics on Plates XXXIV–XXXVII are all Jacquard weaves. They are exceedingly rare in the industry and have therefore been stressed in preference to Dobby weaves, in an attempt to prove the need for giving much greater attention to the versatile Jacquard loom.

1. Linear freedom, the unique quality of the Jacquard, has here been exploited in the manner of a printed fabric and at the expense of that organic relationship which should exist between pattern and fabric. *Designed by Marion Mahler for Edinburgh Weavers.*

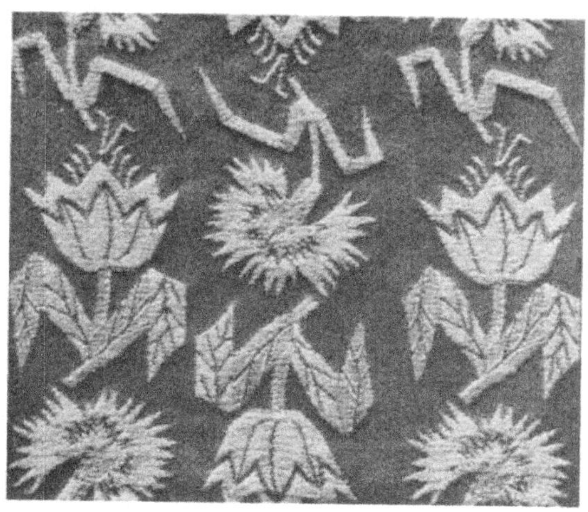

2. Again the linear pattern dominates so that the fabric visually breaks up into separate elements: motif and background. In the hands of a less accomplished designer this approach, general in the industry, can lead to the imposition of all manner of unsuitable patterns. *Designed by Enid Marx for the Utility Furnishing Scheme.*

PLATE XXXV

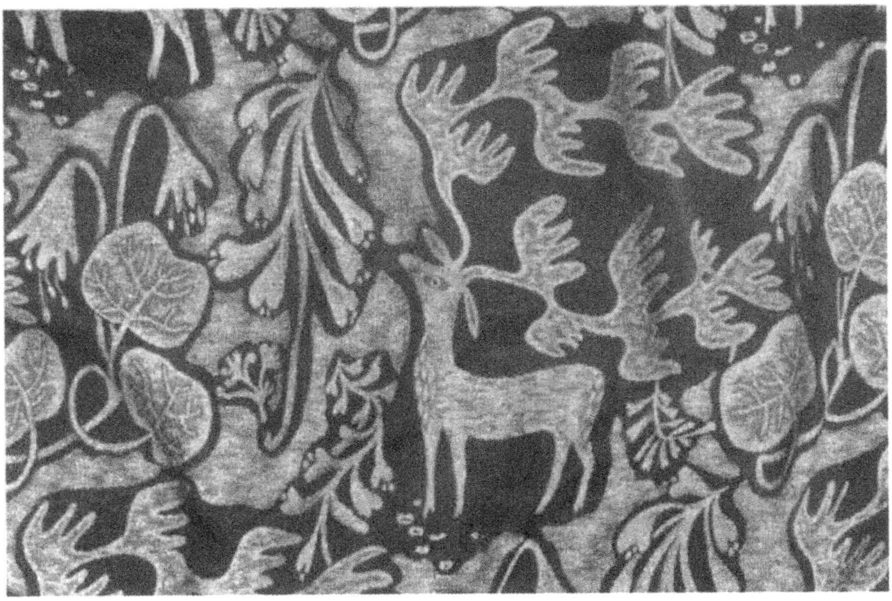

1. This damask is the work of the tapestry artist: though vitally alive it is not essentially a Jacquard design. It is as if the designer drew her pattern on paper, allowing for a neutral ground, and then let the weaver do his best. *Designed by Karin Williger for Edinburgh Weavers.*

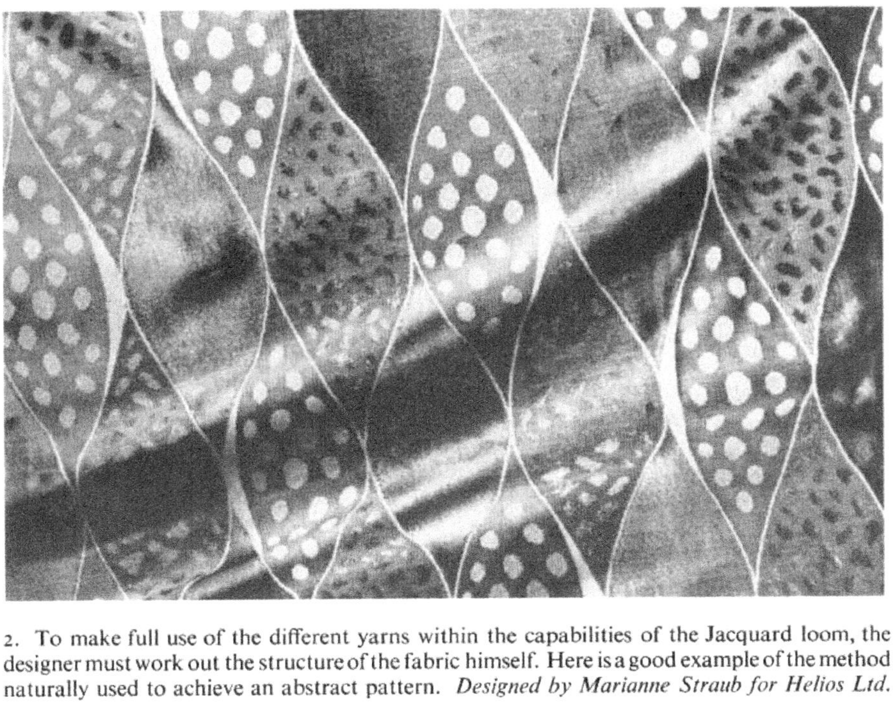

2. To make full use of the different yarns within the capabilities of the Jacquard loom, the designer must work out the structure of the fabric himself. Here is a good example of the method naturally used to achieve an abstract pattern. *Designed by Marianne Straub for Helios Ltd.*

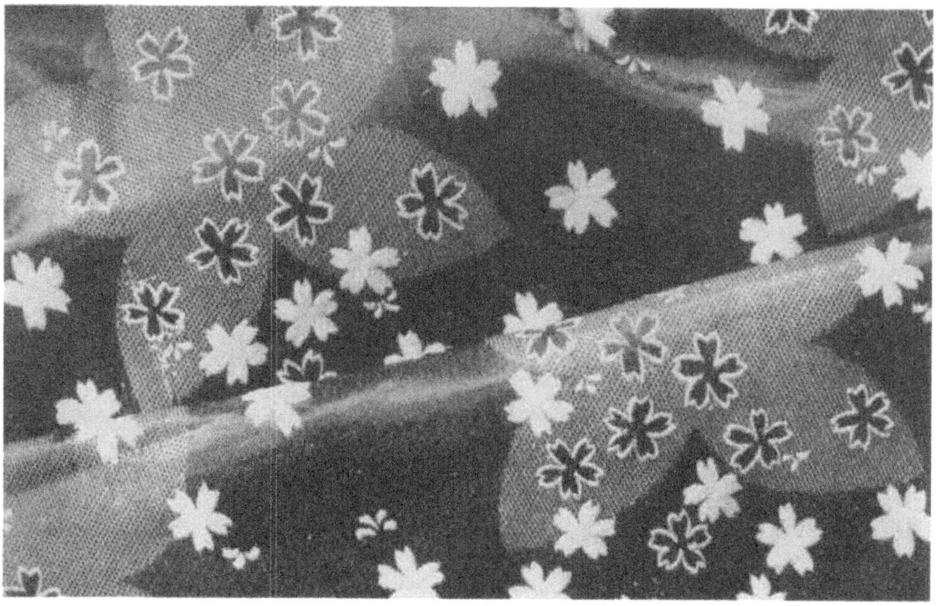

1. Designing a curtain fabric like this is an organic process in which the yarns form the pattern and the pattern enhances each different yarn. The texture has a varied richness giving pleasure to the touch. *Designed by Marianne Straub for Helios Ltd.*

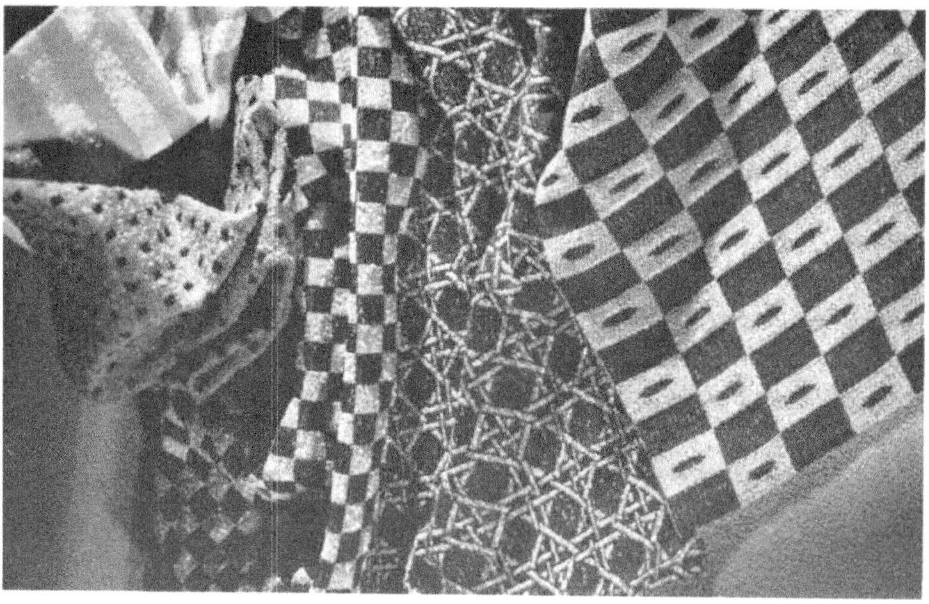

2. The designer of these fabrics spent many years experimenting with yarn textures in the Dobby loom. He came with that knowledge to tackle the increased linear freedom of the Jacquard. These first results are restrained, sophisticated and full of promise. *Designed by Tibor Reich for Tibor Ltd.*

PLATE XXXVII

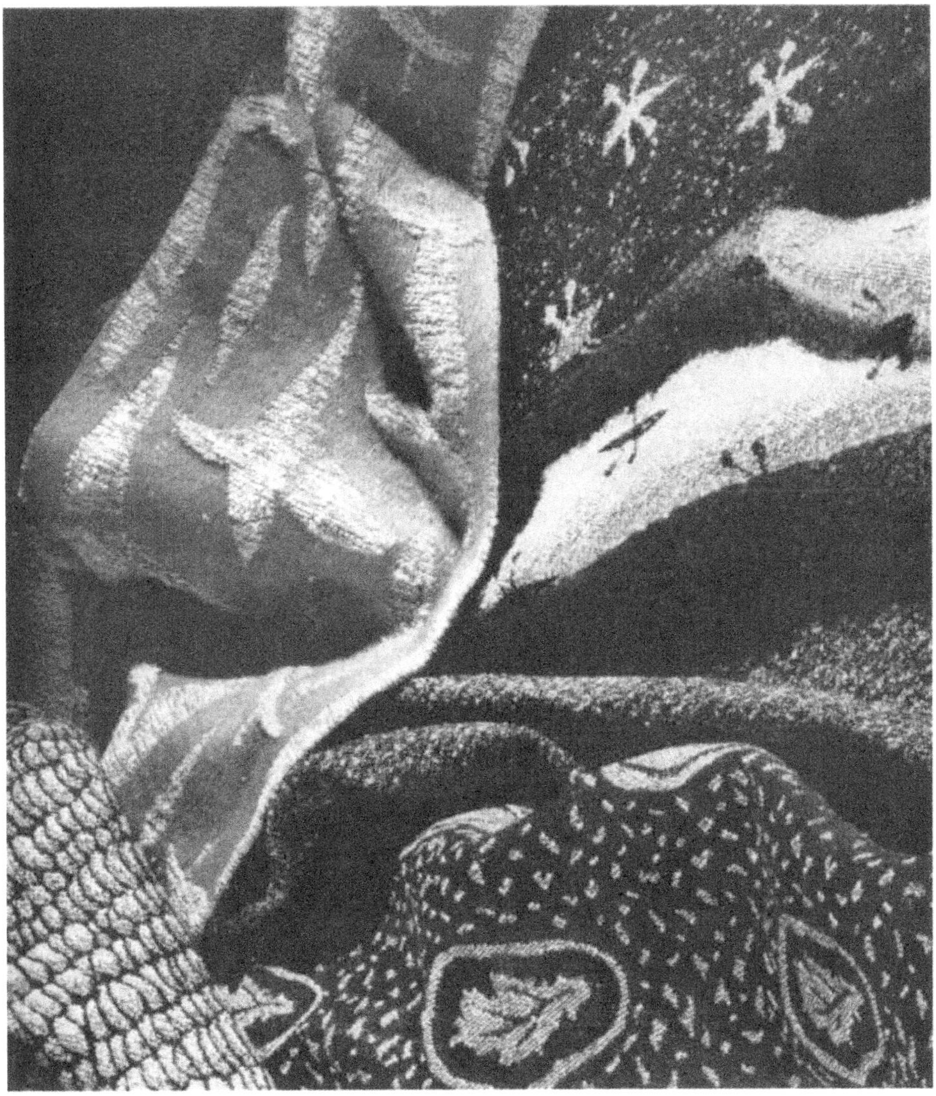

A group of fabrics which combines the linear characteristics of the Jacquard with the intricacies of weave texture, caused by the juxtaposition of variably spun yarns, a characteristic of the best modern Dobby weaves. We get the sensuous, tactile quality of the one, perhaps best described as the presence of the third dimension, coupled with the purely visual interest to be derived from pattern. The fabrics may be appreciated from a distance or studied closely so that the patterns can be seen to resolve into complex texture effects. *Designed by Tibor Reich for Tibor Ltd.*

PLATE XXXVIII

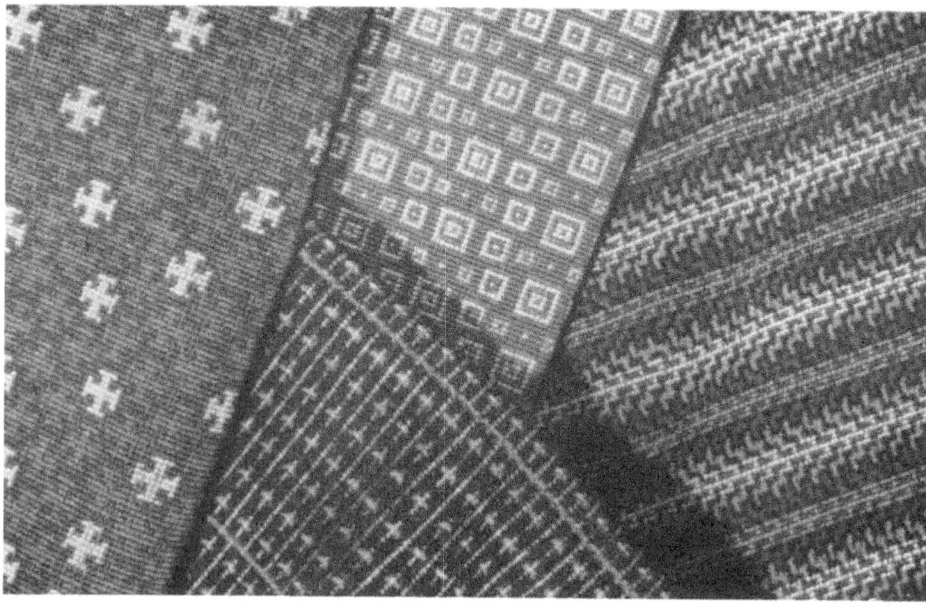

1. British Railways is one of the largest users of moquettes and in recent years has used its influence to set a new and distinctive design standard. The four examples show how the design of pattern and texture is indivisible. *Designed by the late Ian Colquhoun for the Railway Executive.*

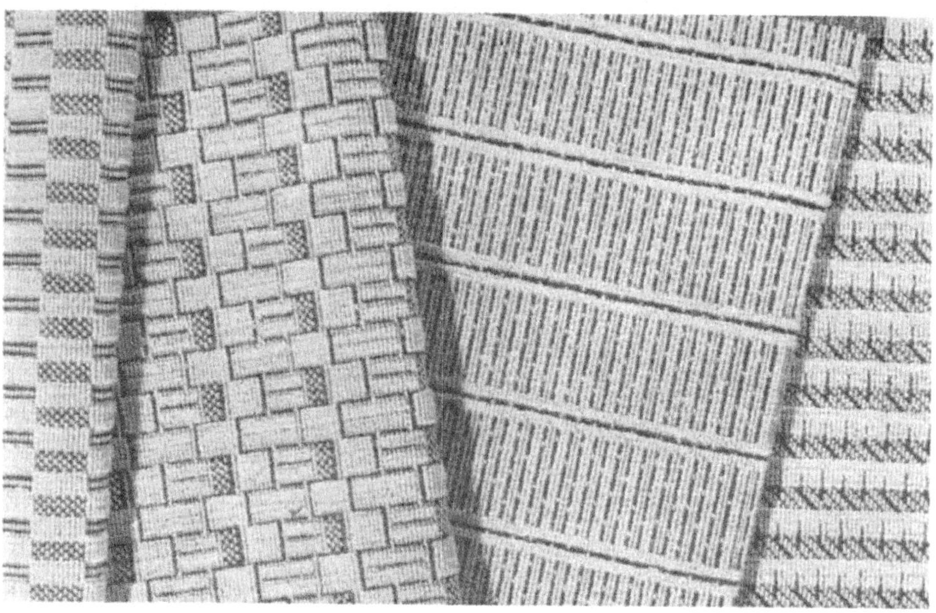

2. By making these cloths manufacturers have been prompted to reassess their own standard ranges: by seeing them in use the public is made aware of designs hitherto almost unknown in retail stores. These examples are for dining-car seats and depend more on varying depths of pile than on linear pattern. *Designed by the late Ian Colquhoun for the Railway Executive.*

PLATE XXXIX

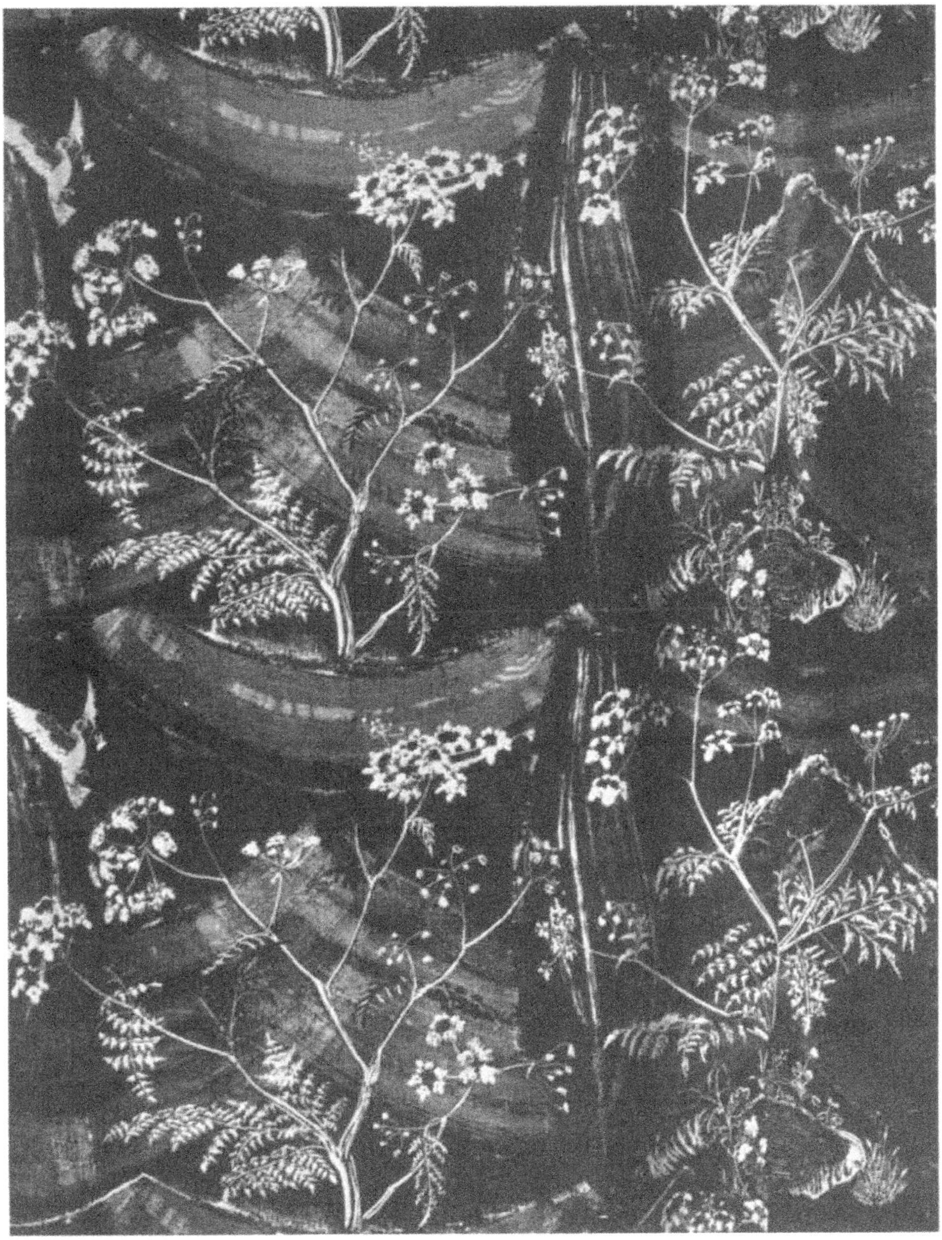

It is a traditional practice in the industry to use free-lance designers for printed textiles, though few of them have the opportunity to transfer their work on to cloth as was the case here. The pattern has been printed by silk screen: it did not previously exist on paper but was drawn on to the screen by the designers. Entirely unsuspected qualities of tone and texture have been introduced by this method which ensures fidelity to the designer's original conception. *Designed by Anne Loosely and John Drummond for Story (Fabrics) Ltd.*

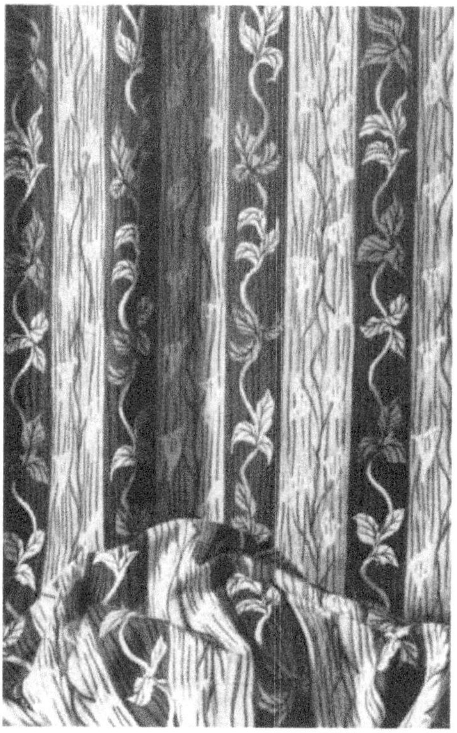

1. Flowers and leaves are stock-in-trade for printed fabrics in all price groups. When carried out with the simple realization that it will spend most of its life hanging in folds by a window this sort of design can be effective. *Designed by Humphrey Spender for Edinburgh Weavers.*

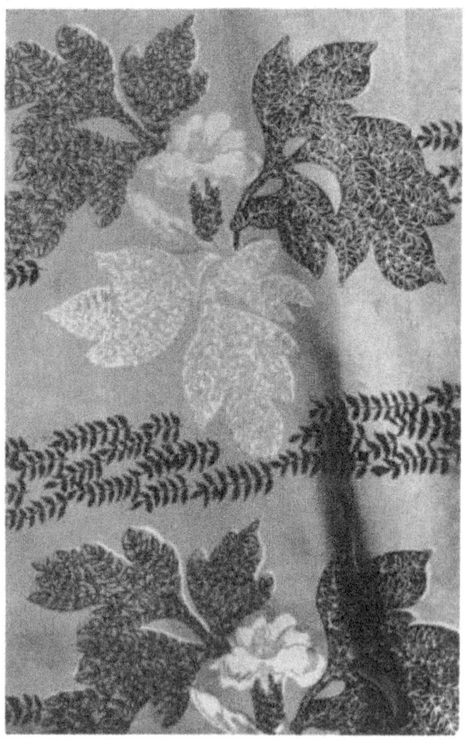

2. The perennial demand is for florals and the designer's main problem is how to make his work entirely distinct from the illustrations in a seed catalogue. Subtle transgressions of Nature's laws have in this case allowed the motifs to be organized into a coherent pattern. *Designed by Sylvia Priestley for Helios Ltd.*

1. A pattern should not lose its character when the curtain is draped. This basic distinction from the creation of flat pattern, which best suits a wallpaper, is often overlooked by textile designers. Here the sprightly design is in danger of becoming disjointed when draped. *Designed by Lucienne Day for Edinburgh Weavers.*

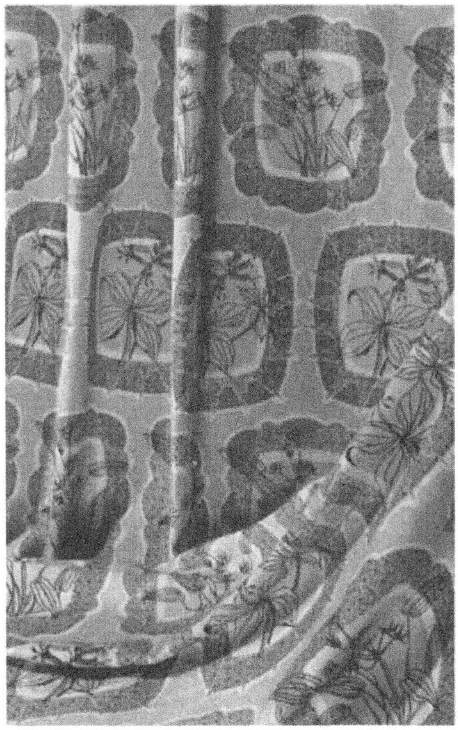

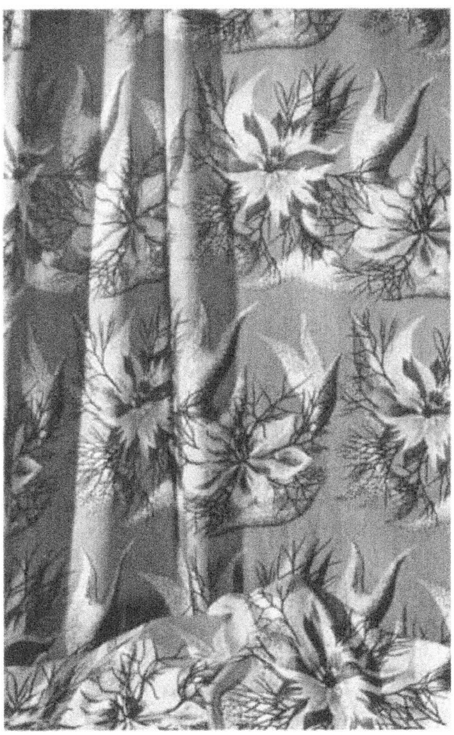

2. William Morris often used flower motifs to give coherence to his designs, however they were hung. Here the effect is the same with the designer exploiting both naturalism and abstraction to keep his conception intact. *Designed by John Barker for Edinburgh Weavers.*

1. Although there is little subject-matter for textiles which is new, its treatment can always be original. This delicately assembled group was engraved early in the nineteenth century and is still being roller-printed. *Made by G. P. & J. Baker Ltd.*

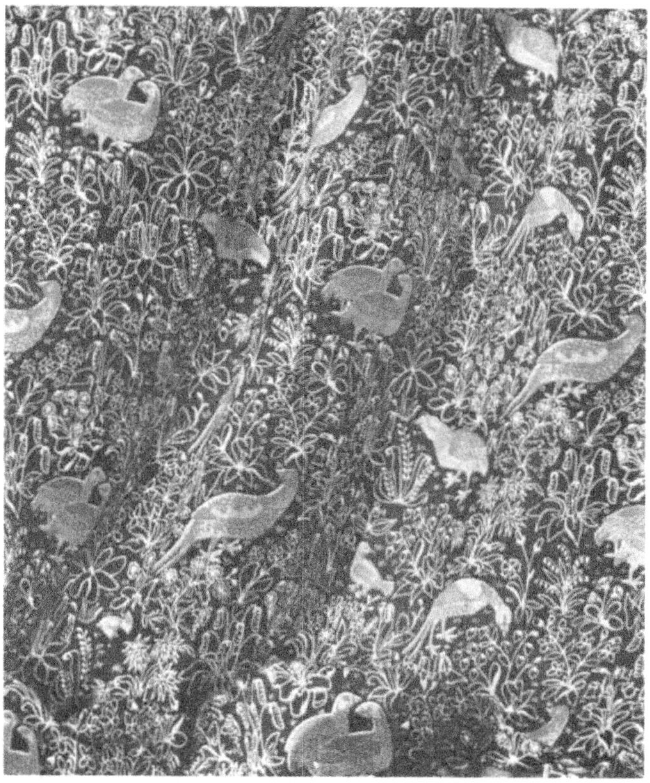

2. Pheasants again, but this time in the modern manner with more than a hint of abstraction. They make a better fabric pattern, but we cannot study them for so long with so much pleasure. *Designed by Sylvia Priestley for Edinburgh Weavers.*

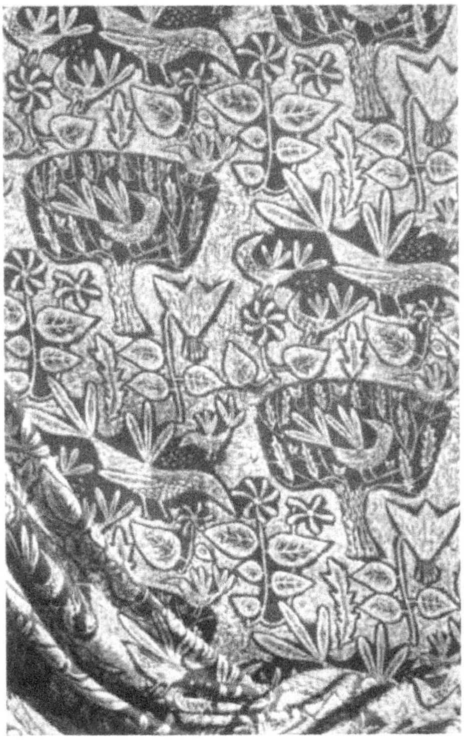

1. An insistence on abstraction with discernible naturalism is characteristic of the most gifted textile designers. While owing a large debt to Paul Klee this pattern gets the best of both worlds with an endearing naïvety. (See the designer's work for woven textiles, Plate XXXV, 1). *Designed by Karin Williger for Edinburgh Weavers.*

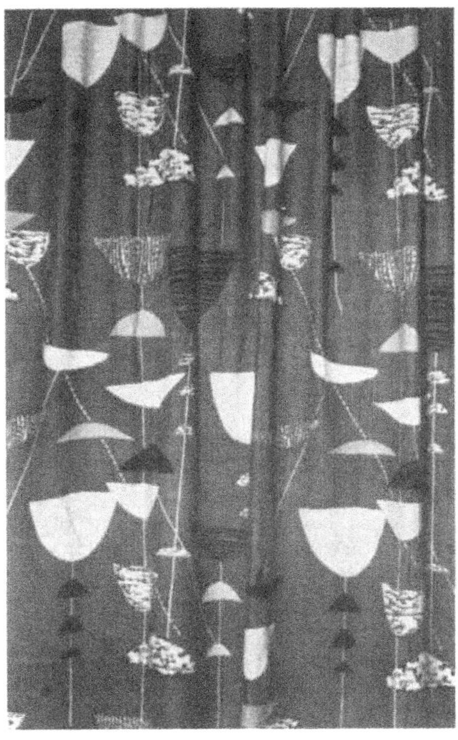

2. The bold bid for freedom in this design encouraged a host of imitations of less merit. The textile is a composition of shapes for the sake of shapes, in colour, to keep the mind alert but the senses unsatisfied. It was judged the best design on the American market, 1952, by the American Institute of Decorators. *Designed by Lucienne Day for Heal's Wholesale & Export Ltd.*

PLATE XLIV

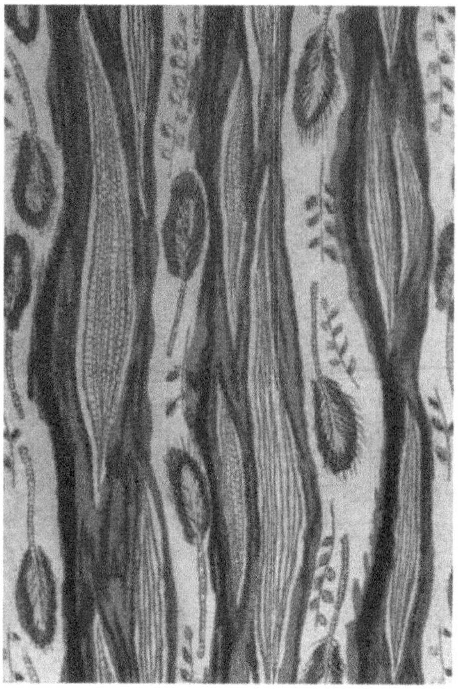

1. Artists skilled in other media are often called on to design for printed textiles. The technique of production is not complicated enough to make their work unpractical. In this case the undulating precision of the drawing shows an unusually intelligent grasp of the essential nature of a fabric. *Designed by Graham Sutherland for Cresta Silks.*

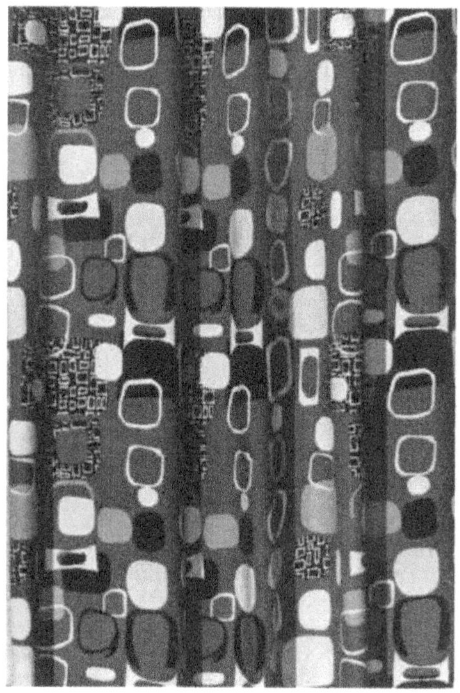

2. Each pattern for printing must repeat at a given interval; in other words it must link up with itself and start again. Diesgners attempt to hide the repeat within the pattern, as in this case, where the motifs appear to go on and on in a visual puzzle of coloured shapes. *Designed by Jacqueline Groag for David Whitehead Ltd.*

experimental loom, that it tends to make him forget about the quality of the individual yarns that make up the woven cloth, and resign himself to drawing on paper patterns which he knows can be reproduced by the Jacquard. Today very few of the best designers attempt to design for the Jacquard. And yet, if they made the effort to master the technicalities involved (and more manufacturers were prepared to risk the relatively high initial costs on each design), the Jacquard could offer a much wider scope for their creative talents. While the organic method already described can only produce the best results on both Dobby and Jacquard looms, we must recognize that, at the present time, the generally accepted practice in the industry is for the designer to remain at all times separate from the looms.

The technical requirements for the designer of printed fabrics are less complicated, and may be dealt with quite briefly. The weave of the cloth used is, as a rule, of no artistic significance, and is considered simply as the structural base on to which the design is printed. Exceptions can be found amongst specially textured cloths for which the designer may or may not be responsible. The three chief methods in use are hand-block-, screen- and roller-printing. Hand-blocks are relatively cheap to produce, but costly to use because of the skilled labour involved. They are confined almost exclusively to expensive fabrics and to designs which are expected to have a limited appeal on the market. Screen-printing is also a skilled operation, but it is widely used in this country, especially where designs are expected to have a short period of popularity. Roller-printing is used for fabrics in price groups where the demand for one design is likely to exceed 10,000 yards. After the copper rollers are engraved and arranged in the printing machine, no skilled labour is required. In hand-block printing the colours are limited only by the cost of producing and using separate blocks for each colour. The same is true of screen printing. But in roller-printing relatively few colours are used as a rule—the possible maximum is twelve. It is, of course, necessary for the designer to appreciate the different ways in which these three methods can be expected to interpret a given design.

In the case of printed fabrics there is no need for the designer to experiment with manufacturing processes. In all the firms I visited the majority of the designs have been purchased from free-lance and commercial designers. Nevertheless, most firms maintain studios of their own, not only for creating new work, but for adapting the designs bought from outside artists. The studios are in the charge of a producer, who is not necessarily a designer. His task is to buy or commission the work of outside designers, or use that of his own studio. The designs have to be suited to the particular quality of cloth on which he wants to print, and to the market to be served. In most firms the producer forms the only link between the designer and the printing works.

With this information in mind we can now deal with woven and printed furnishing fabrics together. In the firms I visited new designs originate in two ways. First, a firm produces a seasonal range and invites buyers to select the designs they require. In most cases this range is divided into three or four groups, each of which contains designs suited to particular markets. The preliminary selection of designs is the responsibility of the managing director and the sales manager. With very few exceptions, the designer is not expected to give his opinion at this stage or later. The result is that the selection depends upon the sales results of the same, or similar, designs which were popular in the previous range. Secondly, a firm would have been commissioned by a wholesale buyer to produce specific designs supplied by him. In some cases the buyer, instead of bringing his own designs, collaborates with the firm's own designer. This contact between the designer and the purchaser generally results in a higher standard of design than is attained when several intermediaries are concerned. Recent striking examples of this are the new moquettes for railway coaches. They were designed by an architect employed by British Railways, in collaboration with the designers of two large mills in Yorkshire.[1]

In the firms where the best designs are produced I found that either the staff designer occupied an influential position, or that the managing director was personally interested in the artistic standard of his fabrics. Generally, the designer's status was unrelated to the size of the firm. In very large firms the executive departments—design, production, buying, sales, etc.—were in a stratified formation and could only correlate their work through the board of directors. In such firms the designer's influence upon standards was negligible. Most manufacturers were indifferent to the designer's need to refresh his mind by visits to foreign countries, etc. For most Yorkshire and Lancashire manufacturers, exhibitions of textiles and other industrial arts in London did not warrant the designer's train fare and lodging expenses. Instead he had to rely on magazine illustrations, the sketches of free-lance designers and what he was told by the managing director and the firm's agents. There were, of course, several exceptions to this. Specially large firms, for instance, did not mind paying the trifling expenses of a designer's journey.

The salary for a fully experienced staff designer is between £750 and £900 a year. As much as £1500 is paid to some designers in the larger firms, particularly those concerned with woven textiles. Boys and girls from grammar schools, or the junior art department at the local art school, enter the studios with a wage of £3 to £4 a week. In some cases where a fully trained student from art school is employed, the starting

[1] See Ian Colquhoun, 'How new moquettes are designed for British Railways', *Design*, February 1953, pp. 22–8.

salary is from £350 to £400 a year. In order to supply print designs a good many of the commercial designers in this country and on the Continent run large studios and employ thirty or more designers, finishers and assistants. I found that the fees which they received from manufacturers varied according to the use to which the design was put and the degree of finish. The same manufacturers were also served by many independent free-lance designers, amongst whom were a few well-known names. Fees range from 5 to 8 guineas for the work of a young designer, and from 15 to 25 guineas for that of an experienced commercial or free-lance designer. Higher fees are paid to exceptionally talented designers, and to those painters and sculptors whose names are likely to bring publicity to the firm. In most firms a free-lance designer is expected to sell an idea, in sketch form, and not a design which is fit to be sent straight to the printing works. It is usual for the colouring of the design and the repeat size to be worked out by the firm's staff designer, although I investigated several cases where the free-lance had been asked to re-scale and colour his design to suit the firm's requirements. In these cases it is rightly felt that the original artist should be responsible for modifying his own work. Successful free-lance designers for woven fabrics are rare, and they command anything between 25 and 50 guineas for a single design.

For both woven and printed goods I was told that fewer new designs were required than before the war. In general, manufacturers were less speculative when purchasing free-lance sketches, and staff designers were kept busy modifying old patterns. Even in firms of a similar size catering for similar markets, I found a wide difference in the number of new designs required. For instance, one firm producing moquettes and tapestries required an average of 120 new designs a year, whereas the corresponding figure in a firm of similar size was only twelve new designs. The only general conclusion which emerged from the information I received is that the demand for new designs is greater at the cheaper end of the trade.

As a general rule furnishing fabrics are selected by the customer in a shop *after* the furniture and the carpets have been purchased. As we have seen, the design of most furniture today is faintly connected with antique styles, and carpets are either Oriental or floral. It is to be expected, therefore, that upholstery and curtain materials will follow, or at least agree with, the dominating styles of furniture and carpets. As a result, more than 90 per cent of all fabrics have floral designs which are either antique or so-called 'modern' in character. The antique floral may have conventionalized leaves and petals, or it may have voluminous bunches of blossom circumscribed by loose indefinite scroll work. The 'modern' floral can be conventionalized or naturalistic, and its intertwining wispy flower sprays tend to run up and down the cloth. There are endless alternatives and there is no need to describe them. In defence of the antique floral it is often urged that many print designs are exact reproductions of period work. In fact a few firms still use blocks and rollers which are well over a century old. These designs are said to have a perfection which is undated, and therefore no amount of argument will prove that they are unacceptable to contemporary taste. But all this is irrelevant when one considers the productions of the majority of firms. There are, after all, only a limited number of period designs to be copied, and the Victoria and Albert Museum has long since been exhausted as a source. It is inevitable, therefore, that modifications must be made to original designs, in order to keep one firm's range sufficiently different from another's. Once this occurs, the worst artistic effects can be expected. With novelty as the only criterion of excellence, a design can be made to combine characteristic motifs from several different periods. In many cases the use of such terms as 'Old Chinese', 'Jacobean', 'Georgian', etc., is considered to be sufficient to remove all doubts about authenticity. There is, of course, no defence for this state of affairs, and still less for period designs which are woven in or printed on low quality or utility fabrics.[1]

The so-called 'modern' floral designs are generally uninteresting because they lack vitality. Designers appear to be too preoccupied with making an exact copy of a flower to remember that their main task is to organize the various motifs into an artistic whole. There is no need to condemn flowers as unsuitable motifs for printed and woven design. What we are concerned about here is that the artistic qualities of a flower should be appreciated in terms of a fabric, and not in terms of a garden. They should be imaginatively interpreted, not slavishly copied. In woven fabrics, the colour and texture of the yarns and the weave structure are fundamental and should be organized by the designer's intuitive apprehension of a flower. In printed fabrics the designer has complete freedom graphically and is not concerned with the structure of the cloth. This freedom is, in fact, a limitation, for the success of his design has to depend upon the spatial relationships of forms and colours in one plane only. Flowers can be depicted with an exactness which asks earnestly for admiration, but representational accuracy is not the criterion which matters. To be successful as a design the flower, in its treatment, colour and arrangement on the fabric, must evoke a sensory and not an intellectual response.

Apart from the many antique and 'modern' florals, there are, of course, a number of exemplary modern designs for woven and printed fabrics. While these are mainly confined to a few outstanding firms, I found that every firm could show me at least

[1] It is worth noting that William Morris conceived his block-print designs in terms of twill cloth. That cloth is not available today, and the firm which owns his blocks has refused to use them on inferior cloths.

half a dozen satisfactory designs. Though in some cases the good designs amounted to only 2 per cent of a firm's range, the fact is that retail buyers were offered some choice. In printed fabrics this choice was fairly wide and the better designs had, in nearly every case, been originated by free-lance designers. In woven goods the general standard was not so high, though each firm had some plain weaves in good, strong colours. As a corollary to this, it should be noted that amongst the few firms where high standards are insisted on, the production of printed designs far outweighs that of woven designs. The lively and enthusiastic way in which manufacturers tend to embark on printed dress fabrics may be responsible for the improved standards of furnishing prints. In one large Lancashire mill a revived fashion fabric department has done much to raise the level of design in the older and more conventional furnishing department. For woven fabrics the need is not for a modern 'style' as opposed to period reproductions, etc., but for a sympathetic understanding of the material, *beginning with the yarn.* This understanding, coupled with a genuine artistic sensibility, would produce designs which could then be looked upon as characteristic examples of modern textile art.

For heavy furnishing fabrics the prevailing colours are rust, dark green and brown. Here again the designer relies on nature, and I repeatedly found fresh spring flowers clothed in the dying shades of autumn. The colours are popular on the market for two reasons. First, there are very few good alternatives which can be chosen, unless the purchaser can afford a more expensive fabric. And secondly, the colours agree so alarmingly well with the drab, dark stain of nearly all the available furniture. For curtains and light covers which are usually printed the situation is much more satisfactory. The four trial colour combinations used by one large firm for all new designs are sufficiently representative to be quoted. For an initial run of 12,000 yards, the quantities of each combination were as follows: brown and fawn 2500, mostly pink 2500, slate-blue 3500, and pastel-green 3500.

It is a remarkable fact that there are no less than two design centres operating on behalf of the industry.[1] In several spheres, notably dress goods, their influence has been appreciable. But in furnishing fabrics it is impossible to say that these centres including the Rayon Industry Design Centre, now closed, have so far been responsible for significant improvements. Rather, the setting-up of high design standards has been accomplished by firms within the industry. And along with these high standards goes the conclusive proof that they are commercially possible. The presence of a few enlightened manufacturers who are determined that their products should be of the highest artistic standard is a matter of vital importance in any industry. While their

[1] The Colour, Design and Style Centre and the Interior Decoration Design Centre (see pp. 223 and 227).

example stimulates other manufacturers to produce better designs, their methods might usefully be outlined for they are too often overlooked.

This is particularly true of the textile industry, and a brief account of the methods used by three firms may be instructive. Sir James Morton established Edinburgh Weavers in 1930 as an independent unit within the big mass-production combine of Morton Sundour Fabrics, with the explicit intention of creating a laboratory for the best modern textile art. He was induced to risk this departure, the financial success of which was inevitably uncertain, by a strong feeling of responsibility for the design standard of his fabrics, which had deteriorated as a result of quantity production. The unit was established by taking over the traditional hand-weaving workshop of Edmund and Alec Hunter, who became the first designers and managers. A few years later a distinctly modern trend in design was developed by Antony Hunt and Alastair Morton so that even before the war, for both woven and printed fabrics, Edinburgh Weavers was one of the most adventurous firms in the country. Another firm is Helios, which was started in 1937 by Sir Thomas Barlow of Barlow & Jones, a Lancashire firm producing furnishing fabrics, dress goods, towels, etc. Within two years the Helios range of inexpensive fabrics of exemplary design was attracting attention at all levels of the retail trade. A wide public was aimed at—a public, incidentally, which had never seen such designs outside the most exclusive shops. The credit for this achievement belongs to Marianne Straub, who has since been working on the same lines for Warner & Sons.

Another method, and one which is comparatively rare nowadays, is for a designer to start a firm of his own. The one outstanding case in recent years is that of Tibor at Stratford-on-Avon. The firm was founded by Tibor Reich, who had studied textile technology and design while at Leeds University. After a short period of stultifying repetitive work in the studio of a large Lancashire mill, he decided in 1946 to begin production on his own. At first there were only two hand-looms, but by 1950 he was employing forty-five weavers for both hand- and power-looms. His fabrics have been used by those furniture manufacturers who insist upon the highest design standards and by architects, etc., for contract work. They can also be seen in many of the better-class retail stores.[1]

Finally, the special scheme to assist a large section of the Welsh woollen industry in the designing of fabrics should be mentioned. Most of the factories are small (five to twenty-five employees) and situated in rural areas. In 1929 the University of Wales appointed an officer to advise the industry in technical matters and to give assistance in the designing and colouring of fabrics. He also attempted to make the industry,

[1] See my article 'Pattern out of Texture', *Design*, April 1954, pp. 12–15.

which was virtually craft-based, more widely known. However, it was soon found that to have one man fulfilling this dual function was not satisfactory. As the funds of the University of Wales were limited, the Rural Industries Bureau agreed in 1932 to appoint a textile designer who was to give full-time attention to the needs of the Welsh woollen mills. The designer, Marianne Straub, produced some outstanding designs for the Bureau before she joined Barlow and Jones in 1937. Although the design scheme closed down in 1939, it has been revived since the war with the appointment of a new designer. Most of the factories are too small to attract more than local retail trade, but if a central sales organization were established the wider market interested in modern textile design could be reached.

From this account of furnishing textiles one very important point stands out. If a satisfactory design standard is to be maintained throughout the industry, then designers must be given greater freedom to experiment and to travel, and more authority to control their work through all stages of production. I am convinced that there are many designers in the industry who would, if given such opportunities, be capable of rising to the highest standards in any other country. Once again, the responsibility for initiating this rests with the manufacturer.

WALLPAPERS

About 80 per cent of the total production of wallpaper in this country is controlled by The Wall Paper Manufacturers, Ltd. The combine was formed in 1899, and employs approximately 5000 persons in twelve mills. The mills range in size from 120 to 850 persons. I obtained information from the head office of the combine, and from the largest of its member firms, producing both handblock- and roller-printed papers. In addition I visited one large firm which, although independent, works in close co-operation with the combine, and one small independent firm specializing entirely in handblock-printing.

As in printed textiles, volume production of wallpaper is based on roller-printing by machine. The process is known as 'surface' printing, and one machine can carry as many as twenty rollers, each of which has a separate colour. Although very few papers have as many as twenty colours, the initial expense of preparing the wooden roller with copper, inlaid with felt, is so high that it would be uneconomical to sell less than 10,000 pieces of one design.[1] Hand-block-printing is used for designs with

[1] A piece of wallpaper has the standard length of 11½ yards.

sharply defined outlines and where the weight of colour required is greater than that obtainable by roller-printing. Hand-blocks are also expensive to prepare (about £20 each), and they demand skilled handling. The quantities produced are small (about 100 pieces is a satisfactory sale), so that hand-block papers are comparatively costly.

For new designs the wallpaper industry draws from staff designers, commercial and free-lance designers. A medium size mill produces about forty new designs a year, each of which has two or three alternative colourings. In big firms the corresponding figure is sometimes much higher. Most of the designs are kept for three years, but it is not unusual to repeat a successful design for five years and more. As in other industries, the tempo of change seems to be slower since the war than it was before. Staff designers are recruited from within the industry (in one firm the head designer was once a salesman) and from art schools.[1] Each firm has its own studio which supplies the bulk of all designs required. The salary of a head designer is between £700 and £900 a year; in one or two cases it extends up to £1500 a year. An experienced assistant designer gets from £500 to £600 a year, and a junior designer about £300 a year. A commercial designer's fees are similar to those paid in the printed textile industry. It sometimes happens that identical patterns are sold in both industries, in spite of the fact that a wallpaper design should be conceived in terms of flat extension, and a design for curtains in terms of creased draping. In addition, free-lance artists are used by a few firms, especially for hand-block-printed papers. This represents a remarkable advance on conditions in the 1930's, when hardly any independent artists or architects sold designs for wallpaper.

In spite of the fact that the general design standard has improved considerably during the past fifteen years, it is still very unsatisfactory. The factor mainly responsible for this is a lack of artistic sensibility amongst the designers. Wallpaper is unique among industrial products in that its only function is to be decorative. A paper is nothing if it is not visually satisfying. The structure and durability of the paper are matters which do not concern the designer, who is left free to work out a two-dimensional pattern in terms of convenient repeat sizes. But decoration, as we have already found, is a quality which is under-developed in modern art. If architects experience difficulty in designing decorative ironwork, then how much harder is the task of the

[1] An important exception to this is the Harold Sanderson Bursary Fund, which operates on behalf of the combine. The Fund opened in 1949 and enrols four students (aged 15–18) a year on a course which lasts for three years in the design studio of a wallpaper mill. Instruction is carried out by a specially appointed art master/designer, and emphasis is given to flower painting and the evolution of pattern from naturalistic forms. The students are well housed, paid a maintenance grant, and a small salary, and given the opportunity of foreign travel. They are not expected to do productive work while on the course.

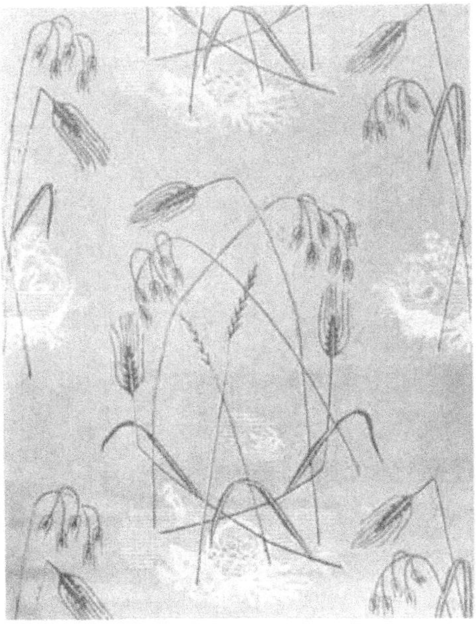

1. Wallpapers are really paintings repetitively arranged and produced in quantity. They have at their disposal the widest range of talent, yet the encouragement given to artists to explore the medium is surprisingly meagre. *Designed by Edward Bawden for Cole & Son (Wallpapers) Ltd.*

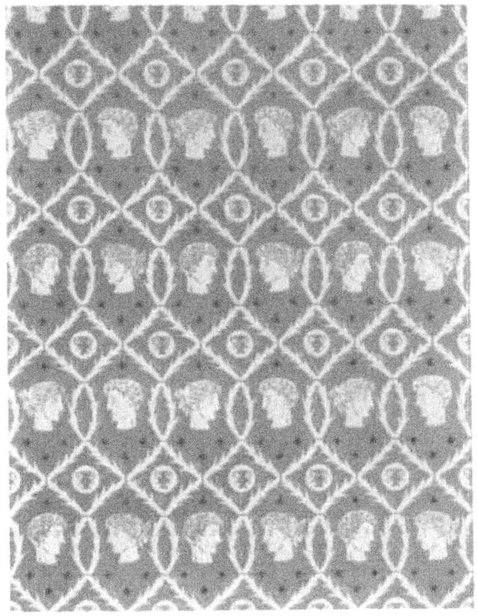

2. The technical limitations for the wallpaper designer are not great, but unlike textiles his work must always be visualized in terms of flat extension. The effect of this dominating pattern is enhanced as it repeats. *Designed by John Minton for John Line & Sons Ltd.*

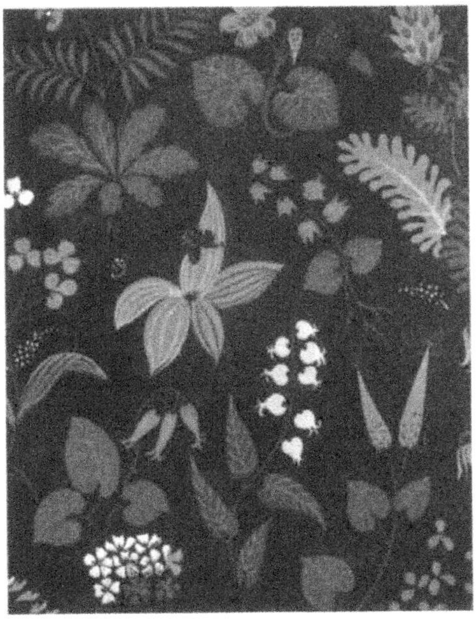

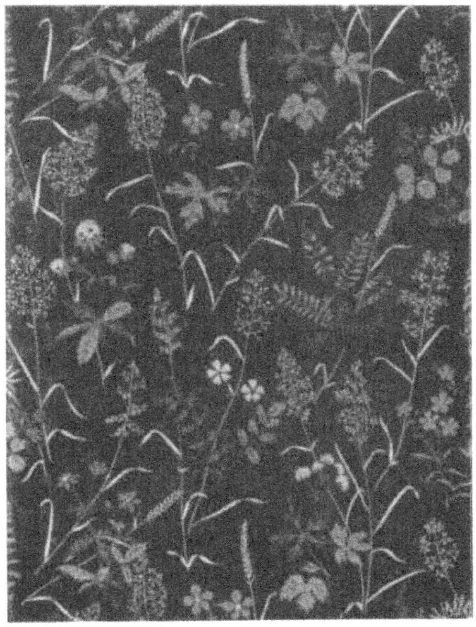

3 and 4. As with textiles flowers are the most popular subject-matter for overall decorative effects. The designer must try to balance the way in which he links his motifs together in a series with the detailed treatment of the motifs themselves. The paper on the left contains elements which are particularized too freely: on the right the pattern is more homogeneous in its organization, while each motif retains sufficient definition. *Made by The Wall Paper Manufacturers Ltd.*

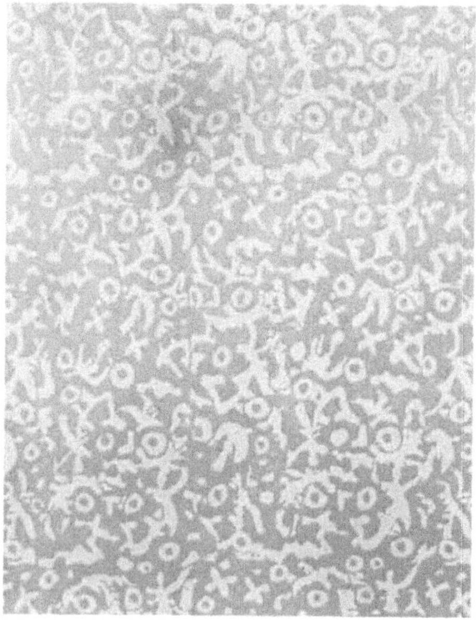

1. Wallpapers' widest use is as a background with low tonal value, leaving other designs in the room to strike a more positive note. Non-representational patterns are often the best for this purpose, especially when they are as restful as this example. *Designed by Graham Sutherland for Cole & Son (Wallpapers) Ltd.*

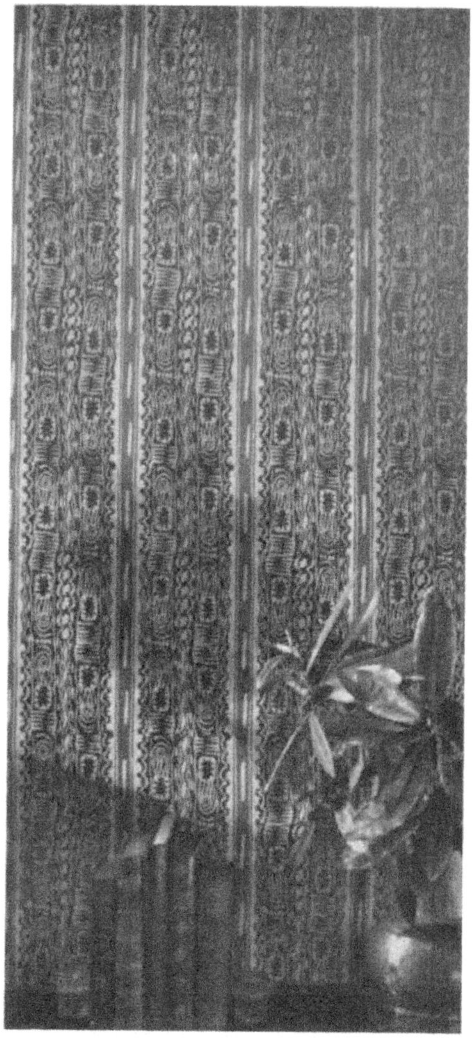

2. *Above.* Used on one wall only some boldly patterned papers give the necessary emphasis to a room. Here the pattern contains the vertical accent most often needed in contemporary houses with low-ceilinged rooms. *Designed by Armfield-Passano for John Line & Sons Ltd.*

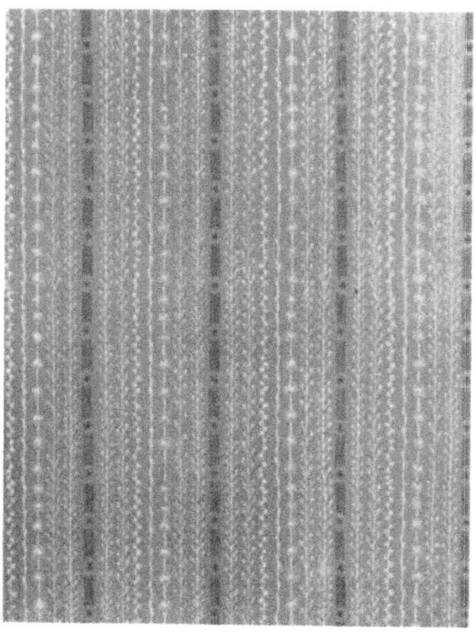

3. The examples on these pages merely indicate the quality of work which *can* be purchased. Although they are very different from the general run of papers in the house decorator's pattern book, they are far less adventurous than we might expect from an industry with such tremendous resources. Left, *designed by Roger Nicholson for The Wall Paper Manufacturers Ltd.*

designer who has to create a flat pattern which must be, in itself, artistically satisfying? This explains the history of wallpaper design. Wallpaper began, like so many industrial products, by clinging to decorative forms characteristic of other materials. Early paper-hangings were looked upon as cheap substitutes for expensive wall hangings. The printed designs on the papers contrived to imitate woven tapestries, embroideries, silk and velvet hangings, painted cloths and tooled leathers. A later development was the flock paper, which copied the brocades and velvets of Florence and Genoa. Throughout the entire development of the industry up to the present time, the majority of wallpaper designs have been imitative of forms and patterns peculiar to other materials.

Alongside and mingled with these traditional imitations, wallpaper design has derived inspiration from other sources. A survey of the dominating styles during the last twenty years would be instructive, chiefly because the industry has been remarkably prone to sudden changes from one fashionable style to another. The ease with which these changes were made suggests that the contemporary negative style of most inexpensive papers will soon be ousted by much bolder patterns and colours.

Modernistic patterns began to be used about 1930. Floral period designs were given up and replaced by riotous pseudo-cubistic patterns derived from German 'Expressionism', but devoid of the expressive qualities which were that artistic movement's *raison d'être*. After a few years colours were softened down, and jagged patterns in pastel shades became the fashion. The next step was the realization that even these appeared too obstrusive. Manufacturers and the public wanted wallpaper to be a background and nothing more, and designs were introduced with an appeal which depended exclusively on broken, stippled effects. The first of the Bauhaus papers were of that kind. In Britain they appeared about 1932 or 1933. Naturally they gave much less scope to the designer than floral or cubistic patterns. Most of the broken effects were the result of experiments with rollers, and the artistic effect could never be exactly foretold. However, it seems as though at that promising moment the disappointed wallpaper designer must have looked for a new outlet for his ornamental fancy, and found it in what is called 'appliqué' work. There had been a vogue during the last years before the 1914–18 war for applied borders, with period motifs, around plain wallpaper panels. They were meant to imitate the appearance of eighteenth-century panelled walls. Early in the 1930's the scheme was modernized and most enterprisingly developed. Instead of the simple and harmless borders of 1914, one was offered clusters of stylized leaves in brown, green and orange (the notorious colours of the modernistic carpet), or clusters of odd stalactite shapes to be stuck along the upper edge of panellings and especially in the corners. Sometimes appliqué grew bolder than this, and life-size hollyhocks or gondolas swimming in a printed and cut-out sea in front of printed and cut-out palaces, rose in profusion from the bottom of the panelling.

The appliqué craze began to wane after 1935, though it still lingered in less wild forms in the cheap trade. The builders of suburban semi-detached houses felt it was their duty to provide at least the 'best rooms' with some appliqué work. In the late 1930's in the medium and high-price trade, plain textured patterns began to outstrip cubistic patterns. By no means were all the textured patterns satisfactory. In fact, many of them were rather crude and thoughtless, and very few had the delicacy of the first Bauhaus designs. Improvement was thus attained by ruling out real pattern effects altogether, instead of by the introduction of better patterns. Although their studios may have lacked creative designers, manufacturers did not use the artists who at that time were serving the printed textile industry. There seems to be only one reason which could account for this. Before 1900, Crane, Voysey and other outstanding designers worked for a few medium-sized firms, the owners of which were personally interested in maintaining a design standard. When the combine was formed in 1899, this personal note of necessity disappeared. Also it was soon realized that there were cheaper methods of obtaining designs than by buying them at 12 to 20 guineas from famous designers. The changed policy did not have an unfavourable effect on sales, and as the policy of the combine meant the policy of almost the whole industry, architects and artists soon found it difficult or impossible to sell wallpaper designs to manufacturers.

Since the war the industry has shown a significant change of policy, which was at first noticeable in the few independent firms. A large number of free-lance designers and some painters are now used, and the best of their work has reached a high standard. However, the best patterns are hand-block-printed, and cost as much as 16*s*. or 30*s*. a piece. This puts their work beyond the means of the majority of persons, who cannot afford more than 5*s*. or 7*s*. for a piece. Despite the fact that these hand-block-prints have sold well, I was told that firms generally are not willing to risk the initial costs of roller-printing designs of a similar standard. For this reason the majority of designs on cheap and medium-price papers are very unsatisfactory. About 50 per cent of these papers have textured effects in beige and cream. Some are embossed to resemble trowelled plaster, or rough cast, others use a faint, confused pattern of leaves. Pale pastel colours dominate all cheap papers. Within the 5*s*.-a-piece price group I looked in vain through the pattern books for an alternative to these devitalized shades. The best-selling patterns of such papers achieve staggering proportions—50,000 pieces of one pattern was considered satisfactory, but

I was shown several others, each of which had reached the amazing figure of 100,000 pieces.[1] Before the war the corresponding figure would have been about 30,000, for firms then had a wider range of patterns. Manufacturers I met instinctively feel that these patterns are too innocuous to find favour, so they confidently suggest that each should be accompanied by a brightly coloured border. The designs of these borders, which are pasted arbitrarily on to each paper in the pattern-book, are the survivors of the pre-war modernistic patterns and colours. As far as medium-price papers are concerned, the spots and stripes which were fashionable with London decorators just before the war are now popular, and there is a small, but consistent sale of imitation marble, wood-veneer and panelling effects.*

At least manufacturers are not unaware of their responsibility to improve the standards of their papers. The exhibition train which toured the country in 1949–50 was a welcome gesture towards the public.[2] It contained several papers of original but restrained design, consisting of stripes, spots and sprigs, for which there has since been an increased demand, partly because the exhibition was seen by 400,000 people. Such direct contact between the manufacturer and the public is needed in this industry, because most people depend entirely upon the pattern-books brought round by builders' merchants and house decorators. The patterns in these books are selected by the merchant from a wide range of manufacturers' samples. To some extent, therefore, the merchant's own taste is responsible for offering the public several large books of patterns, nearly all of which are artistically unsatisfactory.

Nevertheless, it is the manufacturer's duty to provide some designs of a high standard amongst his cheaper papers. This could be done by commissioning the free-lance designers who have already achieved success in block-prints to design for machine-printed papers. There is undoubtedly a public demand for a decorative quality in industrial art today. In other industries we have seen that this demand has entailed a lowering of standards, because of a fundamental misunderstanding on the part of manufacturers. In the wallpaper industry, on the other hand, the more expensive papers have achieved a high standard of decorative design. The main problem now lies with the cheaper papers which have retained a weak style for an unnaturally long time. Indeed, some of the most recent papers seem to be bordering

[1] As in the carpet industry, the social consequences of this are important. For example: if the living-room of a semi-detached house requires about five pieces of wallpaper, then the total sale of 100,000 pieces of one pattern would affect about 20,000 rooms.

[2] The exhibition was staged by The Wall Paper Manufacturers, Ltd. From the public's point of view it was a more effective introduction to modern design than the static Exhibition of Historical and British Wallpapers, organized for the industry by the Central Institute of Art and Design, London, May 1945. See p. 207.

on another wild decorative orgy. There is no excuse for this. Manufacturers must be urged to make much more use of the decorative talent which is now almost exclusively confined to printed textiles and book jackets. Year after year the art schools turn out students capable of modern decorative patternwork. Could not the best of them be encouraged to work for the wallpaper industry?

POTTERY

We have seen in the foregoing chapter that a much wider use of good modern decoration for wallpaper is urgently needed. The same is true in the pottery industry where the many new designs required each year relate mainly to decorations applied to the ware rather than to changes in the shape of the ware itself. But in other respects the two industries are not alike. While it is reasonable to ask wallpaper manufacturers to make greater use of the best students of decorative pattern work from art schools, it would be unthinkable to make the same suggestion in the Potteries. As we shall see later, pottery decoration presupposes that the designer has had a special training and is intimately aware of the complicated techniques involved.

Until August 1952 any discussion of pottery design was of limited use, because no decorated ware was designed specifically for the home market. The Government order prohibiting the decoration of pottery for the home trade was enforced in June 1942. So, for just over ten years, the only pottery permitted in British shops consisted of undecorated pieces in colourless glaze or faulty ware rejected as unsuitable for export. During the year 1951 the total value of pottery sold on the home market was about £12 m. Out of this £10 m. was taken up by white ware and £2 m. by decorated 'export rejects'. Until recently the design of ware for the home market occupied very little of the designer's time, and even now export orders receive first attention in many firms, particularly those producing high- and medium-price china and high-price earthenware.

The preoccupation of manufacturers with export markets, though no longer as keen as it was, has had important results from our point of view. In spite of the abnormal circumstances which faced the industry, especially during the years 1945 to 1952, when the export trade was its chief means of support, the majority of firms are now much more prosperous than they were before the war. There is practically no short time for the workers and most firms, particularly at the expensive end of the trade, and those catering for the American market, have a long waiting list of orders.

Perhaps this general prosperity is best indicated by the fact that nearly all the factories I visited during 1950 and again in 1952 were carrying out plans for improving working conditions, enlarging existing buildings and installing new machinery. For instance, the tunnel-ovens, although known in the Potteries more than thirty years ago, have now become accepted by the leading manufacturers as being superior to the traditional bottle-ovens. They are fired by gas or electricity, both of which permit scientific temperature control. As they are non-intermittent their efficiency increases in proportion to the rate of factory output.

The origin of this prosperous activity can be traced to the export trade, which has been fostered and expanded. A separate range of designs is required for each of the foreign markets; what sells in the United States would not be popular in Canada. While the tendency in all but the American market is to look back to the past for inspiration, and well-known historic styles of decoration are interpreted in order to suit the taste of foreign buyers, the designers now working in the factories have not been neglected. On the contrary, designers today have a status higher than before the war. The foreign markets have insisted on a steady flow of new patterns with traditional characteristics; a demand which forced manufacturers to employ competent and knowledgeable artists. These reasons why the status of designers has improved are very significant, as I do not know of a similar result in any other industry which enjoys prosperous trading conditions. Before the war pottery sold more on price than on design. The cut-throat competition which raged amongst all but the most exclusive firms did much harm to the industry and consequently to its designers. Nowadays, manufacturers have become used to supplying markets which are more concerned with quality than with cost. The importance of this from the point of view of design will be dealt with later.

Ten years' trading in foreign markets has inevitably had a great influence on the types of design now available to the home trade, as it will also affect those being planned. It is axiomatic that every manufacturer likes to cater for a large number of different markets with a limited selection of designs. Wherever possible he will attempt to sell the same design in several markets. But in the Potteries this practice has not been easy and, apart from the few firms with famous names and those others which are able to sell ware of good quality at comparatively low prices, the showrooms in most firms display very many different designs, each one intended for a specific level of taste in a specific market.

In connexion with this one generalization is possible. Manufacturers of medium- and low-price earthenware and low-price china told me that, from their experience of pre-war trade, the taste of the home market is broadly similar to that in the

Dominions. Accordingly, when decorated ware was released, many of the designs originally intended for the Dominions were transferred to the home market. And, although it is still too early to give a reliable estimate of sales results, it appears that the judgement of these manufacturers is proving to be correct. The types of design involved, as far as decoration is concerned, are colourful floral sprays and posies in the old-English garden manner; in general, apart from the Indian Tree, the patterns are naturalistic but pretty. On the other hand, and especially in china, the naturalistic motifs are small in scale, dainty, faint in colour, more tightly bunched and accompanied by gold borders and maroon grounds. Fresh spring flowers have taken the place of the rich juicy roses so popular in the United States until early in 1952.

Although patterns such as these were having a popular reception in British shops, most manufacturers agreed that their success was to some extent due to the impulsive reaction of people hitherto restricted to plain white ware. I also found that the first results of these 'Dominion' designs on the home market revealed a noticeable demand for simpler, quieter patterns. Here again one cannot generalize too freely, for while such patterns on earthenware would probably be cheaper than the former types, on china they might well be more expensive. I shall have to account for this later.

Trading with the United States has produced dissimilar and more interesting results. Here I am concerned mainly with firms producing medium- and high-price china and high-price earthenware. The American market, more than any other, is susceptible to quick changes in taste. In 1950 the majority of firms found that rich rose decorations often coupled with fluted shapes, together with highly coloured floral sprays enjoyed a strong and steady demand. Now, these decorations, although still produced, have given place to others which are free and open in treatment, and in style more informal and more modern.

I say 'modern' here because that is the term used in the trade to describe this new style. Its immediate origin can be traced to designs developed by Californian potters, mainly since the war, although elements in the style can be attributed to Swedish and Danish work in the 1930's. After several years of being rather precious and obscure, the style has made a progressive conquest of markets in the United States from the west to the east coast. A large percentage of American manufacturers have since taken it up because it accords convincingly with the current American practice of 'informal living', a practice which implies that meals, preferably of the buffet type, can be taken in the garden, off the kitchen table or on the living-room floor.

The assertion that this social phenomenon has been expressed adequately in artistic terms is questionable. To account for the new style on this basis would require detailed research in the United States. But what is more important and, as far as this country is concerned, more readily understood is the fact that this modern style is far closer to our best contemporary standards in furnishing than anything yet produced in the Potteries. Although designs attributable to this style are now being made in large quantities by manufacturers anxious to retain their hold on the American market, several of them have appeared in British shops. Again, it is too early to gauge the result of this innovation, so I can only repeat the opinions expressed to me in several firms. One earthenware manufacturer told me that he had noticed in the last six months an increase in 'the public's demand for contemporary furniture and furnishings'. On the strength of this his designer had produced an entirely new range of shapes and decorations to suit both the American and British markets. Another manufacturer said that in china dinnerware of medium price a demand has arisen for 'simple, tailored patterns in positive colours'. While a third, who manufactures inexpensive earthenware in very large quantities, said he was convinced that there had been 'no change in home market taste since 1939' and that, if anything, plain pastel coloured and banded ware would be all the 'modern' design required.

The main part of the pottery industry is situated in Stoke-on-Trent and the five adjacent towns of Burslem, Hanley, Tunstall, Longton and Fenton. Outside this concentrated area there are isolated factories in Worcester, Derby, Dorset, Devon and Cornwall. Altogether about 56,000 persons are employed in 300 or more factories. There are approximately five large firms employing more than 2000 persons each, but the bulk of the trade is carried on by the medium size firms of between 200 and 500 employees. Small firms are described as having 100 workers, or less. About 46 per cent of the industry's total output consists of earthenware, 23 per cent of bone china and 31 per cent of tiles, sanitary ware, electric porcelain, etc.

I visited nineteen firms, seventeen of them in the Potteries. Despite the wide variety of products seen, I was concerned only with tea-services and dinnerware, coffee sets, jugs, bowls and vases. The largest of the firms I saw employs 1800 persons on mass-produced inexpensive tableware. It has an output of 40 million pieces a year. Four large firms, with 1700, 1150, 1000 and 900 employees, make medium-price and expensive earthenware and china. Three firms with 1700, 700 and 550 employees specialize in very high quality china, and three firms, two with 450 and one with 270 employees, make medium-price and inexpensive china. The remaining firms with 600, 500, 75, 70 and 50 employees, are between them representative of all qualities of earthenware. To supplement the information obtained I visited two firms making lithographic transfers and one firm which supplies silk-screen transfers and decorative paints. I saw several studio potters, but I found that their methods of production and standards of design were unrelated to the industry and therefore irrelevant to this context.

In order to know what types and standards of design can be expected from British potteries we should take into account their complex manufacturing techniques, their traditions, and their present-day design policies. One manufacturer I met described his colleagues in the Potteries as 'rugged individualists'. Indeed, by far the greater number of firms are family businesses, a description which does not only imply that an owner's grandfather and his own son have been, or still are, engaged in the business, but that many of his workpeople can trace their families' connexions with the firm through several generations. It follows that in many cases there is not only a strong loyalty surrounding a firm, but a rich assemblage of inherent skills. Girls of sixteen can decorate a plate with a skill not mastered by persons twice their age who were not born and brought up within the area of the six towns.

The pottery trade is essentially based on craft techniques, on hand-operated tools rather than machinery. Basic products are of two kinds, bone china and earthenware. Although amongst firms the proportions may differ slightly, bone china contains approximately 25 per cent china clay, 25 per cent Cornish stone and 50 per cent calcined bones. It is termed English bone china to distinguish it from Continental ware, which contains felspar instead of bone. Both types of china become translucent when fired and are sometimes called porcelain. It is well known that earthenware plays a more important part in English than in Central European pottery. The difference between earthenware and china lies in the clay as well as in the firing. Earthenware consists of approximately 20 per cent china clay and 15 per cent Cornish stone. The rest is ball clay and calcined flint. England's famous potters in the eighteenth century, such as Wedgwood and Spode, produced earthenware and only small quantities of jasper, stoneware, etc.[1] And still today, in terms of quantity, about 66 per cent of Wedgwood's and more than half of Copeland's (Spode's) output is earthenware.

Hand-throwing of hollow-ware on the wheel is now very rare in the Potteries. I visited only one factory (Moorcroft's) where all the hollow-ware is thrown by hand. The shapes were originated by the manufacturer nearly forty years ago and, although reproduced today by several craftsmen, they remain practically unchanged. A high standard of design is upheld because the eye of the manufacturer is constantly at work maintaining the indefinable 'rightness' of shape upon which he prides himself. Nevertheless, even in this factory there is a separation between design and execution—a separation which is inevitable in commercial production. The craftsmen employed in throwing and turning—where the thrown body is thinned-out and burnished on the lathe—are not expected to be creative in their work.

[1] Josiah Spode's formula for bone china was not used for commercial production until 1795.

PLATE XLVII

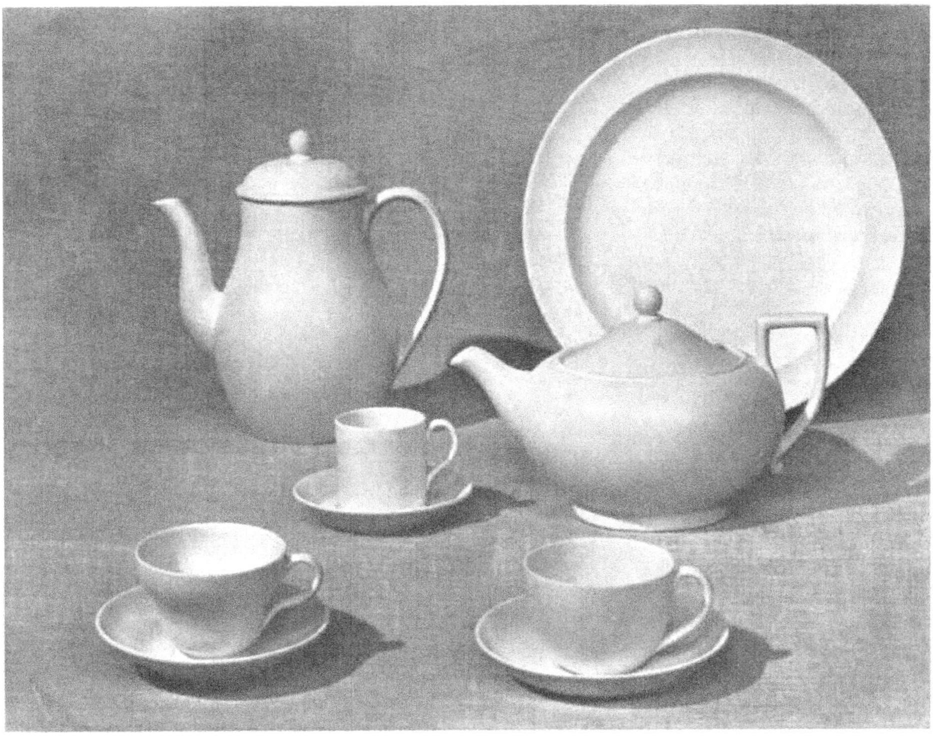

1. These earthenware shapes are perhaps the most perfect examples of 'classic' design still in use. They were designed by the first Josiah Wedgwood in 1770 and are still currently produced by virtually the same methods. With ware of this standard there is no need for changes merely for the sake of having something different. *Made by Josiah Wedgwood & Sons Ltd.*

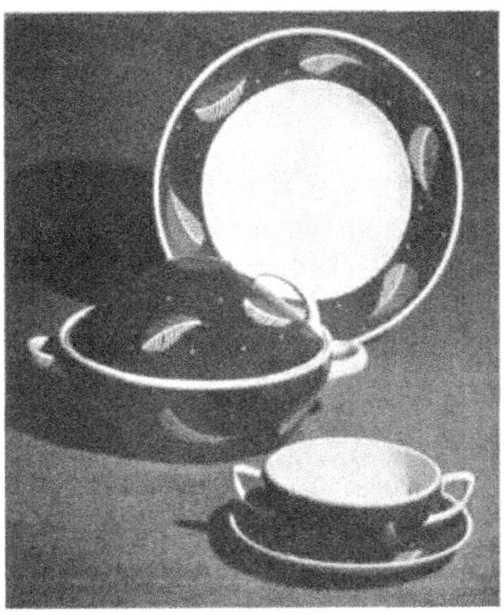

2. No designer in this century has re-valued creatively the basic shapes of pottery as successfully as Susie Cooper. These earthenware shapes are both traditional and modern; the assymetrical handle is the most daring feature. *Designed by Susie Cooper for Susie Cooper Pottery Ltd.*

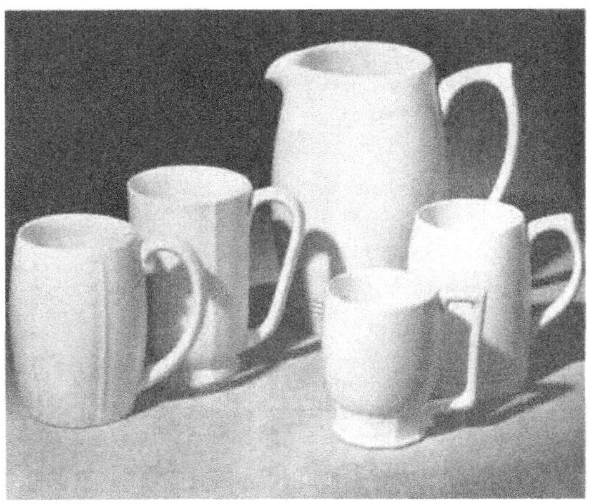

3. The handles in this group serve to emphasize the designer's difficulties when he departs from traditional shapes. While the handles to the jug and right-hand mug are more orthodox, the others appear to be experimental and, when considered in relation to the basically good shapes of the mugs, not equal to the high standards set by this firm. *Designed by Keith Murray for Josiah Wedgwood & Sons Ltd.*

PLATE XLVIII

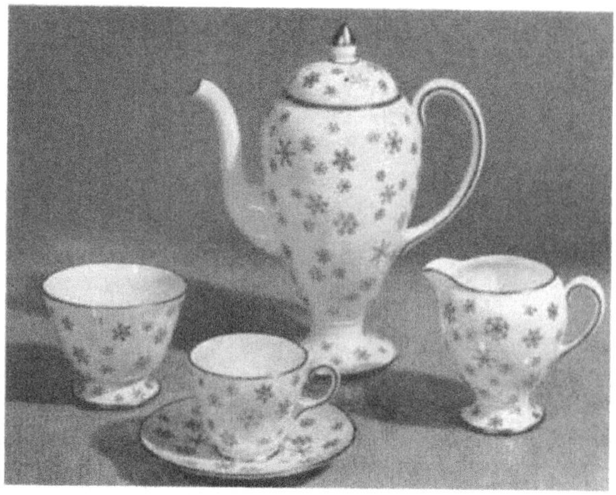

1. The characteristic shapes for bone china are essentially different from earthenware. The patterns, like this one, a lithograph, should also be different, seeking to express the finer body of the ware and its lightness. *Designed by J. F. Goodwin, decoration by Victor Skellern, for Josiah Wedgwood & Sons Ltd.*

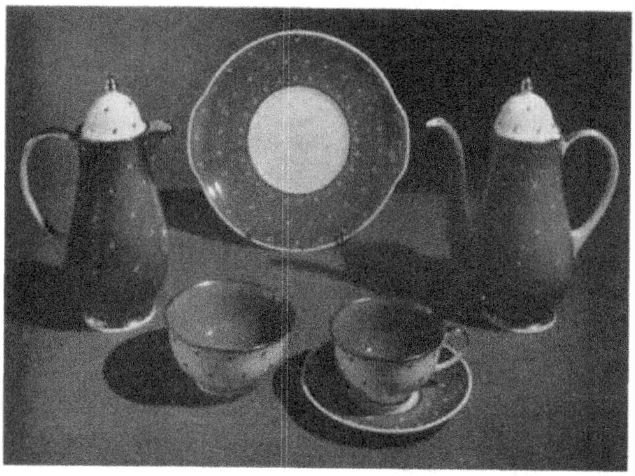

2. The slim elegance of this ware tells you that it is of bone china before you feel it. There was no shape quite like this before: to serve the same purpose it would be difficult to produce another one as good. *Designed by Susie Cooper for Susie Cooper China Ltd.*

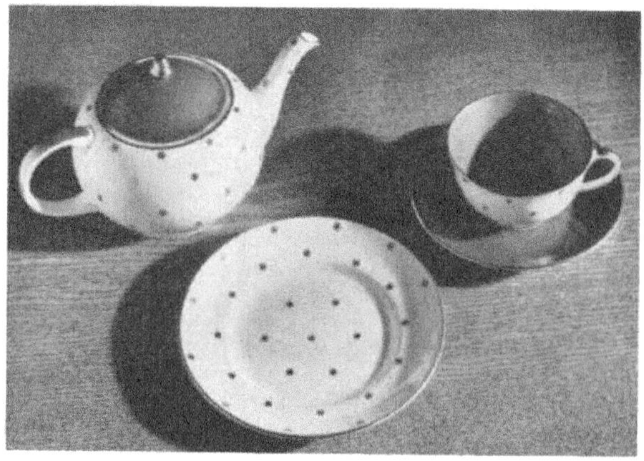

3. Fully rounded contours and sections that taper towards the edge are typical of English bone china. The solid groundlay of colour effectively sets off the pure whiteness of the body. *Designed by Susie Cooper for Susie Cooper China Ltd.*

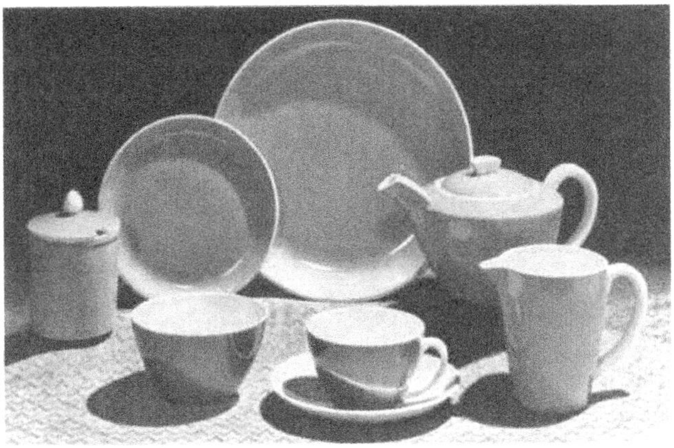

1. New shapes for earthenware, as for china, should be good for they will have to last. The cost of moulds for a complete range makes short runs uneconomic. But small features can be changed and new pieces added, like the coupe shapes in this set. *Designed by John Adams for Carter, Stabler & Adams Ltd.*

2. Robust simplicity in finely moulded forms makes this range of earthenware self-sufficient for the shapes alone. This design is a refinement of shapes traditionally suited to an unchanging purpose. *Designed by Keith Murray for Josiah Wedgwood & Sons Ltd.*

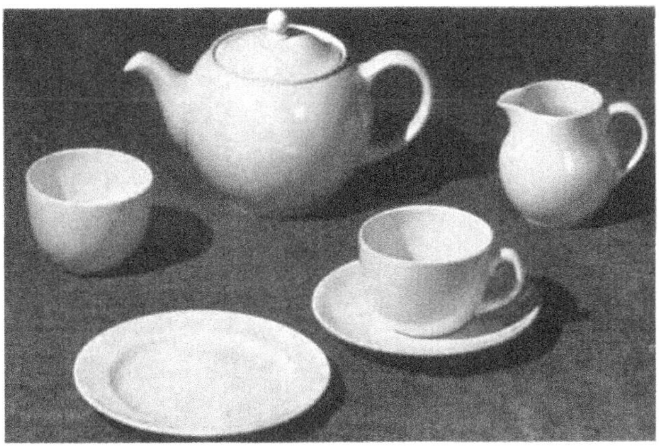

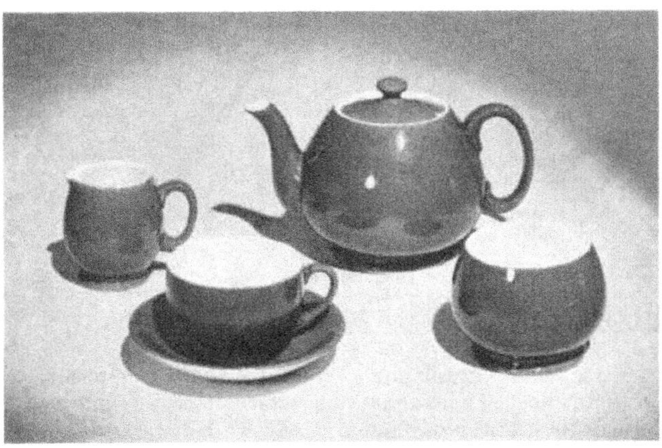

3. These earthenware pieces were hand-thrown on the wheel: a rare practice in the Potteries today. In the distinctive 'powder-blue' glaze they have been popular for more than 40 years, and still look modern and undated. *Designed by William Moorcroft for Moorcroft Ltd.*

PLATE L

1. The widest range of modern decorations is to be found on traditional shapes. Here there is a stronger link with the past for one of the motifs, a border of bells, is similar to a border in the first Josiah's pattern book, 1769–1814. *Designed by Richard Guyatt for Josiah Wedgwood & Sons Ltd. for Liberty's.*

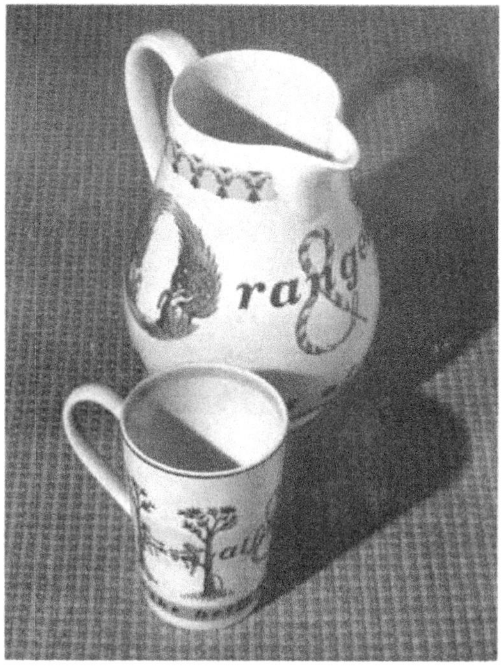

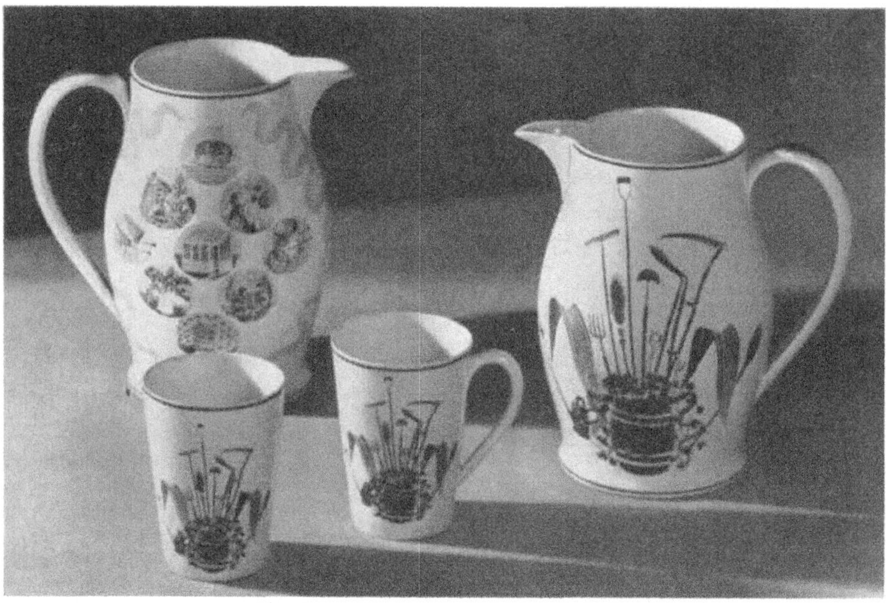

2. The first designer to demonstrate the creative possibilities of modern decoration applied to traditional shapes was the late Eric Ravilious. The shape is late eighteenth-century and the 'Garden Implements' design has made the set the most famous in modern pottery. *Designed by Eric Ravilious for Josiah Wedgewood & Sons Ltd.*

A few firms throw and turn vases, bowls, etc., but the generally accepted method of making all hollow-ware articles in earthenware and china is that of casting the shape in a porous mould. The operations which lead up to the actual casting are highly skilled. In the first place the designer's measured drawing is used to make an accurate full-scale model of the article. In some cases the designer is also the modeller. The plaster of paris mould which is made from the model is not used for reproduction, but serves to make the 'case' mould from which several working moulds are formed. These working moulds are then filled with 'slip' (liquid earthenware or china) and allowed to stand for, say, fifteen minutes. The porous mould absorbs some of the water from the slip and a coating of clay is deposited on the inside of the mould. When a sufficient thickness of clay has been deposited the remaining slip is poured away. While it is drying the clay formation contracts and can be removed when the mould is opened. The article is then ready for the first ('biscuit') firing.

Plates and cups, which undoubtedly form the bulk of any firm's output, are produced by hand-operated machines requiring varying degrees of skill. A plate is made by using a plaster mould which corresponds to the interior shape of the plate. A flat disk of clay is slapped on to the mould which is then revolved by a machine called a 'jigger'. The exterior shape and the thickness of the plate are gauged by a horizontal metal profile, which is lowered on to the revolving clay. Cups are made in the reverse way, on the machine called the 'jolley'. The mould forms the outside of the cup and the metal profile the inside. As the jolley revolves the clay is forced into shape between the mould and the metal profile. Like hollow-ware articles, plates and cups are removed from the mould when dry. Before firing, handles are attached to the cups. Both methods are good examples of mechanized craftwork, and illustrate the essentially traditional nature of contemporary pottery manufacture.

The various methods and temperatures used for firing pottery need not concern us at this point. They are, of course, of vital importance to the designer and to some extent condition his choice of shapes. For instance, when china is fired it becomes almost molten at one stage. The use of thin horizontal or jagged spouts and handles is therefore unwise, because they would tend to fall out of shape. Earthenware, too, has its own peculiarities which must be taken into account by the designer. It follows that there are certain characteristic shapes for earthenware and china articles and they can best be appreciated by studying, for example, the ranges of the older firms, such as Wedgwood and Worcester. The standard techniques of making and firing largely account for the fact that changes of shape occur less frequently than changes of decoration. Cost, too, is an important consideration. I was told that new models and moulds for a complete set of tea, coffee and dinner-ware in bone china would cost at least £1000 and in some cases up to £5000. An additional point is that cups, plates, coffee-pots, etc., are so unchangeable in use that for the sake of obtaining a different design, only small variations of shape can be introduced without impairing their practical value.[1] Thus Wedgwood's is still producing its eighteenth-century shapes, and some of them are so perfect that it would be unjustifiable to abandon them.

Pottery decoration is a different matter. Here the designer has a wide choice of methods for applying decoration to the ware. The decoration of pottery is essentially an applied art, though there are a few important exceptions. Relief ornament, fluting, the Gadroon edge,[2] etc., may be termed organic decoration, because they are cast at the same time and in the same mould as the ware. Another exception is sgraffito decoration, which is incised into the body of the ware before it is fired. In addition, there are two types of decoration which, although applied, become integral with the ware when it is fired. One consists of moulded clay bas-reliefs, which are stuck by hand on to the moistened surface of the ware. The other is called 'tube lining', where slip is squeezed by hand through a small nozzle on to the surface of the ware; alternatively, the slip can be painted on by a brush. In both cases the ware is fired after the decoration has been applied. Lastly, there is ground-laying, where a broad strip of colour is put on the ware by brush. Only the first of these methods, where the decoration is moulded with the ware, and the last, where one colour is applied with a brush, are practised by most firms in the Potteries today. The others are confined to a few firms and are used almost exclusively for more expensive articles.

There are three important types of decoration: hand-painting, printed transfer and lithograph, and they are all applied after the ware has had at least one firing. Free-hand painting is the traditional method of decorating pottery. It affords a wide variety in the choice of patterns, requires small capital outlay and is suitable for short production runs. Free-hand painting is a craft which demands artistic skill, experience and patience, and there are now comparatively few persons who are capable of emulating the work which was done, say, a century ago. Ever since the introduction of ceramic lithography shortly after 1890, the art of hand-painting has been declining. The recent war accelerated this, and few manufacturers can now afford to keep a large staff of creative artists. The most common motifs are landscapes and roses, and many of the artists are proud enough to sign their own work. A somewhat lesser degree of skill is required for gilding, banding, ground-laying and general flower painting. The

[1] It is significant that the majority of the new shapes created by Keith Murray for Wedgwood's (vases, bowls, boxes, etc.) are ornamental.

[2] The Gadroon edge consists of moulded parallel lines or regular indentations, originally derived from silversmithing.

patterns used have been predetermined by the designer, so that the craftsman or, more often, craftswoman is only expected to be technically competent.

The printing of transfers on to pottery is probably the oldest method for reproducing an indefinite number of identical designs.[1] The outline of the designer's pattern is first engraved on a flat copper plate. A type of ink consisting of ceramic oxide and boiled linseed oil is rubbed into the engraved pattern, which is then covered with tissue paper and pressed between two flannel-faced rollers. This action transfers the pattern to the paper, which is then peeled off the copper plate, applied and rubbed on to the ware by hand. When the paper is washed off, the pattern is left firmly printed on the ware. This printed outline is then filled in by hand-enamelling. It is usual for the print to be applied to the ware before glazing in the case of earthenware, which means that the enamel colours have to withstand the temperature of the 'glost' or second firing of over 1000° C. This limits the number of colours which can be used effectively. In china the enamels are applied on top of the glaze, and afterwards fired at a comparatively low temperature (600–800° C.).

Before the war leading manufacturers looked down upon lithography. At that time it was thought that lithography was little more than a cheap substitute for hand-painting, and the general design standard was deplorable. Nowadays, the shortage of skilled labour rather than price is the dominating factor. With foreign markets which insist first on quality (richness of colour) manufacturers have been encouraged to spend more money on lithograph design, with the result that it is at last becoming established as a decorative medium which can exist in its own right. Because of the extensive and complicated apparatus required, lithographic transfers are prepared by specialist firms. Each firm maintains a full-time designer, but for original work many pottery manufacturers prefer to rely on their own design staff. The following is a brief outline of the process involved: the designer's pattern is first copied on to tracing paper, which acts as a key. As a separate lithographic stone, or zinc plate, is required for each colour, the different coloured sections of the pattern must be isolated, and then transferred to the stones or plates. It is usual for this to be done by hand, and great skill is needed to ensure that the designer's pattern is accurately rendered.

As an alternative method photo-lithography may be used. For the general printing trade photo-lithography has been in current use for more than forty years, but it was not successfully applied to ceramic transfer work until 1944. Even now there are only very few firms in the Potteries using the method. Its chief advantage lies in the fact

[1] The process was invented by Sadler and Green of Liverpool, in 1755. Josiah Wedgwood was printing his own designs in 1763.

that the designer's pattern is not redrawn by hand, but photographed, which implies mechanical accuracy. The colours in the pattern are separated individually by colour filters in the camera, a process which is still very much in the experimental stage. Each colour produces a different negative. The negatives are then printed down on to zinc plates, which now correspond to the lithographic stones already described. The image on each stone, or zinc plate, is printed in varnish on to a specially prepared transfer paper. After each printing operation the requisite colour is dusted on to the varnish. Both processes are repeated for each colour, until the complete pattern is built up. It can then be transferred to the ware in one operation.

Lithography is a costly process and can only be justified when a sufficiently large number of lithographs in one pattern is required. Lithographs are sold in editions of 2000, 3000 or 5000 sheets, and unless a manufacturer buys the whole edition he is not safe from competitors. As each sheet contains twenty or more transfers it does not pay to use lithographs if less than 50,000 to 100,000 pieces of ware in one design are going to be produced. The initial expense is considerable, and a manufacturer must be sure that the lithograph is exactly what he requires. In some cases the pattern is re-proofed five or six times before he is satisfied. In addition, there is frequently a delay of one year between the time a lithograph edition is ordered and finally delivered. The cost of a lithograph rises in proportion to the number of colours used. This is because each colour requires a separate stone or zinc plate, which in turn necessitates a complete operation for printing and colouring. For earthenware the average number of colours now used is between six and eight; for china between twelve and fifteen. As an example of the expense entailed, an edition of 5000 sheets of an eight-colour floral spray would cost between £800 and £1000.

Before going on to consider the status and role of the designer, one further type of decoration and one technical point should be mentioned. The silk screen transfer method was first intended for glass decoration, but since the war it has been successfully applied to pottery. The colours in the designer's pattern are first separated, and then individually printed through silk-screens on to a transparent collodion film. The pattern is built up as each enamel colour is printed on the film. When transferred to the ware the enamels stand out in varied relief, so making the result distinctive.[1]

Standards of quality vary, but there is in most firms a fairly high percentage of rejected pieces at some stage in the production process. Flaws may be noticed at an early stage of manufacture or they may only be discovered after the ware has had several firings. The fact that a perfect body with an even glaze is not easy to produce, especially in china, partly explains why it is often economical to apply a decorative

[1] 'Modern Transfers for Pottery', *Design*, August 1954, pp. 21–3.

pattern extensive enough to cover unsightly flaws in the ware. Because of this some sparse patterns, often of modern design, are not popular in the trade.

Except for one case, all the firms I visited employ one or more staff designers. I found that staff designers in the Potteries have in most cases been recruited locally. In the smaller firms and those catering for the cheaper end of the earthenware trade, the designer is called the decorating manager. As his name implies, he is more concerned with new decorative patterns than with new shapes. He is seldom expected to produce original work. For decoration he relies upon the designer of lithographs employed by the printer, or upon old patterns for which copper plates have already been engraved. New shapes are produced by the modeller, who is seldom more than technically competent. The decorating manager is usually a man brought up in the trade, with a long experience of materials and methods, but practically no artistic training. At the other end of the scale, in the medium and large-size firms which are proud of their reputation, the designer, or art director as he is sometimes called, has had a full-time art school training. He is in charge of a studio of three or four assistant designers and he is constantly producing new patterns and, less often, new shapes. Between the two types a variety of designers is to be found. Some have attended evening classes at the art school, some have been released for one day a week, and some have had a full-time training, often at the Royal College of Art. It is noteworthy that design standards improve according to the degree of importance attached by each manufacturer to art education, which to the majority of them means the Burslem School of Art.[1] This has been especially marked since the war. The degree of emphasis now placed on the training of designers is indicated by the fact that the British Pottery Manufacturers' Federation makes an annual award of five scholarships of £150 each to student-designers who are wholly or part-time employed in the industry. The scholarships, which were first awarded in 1950, are intended to be used as travelling grants. In addition, the Federation allocates £250 each year in prizes for original work. Such patronage is almost unknown in other industries.

Free-lance designers have never been widely used by pottery manufacturers. This is chiefly because many of the techniques of making and decorating pottery are complicated, and can only be understood by someone with personal experience. A free-lance would have to gain this experience by working in a pottery; in other words he would, for a short period, become a staff designer. Quite apart from this, manufacturers are satisfied with their own staff and see no reason for employing outside designers as well. Before the war several attempts were made to use free-lance designers for work which demanded a freshness and originality considered to be

[1] For an account of this school's work, see pp. 171–2.

beyond the powers of staff designers. Although many of the attempts were short-lived, it is significant that the most successful ones owed much to the interpretative skill of the staff designer. The ability and the desire of Wedgwood's to call regularly upon free-lance designers are shared by very few firms. Wedgwood's success in this field may be accounted for in two ways. First, the firm has, since the time of the first Josiah Wedgwood, employed distinguished artists in continuous succession. And secondly, it has always been prepared to spend time and money on explaining techniques and encouraging experiments, before expecting a designer to produce suitable work.[1]

Nowadays, apart from Wedgwood's, practically no firms employ free-lance designers. Instead, the staff designer has been given a status and salary much improved on pre-war years. Leading manufacturers expect their designers to visit foreign countries, either alone or accompanied by the sales manager. This practice is intensified at present because of the industry's preoccupation with foreign markets, but at least the designer's experience is not limited to magazine illustrations and agents' reports. In leading firms the head designer gets between £1000 and £1200 a year, and his chief assistant about £500 to £750 a year. Firms of medium size pay head designers and assistants about £750 and £400 a year respectively. Fully-trained art school students start with a salary of £250 to £300 a year. There are, of course, many exceptions to these figures, particularly in cases where the designer or decorating manager is also in charge of production or sales. This is especially true of small firms and those at the cheaper end of the trade.

Faced with this information on the improved status of the designer, it is surprising to find that the number of new designs now required each year is much less than before the war. I was told that the industry as a whole requires about half of the pre-war total number of new designs. This proportion, though only approximate, was confirmed by the information I received from individual manufacturers. Production runs are longer, an average of five years instead of two before the war. In a medium-sized firm a best-selling lithograph design is produced at the rate of 200,000 pieces a year, about four times as many as before the war. But in the largest firm I visited, a million pieces in one design over a period of five years are not exceptional. Engraved transfer prints, made expensive because they are enamelled by hand, do not reach such high figures.

With these facts concerning materials, methods and types of designer in mind, we can now ask the one question of fundamental importance: What will be the standard of pottery design when manufacturers settle down to selling decorated ware on the

[1] In the introduction to his Catalogue of 1787, Josiah Wedgwood wrote: '...no art ever was or can be carried to great perfection with feeble efforts or at small expense.'

home market? The question is worth asking, despite the fact that no conclusive answer can be given. The evidence available may be divided into three main groups. First, the designs sold on the home market between 1930 and 1939 had certain characteristics. Four main types were in current production: all-over engraved prints, period florals and modern florals (in prints and lithographs), and banded ware. All-over prints meant any design of a kind such as the 'Willow Pattern' or 'Blue Italian' or 'College Views', etc. Period florals had conventionalized or naturalistic flowers in sprays or bunches and were 'traditionally English' or Chinese. Modern florals, introduced in the late 1920's, were of two different types: either hand-painted rustic designs, or lithographed (or hand-painted) flowers of a modern commercial character. They were often intermixed with some cubistic elements. Banded ware, introduced about 1930, needs no explanation. The popularity of the four types depended upon the material used (either earthenware or china), and the price. In earthenware, modern florals and cubistic patterns were found in the cheap price range; banded ware dominated the medium, and all-over prints the expensive, price ranges. In both cheap and expensive china the majority of designs were period florals and all-over prints; banded decoration represented about 20 per cent of the total output and there were practically no modern floral or cubistic designs.

Next we need to consider the designs which are now being produced for foreign markets. The first point to note is that all types of decoration which were formerly characteristic of inexpensive earthenware have now been dropped. Banded decoration, too, is practically unknown today. In place of these types naturalistic flower patterns, chiefly roses, dominate all medium price and inexpensive ware for Dominion markets. These patterns owe much to the native skill of the hand-painter, despite the fact that many of them are lithographed. As in fabrics and carpets of floral design, representational accuracy is the only criterion which is acknowledged. Expensive earthenware is printed with all-over patterns and period florals. China is much the same, although a stronger emphasis is placed on traditional conventionalized motifs. As we have seen, manufacturers now find that these 'Dominion' designs are acceptable on the home market. In other industries we have already noted a public reaction against austere or utilitarian designs, and there is little doubt that the long period of plain, white pottery is now having a similar result.[1] Furthermore, manufacturers are not likely to spend money on new designs as long as they can sell all they are already producing.

[1] One good effect of this period of undecorated pottery is that people in this country have become more conscious of the essential difference between earthenware and bone china. The translucency of bone china is an artistic element which has been too often marred by heavy applications of gold and cobalt-blue.

At least from one point of view the situation will not be serious. Vulgar modern florals and cubistic designs are almost forgotten in the industry, and are not likely to be produced again. Before the war, small firms employed semi-skilled labour to paint these designs. Today, that labour is either not available, or requires wages which cannot be proportioned to the value of the work. Although the home market in general cannot afford the quality of ware being produced for foreign markets, there is no excuse for manufacturers to revert to the low design standard of pre-war inexpensive pottery. As we have already seen, one of the most important results of the period of prosperous trading in wealthy markets is that manufacturers have realized that lithography is a decorative medium capable of attaining a high design standard on a mass-production basis.

One further point must be taken into account. The recent change of emphasis in the American market from traditional patterns to those which are more modern in character means that designers now enjoy greater scope for creative experiment. This was clearly demonstrated on my visit in 1952, for in each designer's studio I saw a large number of trial patterns; in five cases I saw the first models of some entirely new shapes. It has already been mentioned that the cost of producing a new shape for a complete range of pieces is very high. With this in mind, the importance of maintaining a hold on the American market can be indicated by the fact that out of the nineteen firms I visited fourteen have started to manufacture the 'coupe' shape. This shape, as far as plates are concerned, is rimless, and therefore unconventional in this country. While it is not new in pottery, it was only rarely seen in the industry until the recent demands of the American market had been felt. As with other designs originally intended for export, the coupe shape can now be seen on the home market, if only in small quantities. Many manufacturers think that it will not be popular because, in dinner plates, it provides no place for the mustard. However, one can only hope that this will be a trivial objection, for the coupe shape offers the designer a much greater freedom for decorative treatment. For technical reasons, it is easier to lay a transfer or lithograph on a coupe shape than on one which has the usual pronounced rim.

Finally, we must recognize that in the pottery industry the word tradition has a real, almost vital, meaning. This is especially true of the older firms where the achievements of the past are preserved as touchstones of artistic excellence. If these achievements encourage a conservative outlook, they do at least safeguard the designer from excesses of novelty. The contents of the Ronald Copeland Art Gallery and the Wedgwood Museum may differ widely from the products in the factory showroom, but they have that spirit which makes the latter so remarkable. The same might be said of Worcester (founded 1751), Minton (1793), and several others, but it should

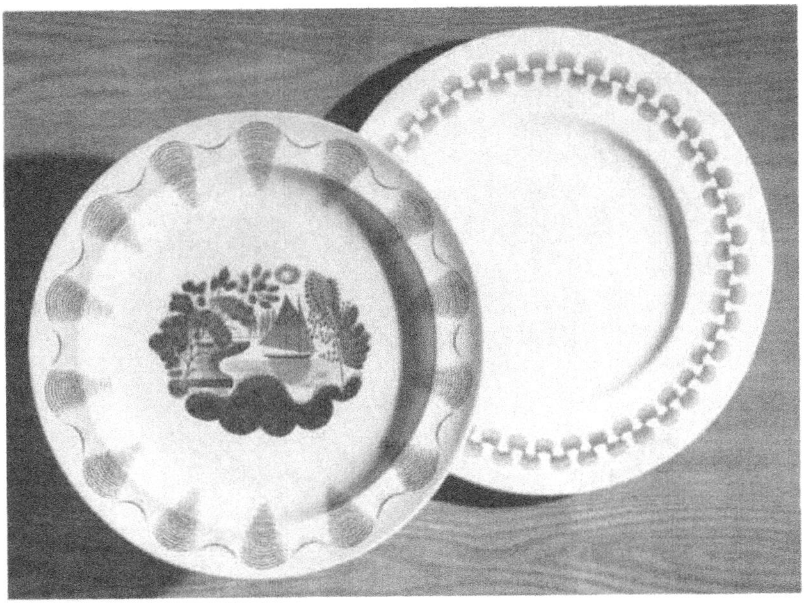

1. Apart from hand-painting, engraved transfer printing is the oldest method of decoration still in use. Two instances of its use on traditional shapes: *left*, on an earthenware coloured body the enamel colours have been applied by hand; *right*, on a china plate the final design has been printed direct in gold. *Designed by Eric Ravilious for Josiah Wedgwood & Sons Ltd.*

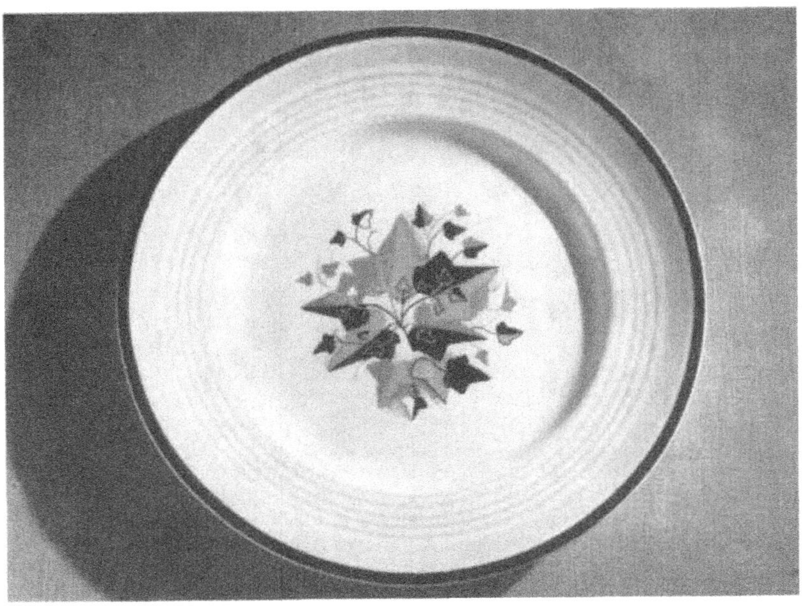

2. The newest method of decoration uses silk-screens to prepare a transfer with the necessary colours that can be printed direct in one operation. Unlike lithography, the solid, flat colours stand out in slight relief. This and the absence of shading render the process unique with peculiarities to be exploited by the designer. *Designed by A. Sayer Smith for Johnson, Matthey & Co. Ltd. for Alfred Meakin Ltd.*

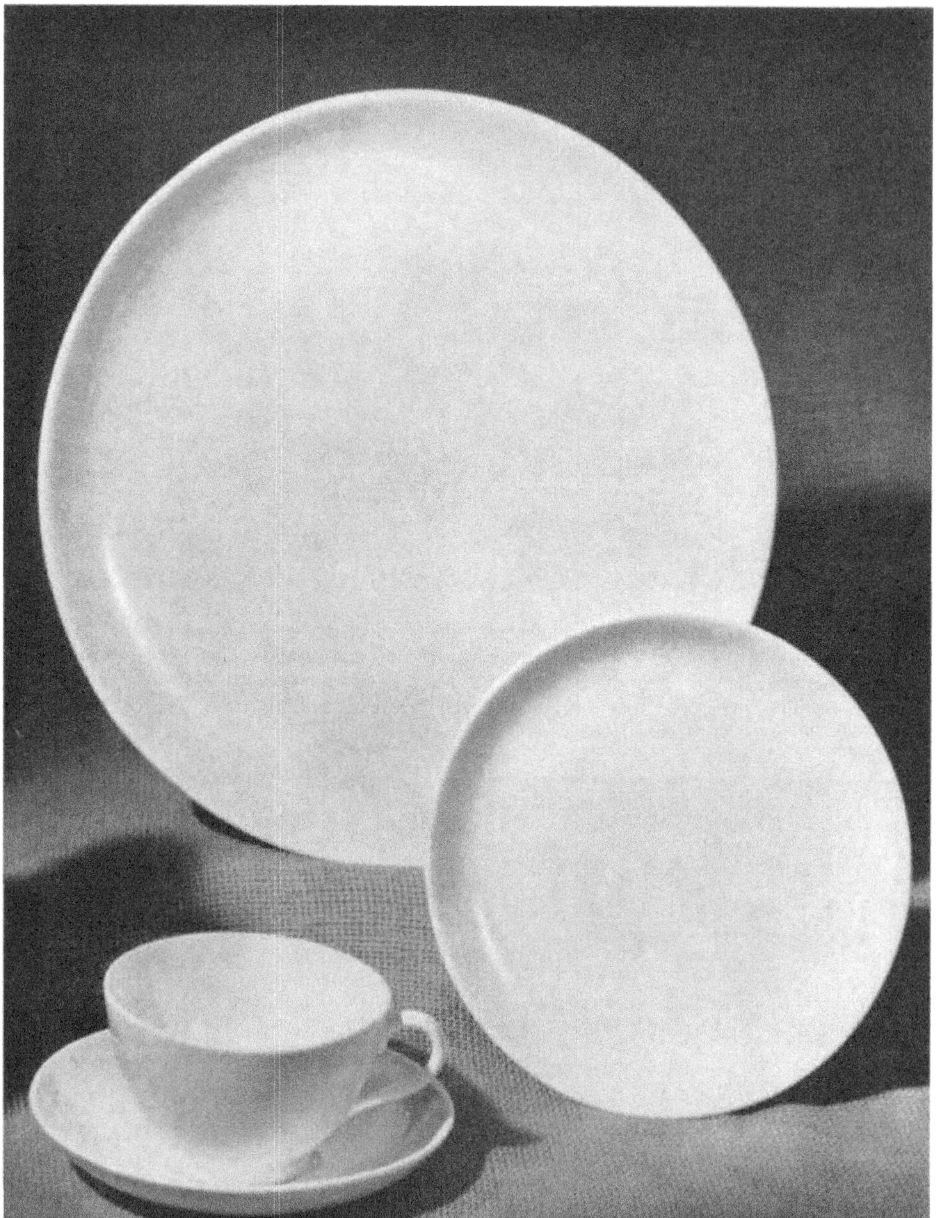

The coupe shape is becoming deservedly popular, for in the hands of a good designer it has that quality of inevitability that makes good, lasting pottery. Although not new it is being made by an increasing number of firms because of the demands of the American market. The shape is rimless and provides for, as well as requires, a new freedom in decoration. It is not easy to make, especially in bone china like this example. *Made by R. H. & S. L. Plant Ltd.*

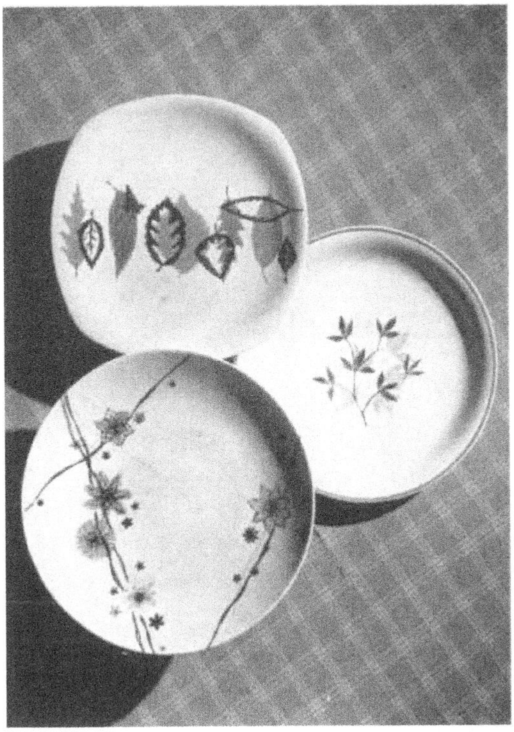

1. The coupe shape gives the designer more freedom to arrange his pattern, and the conventional idea of centre and border, seen surviving in one of these ceramic transfers, can be discarded. The success of the informally arranged pattern depends wholly on the designer's discernment: it could be taken to extremes, like the modernistic patterns of the 1930's. *Designed by Colin Haxby for Johnson, Matthey & Co. Ltd.* (top, *in the range of T. G. Green & Co. Ltd.*).

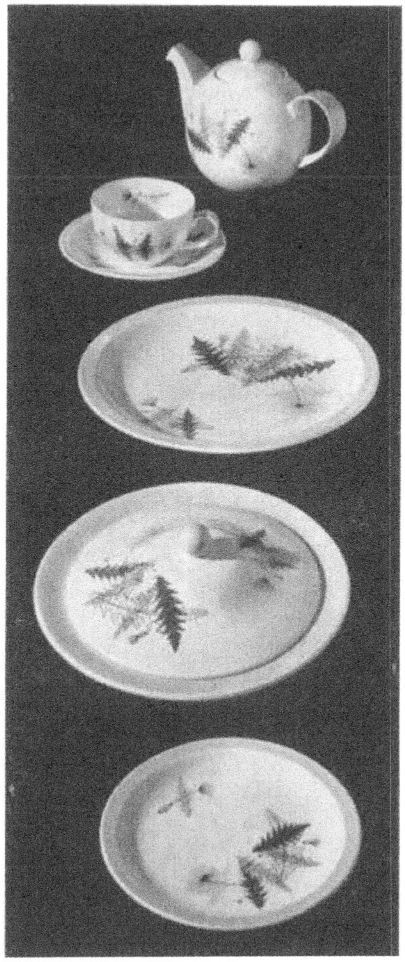

2. A designer's pattern must be capable of reduction or enlargement to suit the different pieces in a range. This silk-screen transfer combines precise drawing with a fluidity that allows it to be used equally well on a variety of shapes. *Designed by Colin Haxby for Johnson, Matthey & Co. Ltd. for Simpsons (Potters) Ltd.*

PLATE LIV

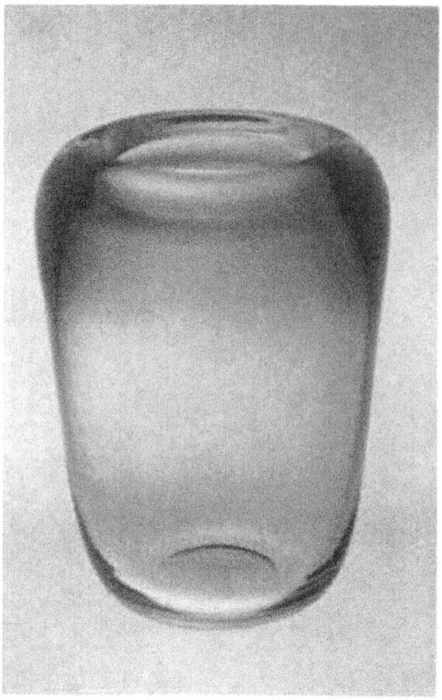

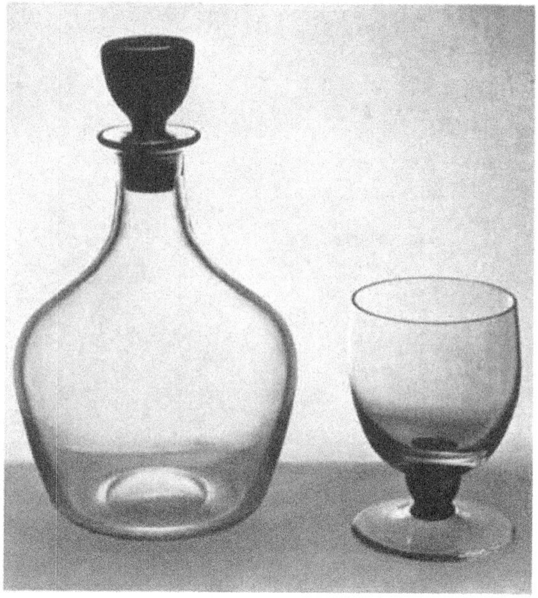

1 and 2. Lead glass is colourless and transparent, two qualities exploited in these designs, which also express the flowing ductility of the metal when it is blown into the initial bubble shape. The lines, neither circular nor straight, are sensuous and yet sharply defined. *Designed by William J. Wilson for James Powell & Sons (Whitefriars) Ltd.*

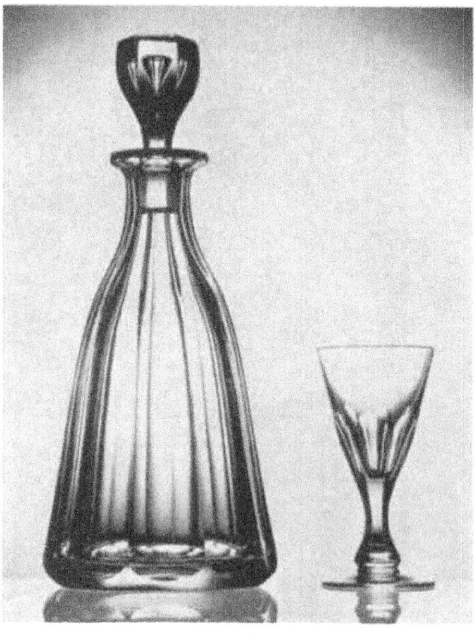

3. The clean, falling lines of this decanter have been achieved by cutting deep grooves in an elongated bubble shape. Wine glasses with a similar instinctive sense of form are rare. *Designed by David Hammond for Thomas Webb & Sons.*

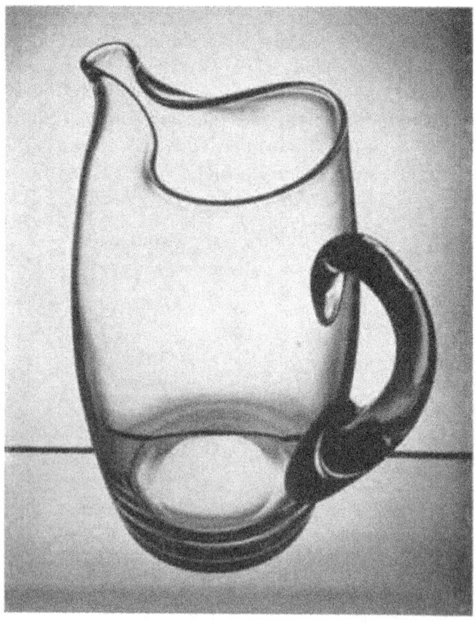

4. For this ice jug it is as if the plasticity of the metal had been suddenly frozen in its movement. Serene formations like this need no decorative embellishment. *Made by The Stourbridge Glass Co. Ltd.*

not be assumed that younger firms are incapable of reaching high design standards. If one accepts the fact that older firms achieve these standards more easily than their younger competitors, then the work of Moorcroft and Susie Cooper[1] is even more outstanding than one would suppose. Many firms which were founded less than fifty years ago, are today very much aware of tradition as it relates to the entire industry. Such is the force of this tradition, that designs which tend to ignore certain fundamental characteristics of shape and decoration are now very rare. A further, and perhaps more important reason why the novel designs of the 1930's have now been dropped is because they were alien to the traditional methods of manufacture.

On the other hand, a tradition is worthless unless it is being revalued continually by designers engaged on contemporary production. This revaluation can be achieved in two ways. Either the designer uses old shapes which are perfectly suited to contemporary needs and redecorates them (some of the shapes first made in the late eighteenth and early nineteenth centuries have been used in this way with excellent results), or he designs both new shapes and new decorations. Neither way offers an easy solution to the problem, but it is undoubtedly more difficult to achieve success by designing the shape as well as the decoration, for when the former is required the designer has to work intuitively in terms of materials and methods which, though inherently limited, have been used to evolve new shapes for well over a century. It is in connexion with these factors, rather than with the styles of old pottery, that the word tradition takes on its full meaning. These materials and methods define the designer's intentions, they limit his choice, but used creatively they will inevitably produce work which is both modern and traditional—modern because it is conceived in the twentieth century, traditional in the sense that it is the outcome of techniques which belong to the past.

GLASS

There are just over 150 firms melting glass in furnaces. Many of these firms have a wide range of products such as glass for lighting, laboratory, optical and other purposes in addition to table-glass. I shall begin with hand-made British crystal glass. There are only eight firms making this type of glass; six of them are situated near Stourbridge, one in Edinburgh and the other near London. Together they represent a small section of the industry, with about 1900 workers. I visited seven of these

[1] Miss Cooper, who was a student at the Burslem School of Art, began to manufacture her own designs in 1933.

firms: the largest of them employs 700 persons, the smallest 100 persons. Table-glass which includes stemware, jugs, decanters, bowls and vases, forms most of the output from these firms. The sale of decorated ware in this country is restricted by a quota.[1] Furthermore, British table-glass, particularly stemware, has a wide market covering many countries and the designer cannot afford to devote too much attention to the artistic preferences of the home market.

British crystal glass, as distinct from 'soda' glass, consists approximately of three parts silica sand, two parts red lead and one part potassium carbonate. The basic requirement is that these elements, when melted and fused together, should constitute a 'metal' which is colourless and transparent. This can seldom be achieved unless a decolouring agent, such as manganese, is introduced to clear away the most usual chemical impurities. The molten glass is maintained at a working temperature of between 1200° and 1400° C., and is contained in a fireclay crucible, or pot, within the furnace. At this temperature it is both viscous and ductile; it is not fluid like molten iron, and cannot be cast in a mould. After it is removed from the crucible, the metal retains these characteristics for a very short time (usually less than a minute) and must be manipulated into shape very quickly. This is done by blowing through a long iron tube so that the 'gob' of glass at the other end takes shape. The initial shape is often achieved by blowing the glass into a mould and then finishing by hand-forging tools. Otherwise it is blown freely, the shape being determined only by the skill with which the glassmaker controls his breath and manipulates his tools. It follows that the characteristic shape of blown glass is that of an elongated bubble. It has neither straight nor circular lines. The further the glassmaker departs from this, the nearer he approaches to the limitations of his material. This is important, for it is perhaps the only criterion in glass design which is universally acknowledged.

At this stage it is possible to decorate a blown shape by applying a long thread or a blob of nearly molten glass. A fusion takes place and the applied glass becomes an organic part of the shape. This is called 'furnace decoration', and is now relatively uncommon in the industry. Otherwise the glass shape is taken to the lehr to be annealed. Here the glass travels along a conveyor belt while it cools, thus exchanging its ductility and viscosity for its generally recognized characteristics of rigidity, brittle-

[1] Since the war the supply of cut glass to the home market has been limited to 15 per cent of the industry's exports to all countries except North America. In relation to North America a rather higher ratio is allowed to enter the home market. Imports of cut glass are banned. There is no restriction on the sale of uncut glassware. All uncut glass for containing food and drink is free of purchase-tax. In September 1952 the purchase-tax on all cut glass was reduced from 100 to 66⅔ per cent. It remained at 100 per cent for all uncut glass not intended for containing food or drink. As from 15 April 1953, all purchase-tax on glass was reduced by 25 per cent.

ness and transparency. The majority of the glass shapes are then transferred to the decorating shop. There are two ways of decorating glass after it has cooled: cutting, and the application of enamels and transfers, a method which, with its hunting scenes and sailing ships, is becoming less popular. The first method is the more important from our point of view, and is far more widely practised than the other. As in the making of glass, all cutting is done by hand. First, the designer's pattern is drawn in red lead paint on to the shape, and then cut (i.e. ground) on a rotating iron wheel fed with water and carborundum. The pattern is given sharper emphasis by the finer cutting possible on a stone wheel. For thin lines and tight corners in the pattern intaglio cutting is necessary. Although a very small wheel is used, intaglio is distinct from copper-wheel engraving, which is now only rarely found on British glass. One general rule may be noted here—the finer the pattern the more expensive becomes the cutting operation, chiefly because of the skill and time required. After it has been cut the glass is polished in acid, which enables the cut portions to regain the clarity of the rest of the surface. It is unusual to find furnace decoration as well as cutting on the same glass vessel.

Before the war, with the exception of three firms, full-time designers were not employed. Instead, one of the directors or the works manager would be responsible for producing new decorations and, more rarely, new shapes. As it was unusual for original designs to be required, the firm's pattern-books (wherein there was a choice of 40,000 to 50,000 patterns) were taken as the basis for all new work. The practice was to design the decoration for a goblet and have a sample cut. This was shown to buyers and, if approved, the foreman cutter would be instructed to reproduce the pattern on all other articles in the set. However, the pattern on the goblet was frequently unrelated to the shapes of the bowl and the jug, and when this happened the foreman was at liberty to arrange the pattern as well as he could. If any blank spaces were left he was expected to fill them in with a few parallel lines or some diamonds. As the foreman cutter was untrained artistically, the majority of pre-war patterns were indefinite and characterless, mere exhibitions of technical virtuosity. Manufacturers appeared to be totally insensible to this, and looked upon the result of the foreman's work as a new design, worthy of a separate entry in the pattern-books. When original work was needed manufacturers called in free-lance designers. Several examples of this work in the 1930's were of a high design standard, but they were not seriously considered from a commercial viewpoint. Instead, they were used for publicity purposes at exhibitions, reproduced in magazines, and displayed in select London stores. Such work was inclined to be more expensive than conventionally cut glass, and that may partly account for the public's apathetic attitude.

I found that since the war there has been a remarkable change in methods of design. In all but one firm a staff designer who has had an art school training is employed; in two cases this training has only been part-time. He is assisted by a draughtsman who is responsible for making copies of his work and keeping the pattern-books up to date. The status of the designer is different in each firm, and in at least three cases I found that he suffered considerable interference from the managing director. These designers do not command high salaries. The approximate scale is from £500 to £800 a year; in no case is as much as £1000 a year paid. Several factors contribute to keep the designer's salary at a low level. The firms are all small, very similar to each other in outlook, range and style of products, and could not possibly absorb a tenth of the potential glass designers available each year. As is to be expected there are no facilities, in individual firms or in the manufacturers' federation, for the designer to travel.

To design glass of a high standard of modern design which can be sold at a reasonable price, requires both technical experience and artistic ingenuity. Like pottery, glass must be designed in relation to methods of manufacture which have not changed for well over a hundred years. The degree of competence required could not be possessed by anyone unconnected with the industry, and manufacturers have realized that original work, to be sold at a competitive price, must be developed inside the factory. A new shape for a vessel may be sketched, but in the initial stages it cannot be interpreted satisfactorily from paper by the glassmaker. The best method is for the designer to instruct the glassmaker verbally when he is actually working the metal. The same applies to cut decoration. Once a sample has been prepared the maker and the cutter will be able to reproduce it without further guidance, though it might take them longer than a piece which is more conventional in shape and decoration. Both craftsmen are paid according to the number of pieces they produce, and they naturally resist the introduction of unfamiliar designs. Finally, for economic reasons the designer must compromise between too much and too little decoration. It is often found that the glass shape has several small flaws (air-bubbles, unfused sand, etc.), which can be removed in the cutting operation. Now it is natural for the cutter to follow a diamond pattern more quickly than a flowing pattern, or one which does not resolve itself into simple geometric terms. If the designer does not want a pattern of diamonds, then he must contrive a pattern which removes the flaws and yet does not require too much of the cutter's time. It is a very great achievement to solve this problem and yet maintain a high standard of design.

When considering the prevailing standard of English cut glass one realizes that so far the new designers have only made a small impression. Apart from a few examples in each firm's showroom, the designs are all traditional and devoid of distinction. And by traditional, one means that they are related to the deeply-cut all-over patterns of the latter half of the nineteenth century. The shapes are unimaginative and the decorations mathematically and not artistically inspired. One cannot blame foreign buyers for this state of affairs, for the design standard in the industry like the methods of manufacture has remained practically unchanged for nearly a century.

Before dealing with possible remedies for this, some conditioning factors must be taken into account. The simplest method of bringing forth the inherent brilliancy in lead glass and at the same time achieving a decorative pattern, is to cut a series of parallel grooves followed by a similar series which crosses the former at right-angles.[1] So the diamond pattern is evolved. It remains popular in the industry for four very important reasons. First, the tradition of cutting glass with flute, mitre, fan and diamond patterns was established early in the eighteenth century. The demand for it has been considerable, though not consistent, ever since. The export markets, with the exception of North America, have always wanted nothing but replicas of old diamond patterns. Secondly, this tradition is served by skilled glass-cutters who comprise a large percentage of those employed in the industry. This is particularly true of Stourbridge, where any reduction in the number of cutters employed would cause immediate hardship. Thirdly, there is a general shortage of glassmakers, mainly because the work is unpleasant, especially in the summer, and because it takes between ten and fifteen years to become a skilled glassmaker. Consequently there is a shortage of men skilled enough to produce shapes which are of a sufficiently high artistic standard to make embellishment by cutting unnecessary. And fourthly, a point already dealt with, it is impossible in a commercial firm to maintain a supply of flawless glass. As the flaws can be removed by cutting there is no need to throw the vessel away, as would be the case if it were to be sold uncut.[2]

It is therefore pointless to condemn the *practice* of cutting glass. Ruskin's well-known remark that 'all cut-glass is barbarous', is unreasonable if he implies that glass-cutting can never achieve a high design standard.[3] The most satisfactory comment on

[1] 'Soda-lime' glass, such as that produced in Sweden, does not possess similar properties and is not subject to so many flaws. It follows that with this type of glass greater emphasis can be given to shapes which need no decoration.

[2] 'This in part explains the paradox of certain articles in crystal, which rely entirely on purity of line and form for their appeal, often being more expensive than pieces smothered in apparently expensive cutting; and also why the latter designs have been popularized.' W.G. Riley, 'Industrial Art and Commercial Design in Glassware', *Pottery Gazette*, April 1946.

[3] 'The peculiar properties of glass are ductility when heated, and transparency when cold...[and]... all work in glass is bad which does not, with a loud voice, proclaim one or other of these qualities. Consequently all cut-glass is barbarous, for the cutting conceals its ductility and confuses it with crystal.' *The Stones of Venice*, Appendix 12 to vol. II.

this remark has been made by the late H.J. Powell. 'A study of old cut glass proves that the object has been to give expression to one of the essential qualities of glass, namely its inherent brilliancy. That this quality should always be coyly hidden under an unbroken surface seems to be a wanton waste of decorative effort. Cutting applied in such a way as to proclaim the brilliancy of glass, without obscuring or cloaking the form given by the glass-blower's breath, helps to illustrate an essential quality of the material, and should no longer be regarded as barbarous. The cutting which most closely corresponds with the spirit of the new law is that of the Romans. Their system of lightly breaking the surface to dispel monotony and obtain flecks of brilliancy is the system from which cutters of modern table-glass should draw their inspiration.'[1] Alternatively, cutting may be performed in a way which emphasizes the ductility in the characteristic bubble shape, maintains the transparency, and yet enhances the brilliancy of the glass. The accompanying illustrations (Pls. LIV–LV) show how this can be achieved by using intelligently the native traditional skill in cutting glass.

Nevertheless, the work shown here is exceptional. The bulk of that produced by Stourbridge firms is richly cut in meaningless designs. All manufacturers have assured me that the geometric patterns of diamonds, stars and parallel flutes are still popular with many foreign buyers because they are 'traditionally English'. Crystal glass, after all, is an expensive product and serves, though not exclusively, that end of the market where design in so many media is judged by outdated standards. However, the outlook is not unpromising. Although in many ways it resembles that in the sterling-silver trade, there is one important difference. In sterling silver it is almost impossible to point to one outstanding design; in cut glass there are several, and the public has at least some choice.

The enthusiasm and initiative required for improving the general standard of design are not characteristic of British glass manufacturers. This was especially marked before the war when modern design was not considered until the influence of Swedish glass had been felt in the home market. The development of the Swedish glass industry is one of the most remarkable chapters in the history of modern industrial art. It was the independent action of a few firms which started the Swedish production of modern glass at reasonable prices. This was mostly 'soda-lime' glass; lead was added for the more expensive pieces. The introduction of the new style in Sweden goes back to Consul Johan Ekman who, in 1913, purchased the small ink-bottle factory of Orrefors. Ekman began to manufacture 'art glass', but this did not advance very far until he called in two Swedish artists. In doing this he followed the suggestion of the Svenska Slöjdföreningen (Swedish Crafts Society), the organization corresponding to the

[1] H.J. Powell, *Glass Making in England*, Cambridge University Press, 1923, p. 152.

German Werkbund and our own Design and Industries Association. The two artists were Simon Gate (appointed in 1916), a portrait painter and illustrator, and Edward Hald (appointed in 1917), also a painter. They were given accommodation near the factory and paid a certain percentage of the total sales of art glass, a minimum salary being fixed. In 1933 Hald was made a director of the works. Neither he nor Gate had done any glass designing before. In spite of that, their careful study of material and processes together with their high talents enabled them to create such outstanding designs, that within a few years the principal European markets were conquered. Orrefors did not limit its production of art glass to 'show pieces' only, but also manufactured a full range of tableware in designs of a high standard.

After a short time other Swedish firms followed Orrefors and shared in its success. It seems that Swedish glass was first introduced into Britain by Sir Ambrose Heal, who had been impressed by the display at the Göteborg Exhibition in 1923. Through the work of Heal's and an increasing number of other shops, Swedish glass became popular in this country and played an important part in making the public aware of modern design in goods for everyday use. At the Swedish Exhibition in London in 1931, there was unanimous enthusiasm for the table-glass shown. It was only then that English manufacturers became aware of the possibilities of modern design. They were too late, however, for the market had been captured by Sweden. Throughout the 1930's attempts were made to compete with Sweden, but only two of them (Keith Murray's designs for Stevens & Williams, and Clyne Farquharson's for John Walsh Walsh) made any impression on the market. The Swedish influence was undoubtedly stimulating while it lasted. Nowadays it is forgotten.[1] It would be unwise to regret this for we do not need to look far for some new influence.

From 1952 onwards manufacturers have been unable to sell sufficiently worthwhile quantities of glass to their established markets in Australia, New Zealand, South Africa and the Argentine. They have had to concentrate more than ever on the Canadian and American markets: since the autumn of 1952 they have been looking for new openings in the home market. In the North American market there is a demand for comparatively inexpensive ware designed so that the shape is more enhanced than obscured by the decoration. This relates to cut glass where the brilliance of the metal has been brought out and allied with the sensuous, flowing shapes that can be achieved by the glassmaker. It seems clear that the design trend in North America is moving towards freer shapes and less conventional decorative patterns. In con-

[1] Nevertheless, the underlying policy at Orrefors, and also at Steuben—the 'art-glass' subsidiary of the large Corning organization in the United States—of using modern design in high quality glass for publicity purposes is now followed more generally in this country. Since the war several manufacturers have used their better designs for illustrated advertisements.

nexion with pottery I pointed to a similar trend and showed how it was likely to find support on the home market. Crystal glass, though it is more of a luxury product and cannot at present compete with pottery on equal terms, is now faced with a similar market, some sections of which are demanding modern designs.

There are at least three firms employing designers capable of setting a new standard —and three out of eight firms would be an encouraging percentage in any industry. These designers appreciate the native tradition of glass-cutting and the economic reasons why it must be maintained. Their task, like that of the pottery designer, is to revalue the tradition from the viewpoint of the twentieth century. It cannot be too strongly emphasized that the necessary creative ability is there, within the industry. How long it is to remain only partly exercised is a matter for the manufacturers alone to decide.

The problems facing the designer of hand- and machine-pressed domestic glassware are fundamentally different from those we have been dealing with. Basically the manufacturing operation is this: a die is cut so that it will form the shape and decorative pattern of the outside of the vessel. A 'gob' of molten glass is then dropped into the die and pressed down by a plunger. After about five seconds the plunger is withdrawn, and the mould opened so that the vessel can be removed. The glass used contains no lead and cannot be polished by acid. Instead, it is 'fire' polished, by playing jets of flame on it. No further process is necessary, and the vessel is allowed to cool in the lehr. It is then packed and distributed.

I visited three firms making tableware: two of them are among the largest mass-producers of machine-made glass in the country. The third firm is much smaller (350 employees) and makes a wide variety of decorative glass, bowls, vases, etc. In addition, I visited one of the two largest producers of ovenware and tableware in heat-resisting glass. The small firm I saw is representative of a large section of the industry, where the standard of design is deplorable. The articles produced are meant to be ornamental, which means that their shapes and decorations are uninfluenced by functional considerations. Few, if any, artistically trained designers are employed and there is apparently no restraint upon the types of pattern used. Design in this section of the industry has been (and to some extent still is) susceptible to modernism in its worst forms. In addition, the characteristic decoration of cut glass, chiefly diamonds, is made use of in the pathetic attempt to imitate the genuine article. The standard of design in the firm I saw is unbelievably bad. The manufacturer told me that his wares reflected the taste of the public. The fact that the public had no chance of seeing anything else of better design in the same price range was a matter of no consequence at all to him. With the present restriction and tax on crystal glass in the home market,

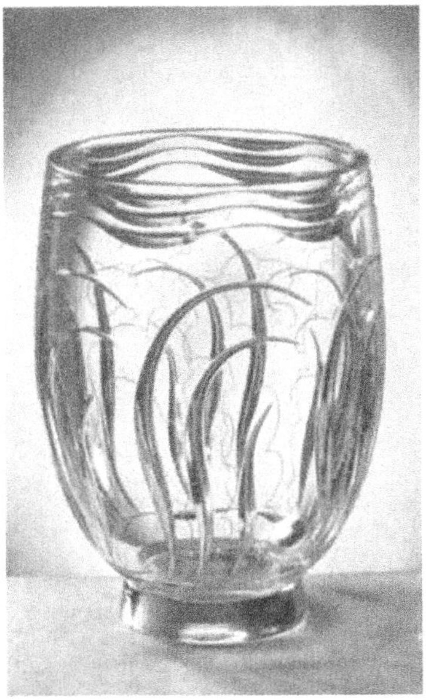

1. Unavoidable flaws occur in crystal glass which must be removed by cutting. Instead of the endless diamond patterns this piece has a relaxed, sensuous design. *Designed by Clyne Farquharson for John Walsh Walsh Ltd.*

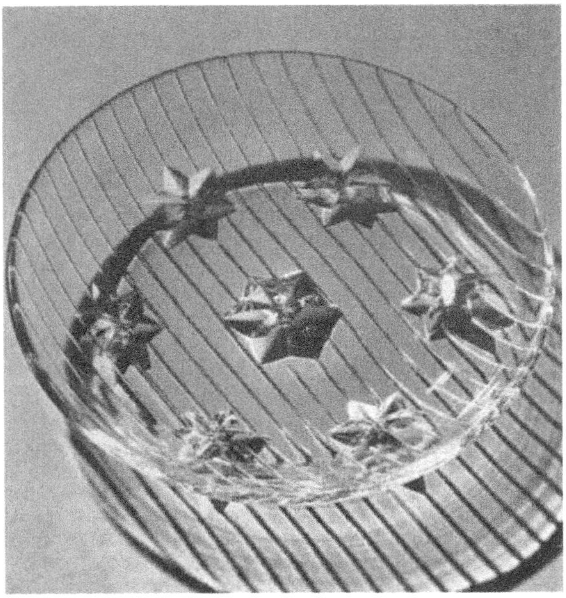

2. The artistic argument for cutting glass is that it brings out the inherent brilliance of the metal. The usual method of achieving this is to rely on harsh, geometric patterns, whereas here the brilliance is enhanced by deep star shapes intersected by thin, lateral lines. *Designed by Irene Stevens for Webb Corbett Ltd.*

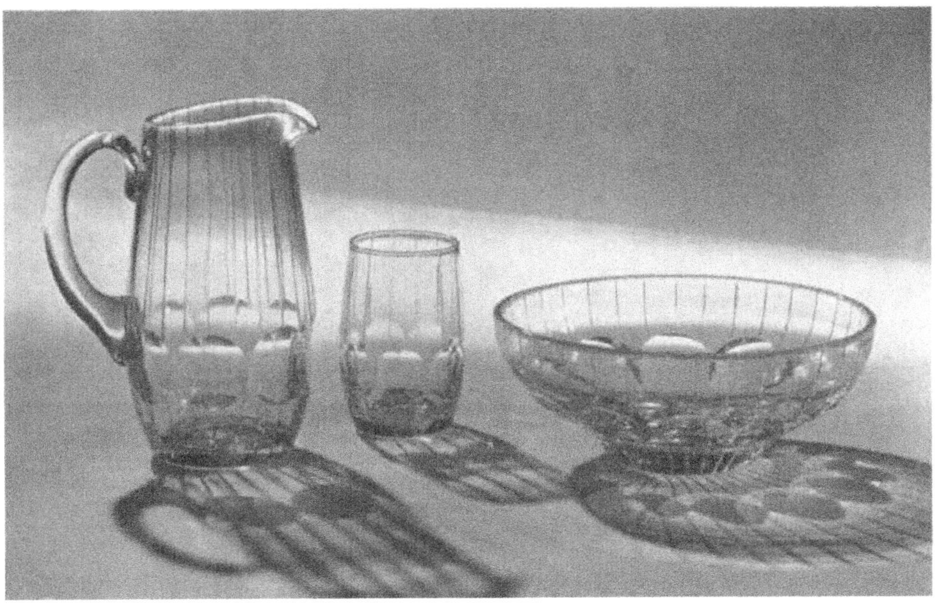

3. The natural flow of the metal when blown has been recaptured here by long slender cuts filling into shallow circles at the widest part. Decoration merges into function for the discs on the tumbler provide a useful grip. It would be difficult to find another example to equal the technique and artistry shown here. *Designed by Irene Stevens for Webb Corbett Ltd.*

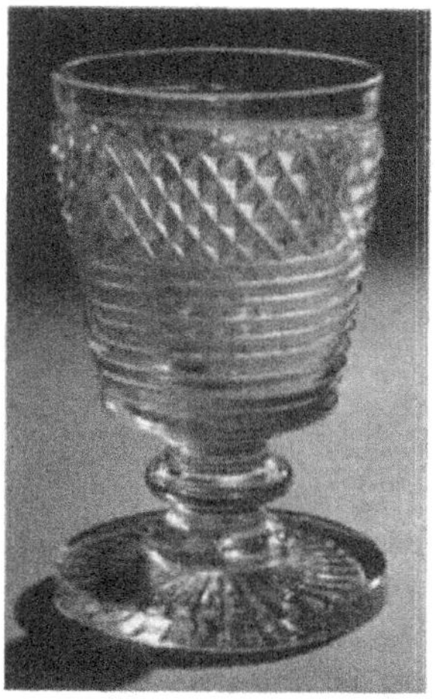

1. Few pressed glass designs are essentially better than this example where the craze for mechanical imitation in one material of the hand-wrought decorations peculiar to another reaches its height.

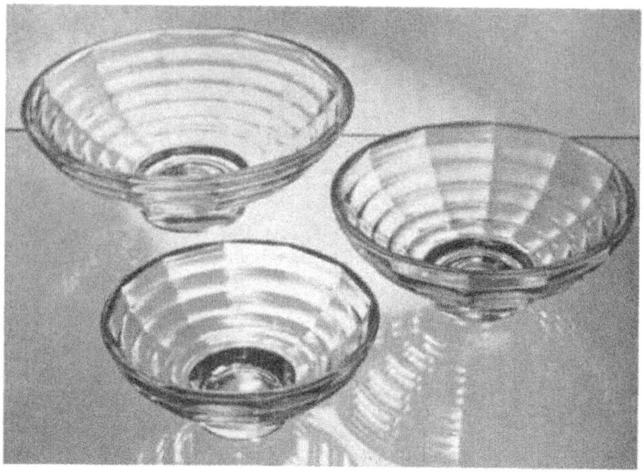

2. The unlicensed artistic excesses of machine-pressed glass can be seen in at least one shop in every town. Creative revaluation has hardly started, yet these bowls indicate that the possibilities are immense. *Designed by R. Y. Goodden for Chance Brothers Ltd.*

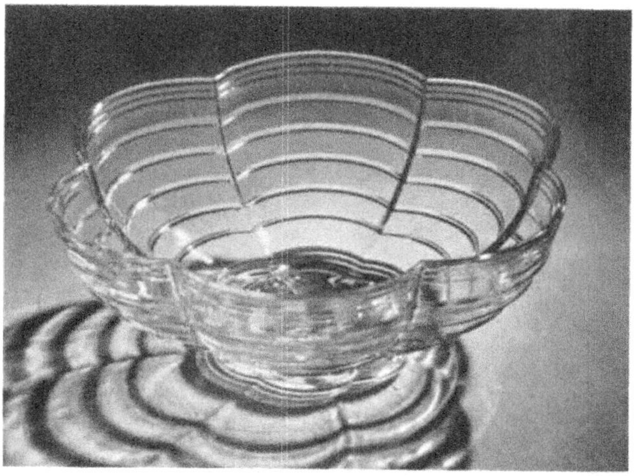

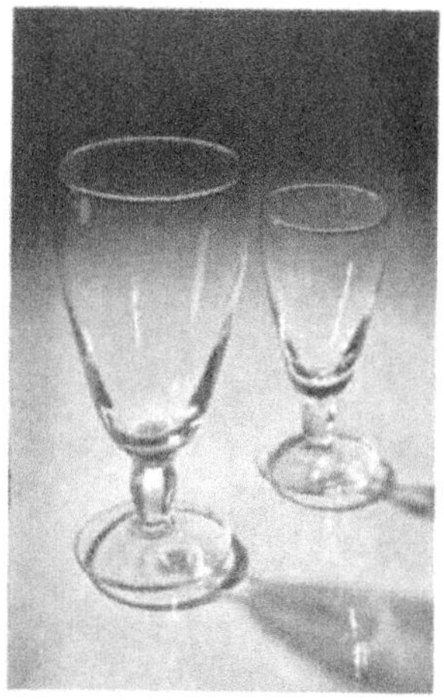

3 and 4. Familiar glasses mass-produced in very large quantities and inexpensive to buy. Machine-pressed glass has its peculiar properties for the designer to explore. The tapering scalloped decoration in the bowl is indigenous and could never be achieved by hand. *Designed by A. H. Williamson for The United Glass Bottle Manufacturers Ltd.*

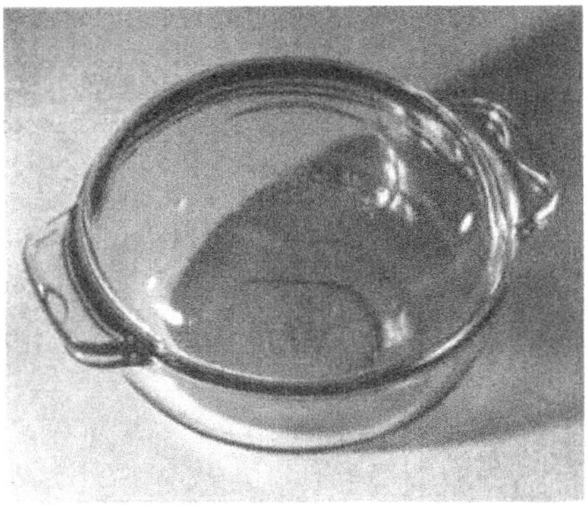

1. Heat-resisting glass poses a special problem. Shapes suited to the kitchen must also be worthy of the table, when taken there direct from the oven: a factor which British firms have only recently considered with any imagination. *Designed (1953) by Milner Gray and Kenneth Lamble of Design Research Unit for James A. Jobling & Co. Ltd.*

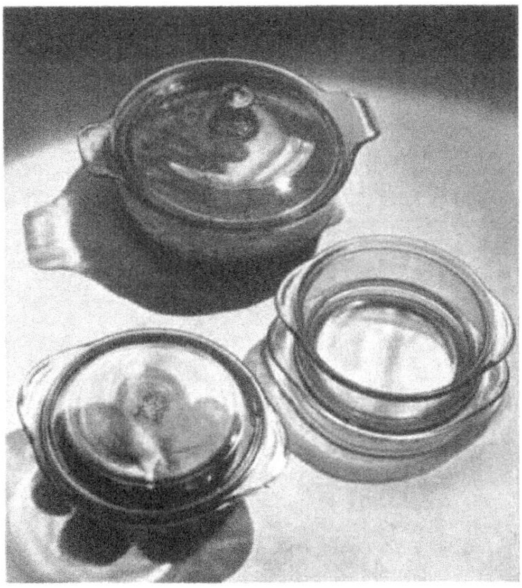

2. German 'Pyrex' dishes which so well exploited the new medium that they have not been changed for 19 years. *Designed (1935) by Wilhelm Wagenfeld for Jenaer Glaswerk Schott & Gen.*

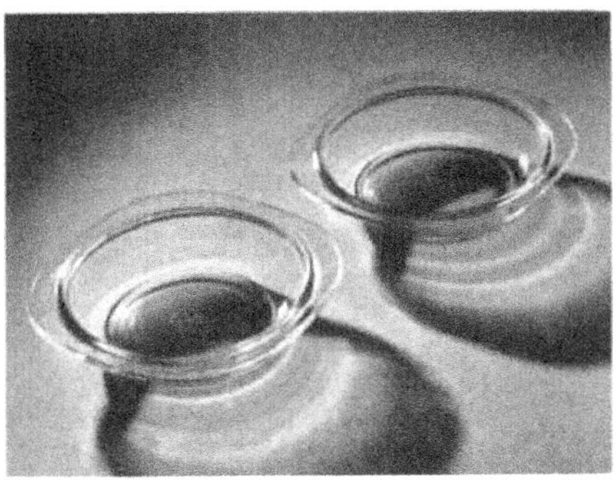

3. *Above.* Two examples from the new range of British 'Pyrex', which ingeniously uses a limited number of shapes to make storage easy and a table-setting attractive. *Designed (1953) by Milner Gray and Kenneth Lamble of Design Research Unit for James A. Jobling & Co. Ltd.*

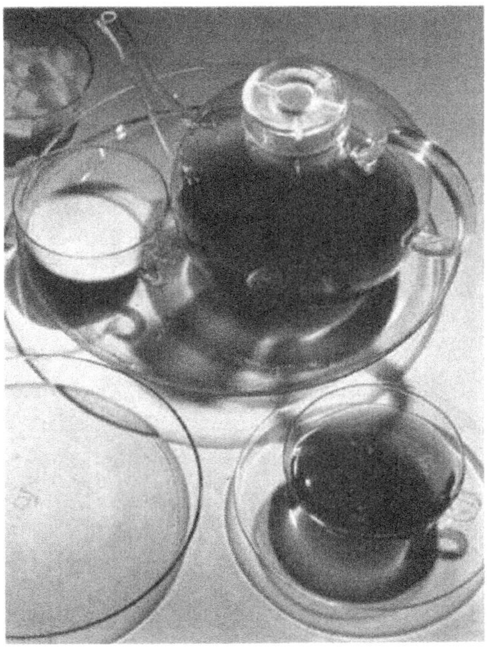

4. *Right.* The tea is strained by a central diffuser and not by the usual perforations in the base of the spout. *Designed (1932) by Wilhelm Wagenfeld for Jenaer Glaswerk Schott & Gen.*

PLATE LVIII

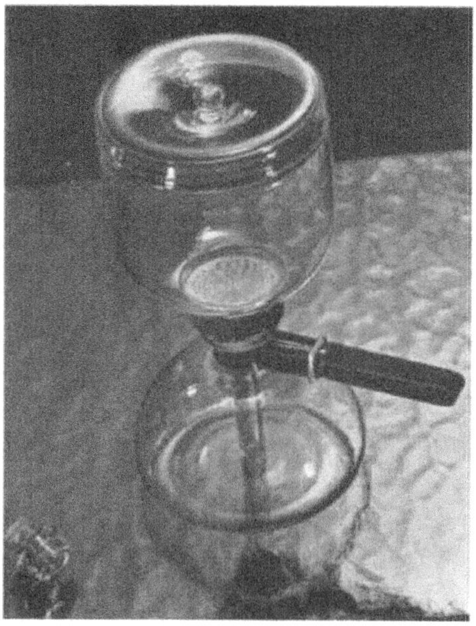

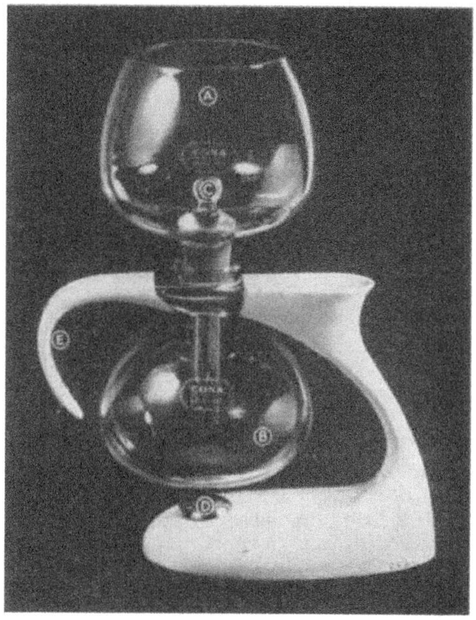

1 and 2. Allowing for the difference in date (*left*, 1934 and *right*, 1949) these percolators show how designers from different cultures react to a similar problem. On the left Wagenfeld, having worked at the Bauhaus, attends strictly to functional requirements and produces strong, full shapes, undecorated. Abram Games, on the right, seizes the opportunity for abstract sculpture and evolves an original relationship between stand and handle. Which is best for the job: the functionalist or the humanist? *Designed* (left) *for Jenaer Glaswerk Schott & Gen.*, (right) *for Cona Ltd.*

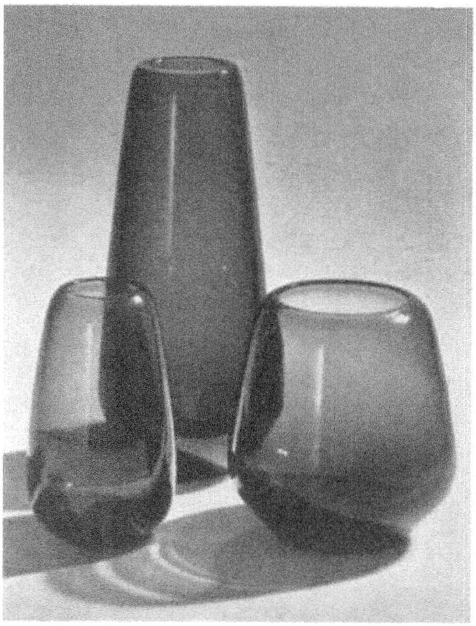

3. These essentially blown glass shapes were worked out with models and produced from measured drawings. *Designed by Wilhelm Wagenfeld for Württembergische Metallwarenfabrik.*

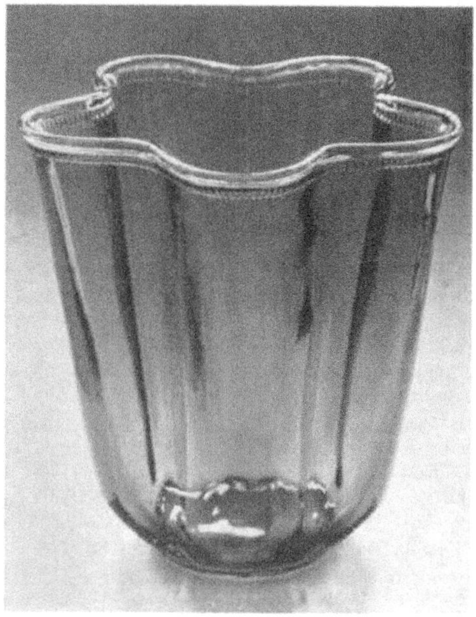

4. Typical pressed glass formation, with integral ribbed decoration to lend interest to the lip. *Designed by Wilhelm Wagenfeld for Vereinigte Lausitzer Glaswerk.*

the manufacturer of pressed glass (which is neither taxed nor restricted) is in a very responsible position. His products are cheap, but at the same time they are the only glass articles which, for decorative purposes, are generally available in this country.

In the two large firms the situation is different. All their products, also inexpensive, are designed for specific utilitarian purposes, and consequently there is a restraint upon artistic excesses. Each design is expected to have a long production run. In one firm I was told that 250,000 of one design had to be sold to pay for the high cost of moulds in fully automatic machines. If one million were sold, the design would be called a best-seller. In the other firm the corresponding figure for best-sellers was between six and eight million. These figures related to such things as tumblers, jugs and fruit-bowls. Each firm employs a staff of draughtsmen to produce these designs. Apart from the many imitations of cut glass, the bulk of the work turned out is free from modernism or other more recent decorative crazes. It is not possible to argue that diamond, star and flute patterns are unsuited to the manufacturing methods of machine-pressed glass. But it was explained to me that the complexity of the mould required for a cut glass pattern caused more rejects than the mould used for a modern pattern. However, this economic fact matters little when the manufacturer feels assured that his public wants imitation cut glass. On the other hand, there are certain shapes which are characteristic of pressed glass, shapes which could not be produced by hand. They offer rich scope for design experiments which might well be undertaken by the larger firms in the industry. Such experiments could only be profitable by employing a staff designer with a studio on the premises.

Apart from its staff of draughtsmen, one of the firms relied for a number of years on an architect. He was paid a retaining fee and produced some outstanding designs. He parted with the firm because he was not prepared to make use of cut-glass patterns in his work. The other firm had employed a consultant designer since the war for a small section of its range. I was told that there had been a recent improvement in the taste of the public and that the simple, straightforward designs of the consultant had been very popular. Both designers had been trained by the firms to appreciate the possibilities and limitations of pressed glass, none of which is as complex as those confronting the designer of hand-made crystal glass.

Ovenware and tableware in heat-resisting glass produced by the firm I visited was satisfactory from a functional viewpoint, but artistically undistinguished. No designer was then employed and new shapes were evolved by a draughtsman. Many of the standard patterns had been in production for more than twenty years. More recently, however, a team of consultant designers was called in and the first results of this collaboration are extremely encouraging.

As long as it is accepted that a high standard of design is required in cooking utensils, it follows that the quality of 'fitness for purpose' alone is not enough. An artistic quality is necessary; a quality at last being appreciated in English oven-glass. In Germany it has been different, and some details concerning the career of one German designer of pressed glass for all domestic purposes, might help to make it clear that the employment of a trained designer is a matter of fundamental importance. In 1930, Dr Erich Schott of the Jenaer Glaswerk engaged a designer who had spent three years specializing in metalwork at the Bauhaus, and was then teaching at the Bauhochschule, Weimar. The designer, Wilhelm Wagenfeld, worked part-time at the factory and produced a wide range of Pyrex glassware. His designs represented a striking advance on previous products, and have been so successful that they are still being used today.[1] Their success is reflected by Wagenfeld's salary, which increased by more than 300 per cent during his service with the firm.

In 1935 Wagenfeld became head staff designer at Vereinigte Lausitzer Glaswerk, which then employed 3500 persons. Instead of improving existing designs and gradually replacing them with new ones of his own, he began by concentrating on new designs in glassware of high quality. His object was to produce pressed glass which would be cheap enough for the workman and good enough for the wealthy classes. With the help of Dr Mey, the chairman of the company, Wagenfeld established a design department with a staff of ten, equipped with studios, modelling rooms and a grinding shop.[2] The necessary co-operation between the department and the already existing scientific laboratory, when developing new projects, was greatly assisted because the departments were placed adjacent to each other. For some time Wagenfeld's new designs met with stiff opposition inside the firm. It was said that the first set of glasses (Oberweimar) would be of interest to a few aesthetes and writers only. In fact this particular set later reached the highest production figure of any set of glasses ever produced. Similar results followed and in the end the high quality department was so successful that its designs affected adversely the sales of the company's dubious products. Since the war one of the company's factories has specialized entirely in high quality products originated in the department. Wagenfeld's salary rose steadily until it was equal to that of the company's directors. After the war he was paid on the basis of a royalty of 1¾ per cent of selling prices. He was also allowed to work privately for other firms, provided they were not in competition with V.L.G. Journeys for purposes of study were paid for by the company.[3]

[1] The 'Pyrex' glassware made in this country has not been designed by Wagenfeld.

[2] This was in 1936, and was perhaps the first of its kind in any firm of this type.

[3] I am extremely indebted to Professor Wagenfeld for allowing me to publish the above facts and figures relating to his career. See also p. 265.

There is no parallel to this state of affairs in British pressed glass factories. As far as hand-made crystal glass firms are concerned, the employment of a staff designer has only become a commonplace since the war. My impression is that the industry as a whole is not concerned with the maintenance of a satisfactory, let alone a high, design standard. The initiative required to undertake an experiment similar to that just described appears to be lacking. The position is most serious in the section of the industry making cheap pressed glass, where the consequences of one bad design can be very far-reaching. In two of the pressed glass firms I visited free-lance designers are only employed occasionally for one or two items. Clearly they have no chance to keep up a continuity in design development, and so make a convincing impression on the market.

LEATHER AND ALLIED GOODS

In every section of the leather trade I found that manufacturers associated form with function, function with form, and called the result design. Their attitude of mind was fostered by nothing more than commonsense allied to a traditional and still living craft. The craft of working leather, unlike hand-weaving, has never died out of commercial practice. The great advances in mechanization during the nineteenth century had little effect on the leather trade. Although in recent years several new machines have been introduced, there are still many operations which can only be performed by hand. Furthermore, most manufacturers and employees still possess a sympathy for, and an understanding of, the material they use. These factors, together with the pride in craftsmanship, or what William Morris called 'the joy in making', are the main reasons why the general standard of design in the trade is maintained at a satisfactory level. In one other respect the trade is remarkable. Many industries and trades have been the subjects of books where the accent has been placed on design; but how many of the manufacturers concerned have read them? As far as the leather trade is concerned, the one recent book on the subject is familiar to most of the manufacturers I met.[1]

I visited eighteen firms in London, Walsall and Birmingham. For simplicity they may be divided into three main sections of the trade: travel goods, handbags, and

[1] John W. Waterer, *Leather in Life, Art and Industry*, Faber and Faber, 1946. This work is a comprehensive account of the leather trade from the earliest times to the present day. The meaning and importance of design are explicit throughout and a detailed analysis is supplied of every process which concerns leather right up to the finished article. There is no doubt that the book has had noticeable influence on the trade.

fancy leather goods. Before each section is considered in detail there is one general point to make. A very small proportion of the travel goods and handbags sold on the home market are made of leather. One reason for this is the high rate of purchase-tax attached to leather articles, apart from briefcases. For example, the price of a 26-inch leather suitcase would today be about £25. The price includes a tax of 42 per cent and the retailer's profit of about 40 per cent.

From as early as the fourteenth century London has made travel goods in leather and today it is the acknowledged centre of the trade. As John Waterer has pointed out there are still a number of firms operating within the City limits.[1] I visited five firms in or near London, and in addition I saw one in Cambridgeshire and one in Birmingham. The largest firm employs 500 workers and specializes in mass-produced inexpensive luggage made entirely without leather. The smallest employs fifty workers and makes all-leather luggage. A firm of average size, say 250 workers, produces a thousand cloth-covered articles a week but only thirty-five leather cases. In nearly every firm the production of fibre-board and synthetic leather luggage is greater than that of real leather. When leather is used as the covering for a suitcase it is seldom more than 2 mm. thick, except at the corners and edges where its function is to add strength. The days when a trunk or a suitcase was made of solid leather are gone. The weight of such an article alone would today make it unsaleable and the cost of producing it would, of course, be enormous. The usual method of constructing a suitcase is to bend thin plywood sheets into the required shape, which is then strengthened with strips of sheet metal. A cheaper method is to use fibre-board for the basic construction and strengthen it with a wire frame. Two recent developments must be noted. One is the use of a moulded plastic shell, which is stronger than plywood because amongst other reasons it has rounded corners and is unjointed. The other is the use of aluminium, probably the strongest and lightest suitcase construction known.

Of whatever material the case may be made, it has to be covered. For leather covering cowhide is invariably used. Before it reaches the manufacturer a hide goes through several processes, each of which will affect the design of the suitcase. After the hide has been tanned it receives an aniline stain, which rarely produces an even overall colour; it also leaves any scars or growth marks on the hide plainly visible. In this country, but not in the United States, the retailers and the public appear to think that a hide which shows such marks is faulty and therefore undesirable. They forget that leather was once the skin of a living creature and subject to the natural wear and tear associated with animal life. Instead of rejecting such leather they should appreciate that the marks on it prove that it is 'top-grain', or the hardest wearing

[1] *Leather in Life, Art and Industry*, pp. 220–1.

portion of the hide and not the second layer, or 'leather-split', which is not so durable. One old manufacturer I spoke to remembered that, when he was young, it had still been a source of pride to a craftsman if he could display the pronounced growth marks of the beast's shoulders on the panel of a suitcase. To satisfy the public, leather dressers use a dope or pigment which imposes an even, shiny texture on the aniline stain and so robs the leather of its characteristic depth of colour and richness.

The use of pigment will obscure some growth marks and leave the leather with a plain, characterless surface, but it will not mask every so-called fault. If a suitcase is to be made out of such leather and yet not reveal any marks, then a selection from five or six hides may have to be made. This is, of course, extremely costly and is limited to expensive articles. There is a cheaper method known as graining or embossing, which is the application under heat of an engraved steel plate which forms a regular pattern, in relief, on the surface of the hide. The most common pattern used is called the 'oat grain'. It has a distinctly leathery character and is the least objectionable of the patterns available. Another frequently used pattern is the 'pigskin grain' which, when used on cowhide, creates a very misleading impression; while other patterns either distort or magnify the natural grain of the hide (one is called 'elephant grain'). Apart from covering growth marks, this process provides the leather dresser with an opportunity of passing off second-quality hides with little risk of being detected, at any rate by the retailer and his customer. There are, however, certain occasions when an embossed leather is desirable. In the manufacture of some suitcases or portmanteaux it is necessary to stitch the leather on the inside. After stitching, the case is turned right side out and, if the leather has not been softened by embossing, it will crease badly. Leather may be softened by rubbing the flesh side with a cork board, but embossing it with a pattern is much more common and less expensive.

For making travel goods there are several alternatives to leather, such as plastic and woven cloths available in wide varieties. Some of them do not imitate leather and their choice of markets is therefore considerably reduced. The majority imitate leather to a greater or lesser degree, although in travel goods there are few really objectionable examples. Leathercloth—the term itself adequately suggests its intention—with a 'pebble grain' or an 'oat grain', is found in all but the more expensive price groups. Such plastic cloths are frequently made to imitate the costliest of leathers, a dark shiny crocodile being very popular. Synthetic leathers have several useful qualities. They can be obtained in unlimited lengths with no variation in colour or texture so that, unlike leather, little is wasted in the cutting-up operation. Comparative cheapness added to variety and evenness of colour are considerable assets to the manufacturer who wishes to keep his products in fashion. While this is chiefly relevant where handbags are concerned, there are also several manufacturers of travel goods who follow fashionable colours with close attention. There is little doubt that leather is more durable than leathercloth, but if the suitcase or hatbox is only wanted for one season, it seems pointless to pay for the pleasure of possessing a leather article in a fashionable colour.

The introduction of the zip-fastener in the early 1930's was revolutionary in this section of the trade. Many firms made use of it and cumbersome locks and heavy metal fastenings were discarded. The merits of the 'zip' were soon appreciated and in a very short time it had worked its way round the three sides of many suitcases. It is light and seemingly efficient and for many years has held an unrivalled position in the trade. But for some firms it is no longer so popular. Its one major fault, of being irreparable if once strained, has induced manufacturers to replace it with locks of very light construction and of increased efficiency and ease of movement. This quest for lightness of weight is connected with air travel. More than twenty million people in the world make an aeroplane journey every year. The overall weight of their luggage is limited, depending on the length of the flight, and in no case does it exceed 90 lb. A suitcase must therefore be as light as possible, but it must also be strong, for air travel does not necessarily mean that luggage will be handled with special care. Suitcases for this purpose weigh between 5 and 9 lb., according to the permanent fittings installed. It makes no difference to the weight if leather or another material is used; the conditioning factors are the internal construction and the fittings. Luggage suitable for air travel constitutes a challenge which designers in this country have not yet fully appreciated.

From our point of view the suitcase fitted with coat-hangers, special devices for folding dresses and suits, lined pouches for toilet articles, etc., is of special interest. It is the logical development of the wardrobe-trunk which originated in the United States. One manufacturer I met said that he had introduced the fitted suitcase to the British market in 1936. Its popularity had increased steadily since that date and many improvements had been incorporated. It had great appeal for the person who wished to pack for a journey at the last moment for he could, with the minimum of meditation, pack all he required in two minutes. His suits folded automatically into a very small area. His shoes found themselves a place well clear of his shirts. For the achievement of this excellent design no one person was responsible. It was the work of the management of the firm during the holidays of the last fifteen years.

In this country it is quite common for the purchaser to buy his luggage in a suite of three pieces, whereas in the United States he would require six or seven pieces. This is an important factor for the designer, who must evolve a pattern and arrangement of the several parts of a suitcase so that they will agree proportionally with the rest of the suite. Two examples of luggage are illustrated here and it will be noticed that their designers have had, apart from each separate item, the ensemble in mind. Close attention to details such as locks and linings is important. I met only three manufacturers who had taken the trouble to have locks made specially to suit a particular design. To do this a manufacturer has to pay the lock-maker the initial tooling costs, about £150, and to undertake to purchase a very large number of the locks. It is thus not surprising that most manufacturers prefer to use standard locks for their new designs. There are very few alternatives to the 'watered moiré' artificial silk linings. They are not unattractive and would be ignored in this context were it not for the fact that one manufacturer had commissioned a well-known textile designer to create some new patterns for him. The result greatly enhanced the appearance of his suitcases.

It is usual for firms to produce about five new lines every year. The designs for them are obtained in several ways. Two of the seven firms I visited do not employ designers. Of the other five, one manufacturer is himself a designer, while the remainder all employ consultants. Two of these firms are outside London and their object in employing consultant designers is to ensure that their products possess the required metropolitan flavour. In one of them the designer was given a complete suite of luggage to design and was able to carry his ideas right through to the last stage of production. The result was very satisfactory from the artistic and from the commercial points of view. The consultants to the other three firms are all retained for the criticisms and suggestions they make in connexion with the firms' products. In no case are they asked to take charge of the design in its early stages. While the results obtained are not up to the highest standard, they could not in any way be called unsatisfactory. To understand this it is important to remember the innate sense of fitness which exists in the minds of most manufacturers and craftsmen. This characteristic quality in the trade is the strongest possible insurance against a low standard of design. The consultant designers are also expected to advise on colour, although the most popular colour for a leather suitcase has always been 'London brown'. I know of only one case where royalties are paid to the designer on an increase in the firm's sales. The majority of manufacturers consider that the responsibility for a design rests with several members of their staffs, including themselves, and they see no reason why the designer should be singled out for special reward. They are inclined, however, to think differently if the design happens to be a failure. The retaining fee paid to one of the consultants is £200 a year. Both free-lance and staff designers are comparatively rare in this section of the trade. Of the former there are no more than six in England, while the latter become confused with the pattern-cutter or the managing director.

In John Waterer's book there is an excellent account of the recent history of the British handbag trade.[1] The following is a brief sketch of the outstanding points supplemented by material obtained from five manufacturers. The handbag as we know it today was first made in the early years of this century. Handbag manufacture rapidly became a specialized craft and took up its allotted position as a follower of fashion. By 1914, Offenbach-am-Main in Germany was the largest centre of production, taking its designs from Paris and Vienna. The British trade began to gain ground during the First World War, but it was not until the late 1930's that it achieved results comparable with those of its Continental rivals. There were no trained designers in this country and no steps were taken to remedy this deficiency in the art schools. Two possibilities were open to the British manufacturer. He could, as in one case which I investigated, employ as many as six German designers at one time to provide him with a yearly output of 400 different handbags. Another firm employed an Austrian consultant designer who came over twice a year. The alternative method was to purchase handbags in Paris or Vienna and produce copies of them for sale in Britain. This method, augmented by magazines such as *Schuh- und Ledermode* from Vienna or *Handbags Illustrated* from New York, still provides the majority of designs in the British trade. The case of one firm producing leather handbags of medium quality is typical of many others. Its designer, a man of considerable talent, is employed for the purpose of toning down the vagaries of Continental fashion. The firm's best-seller (1000 sold at £4 each) was a commodious, solid-looking bag with little character.

I visited three more firms where staff designers are employed. One of them makes cheap leather and plastic bags at the rate of five new designs each week. The designer is untrained artistically, but possesses an intimate knowledge of handbag construction. His work has many utilitarian virtues but is totally uninspired. It was evident that he had placed too much faith in magazine illustrations, and not paid enough attention to the more immediate signs of fashion—the season's new millinery. In the other two firms the designers were able to anticipate a new vogue. Each of their products had the mark of being first-hand. In addition both are interested in new materials. One specializes in a pre-formed plywood frame for a small portmanteau, for which the clasps have been specially designed. The other, working in many different leathers such as calf, pigskin, crocodile and ostrich, has developed his own technique of construction without the aid of metal frames. His best-selling design is a handsome, top-opening bag in calf, of which 700 have been sold at £9 each. In this class of bag the most popular colours are black, nigger, navy and tan. The remaining firm to be con-

[1] *Leather in Life, Art and Industry*, pp. 224–6.

PLATE LIX

1. It is difficult to find a badly designed leather article. Manufacturers in general produce good-looking useful shapes instinctively so that, unlike most new products in other trades, a distinguished design like this is not exceptional. *Made by Papworth Industries.*

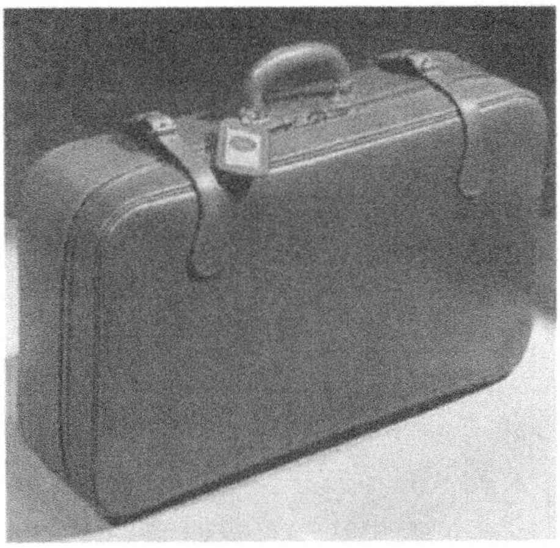

2. Although traditional skills are still active in the trade, manufacturers are keen to experiment with new materials, particularly where greater strength and lightness are to be gained. Here the hide covers a moulded plastic shell. *Designed by John W. Waterer for S. Clarke & Co. Ltd.*

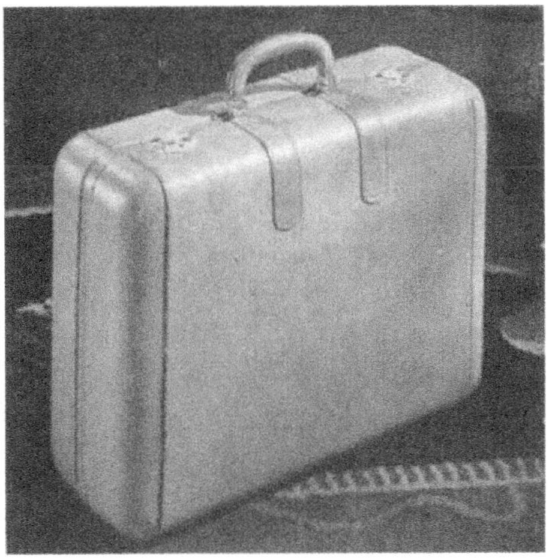

3. Different cases produced as a suite have a growing appeal, particularly on the American market. This is the most adventurous example to be produced since the war: special locks and carrying handles are carefully considered features of the design. *Designed by Scott-Ashford Associates Ltd. for Papworth Industries.*

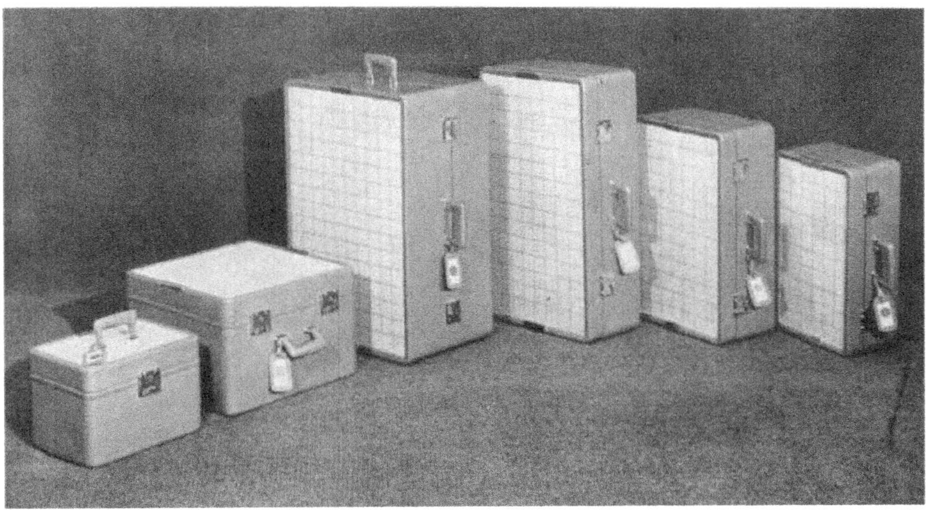

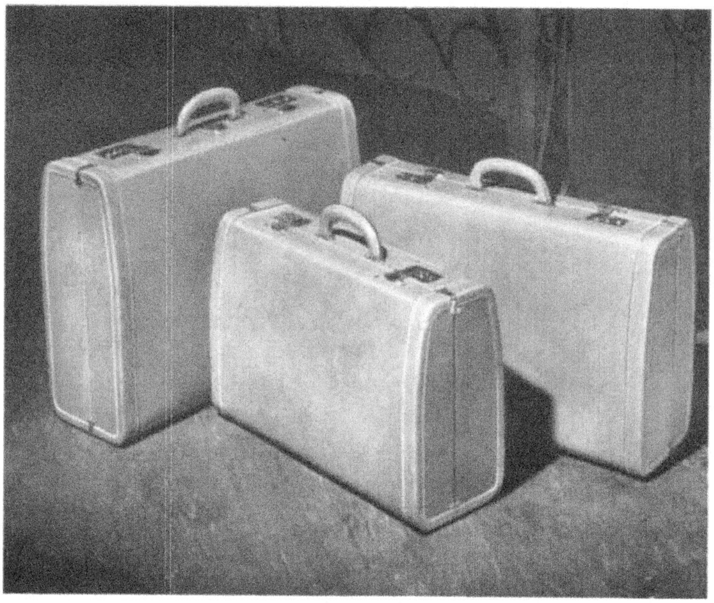

1. The weight restrictions for air travel provide a challenge for the trade. These leather covered cases have soft sides to save weight and a wooden frame that curves towards the forward edge for strength and ease of carrying. *Made by Parker, Wakeling & Co. Ltd.*

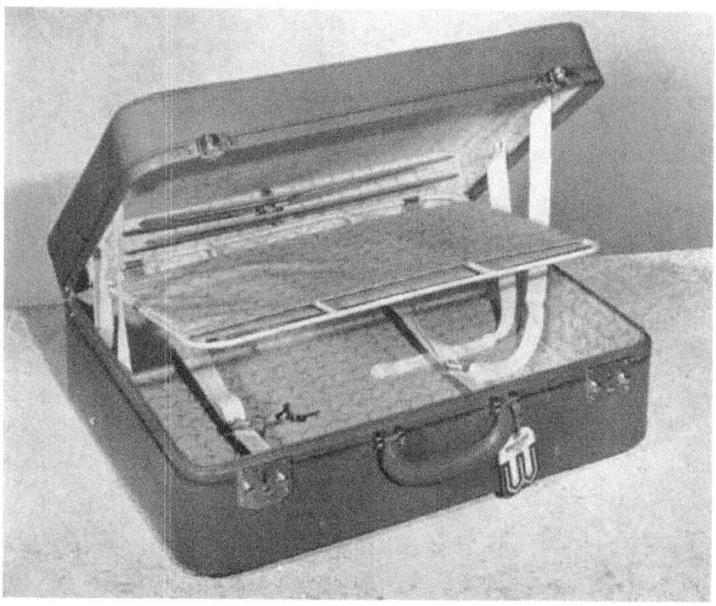

2. There are many ingenious solutions to the fitted suitcase problem, though few are as simple and well detailed as this, where two suits can be folded in the lid. *Designed by John W. Waterer (lining by Enid Marx) for S. Clarke & Co. Ltd.*

PLATE LXI

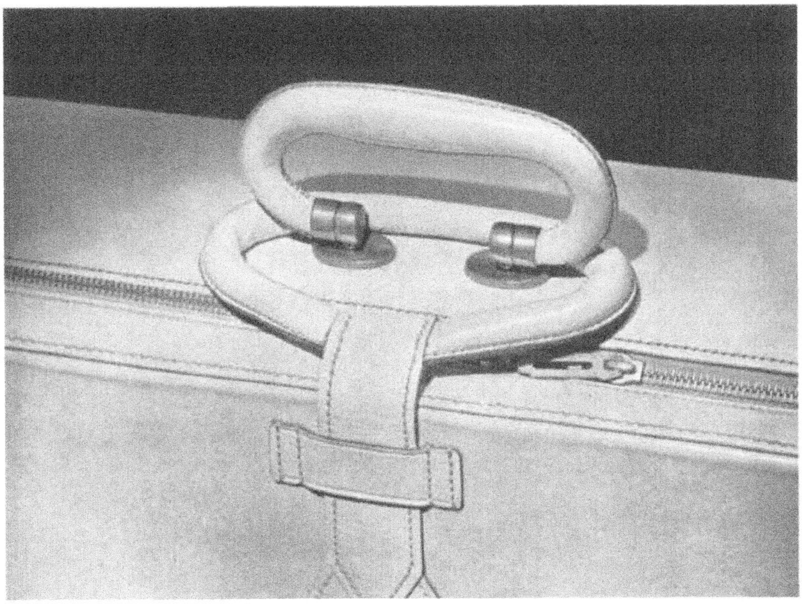

1. Leather itself is an immediately attractive material when undisguised by false grains and pigments. The fittings that go with it are equally important. This light alloy handle with leather covering would be a pleasure to use. *Designed by John W. Waterer for S. Clarke & Co. Ltd.*

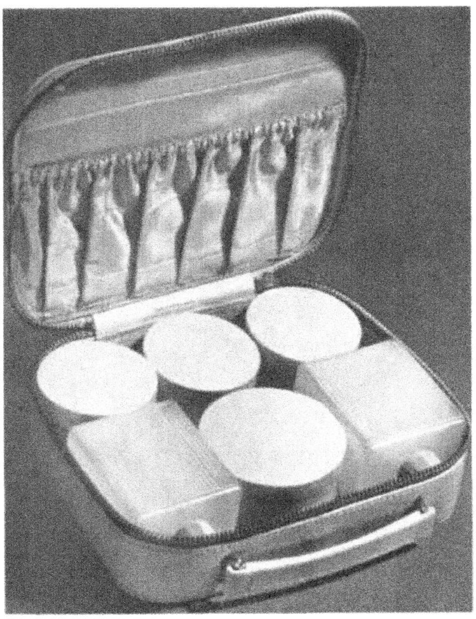

2. The zip-fastener has modified the shape of many leather articles for it dispenses with relatively heavier clasps and locks. For this 'beauty-case' the plastic containers were specially designed to fit together without straps. *Designed by T. C. Maylor for W. A. Maylor Ltd.*

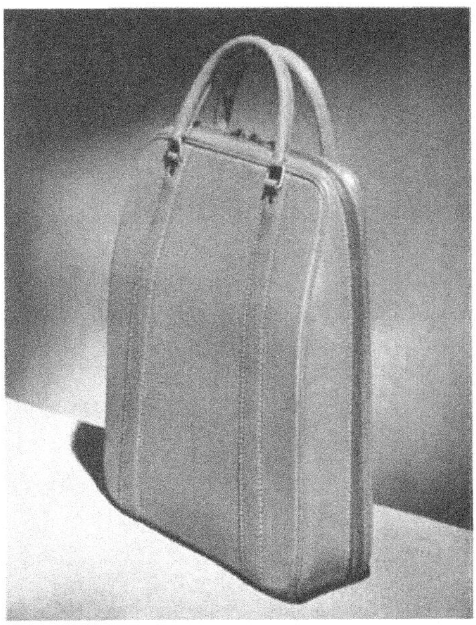

3. Here the zip-fastener was the most efficient solution to a particular problem: a carrying case for the Olivetti portable typewriter. English taste for leather craft goes well with this famous Italian machine. *Designed by John W. Waterer for S. Clarke & Co. Ltd.*

PLATE LXII

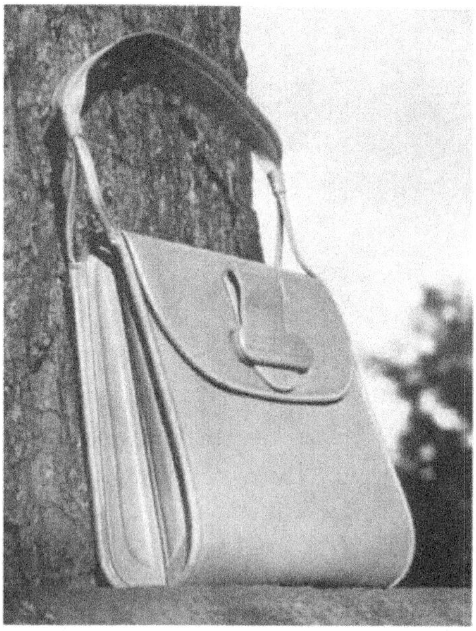

1. Changing fashions predict the styles for handbags, but beneath the superficiality there is, in the best work, a traditional sense of fitness which conveniently provides a lasting quality in a new design. *Made (1949) by Bagcraft Ltd.*

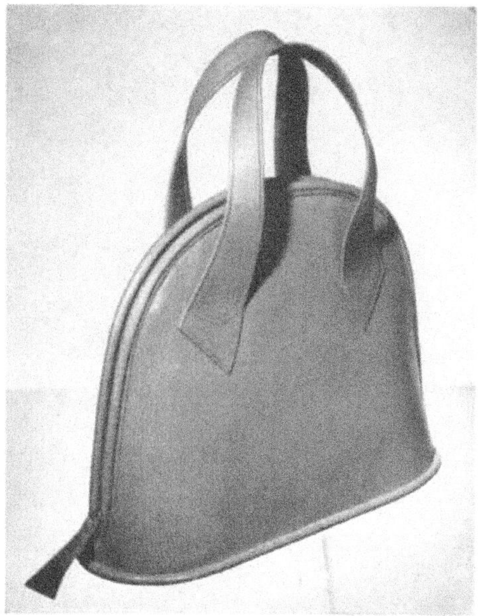

2. Often clumsy clasps spoil the best shapes for leather handbags which must, in any case, compete with cheap imitations in plastic. Here a bold treatment has made the zip-fastener the co-ordinating feature of the bag. *Designed by F.M. Jennings for J. Jones Sons (Birmingham) Ltd.*

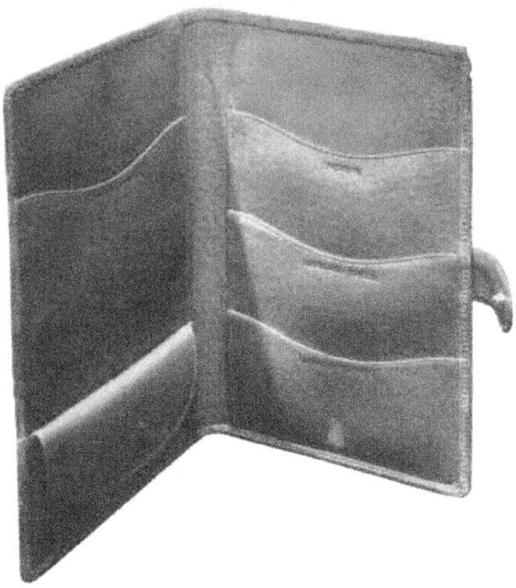

3. Fancy leather goods are too numerous to illustrate, but again there is an underlying sense in the trade of what looks best and lasts longest. This pigskin wallet is unusually restrained and refined. *Made by Mark Cross Ltd.*

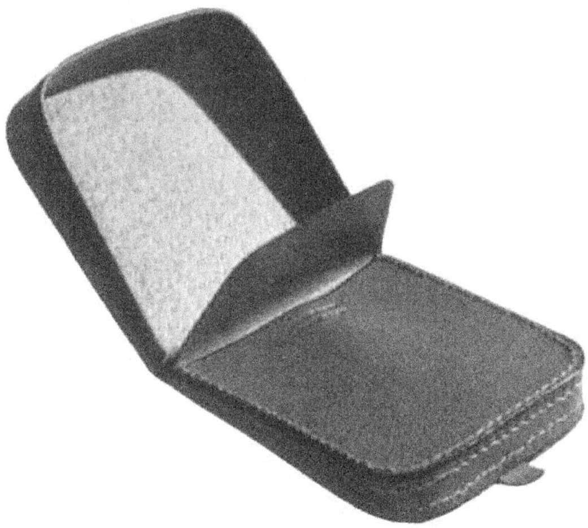

4. Walsall is the centre for the fancy leather goods trade and much of the work has a characteristic appearance: for instance, the robust stitching on this pigskin purse which secures the cut edges. Agricultural harness is not far in the background. *Made by George Sheldon (Walsall) Ltd.*

sidered employs free-lance designers only. For a sketch anything from £5 to £50 might be paid, according to the status of the designer. The manufacturer told me that he was convinced that this method was the only one possible, if a constant stream of fresh and original ideas was to be maintained.

The leather handbag has to withstand great competition from bags made out of cheaper materials, such as knitted fabrics, felts, leathercloth and other plastics. It is possible to purchase for 35*s*. an attractive plastic handbag which will last out a fashion period and then be discarded. Plastics, as mentioned earlier, have several useful qualities—wide colour range, flexibility, and constancy of composition which enables them to be handled successfully by semi-skilled labour. Above all, they can be made to imitate leather with little difficulty. One typical trade announcement states that its product, a plastic handbag, 'captures the beauty and finish of real leather'. There are imitations of calf, pigskin, pinseal, morocco, alligator, lizard, python and crocodile. They imitate real leather, and in some cases great skill and ingenuity are evident, but they rarely deceive.

The production of fancy leather goods is centred in Walsall. A tradition in the town for making leather articles goes back for nearly two hundred years. Walsall was then in the midst of an agricultural area which it supplied with leather goods, such as saddles, harness, belts, straps, etc. With the growth of industrialism the leather trade began to decline as more and more factories spread over the land. In the early years of this century the decline was hastened as horse transport was replaced by the motor-car. Between the years 1916 and 1918, most of the firms which were still making harness and saddles changed over to those smaller articles nowadays known as fancy leather goods. The term denotes wallets, purses, billfolds, notecases, watch-straps, cigarette-cases and many other items. A few years later Walsall added sports goods to its range of products. Before the war the standard of craftsmanship was very high, and even now there are a few firms with unchallengeable reputations. But Walsall leather work is no longer as fine as it was. The loss of skilled labour during the war and the mechanization of hand-work partly account for this, but the chief reason why craftsmanship is declining is the use of cheap, synthetic materials.

Nevertheless, the original technique of Walsall workers is still apparent. In the town there remain five firms which specialize in harness and saddles, and their methods of working find many echoes in the rest of the trade. There is what is known as the 'Walsall touch'. It shows most clearly in the heavy and secure stitching on many articles. The town is also known for its 'cut edge' work, or the stitching together of two pieces of leather so that the raw edges are visible. This traditional familiarity with stout leather articles contributes to the character of Walsall fancy goods, which are inclined to be heavier, more robust and more practical than the corresponding London products. Altogether there are eighty-seven leather factories in the town. They range in size from five to two hundred employees, with an estimated average of twenty-five. Just as in the jewellery trade, the low cost of plant has favoured the development of small manufacturing units.

The markets served by Walsall range from Brighton Pier to exclusive London stores. In both there is a quest for novelty, and manufacturers have each year to modify their designs slightly. However, quasi-artistic motifs are rarely used, and most manufacturers obtain new effects by juxtaposing different kinds of leather in the same article. That cheap pouches and purses look cheap cannot be helped. In leather more than in many other materials, such as pottery, the charm of the finished product depends on the quality of the material used, its grain and texture. But there is a difference between looking regrettably cheap, and looking cheap and nasty; and nasty-looking fancy leather articles are exceptional. Several manufacturers told me that they did not know of one full-time designer in Walsall. I did not find one, except in a handbag factory. The new designs are produced by the pattern-cutter, who usually has the duties of works foreman as well. He is little more than a medium through which the manufacturer's own ideas take a practical shape.

This brief survey of the leather trade does not pretend to be complete. As there are numerous uses for leather other than those mentioned, it is strange that openings for designers in the trade are so very limited. Why, for instance, is there not more co-operation between makers of handbags, gloves, shoes and fancy leather goods? Could not one designer produce, and thereby correlate, the designs of all four manufacturers? In this way his fees could be shared amongst the members of the group. As John Waterer points out, there are co-operative systems of this sort successfully at work in the United States and Germany.[1] It is plain that the native craft tradition in the trade does maintain a very satisfactory design standard, but if real distinction is to be achieved in British leather goods this tradition must be used more creatively.

PLASTIC MOULDINGS

Plastics form a group of organic basic materials derived from natural substances and chemical compounds. They are stable in use at ordinary temperatures, but become plastic at some stage in their manufacture and can be shaped by the application of heat and pressure. They may be used on their own, or in combination with wood,

[1] *Leather in Life, Art and Industry*, pp. 288–9.

metal or glass. Apart from artificial horn, cellulose nitrate is the oldest plastic material. It is known as celluloid, and a patent for its commercial use was taken out in 1865. Since 1877 the British Xylonite Company has specialized in the various applications of celluloid. As it is highly inflammable, it is not generally used as a moulding powder. The next type to be introduced was casein, a protein plastic which originated in Germany in 1899. It is used mostly for buttons, jewellery, etc., where bright colour is necessary. Following this was cellulose acetate, which appeared during the First World War. It was used for rayon, film, and later for a wide range of toilet articles formed by injection moulding. All these early plastics are derived from natural substances.

The next step was the production of synthetic resins from simple chemical compounds. The first of these was 'bakelite', so-called because it was developed into a commercial material by Dr Leo Bakeland, who filed his patent in 1907. Commercial production did not start in this country until 1916. This plastic is formed by the reaction of phenol (a coal tar product) and a solution of formaldehyde (derived from carbon monoxide and hydrogen). It differs from earlier plastics in that it is thermosetting, i.e. it hardens when heated and cannot be softened afterwards. With the invention of bakelite the plastic compression moulding industry began. The other important synthetic resin is urea-formaldehyde, called 'beetle', which was developed by the British Cyanides Company (now British Industrial Plastics), and produced commercially in 1928. Unlike bakelite, it has a wide range of colours. Of all plastics, phenol- and urea-formaldehyde have the widest industrial and domestic uses. Since 1928, ten new basic plastics have been introduced. Most of them are in the thermoplastic group of materials which, once they have been formed into shape, can if necessary be softened by a further application of heat.

There are three chief methods of forming articles from plastics. The oldest method for many thermo-plastics is to cut the required shapes from plastic sheet. A typical product is the knife-handle. Second is compression moulding, by which a thermosetting powder is pressed in a mould by a plunger. The mould, which is heated, should not be opened until the plastic article has been cured, i.e. set in a rigid shape. Most hollow-ware and three-dimensional articles and those requiring rigidity are moulded in this way. Thirdly, there is injection moulding, which nowadays is a fully automatic and continuous process, whereby heated thermo-plastic powder is forced through a small hole into a cold mould. It is usual for comparatively small articles to be produced in this way—radio-knobs, combs, toys, etc.

The plastics industry is a development of the late 1920's. At that period nearly all its products were accessories used in the electrical industry, notably in the manufacture of radio components and the telephone.[1] Nearly all the early plastic articles copied articles in metal, wood and other materials, in order to replace them. Comparatively few of them were complete in themselves. Nevertheless, the moulded telephone receiver hand-set (first made in this country in 1929) is, perhaps, mainly responsible for the steadily widening acceptance of plastic as a basic material for domestic articles. It did not imitate the metal shapes of earlier receivers, but was designed to evolve naturally from the material and the moulding process. Since the late 1920's, and particularly since the war, the industry has expanded at a phenomenal rate. For the year 1929 the production of plastic materials amounted to 8000 tons. In 1939 the figure was 30,000 tons, and in 1948, 110,000 tons. Much of this expansion is due to the industrial uses of plastics, but throughout the 1930's and since the war there has been a steady increase in the demand for domestic articles.

It is important to stress the technical and commercial development and the expansion of the industry. They are some of the factors which must be taken into account when assessing the design standard of plastic articles. I visited eleven firms. They may be divided into two groups: firms which mould and sell their own products, and firms which mould articles for other manufacturers (e.g. radio cabinets). No clear-cut division can be made, as most firms which sell their own products also mould some components for manufacturers in other industries. In the first group there were five firms. Two of them were started as subsidiaries of companies which supply raw material, in order to demonstrate the possible commercial uses for new types of plastic. One of these firms employs 1000 persons, and specializes in celluloid toilet articles. It also has a yearly output of ten million table-tennis balls. The other, 675 employees, moulds tableware and electrical components, chiefly in thermo-setting materials. Two firms are established brass-founders and employ 100 and 150 persons respectively on plastics production. One firm makes plastic door furniture to replace its cheaper lines in brass, while the other has an independent range of picnicware. The remaining firm in the group is small, and remarkable for its design standard in catering equipment and tableware. Four firms fell in the second group. Three of them (275, 550 and 800 employees) mould light fittings, radio cabinets and electrical and mechanical components. The other firm (E.K.Cole) has a similar range of products and has already been dealt with in connexion with radio cabinets. In addition, I visited two of the industry's chief suppliers of raw materials.

As in most of the recent industries, only mass-production methods are used. The relation between number of workers and volume of output is totally different from that in old trades such as jewellery, silversmithing and pottery. For each moulded

[1] The first plastic radio cabinet was the Marconi 'Baby', 1924.

article special tools have to be made. The considerable expense which this represents before manufacturing begins may be indicated by two examples. To mould a cabinet for a table radio the tools required cost between £800 and £1000, which means that at least 10,000 cabinets have to be ordered before any profit is made. In this case the financial risk is borne by the radio manufacturer, who is bound to be hesitant and conservative with regard to the design unless he is sure that it will sell successfully. The same applies to all manufacturers who commission mouldings, though, as we have already seen, artistic judgement in the radio industry is notoriously unstable. Secondly, in firms which mould and sell their own products it is the sales manager who has to be convinced that the proposed manufacturing costs will be justified. The large quantities which must be sold give the sales manager more authority in the early stages of a project than is usual in other industries. I was given details of the risk involved when a design for a condiment set, consisting of a stand and two containers, was accepted for production. Although the set was intended for one of the leading chain-stores, its design was outstanding and in advance of its time. This last factor is a necessity for firms which expect to run the same design for at least five years. The cost price of the set was 11*d.*, of which 2*d.* was needed to pay for the powder. The remaining 9*d.* had to pay all expenses, including tools costing £1450. It was necessary, therefore, to sell at least 38,000 sets in order to recover initial costs. In this case the designer and the sales manager were justified in their enterprise, for within a year 560,000 sets had been sold. However, it should be emphasized that the risk involved by high tool costs, partly accounts for the fact that modern design in plastic articles has had a relatively slow evolution.

For many industrial and domestic purposes plastics have been considered as substitutes for other materials. To establish its position the industry has been forced to produce in plastics articles formerly made of metal, wood, glass, etc. Plastics' chief claim is cheapness, for if a sufficiently large quantity of one design is required the article will cost less in plastics than it would in metal. In addition plastics manufacturers can offer articles which are comparatively light and strong in a wide variety of colours which will not tarnish. In very many cases the plastics manufacturer is asked to copy the form of a metal article. I was shown two recent examples of this. One was the housing for a small weighing machine, where imitation screw-heads had been moulded along the edges to recall, apparently, the method whereby the former metal housing had been held together. The other was a camera case which had a moulded stipple pattern on the outside to resemble leather.[1] Although it is often

[1] It is interesting to note that in this case, after several hundred mouldings had been made, the stipple pattern wore off the mould owing to the abrasive action of the powder.

impossible to copy the form of a metal article in plastics, the designer in such cases rarely has the opportunity of considering afresh the purpose for which the article is required in terms of moulding technique. This is the second important point which affects design for plastics. Like the high cost of tools, it partly accounts for the industry's slow progression towards an artistic style of its own. Moreover, it is chiefly responsible for the widespread opinion that plastics are 'ersatz' materials, and articles made from them are merely the next best thing to articles in wood, metal, glass, etc.

It is in the firms which mould and sell their own products that the best design standard is achieved. These firms are concerned to put on the market new products which are made, in many cases, entirely of plastics. In other words the attempt is made to serve old purposes by using new materials. By no means all the work turned out by these firms is satisfactory from our point of view, but a high proportion of it reaches an excellent design standard. Plastic materials and manufacturing methods exercise a control over the designer's choice of shapes. A typical plastic shape has generous radii at corners and edges. Vertical flat surfaces are excluded wherever possible. A convex shape allows the powder to flow evenly into place as the plunger descends and also enables the article to be extracted easily from the mould. Less powder is required to achieve a given strength in a rounded form than in an angular one, which means a saving in cost. Finally, a convex surface readily reflects light and so gives an added lustre to unpolished plastic. When a firm is wholly responsible for the design of its own products, these factors can be given their right degree of significance.

All the firms I visited in this group employ designers. Three of them have staff designers, two use consultants. Two of the staff designers interviewed have had a full-time art school training in metalwork and modelling, which has enabled them to bring a fresh approach to plastics. Both have studios and facilities for model-making. They are sent abroad frequently to study design and production methods, particularly in the United States. When engaged on contract jobs their status enables them to work out the most suitable design in collaboration with a client. Of all the private consultants used, I know of only one who has had steady success. This designer is retained by a firm which produces catering equipment and tableware, in plastics, metals and glass. Plastics are treated as the basic materials for all articles; metal and glass are supplementary and only used where their qualities are superior to those of plastics. A very high standard of design is maintained throughout the firm's range. In the other group of firms, which mould components for manufacturers outside the industry, only one artistically trained designer is employed. He joined the firm as a commercial artist and gradually took over product design. The influence which he should be able to exert over the taste of his firm's clients is not encouraged.

Apart from artistic training a plastics designer must have technical experience. There are at present fifteen basic plastics and each one has its own peculiar properties. In other words each type of plastic must be fabricated in a specific way to do a specific job. For example, in the make-up of a certain brand of electric shaver now on the market no less than seven different plastics are employed. One is included for its insulating property, another for rigidity, another for strength and colour, etc. Moreover, new technical developments involving new possibilities for plastics are an almost daily occurrence in the industry. It is for this reason that free-lance designers, and even consultants, are rarely used by plastics firms. The case referred to above is exceptional, though it proves that a close relationship between consultant designer and manufacturer can be successful.

The two suppliers[1] of raw materials which I visited have, since the war, opened design studios. Although these firms do not mould their own powders, they feel that they are in part responsible for the way in which their materials are used by the industry. Since the 1920's both firms have had to maintain a staff of technicians who could visit a moulding firm to give advice. It was in their interest to see that their materials were used correctly. The service, which was free, involved the design of tools. Since the war both firms have added product designers to this service and have set aside special studios for their work.[2] The designers operate as consultants within the industry and they are both artistically and technically capable of designing the product and the mould.

The method adopted by one firm (British Industrial Plastics) is worth close study. Its product and mould design unit consists of a designer with more than twenty years' experience, an assistant and four draughtsmen. The unit sets out to do three things: first, to approach the buyers of chain-stores, large public corporations, wholesale houses and the bazaar trade, and offer to interpret their ideas for new products in terms of plastics. It is rightly assumed that these retail buyers cannot be expected to discriminate between suitable and unsuitable applications of plastic materials, unless they are given the necessary knowledge and assistance. Secondly, the unit offers to design or re-design the products and moulds of manufacturers who do not employ designers. A brochure has been compiled which contains photographs and drawings of articles of modern design. In cases where the firm's representative, instead of the

[1] Bakelite, Ltd., and British Industrial Plastics, Ltd. See *Design*, December 1953, pp. 11–17.

[2] The origin of this new departure may be traced to the leading suppliers of plastic materials in the United States.

designer, meets the manufacturer, this visual evidence can be of great assistance. And thirdly, the unit is able to suggest suitable applications for the company's new technical developments. With the aid of a market research department, the unit may, for instance, design comprehensive collections of bathroom and kitchen fittings, etc. These would be shown to the industry and the public as illustrations of a line of thought which can be progressively developed.[1]

It is particularly important to note that both firms started this design service at a time when there was a shortage of moulding powders. It follows that the effect will be far-reaching. By the time that full competition and price-cutting return, the discriminating public which is being built up will perhaps force manufacturers to maintain a reasonable standard of modern design. Although both firms specialize in thermo-setting materials, there is no suggestion that the designers should be limited to these materials. As consultants these designers possess an enviable status, for they represent two of the most influential firms in the industry. I do not wish to imply that this design service is altruistically inspired. No doubt it will increase the sales of thermo-setting materials and the prestige of the firms concerned. Nevertheless, it indicates an enlightened attitude towards design and a feeling of moral responsibility for the form in which the raw material will eventually be judged by the public. It is significant that this attitude to design is to be found in two of the largest firms in the industry—not two of the smallest, as is the case in so many other industries.

The plastics industry may be usefully compared with the radio industry. Both are new and very much dependent upon new technical developments, though only the plastics industry has realized that the commercial applications of these developments should be evolved without help from traditional methods of working. As we have already seen, the radio industry has largely failed to produce cabinets worthy of the advances which have been made in electrical engineering. It has preferred to adopt the forms, both period and modern, used by the furniture trade. The enthusiasm which the industry displayed in the 1930's for the modernistic style—at least a sign of vitality—has now been dissipated. The majority of radio cabinets today are characterless boxes, devoid of artistic significance. The situation in the plastics industry is remarkably different. Although plastics were also affected by modernism in the early 1930's, several manufacturers soon appreciated that the jags and angles which belong to the style were fundamentally unsuited to compression or injection moulding. New forms, peculiar to the materials and the manufacturing methods, began to appear in the late 1930's. The fact that not more than five artistically trained designers were employed at that time, makes this achievement remarkable. Apart from tortoiseshell

[1] Cf. the policy of R. Greg & Co., p. 79.

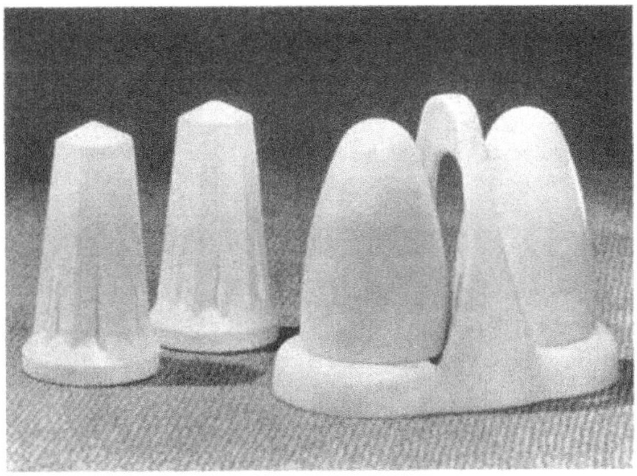

1. Old purposes are often served better by plastics, a group of new basic materials. Design progress in the industry may be indicated by these best-selling condiment sets. *Left*, 1932, with decorative dirt-traps; *right*, 1945, with smooth, clean contours. *Designed by A. H. Woodfull for The Streetly Manufacturing Co. Ltd.*

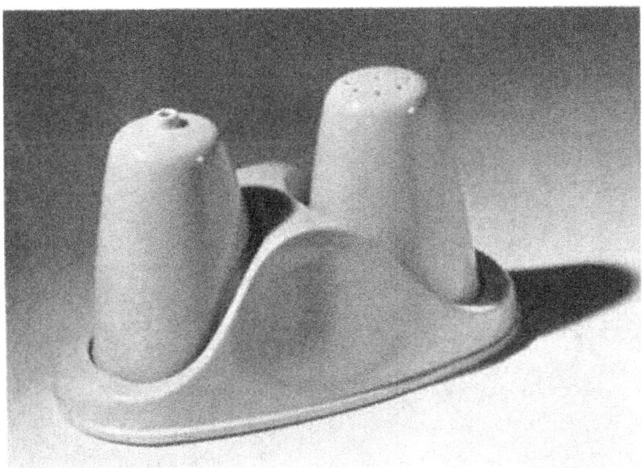

2. To their detriment extravagant claims have been made for plastics; yet in the right context they remain stronger, lighter, cheaper and more flexible than other materials. These are typical mouldings distinctive in design. *Designed by W. Bruce Brown for Halex Ltd.*

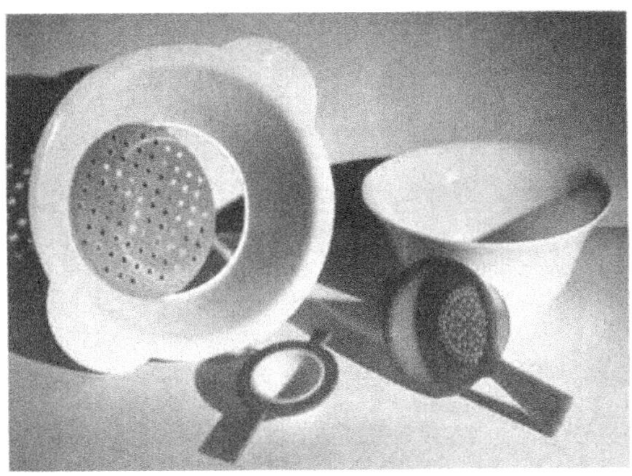

3. The discovery of new possibilities for plastics is relatively slow for economic reasons. When launching a new moulding the manufacturer must be sure of selling enough to repay the high tool costs. Here older materials have been successfully replaced without imitation. *Designed by Gaby Schreiber for Runcolite Ltd.*

1. Rightly the public has conservative views on cups, but habits can be modified when fresh advantages are offered. These articles were moulded to weigh the same as earthenware, but they are far stronger. *Made by The Streetly Manufacturing Co. Ltd.*

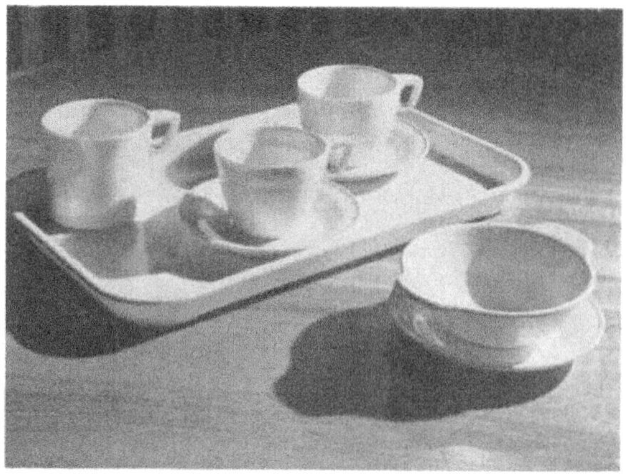

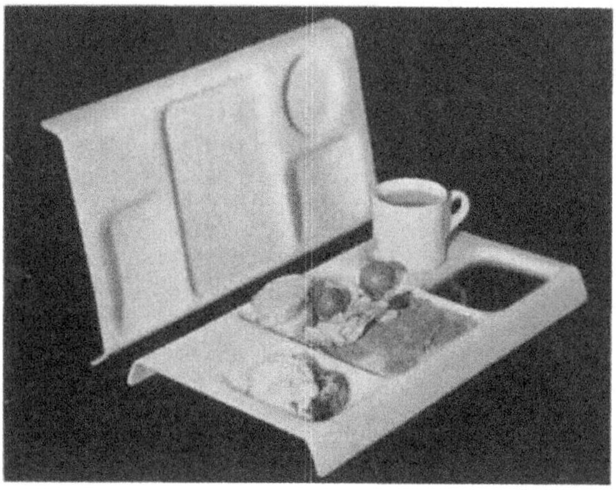

2. With varying ingenuity plastics manufacturers have competed with more established materials to satisfy old markets, but to survive they must also create new uses for their products. The simplicity of this meal tray helps to make its originality more palatable. *Designed by Gaby Schreiber for Runcolite Ltd.*

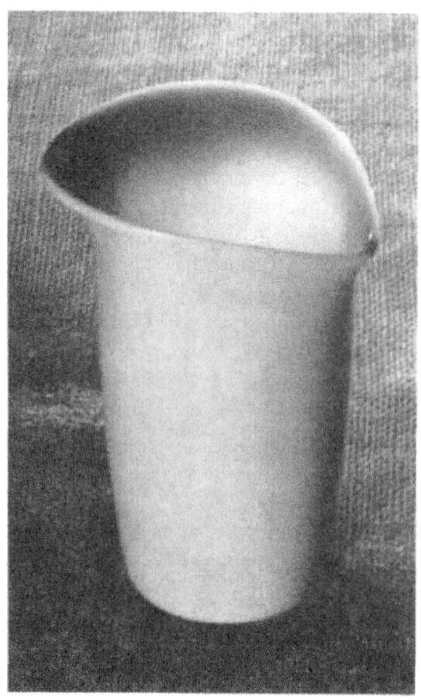

3. A fine sense of form is integrated with the function of this milk jug which can be held without a handle. In manufacture the design is straightforward as the powder in the mould can flow smoothly to form the shape. *Designed by R. E. Brookes for Brookes & Adams Ltd.*

PLATE LXV

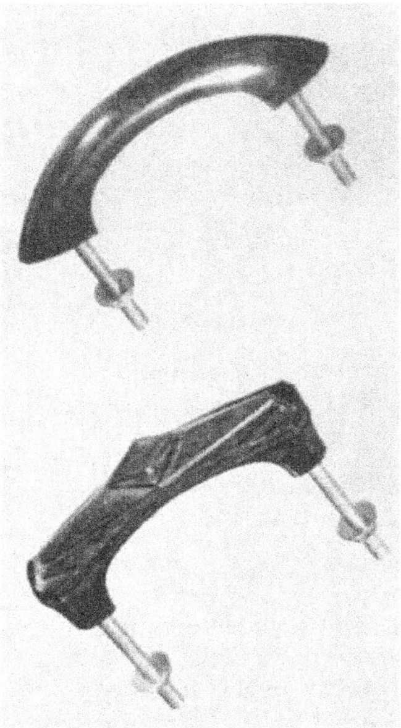

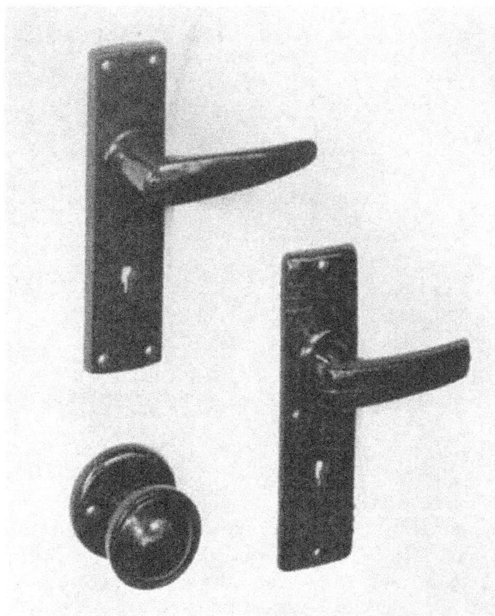

1 and 2. A great many plastic mouldings are small accessories, like these handles for fitting to doors, cupboards and drawers. Although relatively unimportant they can make no useful contribution to an interior design scheme, unless they are good designs themselves. In both photographs the top examples, when compared with the lower ones designed in the early 1930's, indicate the trend in the industry towards well-proportioned shapes. The knob, now 25 years old, is a best-seller. *Made by Evered & Co. Ltd.*

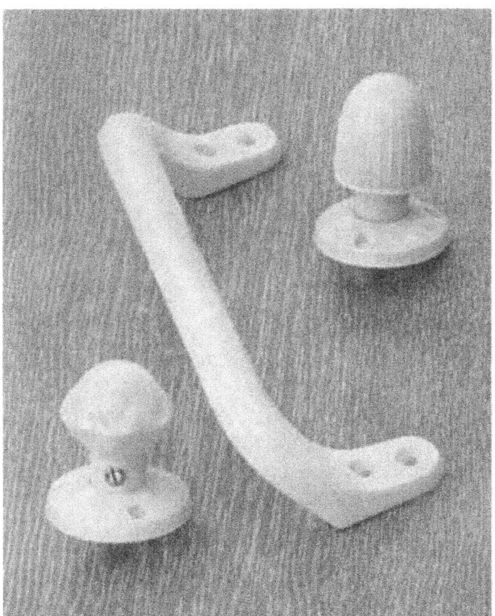

3. Three cases where functional and artistic standards are in balance. The top knob is ribbed for gripping; the other has indentations on a shape for turning and pulling; and the door grip is offset to keep the hand clear of the joint. *Designed by Rodney Hooper for Lacrinoid Products Ltd.*

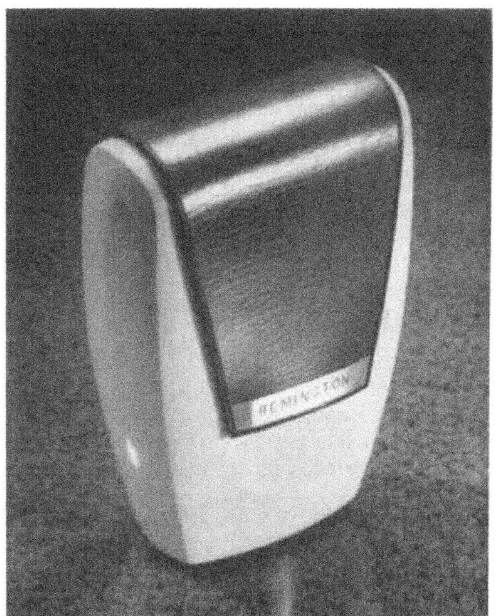

4. The tapering shape of this razor case enabled the designer to exploit a crispness of line without impeding the moulding process. Leather brings out the texture of both materials and solves the difficult problem of a hinge in thermo-setting mouldings. *Designed by A. H. Woodfull for The Streetly Manufacturing Co. Ltd.*

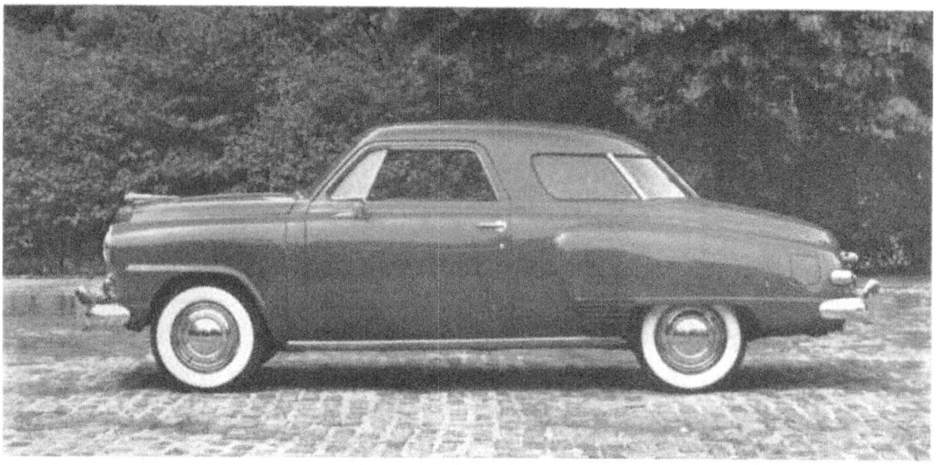

1. American. Revolutionary post-war mass-production motor-car to a design conceived three-dimensionally. It indicated the trend now followed by progressive British designers. Seats are slung between the wheels, the area between roof and slab-side panels is mainly filled with glass, and the luggage compartment balances the bonnet. The form alone is evocative of smooth, forward movement. *Designed (1946) by Raymond Loewy for The Studebaker Corporation.*

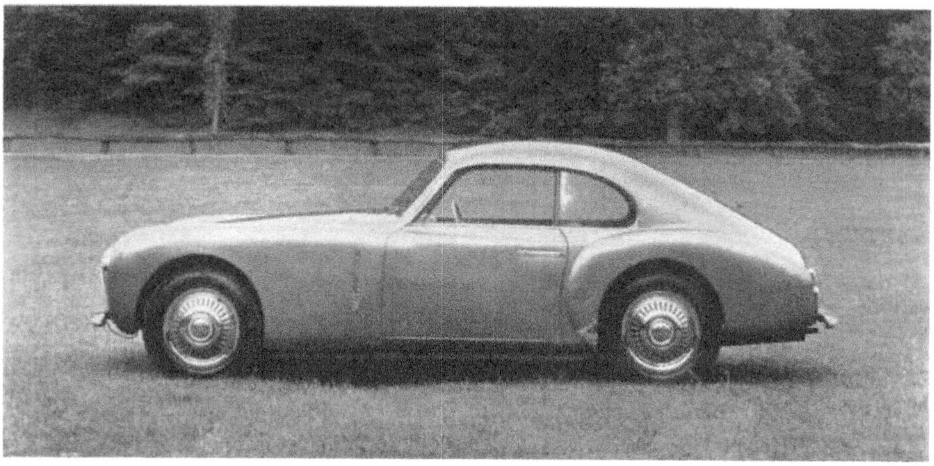

2. Italian. Expensive car produced in small quantities on a craft basis. The body consists of a tight metal skin cut away to reveal only the wheels. Enveloping form is all-important here, for it presents the body as a three-dimensional design with distinct elements flowing imperceptibly into each other. *Designed (1946) by Pinin Farina for Automobile Cisitalia.*

decoration—now comparatively rare—there are practically no moulded plastic articles which imitate the outward appearance of other materials. In particular, period styles, such as the Baroque-type scroll, are almost unknown. Since the war a few more designers have been added to those already employed, and it appears that the cumulative influence from their work is being felt throughout the industry.

In less than twenty years, plastics have almost entirely ceased to copy the characteristic forms of the materials they have replaced. Design in the industry has reached the stage where it can be dependent only upon the materials available and the artistic ability of its designers. Only one other British industry (motor-cars) has put itself into such a position of advantage in so short a time. The reasons why the artistic development of the industry is encouraged to keep pace with its technical progress are worth repeating. First, the leading firms, including two of the suppliers of raw materials, wish to set a design standard. Amongst moulding firms this standard is highest in those which are responsible for selling their own products. Secondly, the artistic judgements made by manufacturers—most of them are professional chemists or engineers—are remarkably sound. And thirdly, it is realized in at least five firms that the designer should possess the authority to meet buyers in order to work out the best solution to their needs. As might be expected, there is abundant evidence to prove that the public, if given the opportunity, will demand a high standard. We have already dealt with one condiment set of which more than half a million were sold in twelve months. In another firm which also supplied a leading chain-store, I was shown a condiment set of excellent design of which 1,340,000 were sold in eighteen months. It is especially significant that much of the commercial success of modern design in plastics does not depend upon the patronage of exclusive West End shops, or extensive publicity campaigns, but upon chain-stores. With a responsive public and a progressive industry the standard of design in plastics could be as high as that in leather goods and for similar reasons.

MOTOR-CARS

Several points about the motor-car industry are of special interest in our connexion. For one thing the motor-car is a creation of the twentieth century. The first patents were taken out by Daimler and Benz in 1885 and 1886, the first cars were built in the 1890's and the industry began to grow rapidly shortly after 1900. Another noteworthy fact is that motor-cars hold a central place in popular interest.[1] Changes in their

[1] The Motor Exhibition in London in 1948 (the first for ten years) attracted 562,000 people.

appearance are discussed everywhere and seem to concern everybody, while other objects of industrial art steal on to the market almost unnoticed. The ambition to recognize the various makes of car in the streets has infected all boys from four to seventy. To have a car is the longing of clerk and shopgirl; to have a smart car the longing of the wealthier classes. Furthermore, motor-car design is one of the most controversial subjects outside politics, and receives more attention in the daily press than all the other industrial products included in this book.

The body construction of early motor-cars consisted of ash supports with thin sheet steel and wooden panels. Apart from the chassis there was comparatively little standardization of parts, and people were free to choose their own style of bodywork. The engine and chassis were made by motor firms, the body by coach-builders, and it is well known that early cars were designed to emulate coaches and carriages. The evolution of a design for cars which was independent of earlier forms of transport occurred during the First World War. As firms increased in size, standardization and mass-production led to cheaper cars and by the early 1920's an entirely new market was being served. Typical of this trend was the introduction of the 7 h.p. 'Baby' Austin in 1923. Except in cars with individual coachwork, ash supports were gradually replaced by steel. The all-steel frame was first introduced in the United States in 1914 (Dodge). It led naturally to the all-steel body appearing in the United States in 1925 (Willys-Overland). Britain followed soon after, and in 1932 the complete side of a Hillman Minx was pressed out of one steel sheet.

The introduction of the all-steel body is important for two reasons. First, the manufacturing method demanded a new approach to design, in which the supports and the panels were integrated to form the construction of the body. And secondly, it meant that each new body design required a new set of tools. Today, the tool costs for a four-door saloon can amount to as much as £500,000. The steel body maintained the traditional idea that the chassis should be designed first and the body imposed upon it afterwards. Although in 1938 a car with the 'monocoque', or integral chassis and body, construction—a method which is steadily becoming more popular—was produced,[1] it is still generally thought that the chassis should take precedence and be separate from the body. Until the war the bodywork of all but the most expensive cars was designed by the technical engineer. Apart from seating two or four persons as comfortably as possible, his main considerations were cheapness and simplicity of construction within the limits of a more or less standard chassis. The design of the body was conditioned by these functional considerations, and the engineer's artistic judgement was only unconsciously involved. Despite the fact that the engineer

[1] The first American car with 'monocoque' construction was the Airflow Chrysler, 1934.

designed in terms of steel pressings which must be curved for structural reasons, it is apparent that he evolved his bodywork two-dimensionally. In other words, the body was conceived in independent parts, one relating to the plan view, another to the side elevation, another to the front elevation, etc. The body was simply a series of lines on the drawing-board, and consequently when it was constructed it only looked satisfactory from a limited number of viewpoints and had no integrated form. Although scale models were made in some factories, the artistic significance of volume in pre-war bodywork appears to have escaped notice.

Nowadays, most car bodies in this country show that this approach has been dropped. To account for this we must turn first to the United States, and afterwards to Italy. In the United States the mass-production of all-steel bodies developed at a much greater rate than in this country. Backed by a demand for huge quantities of one design, manufacturers were able to afford larger tools. For example, a typical pressing was made to include the side of the body and part of the back as well. This immediately involved the third-dimension and made it a factor to be considered from the start. Later on the so-called 'slab-sided' body was developed, which meant that there was no longer any need for separate mudguards, or a bonnet which was constructed independently out of four flat panels. It was found that sufficient strength could be obtained by incorporating the mudguards and bonnet in a minimum of three pressings, where six had been needed before. A simple pressing is one which maintains a gradual, constant curve with a depth that can increase in proportion to the total area. The characteristic low bonnets of American cars are dependent on this mechanical principle.

There are, of course, other considerations which have determined the typical shape of American cars. A slab-sided body affords greater seating width and a low bonnet greater visibility. But more important is the fact that artistically trained designers (called stylists) are used freely by American manufacturers and their effect upon car design has been decisive. In the United States the appearance of a car is a factor of such great importance for the salesman, that the stylist often designs the body first and the chassis is fitted in afterwards. Certain trends in design have become apparent —especially since the war—and they are worth discussing because of the effect they are having on English cars. By using the slab-sided construction, American stylists are able to extend the rear portion of the body. While this provides increased space for luggage it also allows the rear wheels to be moved further back. Instead of placing the rear seat in its usual position on top of the back axle—a most uncomfortable position only alleviated by elaborate springs—it is possible to sling both front and rear seats between the axles. In this way, too, the projection caused by the luggage compartment is made to balance the projection of the bonnet. The system of suspending the occupants between the axles is one of the inevitable developments in car design, despite the fact that it was anticipated centuries ago by coachbuilders faced with a similar problem. Another point is the American conception of design in relation to an object which moves horizontally. Vertical features are obviously out of place on a moving vehicle and recent trends show that horizontal lines have been given increased emphasis. The most marked of these is the line which extends along the length of the body, passing just below the windows. Initial prominence is given to this by the head lamp which is embedded in the top of the front mudguard where the line begins. A complementary line is formed by the roof, which is suspended on light metal struts, the intervening space being filled in with as much glass as possible. Again, the logical nature of this development appears to make it inevitable.

At this stage one more aspect of motor-car design must be considered—streamlining. Objects which move, such as birds, aeroplanes and cars, can be shaped in a way which allows the air to flow as easily as possible over their surfaces. Any interruption to the flow of air will build up resistance in front of a moving object and cause turbulence, or drag, behind. This shaping of an object is called streamlining, and for our purposes we may take it as a fact that streamlined objects move more easily, and therefore faster, than objects which are not streamlined. The proper degree of streamlining for a given object has to be calculated scientifically. Aeroplanes receive this sort of consideration, but commercially produced cars do not at present merit the time and trouble which would be involved. On this all manufacturers whom I have asked were agreed. Only at speeds above 45 m.p.h. does streamlining become effective. From then on its importance increases in proportion to the speed. However, streamlining can never be fully effective when applied to moving objects, such as cars, which remain in contact with a stationary surface. As a car moves forward some of the air which is displaced passes under the chassis and is partially arrested by the road, thus causing turbulence. The dragging effect of this turbulence on the car cannot be eliminated completely, though it can be reduced. If the underside of the chassis is totally enclosed by a smooth metal plate (a costly procedure), and moreover, is raised as high as possible above the road, much of the turbulence will become ineffective. Admittedly these are factors which are only related to streamlining, but they help to explain why the application of aerodynamic principles to cars is limited. Another factor which is frequently overlooked is known as 'tyre-rolling-resistance'. As the speed of a car increases its tyres lose resilience and become soft. Instead of remaining circular they tend to become oval and, with each revolution of the wheel, a section of the soft outer rubber is thrown forward to flap upon the road. It is interesting to note that the energy required to propel a car at 90 m.p.h. is consumed equally by the turbulence of the air and the rolling-resistance of the tyres.

The pioneers of streamlining[1] in motor-cars intended to apply the aerodynamic principles honestly and scientifically. They were Dr H. Rumpler, who built his 'drop-shaped' car in 1921, and Sir Dennistoun Burney, who followed in Britain in 1930. Their models had their engines in the rear. Both men were not at once, nor ever fully, successful. Manufacturers, in spite of Sir Dennistoun Burney's passionate propaganda, were not convinced that either streamlining or the rear engine merited a change in their attitude towards design.

To some extent their hesitancy was justified, because at that time it was thought that a streamlined shape could only be obtained if the engine were placed in the rear. And the disadvantages of having a rear engine are many.

A motor-car engine may be described as a long narrow box. If it is placed in front it allows the wheels sufficient space in which to swivel, either right or left. If the engine is taken out and placed above the back axle it pushes the rear seat forward, thus reducing interior space and practically eliminating luggage accommodation. The space which is left between the front wheels is wasted, because it is neither sufficiently large for luggage nor wide enough to take the legs of the driver and passenger. Again, if an engine of more than 10 h.p. were placed in the rear, its weight would have to be balanced by an equivalent dead-weight in the front, or the car would become unstable at high speeds. Alternatively, a stabilizing fin could be erected. The elimination of the transmission shaft made possible by the use of a rear engine, would be a trifling advantage in the face of this rearrangement. Consequently only very few manufacturers took to the rear engine in the years before the war.[2] Their results to date, however, leave the difficulties of reduced seating, leg and luggage space unsolved.

Streamlining, of course, does not depend upon the position of the engine, a fact which was soon appreciated in the United States. The economic crisis of 1929–33 in that country made car manufacturers eagerly design-conscious. They seized upon some of the more obvious characteristics to be found in Dr Rumpler's and Sir Dennistoun Burney's experiments and applied them to their own cars. Slanting windscreens, slanting radiators and outcurved backs became the fashion. On the Continent streamlining began to be applied about 1931. The first year in which the new shapes

[1] Incidentally, the word 'streamlined' was already used by Frank Lloyd Wright in his lecture on 'The Art and Craft of the Machine', delivered in 1903 (F. L. Wright, *Modern Architecture*, Princeton, 1931, p. 19).

[2] They were (in chronological order) Trojan 1923, Crossley 1932, Mercedes 1934, Tatra 1935, Volkswagen 1938. After the war a few firms mass-producing specially light cars began to make use of the rear engine. The most familiar example is Renault in 1948. In Britain the Rover firm in 1950 introduced a rear gas turbine engine, lighter by at least 200 lb. than the conventional engine of similar power.

could be seen in Britain was 1934, and by 1935 only a very few new models on the Continent were not streamlined in some way. There were considerable differences in degree, but the general tendency was obvious. Britain, averse to novelty and artistic adventure, was much slower in taking up the new idea than the United States, France, Germany and Italy. Amongst the three largest mass-production firms in the country at that time, the two wholly British ones were extremely cautious, and it was Ford, the American firm, which pushed streamlining in cheap cars. So did Citroën in France and Opel in Germany, the largest firms in those countries. Britain's reluctance to follow the trend was not confined to the inexpensive models. The same attitude seems to have predominated in the upper- as well as in the middle-price groups.

It will have been noticed that the word streamlining has been used freely in the preceding paragraphs, although it has been explained that the term cannot really be applied to commercially produced motor-cars. However, the term is popularly accepted, and has its value, as long as it is realized that it does not refer to shapes designed to *be* faster but to *look* faster. It is a case of evocative rather than functional values, and possesses as such considerable significance in connexion with the problems on which this book is engaged. There is, of course, nothing wrong with the application of evocative patterns to an object as long as they do not interfere with use, and this in the case of the car they rarely do. What makes the design of most American cars of today and also so many American railway engines and carriages, etc., so objectionable is not that the parallel lines and curves derived from streamlining are chosen as a means of decoration, but that they are so often overdone. How many chromium bands along the sides of car bodies stuck on as a substitute for something more daring, how many freakish radiator grilles do we see in the streets everyday?[1] In the United States this style is known as borax. Borax usually appears as a stripe formation possessing the symbolic significance of streamlining. Thus it is applicable to moving objects, such as cars, and irrelevant when applied to static objects like radio cabinets. As a type of ornament borax lends itself to artistic excess and in this way it has become disreputable. Often borax indicates the wish of designers to obtain results quickly and their readiness to let their standards be swayed by expertise in salesmanship. From another point of view, one which is purely associational, borax represents the noisy and self-assertive element in the American character. And from yet another, it can be used to keep a particular make of car always new and different. Changes in chromium decoration are less costly than structural alterations to the bodywork. The intrusion of borax into car design is particularly unfortunate; if most American cars

[1] On one well-known American car there are 182 pieces of chromium and stainless steel; only eleven of them are attached to accessories such as lamps.

were stripped of their chromium blobs and bands, it would be seen that the bodywork alone is a sufficiently convincing symbol of speed.

American bodywork is, as we have seen, appropriate to the manufacturing methods and the purposes for which it is intended. It is a three-dimensional conception of a horizontal moving object in which one can have a comfortable ride. Moreover, it is evocative of speed. It presents, however, a difficulty which American stylists have not yet overcome. This is the necessary use of a horizontal instead of a vertical radiator. The radiator on a car should be placed where the air pressure is greatest. And on American cars with their low, flat bonnets, the pressure is greatest over a horizontal, oblong area between the front wheels. For some unaccountable reason the radiator opening presents a problem to American stylists which none of them has managed to solve. The usual result, which is popularly compared with a mouth-organ, is abominable and specially distressing as viewed from this country where it has had a great deal of influence. It is said in Britain that this influence was a necessity because of the needs of the export market. However, the answer to this argument is that the British mass-produced car, if only for reasons of size, can in any case not compete with the American on its own ground. British cars will have to show certain qualities absent in American cars, and one of these might well be a more sensitive approach to body design. In fact Britain, for inspiration in this respect, is at liberty to draw on the best of Continental designs as well.

The most interesting example here is Italy. In Italy, the influential section of the motor industry may be said to be craft-based. Although steel pressings are used by some firms in this section, no attempt is made to produce in large quantities. Italian designers have used their tools to produce bodywork which is unrivalled elsewhere. With short production runs which give flexibility in manufacturing methods, relatively unlimited time to spend on details, a small, wealthy and enterprising market (which makes cost unimportant) and the services of brilliant designers, Italian firms have, especially since the war, set a very high standard. Take, for example, the Cisitalia (Pl. LXVI. 2) designed by Pinin Farina. All working parts, except the wheels, are enclosed in a tight metal skin. It has a slab-sided construction which gives full seating width and allows the line from the front headlights to flow horizontally towards the rear. After the door the line is subtly eclipsed by the rear mudguard which is humped up and forward to return the eye along the directional, forward emphasis of the body. This shaping of the rear mudguard, which is helped by the lifted line of the cantilevered luggage compartment, is indicative of the car's movement stemming from the rear wheels. Also expressive of forward movement is the bonnet, which slopes towards the front to sink below the adjoining mudguards. Several other elements in the design are worth notice. The forward pointing side-window, the emphasis on stability by thickening the bodywork between the front wheels and the window, and the radiator with its edge rolled inwards to belong to the rest of the body. These points, suggested by aerodynamic principles, have made the design outstanding.

The best-designed British cars certainly owe something to Italian practice. The British car industry, though considerably larger than the Italian, comprises several firms with methods similar to those used in Italy. In fact it is the peculiar nature of the industry in Britain that its manufacturing methods can be compared with those in Italy as well as in the United States. With the mass-production methods of the United States on one hand and the artistic standards of Italy on the other, manufacturers in this country are in a fortunate position. They can be stimulated by what is best in both countries and be able to apply it imaginatively to their own peculiar scale of production which, though not craft-based, deals in much smaller quantities than do American manufacturers.

Since the First World War the British motor industry has expanded at an increasing rate. In 1922 Great Britain had 875,000 cars on its roads, 1,475,000 in 1935 and more than two million in 1950. For the year 1935 the industry's production was over a quarter of a million vehicles; for 1950 over three-quarters of a million.[1] About 80 per cent of the cars produced have a capacity of less than 2½ litres, which accounts for the characteristic small British cars. There are thirty-one firms making motor-cars and numerous factories which supply pressings, engine parts and accessories; the total number of employees is approximately 750,000. Thirteen of the firms are controlled by six large companies (known as the 'Big Six') and together their production represents 90 per cent of the output of cars. The majority of car firms, including three of the largest producers, are located in Coventry and Birmingham.

In terms of production methods and quantity of output, the industry may be divided into three main sections. In the first section, where only mass-production and assembly-line methods are employed, I saw five firms, three of them members of the 'Big Six'. One firm has an average yearly output of 140,000 vehicles and employs 18,000 workers; this firm makes four main types of car in the low- and medium-price groups. Two other firms, both with 10,000 employees, have an annual production of 100,000 cars each. The two remaining firms make pressed-steel bodies, and now more frequently the integrated chassis and body, for motor-car manufacturers. I visited these firms for information on production methods; neither is responsible for the

[1] Although England has the second largest motor industry in the world, it is much smaller than the American industry, which in 1950 produced more than 6½ million vehicles. About half of these came from the General Motors combine, which employs 516,000 workers.

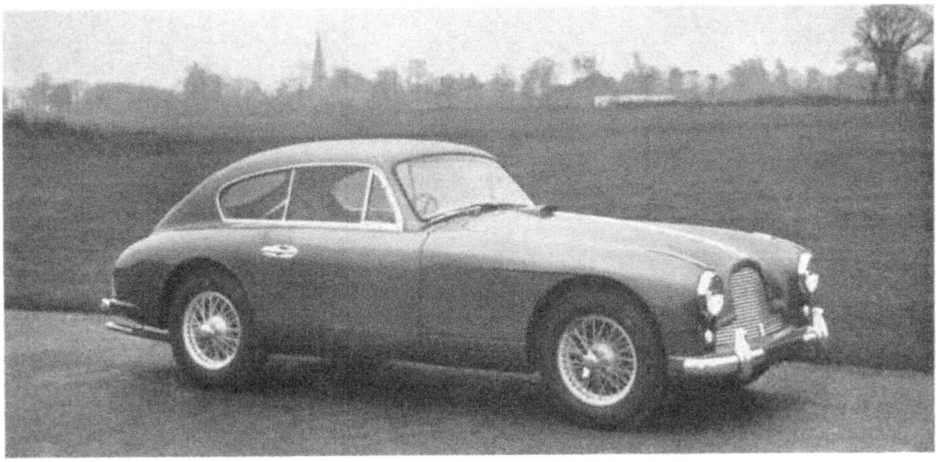

1. Reduced air resistance by streamlining is worth achieving for any car that travels frequently at more than 45 m.p.h.: for this car with speeds over 100 m.p.h. it becomes a vital factor to be determined by wind-tunnel tests. Influence from Italy blends here with the maker's 'family likeness'. *Designed by Frank Feeley for Aston Martin Ltd.*

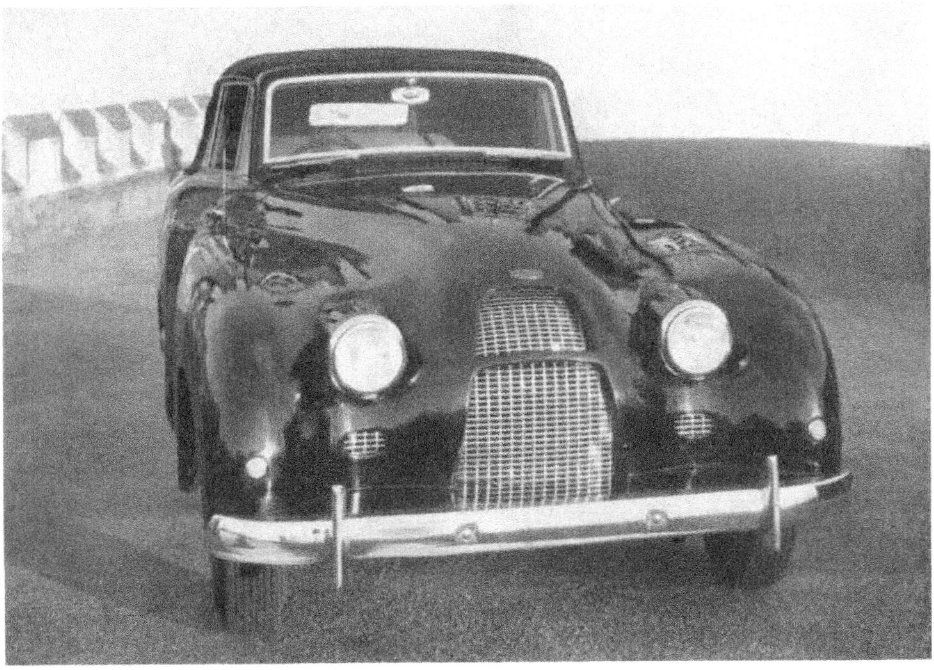

2. The air intake appears to present such a problem to American designers that they prefer to clutter it up with chromium plate. With the virtues of the American inspired slab-side this car, following an Italian lead, treats the radiator as simple, textured openings: a foil to the broad, flat bonnet. *Made by Allard Motor Co. Ltd.*

PLATE LXVIII

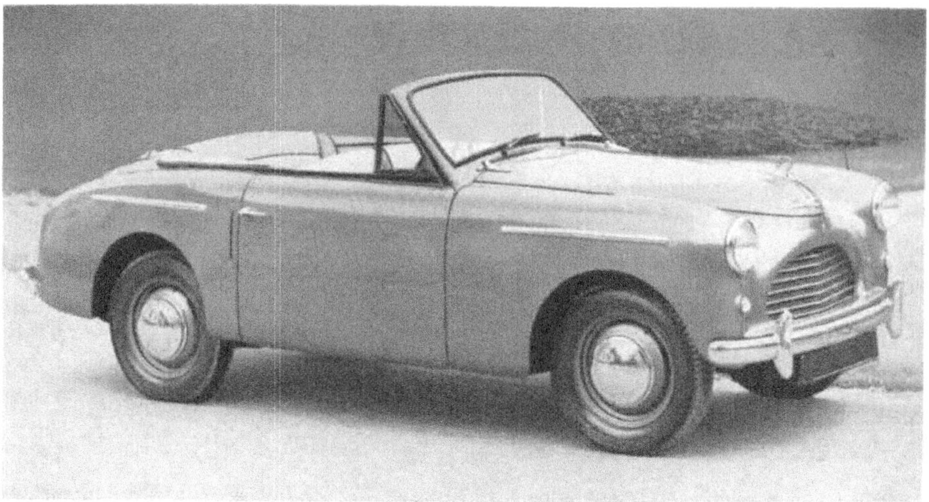

1. Horizontal emphasis is fundamentally suited to a car, and with the typically British short wheel-base it can more easily be achieved in sports models where the designer does not have to contend with a permanent roof high enough to give head-room. *Made by Austin Motor Co. Ltd.*

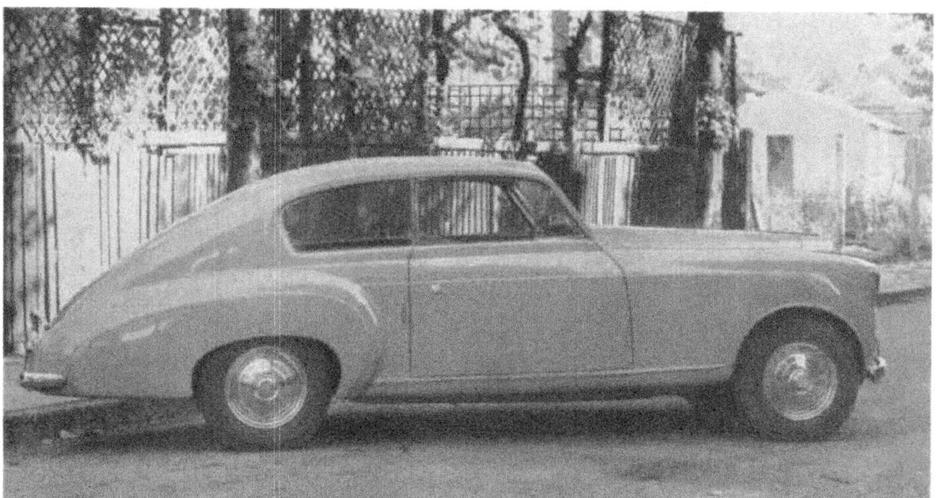

2. In this class of car the length and height of the engine mean that the bonnet must dominate the design. To counteract this the rear of the car is carried up and forward along the line of motion, by shaping the mudguard which protrudes from the slab-side. Chromium details have been tucked into the envelope. *Made by H. J. Mulliner & Co. Ltd. on a chassis by Bentley Motors (1931) Ltd.*

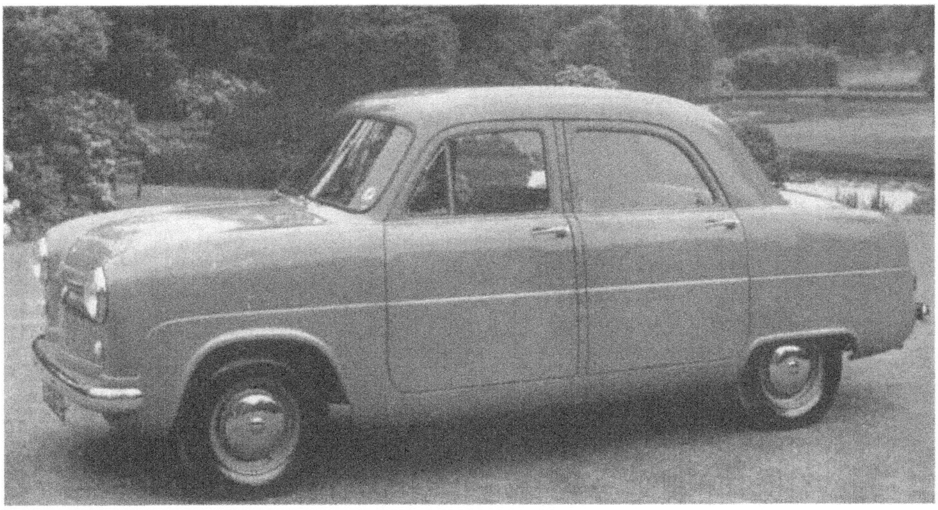

1. The first car on the British market to appreciate fully the advantages of the slab-sided monocoque construction. The occupants ride between the wheels on wide seats, have good visibility over a low bonnet and a large luggage compartment in the rear. On the characteristically short British wheel-base this design, only 13 feet 7 inches long, is a very great achievement. '*Consul*', *made (1950) by Ford Motor Co. Ltd.*

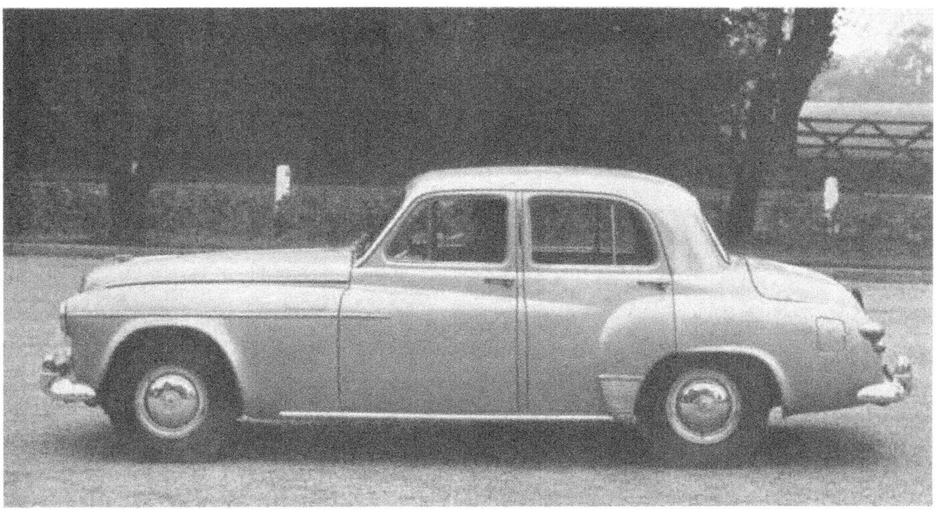

2. The large car has a better opportunity to achieve a good proportion between height and length, and there is little temptation to supplement the body with ornamental streamlines or borax. Like the Bentley opposite, a vertical radiator survives to contradict the modern trend. *Made by Rootes Ltd.*

PLATE LXX

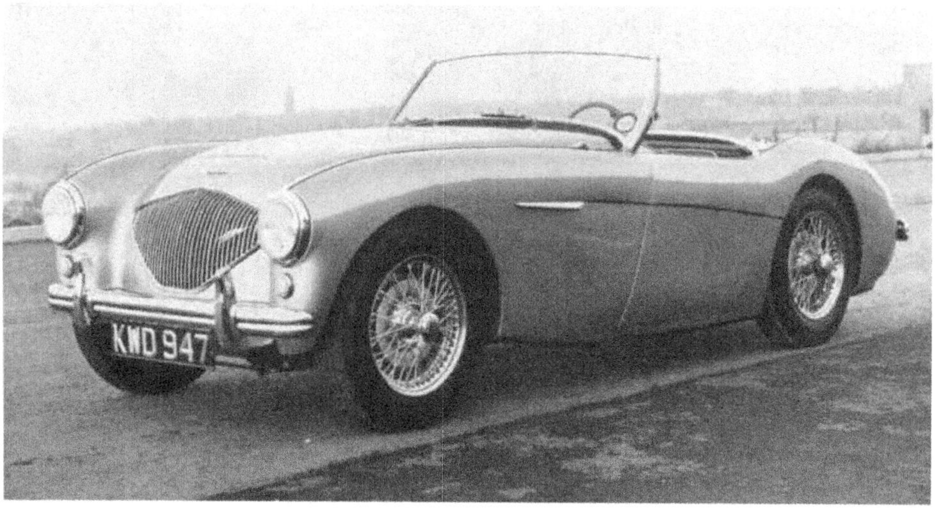

1. Much can be learnt by the British motor industry from design and construction developments in Italy and the United States. This is a mass-produced car having implicit American characteristics combined with the Italianate stressed metal envelope. Yet it is one of the best examples of the modern British sports car. *Austin-Healey 'Hundred', made by Austin Motor Co. Ltd.*

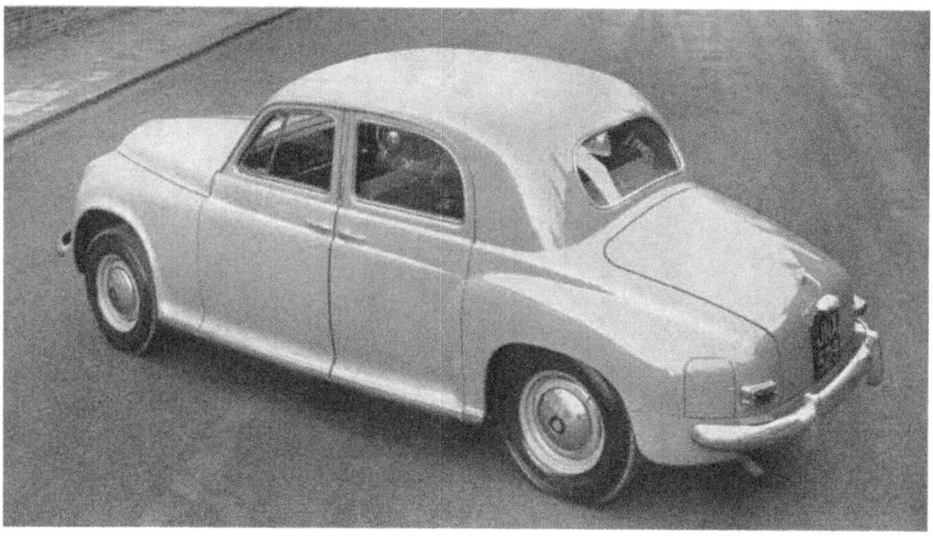

2. A car can be seen from all angles, not least from the rear. This model indicates both the complexity and the economy of large metal pressings held in tension to form the total shape. The body becomes a steel sheath composed of standardized parts to cut down tool costs. *Made by Rover Co. Ltd.*

design of the bodies. The largest of them employs 10,000 workers and turns out 1100 bodies a day.

In the second section I saw three firms, employing an average of 1500 workers. The batch-production methods used, together with small quantities (about 7000 medium-price cars a year) give greater flexibility, and more scope for individual skill and craftsmanship. These firms are particularly interesting for us because their size and relatively small production enable them to risk more enterprising designs. To some extent firms in this section set the design and engine performance standards which are followed by the rest of the industry.

In the third group I visited two firms. One of them (3000 workers) is responsible for two of the most expensive makes of car in the industry. Before the war engines and chassis were supplied to coachbuilders who were responsible for all body design. Since that time the firm has produced its own pressed-steel bodies. It has a yearly output of about 1000 cars, including the chassis that are still supplied to coachbuilders. The other firm is a coachbuilder with 270 workers. The expansion of the mass-producing firms since the First World War has been accompanied by a steady decline in the coachbuilding section of the industry. In 1918 there were more than ninety coachbuilders—at least one in every large provincial town and several in London. At the present time there are twelve, of which no more than five are of importance. It is usual for the coachbuilder to have a standing agreement with one or more manufacturers of engines and chassis. Some of the London coachbuilders have an international reputation, but only one or two of them are able to undertake individual commissions. The remainder carry out orders for which twenty to fifty bodies of standard design can be supplied. Clients who wish to own an individual car are encouraged to indulge their fancy in such things as colour, woodwork, internal fittings and trimmings. Finally, I saw two large and important producers of car accessories, such as lamps, horns, traffic-indicators, speedometers, gauges and clocks.

Certain factors in motor-car design are common to all firms. Standard chassis and engines that have been evolved after many years of research and experiment tend to dominate, and so impose a limitation on, body design. As the use of the integral chassis and body becomes more frequent, this difficulty will cease to exist. Nearly all bodies are hampered by the peculiarly short wheel-base demanded for British cars. Except in coachbuilding works, it is no longer usual to conceive a new body in terms of lines on the drawing-board. All firms make at least a quarter-scale model; the largest firms make full-scale models in wood and plasticine. In this way new bodies begin and are developed in three-dimensional terms. The managing directors in all firms I visited make themselves responsible for, and in some cases initiate, new designs.

All cars, including those produced by coachbuilders, are designed in terms of steel pressings. For light cars and some of the more expensive types, built by coachbuilders, aluminium instead of steel is being used for parts which are not structural. In addition there is the experimental use of plastic reinforced with glass fibres.

Finally, I found that manufacturers depended almost entirely upon the stock accessories supplied by three or four leading firms. This is an important point to bear in mind when assessing the design standards of British cars. For example, if the glass for a windscreen were offered more cheaply in one shape than in another, it would have to be accepted by the car manufacturer and worked into his design. If a new type of speedometer were required, the manufacturer would have to pay a tool cost of between £4000 and £5000 before the accessory firm could agree to produce it. While only the largest firms can afford this, it is comparatively simple for the designer to impose his own dial shapes upon stock instruments. Similarly, car manufacturers prefer to use dashboards designed by their own staff. Dashboards, incidentally, are one of the worst features of British cars. The lettering (described in one firm as 'Arabic styled Roman') is particularly thoughtless. Although cramped and badly arranged, many dashboards may be described as pathetic copies of the lavish instrument layouts in American cars. In general the design of instruments and lamps maintains a satisfactory standard; if any really bad examples are seen they may be traced to the car manufacturer who has attempted to improve them. Nevertheless, it must be noted that the design of lamps, especially rear lamps, often detracts from the artistic value of a car. In one accessory firm I found lamps which had been designed twenty years ago. They were being sold to the smaller manufacturers who could not afford the tool costs of a lamp which would be more in keeping with their own standard of bodywork design. Although the design of accessories progresses slowly, two recent trends may be noted. Circular rather than square, oblong and hexagonal dials are preferred for instruments; pastel coloured calibrations are more popular than those in black and white. Headlamps are now inset or half-inset in the mudguards. This is an obvious improvement, especially when used in slab-sided bodywork.

In the three largest firms I visited completely new designs are very rare. This is mainly because a production programme, involving special tools and specific numbers of skilled workers, is in constant operation. Every alteration to the programme means a delay at some point along the production and assembly lines, where the speed of operation is, in any case, dependent upon the slowest worker. Once a certain type of car has been accepted for production in these firms, it remains basically unchanged for five and sometimes seven years. If the design is satisfactory there is no reason to regret this. However, it is usual for some superficial changes to be made from year to year, so that the car maintains the *appearance* of being a new model. Such changes are commonly confined to the radiator, or the contours of the back mudguards and luggage compartment. These changes are justified economically as they do not involve the expense of large pressing tools. Artistically they should be condemned because they disrupt the proportions of the original design. It is worth pointing out that slab-sided bodywork could not be treated in this way. Where large quantities of three or four different types of car are in production, an entirely new design will be introduced only once in two years. The development of new designs and of modifications to existing models is in the hands of the managing director and the chief chassis and body engineers. The three basic factors to be considered are dependability, running cost and performance, and they are all controlled by the chassis engineer. Chassis and body engineers must work in terms of specific production methods, and include nothing which will increase the proposed maximum cost. Within these limitations the body engineer or designer must attend to the comfort of the occupants and the general appearance of the car. From this it will be seen that the application of a preconceived artistic body form is impossible. As with design in all other products, the shape of the body must evolve naturally out of the engineering and production methods, which are, of course, only partly controlled by the body designer.

The work of the body designer is often marred by the fashionable styling clichés requested by the sales manager. Furthermore, the artistic judgement of the managing director is decisive. This is understandable once it is realized that a new design takes about two years to develop and must remain in production for a further five years at least. This means that the managing director must decide what type of design will best sell his car in seven years' time. If he thinks that the appearance of a new model could be improved within the limitations already settled on, he may call in a consultant designer. This was the practice in one large firm I visited, where an American designer was retained as a consultant. Unfortunately, the influence which he was able to exert was merely superficial. In the other two large firms the design of new models is carried out differently. I found that the body engineers in both firms have sufficient authority to build their own prototypes. As they possessed the production experience and technical ability to work out their own ideas, they were able to present a convincing case to the chassis and production engineers. In this way the artistic standard they desired has a much better chance of survival. But I was also told that in large firms there is little sympathetic collaboration between the technical, production and sales departments. This leads to a narrowness of outlook which, from the design point of view, can be harmful unless the body engineer is either a director, or able to develop his work to a stage where it can withstand superficial criticism. Not one exceptionally gifted man but a team of experts makes a motor-car, and the best motor-cars are made by experts with equal authority. The technical engineer wishes to make full use of his experimental department, the production engineer wants to keep his assembly-line inflexible and so maintain maximum output, while the sales manager has to meet the varying demands of a large market organization. But what does the body designer represent? What section of the factory can give him an authority to match that of the other members of the team?

In the three firms where the batch production of motor-cars is carried on, I found that the bodywork is more important than the chassis. Unlike in the largest firms, cost is not a deciding factor for it is realized that cars in that class sell more readily upon engine performance and appearance. In two firms the body design is in the hands of the managing directors. The other firm employs a body designer with a status equal to that of the technical engineer. Because the cars are priced to carry a comparatively high tool cost, very long production runs are unnecessary. This means that steel pressings can be used more adventurously; a fact which is especially true of coachbuilt bodywork. Here the body engineer has so much freedom that his work, considered as a design influence upon the mass-production side of the industry, almost ceased to be relevant. This need not be regretted. The coachbuilder serves a small, wealthy public which is antipathetic to modern design trends. Although steel and aluminium pressings are used, they are thought of merely as substitutes for hand-beaten panels. The design style of coachbuilt cars belongs to a bygone period of staid, upright elegance.

All the body engineers I met are encouraged to visit the Continent and the United States frequently. Their salaries vary according to the importance which is attached to their work, and not the size of the firm. The salary scale ranges from £1000 to £1500 a year. I was repeatedly told that there is no accepted source of trained body engineers.[1] The usual method is to promote them from the drawing office or obtain them from other firms in the industry. All but one of the designers I interviewed had begun their careers at a technical college and continued them as draughtsmen, while gaining experience of production techniques. The exception is a designer in one of the largest firms. He had been trained at an art school and afterwards apprenticed to a coachbuilder. It is recognized throughout the industry that body design has international ramifications. Although American and Italian sources are the most important ones, recent work in France and Germany is not overlooked.

[1] The Institution of Body Engineers, formed in 1948 to give motor-car designers a professional status, has devised a three-year course of training for students. The course is held at the Polytechnic Institute, London, but it is hoped that similar courses will soon be started in Coventry, Birmingham and Oxford, the chief production centres for motor-cars.

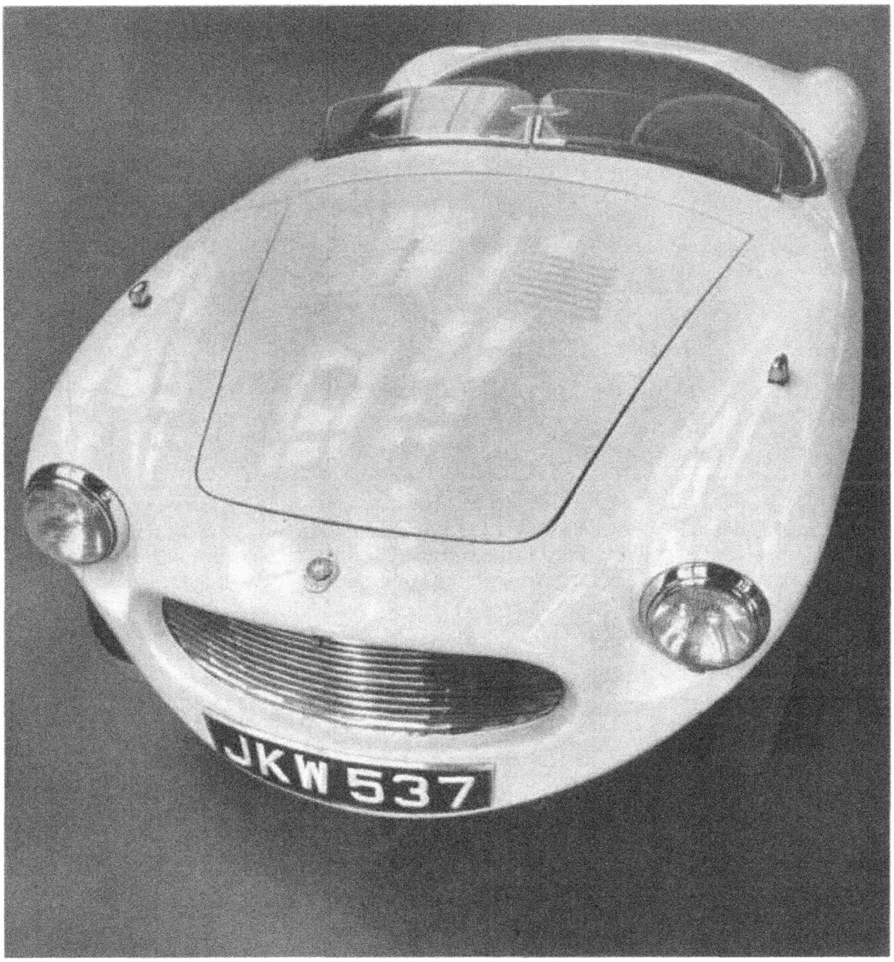

A dramatic view of a car using a material which may fundamentally alter the appearance of bodywork. The broad bonnet and wings are made of laminated plastic with woven glass cloth for reinforcement. While giving lightness and strength, and in some circumstances low cost, plastic bodies will bring with them the opportunity of designing unusually complex forms. *Made (1953) by Jowett Cars Ltd.*

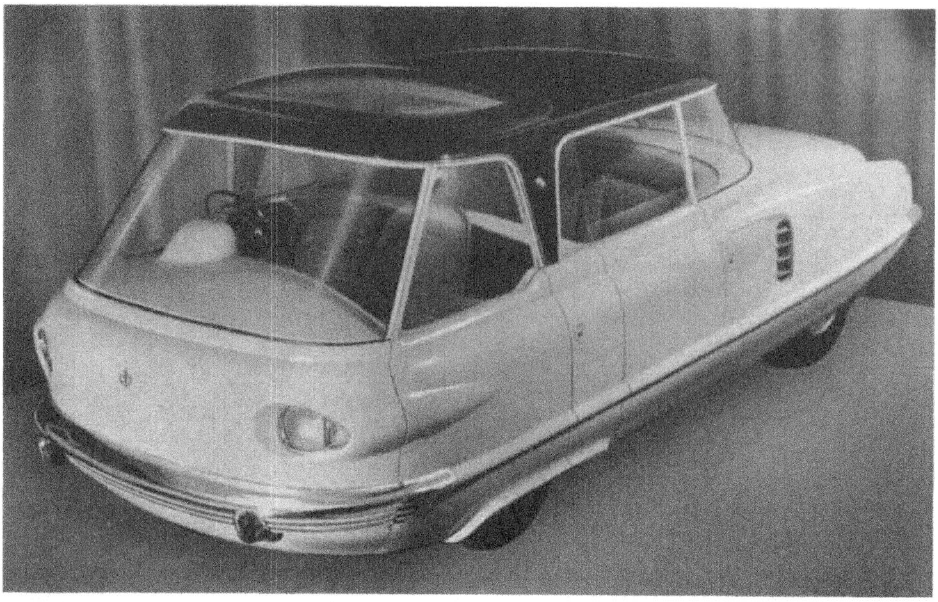

1. 1951. 'Car of the Future': a scale model for the Festival of Britain. The designers were asked to indicate a possible car for 1965, using a gas turbine in the rear. Lack of data prevented the effect of engine weight on stability from being considered. Integral chassis and body construction, waist-high side-panels and good forward visibility epitomized current trends. The unusual width of 6 feet 3 inches, to seat seven persons, is brilliantly disguised by crisp body lines. *Designed by E. G. M. Wilkes, Hugo G. Poole and Peter J. Ashmore for the Council of Industrial Design.*

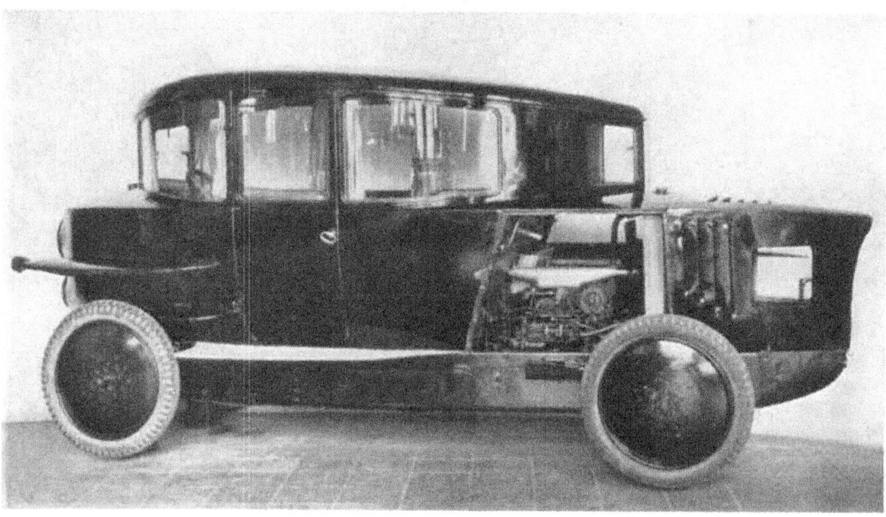

2. 1921. The first streamlined car ever designed was an attempt to apply aerodynamic principles to cars scientifically. The car was shaped like a boat on a four-wheeled carriage. At that time it was thought that a streamlined shape depended on having the engine in the rear. *The 'Tropfenauto' designed by Dr H. Rumpler.*

MOTOR-CARS

During the war practically no motor-cars for civilian use were produced. In 1945 when the industry had once again to sell its products on the domestic market, the only tools available belonged to the 1939 models. At the same time there was a heavy demand for cars of all types, and production began with whatever was available. By 1947 most firms, including the 'Big Six', were producing cars based largely on the pre-war models, but incorporating new bonnets and new rear sections. This was the interim period of car design, and it was not until 1948, at the Motor Exhibition, that entirely new models were introduced. These post-war phases of car production are important, because they partly account for the three main types of design apparent in the industry today. The first type includes large, medium- and small-size cars, in all price groups, which are said to be 'traditionally British'. In style they belong to the typical motor-car of the 1930's, a car which had the so-called traditional features of a vertical radiator grill, mudguards attached as separate units and headlamps poised on chromium columns, vertical lines being everywhere accentuated. The style of the second type of car is derived from the interim period. There has been an attempt here to maintain an upright, British dignity, while allowing certain American features to intrude. The designs are necessarily half-hearted and reflect the efforts of mechanical engineers to do something artistic. Instead of attempting to make the *whole* body symbolic of speed, they prefer to stick on little bits of chromium here and there. Again, these cars show the disastrous result of trying to suit the tastes of too many markets at the same time. For instance, the bonnets and mudguards remain divided into six or more separate pressings, but the radiators are half vertical and half horizontal. Again, the luggage compartments are extended, as in American cars, but the rear seats remain fixed above the back axles. These cars are full of such contradictions, and have done more than anything else to prejudice the public against a genuine modern style.

Finally, we come to the third type of car, represented by two of the largest and several medium-size firms. In this group there has been a serious attempt to evaluate what is best in recent American and Italian achievements, and then relate it to a British scale of production. The body designers responsible have been able to work out their ideas in collaboration with, and not under the direction of, the technical engineers. The following points are characteristic of the group. Both front and rear seats are slung between the axles. Slab-sided construction is used to give an increased seating width and a unified exterior form. This, together with horizontal lines emphatically carried out, has been related to streamlined contours. The remarkably short wheel-base of most British cars has made it difficult for designers to obtain the right degree of crispness in pressed steel forms. There is a tendency, especially in small cars, for the radii in the bodywork to become too full, resulting in an inflated appearance. Although symbolic ornament, like borax, can be legitimately used to replace this loss of horizontal emphasis, the problem of evolving a satisfactory design related to the short wheel-base remains. Out of the various attempts to solve this problem, a characteristically British style will emerge. There need be no loss of individuality amongst well-known makes of car if this is tackled intelligently. Britain has proved in the last fifteen years, in such cars as the Rover, Armstrong Siddeley and Triumph, that designers can achieve artistic results of a high order. We know now that they were achieved on inadequate premisses. The slab-sided construction has so much to recommend it that, in the near future, most manufacturers will adhere to it. Some cars such as the Ford Consul have on these new and improved premisses already achieved a remarkably sound appearance. But more is necessary, and looking back on the past one can say with every confidence that it will be achieved. It is likely that body designers will be encouraged to concentrate upon this problem because manufacturers have realized that facts concerning engine performance and reliability are no longer sufficient to ensure the sale of mass-produced cars in competitive markets.

The outlook for design in the motor industry appears to be very promising. It is a prosperous, enterprising industry, not afraid of taking risks. It is keenly competitive and many of its leading firms have survived by innovations, not retrenchment. It has the benefit of a critical public, the real force of which has not been felt since the war. The public knows what it wants because it thinks more about motor-cars than about electrical fittings or carpets—cars after all need not be related to a furnishing scheme. The manufacturer, for his part, is aware of his responsibility as the initiator of designs and is very proud of his products. Both parties to the design process, producer and consumer, should be able to work reasonably and interestedly towards a high standard of modern design.

THE PLACE OF THE CRAFTSMAN IN INDUSTRY

In this section I am not concerned with the routine handwork of craftsmen in firms making pottery, glass, hand-made furniture, leather goods, jewellery, block-printed textiles and wallpaper. This is not creative work, for manual dexterity has no artistic significance. But where good design results from constant and intimate contact with the production methods—methods which change and develop and methods which are complicated—an experimental craftsman is needed to aid the designer. His task is to work on machines similar to those involved in the production process. He may also use hand-tools, but their flexibility can become a menace to design for mechanical production unless it is at all times related to machinery. The craftsman may begin his experiments with new materials or new techniques with hand-tools, but his object is to decide, with the engineer, if a modified or entirely new machine can be built to do the work. In cases where suitable machinery cannot be evolved there is no point, for a firm with its costing system based on mass-production, in carrying out more research. The second use of the craftsman is in a firm wanting to try out new designs on the market. In nearly every case, this implies that the craftsman's tools are used to reproduce the machine's work in quantities so small that mechanical production would be uneconomical. In both cases craftwork is the experimenting ground for the designer: first with new processes, second with new designs. It can perhaps be added that it is a great advantage to design in association with a craftsman whose intimacy with materials and elementary functions endows him with a freshness of approach often lacking in the staff designer for machine production. I am, of course, aware that in some firms the designer is himself a craftsman. This is especially true of woven textiles, leather goods, furniture, silver and pottery.

EXPERIMENT WITH NEW PROCESSES

In this book there is no place for adequate discussion of even the most important new materials and processes which have been developed for industrial use since 1935. While I have seen many of them used in factories I have also seen others, potentially useful, but neglected or misused by manufacturers. New processes, and new purposes as well, require new forms and patterns. These can only be achieved when a designer and a craftsman are employed and given freedom to experiment. The designer's scope in this field is very wide. Apart from developments in his own firm he can draw upon the work of universities, the Department of Scientific and Industrial Research, the National Research Development Corporation and other public and private bodies, such as the material suppliers (Courtaulds, British Nylon Spinners, Imperial Chemical Industries, Bakelite, British Industrial Plastics, the British Rubber Development Board, etc.).

In the textile industry a number of new materials (nylon, ardil, terylene) have not yet been exploited by designers of furnishing fabrics.[1] An example of the process which should be followed is provided by Fothergill and Harvey Ltd., who commissioned Margaret Leischner to direct the weaving of a new plastic yarn in experimental designs. Miss Leischner's work in fancy yarns for R. Greg & Co. has already been described. In furniture there are several new materials and techniques available, but

[1] At the British Industries Fair in 1953, I.C.I. showed a wide selection of experimental fabrics of ardil and terylene designed by some leading textile designers.

the industry has shown little interest in them. Moulded plywood is one example,[1] and craftsmen and scientists of the Forest Products Research Laboratory have demonstrated how thermo-setting glues and simple jigs can produce new constructions which, if used in commercial production would, for several purposes, be more economical than the conventional woods bent in the solid state. Some few instances in this country, but far more in Finland, Denmark and the United States, suggest remarkable possibilities for furniture designed for laminated construction. Again, the science of anthropometry, concerned with analysis of the dimensions of the human frame, seems to have left British designers and manufacturers singularly unimpressed.[2] While one has to look abroad for the truly creative use of metal furniture (solid, strip and tubular steel), British furniture designers and small firms have shown what can be done with cast and sheet aluminium, moulded plastic sheet and sponge rubber.

In connexion with rubber, the designs by Dennis Young for the British Rubber Development Board should be mentioned. As one of its activities, the Board promotes Latex Foam, a compound of natural rubber, which can serve as an alternative to metal springs in many types of upholstery. As the Board was anxious to demonstrate its product to furniture manufacturers, designers and architects, Dennis Young was appointed, in 1951, as design consultant and lecturer. His lectures are given in furniture schools and colleges and are attended by local manufacturers and retailers. His design work, so far, relates mainly to two experimental chairs, using metal and wood as the structural components. In both chairs the upholstered units incorporating Latex Foam have been prefabricated by hand. The peculiar versatility of the material, and the ease with which it can be prepared for use, have been emphasized by the unconventional shapes achieved.[3]

In pottery there is photo-lithography, in jewellery investment casting, in light fittings the 'Chrysaline' lamp shades of liquid plastic,[4] in crystal glass a specially adapted process of sandblasting. These four processes have already been taken up by commercial firms. The majority of the others have been neglected. They have reached the stage of being developed experimentally by private craftsmen and designers, but without the assistance of industrial firms. This means that they are still in the category of hand-made articles and lack the discipline of design for mass-production. Moreover, they are unnecessarily difficult and expensive to buy.

[1] See experiments by Michael Thonet in 1836 (Gordon Russell, 'Design Policy for New Techniques', *Design*, December 1953, pp. 7–10). Alvar Aalto has designed for moulded laminates since 1928.

[2] See Gordon Logie, *Furniture from Machines*, London, 1947, chap. 1; Bengt Åkerblom, *Standing and Sitting Posture*, A.B. Nordiska Bokhandeln, Stockholm, 1948; Brigid O'Donovan, 'Sitting in Comfort', *Design*, July 1954, pp. 17–21.

[3] See Dennis Young, 'Experiments with Latex Foam', *Design*, January 1953, pp. 20–2.

[4] For the lamp shades, see Pl. XXVI, 1 and 2; also *The Architectural Review*, May 1951, p. 321.

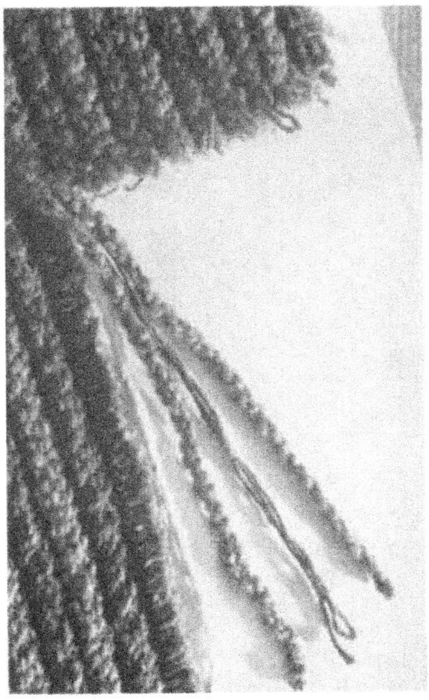

1. Three yarns from the range of R. Greg & Co. Ltd., which is shown to the trade in sample lengths of possible combinations woven by Margaret Leischner (see 3).

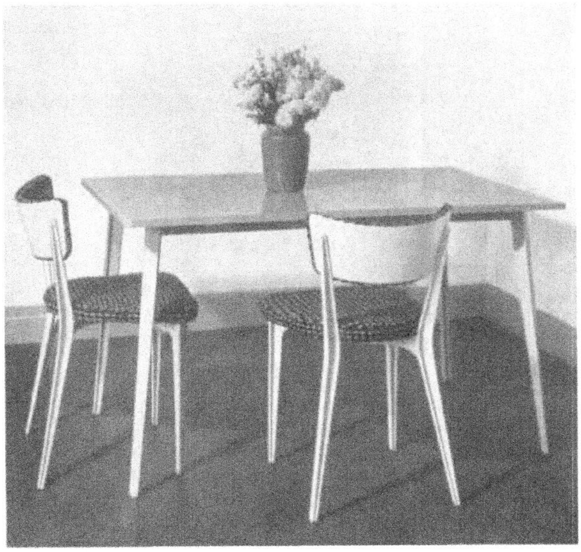

2. For lightness and strength the chairs' and table's supports are made of aluminium castings. An unfamiliar material for furniture, naturally employed, required new forms and opened a new market. *Designed by Ernest Race for Ernest Race Ltd.*

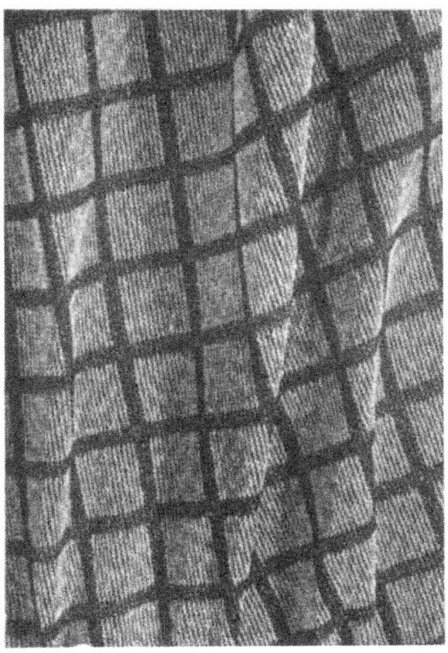

3. Margaret Leischner creates new yarns, as well as weaving others in Greg's range into finished patterns, for customers to see how to exploit them without copying the designs.

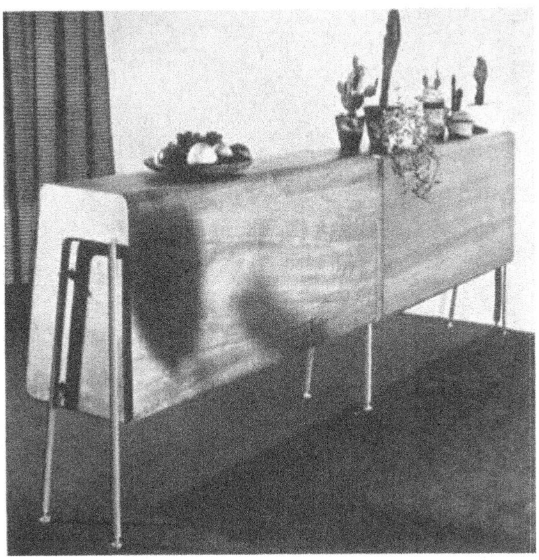

4. The stimulus of a specialized furniture competition in New York prompted this winning entry, which explored in prototypes the possibilities of bent plywood sheets and steel supports. (View from rear.) *Designed by Robin Day and Clive Latimer.*

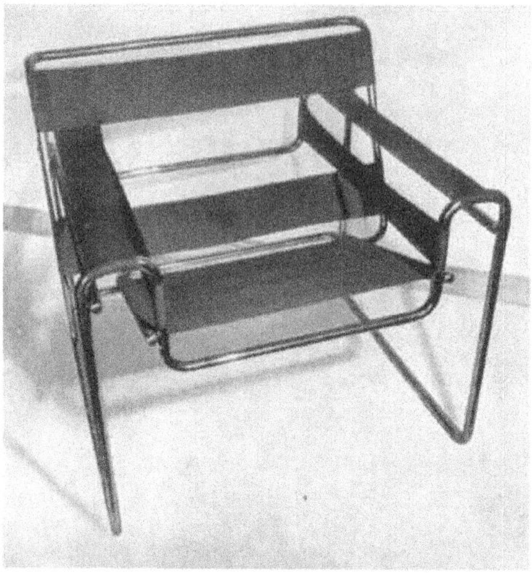

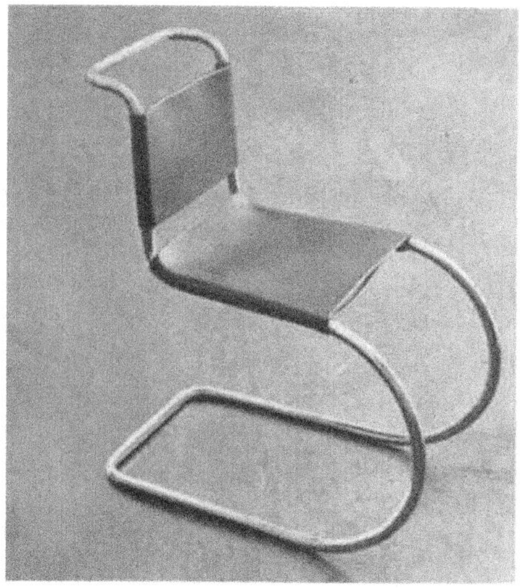

1 and 2. The original tubular steel chairs were developed at the Bauhaus: they result from designers who were helped in the initial stages by workshop craftsmen producing on a small scale. *Left*, the first tubular chair, designed in 1925 by Marcel Breuer. *Right*, the first chair to exploit the resilience in steel, designed in 1927 by Ludwig Miës van der Rohe.

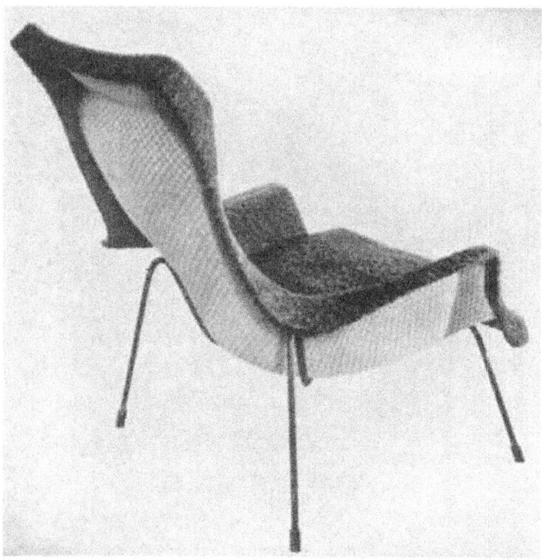

3. This chair is basically a moulded plastic shell slung in a cradle of steel rods: a series of shells and rods could be stacked inside each other for packing. The great possibilities here have not been taken up by the trade. *Designed by Dennis Young.*

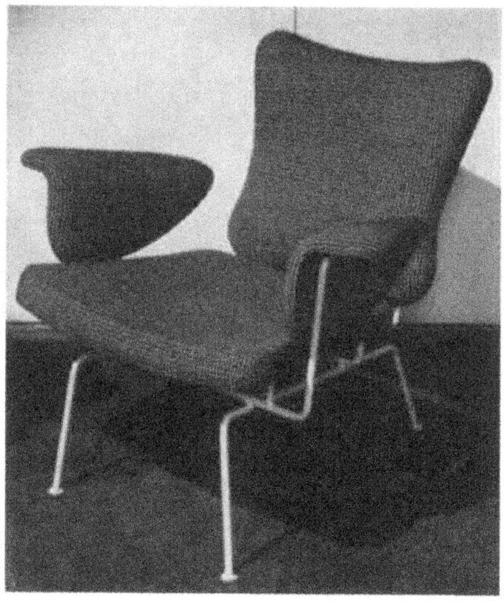

4. To demonstrate the uses of Latex Foam the British Rubber Development Board commissioned this hand-made chair. The material presupposed an unconventional shape. *Designed by Dennis Young.*

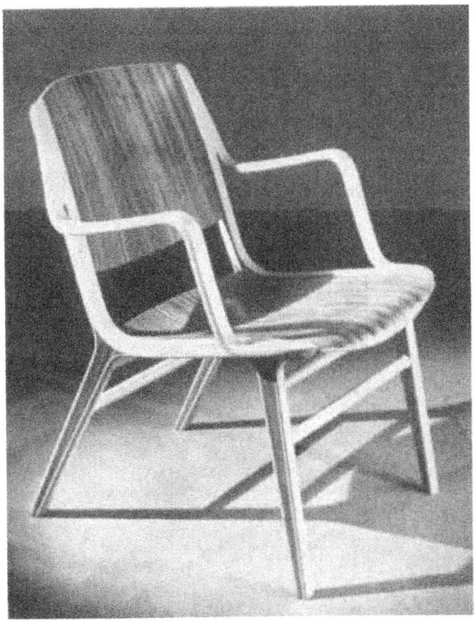

1. A long and costly experimental period was needed before this remarkable design for laminated structures could be produced. It is a best-seller. *Designed by Peter Hvidt and O. Mølgaard Nielsen for Fritz Hansens Eftfl, Denmark.*

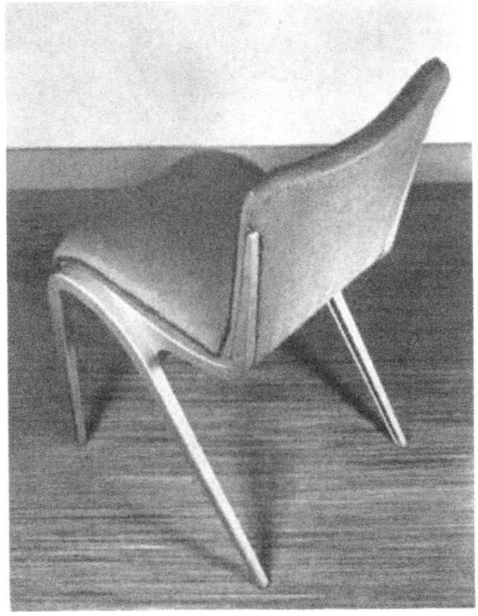

2. Freedom to experiment with new processes is lacking for designers, especially in the furniture trade. This stacking chair exploits light alloy die-castings. *Designed by J. W. Leonard for Educational Supply Association Ltd.*

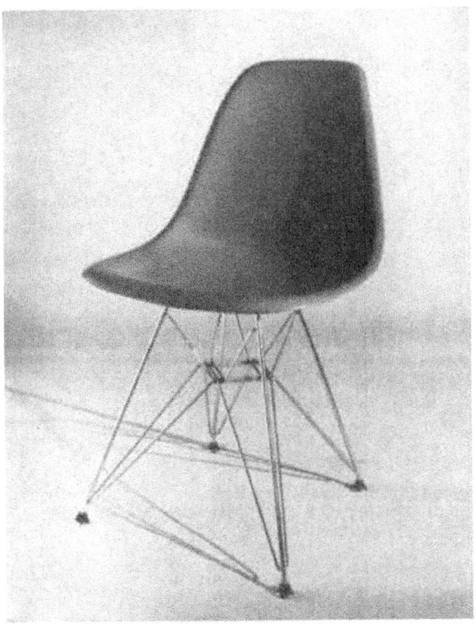

3. New shapes possible with plastic reinforced glass fibres have been developed in the U.S.A. The shell here is $\frac{1}{8}$ inch thick and flexible for comfort. *Designed by Charles Eames for Herman Miller Furniture Co. U.S.A.*

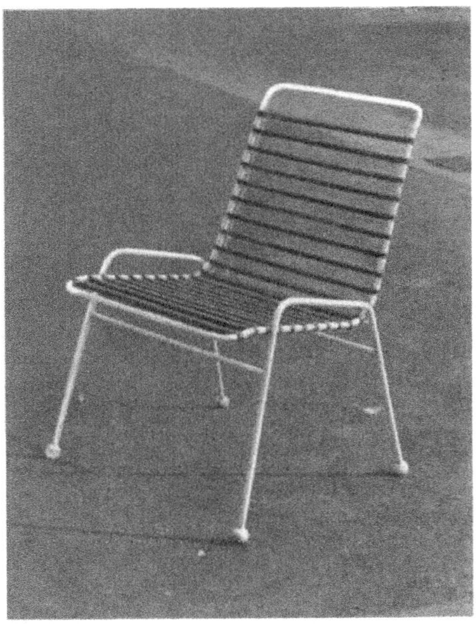

4. The Festival of Britain was one of those useful occasions for introducing unfamiliar designs. This chair of steel and plastic was admired on the South Bank and later bought in quantity. *Designed by Ernest Race for Ernest Race Ltd.*

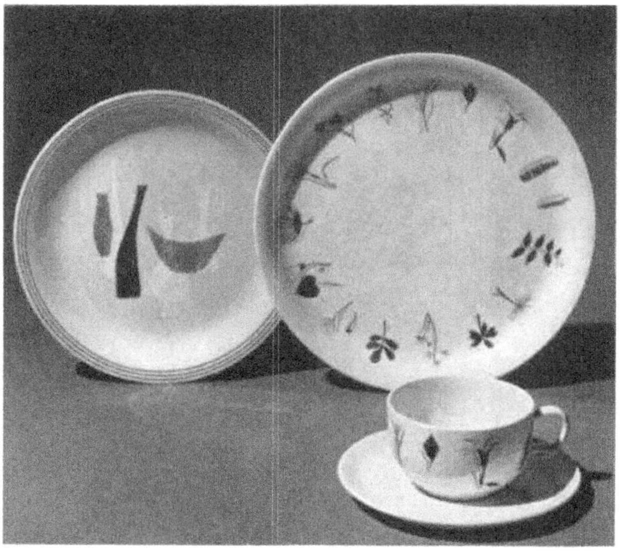

1. To avoid initial misunderstanding the inventor of a new process should experiment with small-scale production. Customers can appreciate the unique nature of the silk-screen ceramic transfer process by handling these designs. *Designed* (left) *by Colin Haxby*, (right) *by Jacqueline Groag for Johnson, Matthey & Co. Ltd.*

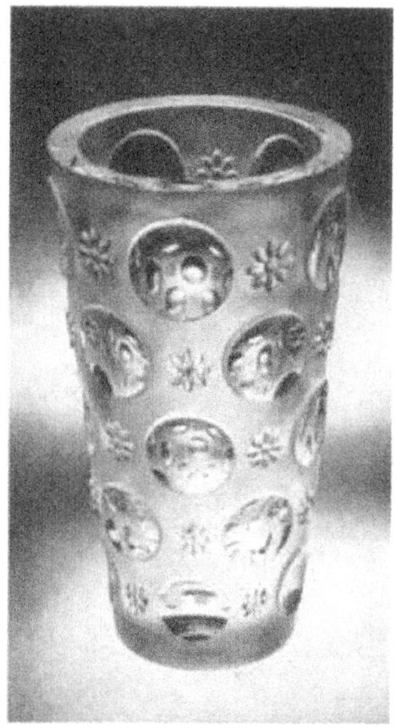

2. Working through the medium of a craftsman, designers can develop new techniques. Though for crystal glass sand-blasting may never become a mass-production process, it widens our knowledge of pattern work. *Designed by Irene Stevens for Webb Corbett Ltd.*

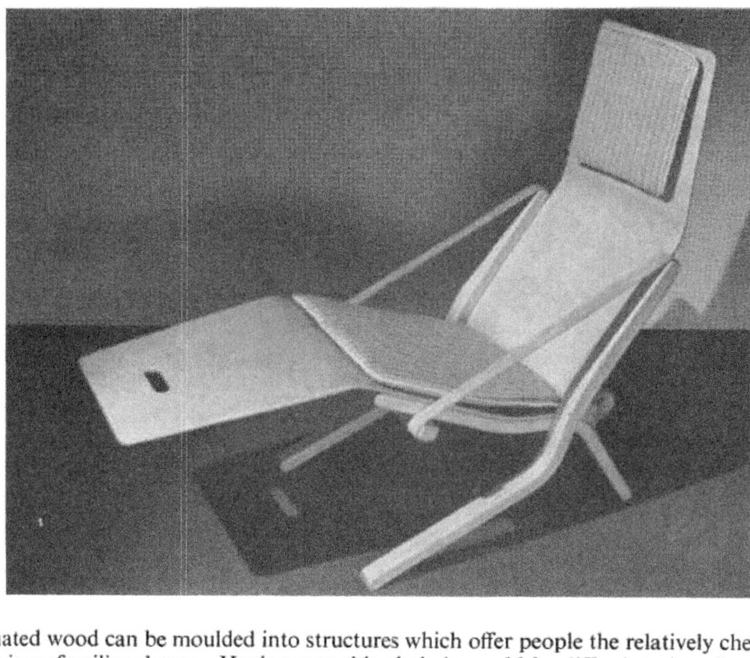

3. Laminated wood can be moulded into structures which offer people the relatively cheap articles they want in unfamiliar shapes. Having seen this chair it would be difficult to imagine an equally good solution carried out in a different way. *Designed by Ernest Race for Ernest Race Ltd.*

There are several instances of the successful application of new processes and materials by designers aided by craftsmen in mass-production firms. Some of them have just been noted, while others, such as stainless steel for tableware, fluorescent lighting, the baffle-board radio receiver, moulded plastic shells for suitcases, have been described in earlier chapters. Two more examples can be given. The first comes from Germany and concerns tubular steel chairs.[1] The first piece of steel furniture was patented by an American in 1904. It was a divan constructed not from steel tubes but from steel rods. About 1924, another patent was taken out in the United States referring to a chair without separate back legs which was also made from steel rods.[2]

The first chairs using steel tubes were created by Marcel Breuer, who was then in charge of the cabinet-making class at the Bauhaus. That was in 1925.[3] In 1925–6 he provided the new Bauhaus buildings at Dessau with his furniture. The designs were carried out at a small locksmith's workshop at Dessau. Production amounted to about 500 pieces in 1926 and 1000 to 1500 pieces in 1927. In 1928, A. Lorenz in Berlin took over the production, and thus became the first real manufacturer to turn out tubular steel furniture on a commercial scale.

Ludwig Miës van der Rohe, the last principal of the Bauhaus, was the first to realize the great possibilities of steel construction for giving chairs lightness and resilience. His solution of the problem, which was the conclusive step in the development, was patented on 24 August 1927. The manufacturing was for the first few years done in the workshops of a firm called Berliner Metallgewerbe Jos. Müller. But since 1929 Gebrüder Thonet, the famous mass-producers of bentwood chairs, who some time later took over the Lorenz firm, started production on a larger scale. In this country the first steel chairs were put on the market about 1930. Since then several other manufacturers have followed. As for the designs now on the market, they do not compare well with those of Rohe and Breuer, which have been often copied or imitated with slight and artistically unnecessary alterations.

The second example concerns laminated plywood furniture and comes from this country and the United States. In 1948 the Museum of Modern Art, New York, held an International Low-cost Furniture Competition. One of the first two prizes of $5000 was won by two British designers, Robin Day and Clive Latimer. Their designs were for cabinets made from sectional units using a laminated plywood body wrapped round solid end-pieces. The basic design could be adapted to form a sideboard, book-

[1] The following account is taken from Nikolaus Pevsner, *An Enquiry into Industrial Art in England*, and is based on detailed letters to Dr Pevsner from W. Gropius (22 September 1934), Marcel Breuer (3 October 1934), Miës van der Rohe (7 May 1935) and A. Lorenz (12 June 1935).

[2] H. E. Nolan.

[3] See *Das Werk*, vol. VII, 1926, p. 210.

case or secretaire. The plywood technique was first developed by James Henderson, the managing director of a small Scottish furniture factory. It enabled the units to be constructed with only two joints while using a comparatively small amount of wood.[1]

Although the first models were constructed by hand the designs were specifically intended for mass-production. This complied with the aim of the competition which was to interest American manufacturers in the winning designs. But when an American firm, Johnson-Carper Furniture Co., started to produce the sectional units the original design was almost entirely abandoned. The laminated process gave place to conventional solid wood and joints construction. The tubular steel support frames were replaced by separate legs, the sliding doors by the more common hinge type. Standardization and flexible interior arrangement were both dispensed with and the units became little more than a series of chests of drawers.

This is a particularly vicious example of what might easily happen in this country if the original designer and craftsman are not able to supervise all stages of development within the firm concerned. On the other hand the story of tubular steel chairs shows how admirable results can be achieved, when both persons are implicated in all stages. It also shows how the hand-craftsman is the necessary medium for a designer's creative experiment before a new process is fit for mass-production. In addition, the point should be made that in nearly every case the initiative for using new processes comes from outside the industries most concerned. If British manufacturers were not so notoriously unenterprising, many more inherently valuable results of scientific and workshop research would now be available on the market at reasonable prices.

EXPERIMENT WITH BETTER DESIGN STANDARDS

Craftsmen are also used by manufacturers who wish to try out new or better designs. In this context I am only concerned with the latter. As we have seen, manufacturers are naturally unwilling to risk the introduction of an improved design in large quantities before the public has indicated its response. In many firms, of course, no improvements have been attempted; and yet, in all industries but those concerned with linoleum, jewellery and electro-plate, some examples can be found where there has been a serious endeavour to improve design standards. The usual method is to set up within the firm a separate department where small quantities of designs are made up by hand or machine. The techniques used are not necessarily related to mass-production and often better quality materials are used. Both circumstances mean, inevitably, that an unnecessary premium is put upon the better designs with the result that their market is limited. Nevertheless, it sometimes happens that the sales of the

[1] See *Design*, May 1950, pp. 17–19.

better quality article will encourage the firm to introduce improved designs into its mass-production business. This has happened at Morton Sundour Fabrics after the experience of Edinburgh Weavers, at Orrefors after the success of its art glass, at the Vereinigte Lausitzer Glaswerk after Wagenfeld's high quality glass had established itself. The same thing has happened at the Corning Glass Works, in the United States. For instance, it was decided to improve the artistic element in its designs for mass-produced oven-ware, after the success of high quality glass, made by its subsidiary Steuben Glass, had been proved.

There is, however, a different, and I think preferable, arrangement. In outline it is this: The department producing better designs in small quantities would be only physically, not financially, separated from the main factory production. The craftsmen employed under the designer would use methods which enable the same designs to be mass-produced without alteration. Furthermore, the department would be subsidized so that its products could be sold at a competitive price. When a successful design is found under this system it could be transferred directly to the machines and produced in large quantities. The use of high quality materials is irrelevant to good design. If they are introduced at all they should be confined to the mass-production process where large sales of one design warrant an improvement in the materials used. The ultimate object of the department is to merge gradually into the supporting firm whose mass-produced designs are one by one improved, until all of them maintain a high standard. But, no matter which system is adopted, it is essential for the manufacturer to insist upon the collaborative goodwill of his production and sales staff. Without their help any project will fail.

To sum up: the craftsman is the necessary interpreter of the designer's ideas during the experimental stages of a new design. With new materials and techniques his job is to exploit their possibilities, often by using hand-tools, and then to relate his research to mechanical methods of production. His other job is similar. By conforming very closely to the production methods of his firm, he provides the designer with a practical trial ground for better designs. He then makes up small quantities which are placed before the public.

Far too few manufacturers are aware of the *essential* function of the craftsman in firms specializing in quantity production. And yet in nearly all firms I visited one or two men can still be found who are skilled in the use of both hand and mechanical tools. Instead of routine repetitive work on the production line they could, in many cases, be far more profitably employed as experimental craftsmen under the designer's supervision.

CONSULTANT DESIGNERS

Consultant designers have been already mentioned, without much explanation, in these industries: furniture, carpets, metalwork, domestic equipment, light fittings, radio cabinets, leather, plastics and motor-cars. The consultant designer is distinct from the staff designer, the free-lance and the architect. He is an independent specialist in design for firms in one or a group of industries, able to carry out research work in materials, production methods and the market. He can also follow a design from its inception through models, mock-ups and prototypes—made by him—until it is in production, without relying altogether on the skill and machinery possessed by a manufacturer.

For mass-production in industry the idea of the consultant designer was first developed in the United States in the late 1920's by men such as Norman Bel Geddes, Walter Dorwin Teague, Henry Dreyfuss, Harold Van Doren, Egmont Arens and Raymond Loewy. Mr Loewy's career, which began in 1927 with the designing of a duplicating machine,[1] is the most remarkable, as it is also the best known.[2] In the middle 1930's Loewy, in association with Carl Otto, started to work for British manufacturers and he opened a London office in 1936, but this was closed down in 1952.

The question of receiving and paying for advice on design from somebody outside industry was entirely new to American manufacturers. Mr Loewy's early experiences illustrate the point. For instance, in connexion with a contract with the Hupp Motor Company at Detroit, Loewy patented, in 1930, a very advanced car body design. To get it even considered by the Company he had to build a full-size prototype himself. During the slump, when manufacturers were looking for novelty easily applied to their products, he had surprisingly few commissions. In the autumn of 1933 he had only one account. His first big opportunity came in 1934 when Sears, Roebuck & Co. engaged him to redesign a refrigerator.[3] Mr Loewy's expenses for making models, prototypes, etc., amounted to nearly three times the fee he received—$2500. But the demand for the refrigerator doubled, and for his next design he received $7500. In 1936 he was paid $25,000 for similar work. By this time Loewy had formed his own company, with three managing designers, a business manager, and a publicity and

[1] For an account of the original Gestetner duplicating machine and subsequent modifications, see *Design*, April 1952, pp. 12–14.

[2] Raymond Loewy, *Never Leave Well Enough Alone*, Simon and Schuster, New York, 1951.

[3] This was not the first case of a refrigerator being designed by a consultant, as Mr Loewy suggests in his book, p. 125. In 1933 the Norge Company engaged L. V. A. Guild to design a complete range of refrigerators, two of them embodying an entirely new pressed-steel technique.

public relations department. In 1937 his staff included thirty-eight architects, technicians and engineers. By 1939 it numbered 100, by 1951, between 140 and 150. Apart from his office in New York he has two branches in the United States. His work is divided into designs for domestic and commercial products, transport, packaging and specialized architecture. His most spectacular commission was with the Pennsylvania Railroad Company when the steel plates for fifty-seven locomotives were welded on his suggestion and to his design. Mr Loewy's experience of motor-car body construction enabled him to suggest this alternative to riveting which made production less costly. Since 1943 he has been retained by The Studebaker Corporation, and has been responsible for all new models since the war. His design for the 1946 model was particularly progressive and set a very high standard.[1]

So much for the American example. In this country before the war any designs obtained outside the factory came from architects and free-lance designers. Only in the printed textile industry were independent commercial designers in operation. Apart from Carl Otto and Raymond Loewy, the consultant designer capable of handling a new design for a domestic product through all its trial stages was unknown.

In 1942 the Design Research Unit was formed by Herbert Read to provide a link between designers and manufacturers. It had no permanent designers on its staff, but in 1945 it was joined by Misha Black and Milner Gray. These two designers had originally practised together as Industrial Design Partnership since 1934. Since the war the Unit has increased in size and now consists of twenty-three architects and designers. In normal times architecture, exhibition and interior design and packaging design provide most of the work, industrial design of commercial and domestic products least. I interviewed this organization and in addition I saw four separate firms of design consultants, two in London and two in the Midlands. Two of them started work in 1945: the others in 1947. One firm specializes in kitchen and catering equipment in plastics, but carries on an extensive practice in design for commercial interiors. A staff of ten designers and craftsmen is employed. Another firm, with three head designers (two with engineering degrees, one with art school training) and over twenty engineers and draughtsmen, is concerned with light metal products, chiefly for commercial use. In the third firm a wide range of domestic and commercial products has been produced by two designers (both trained in art schools) with a small staff. The remaining firm, with a staff of seven, deals with a limited range of light metal and plastic articles for commercial use.

With small variations the firms employ the same working methods. After an inquiry from a client has been received the head designers will estimate the scope of

[1] See Pl. LXVII.

their work and the time required. Whether the design is to be entirely new or an adaptation of an existing article, the first necessity is research. The success or failure of previous designs will be weighed against those of competitors. The client's manufacturing methods will be studied in case new uses of machinery will produce better designs or lower the cost. Market research is not usually conducted by these firms. The cost of packing, storage, repair and maintenance must also be considered. With this information the firm will prepare several rough drawings of alternative designs and discuss them with the client's production and sales staff. Normally a scale model would follow, and if approved the firm would construct a mock-up. If a prototype were necessary it would, in the case of the firms I visited, be constructed by the client. The final stages of work by the firm would entail fully detailed working drawings and supervision of the production trials. In some cases the firm will design the carton or package and suggest methods of presenting the article in retail stores. Payment for work can be made in several ways. A retaining fee assures the client that his consultant designer will not work for competitors. Payment can also be made for each piece of work, either on completion or after each stage in development. Or, very rarely, a royalty is paid, consisting of a percentage of the sales of the new design. For exhibition work it is usual for payment to be based on a percentage of the total costs.

The consultant designers I have interviewed were concerned mainly in work of a satisfactory, often very high, artistic standard. As might be expected none of them has been outstandingly successful with manufacturers. But unlike the staff designer the consultant is an independent expert called in to perform a specific task. If he is well known and has an experience of methods in other industries he should obtain the respect of his client's managers and foremen, upon whose collaborative goodwill his success depends. As far as product design is concerned, the consultants I met have not had time to secure this necessary status. In addition they have comparatively small staffs with a limited knowledge of current production methods in the industries for which they work. For instance, it would require a very large firm of consultants to design competently for the light metal industries alone. And at the present time there are several cases where a consultant, in order to satisfy his client's request for better appearance (the term is 'eye appeal'), has patched up an inherently bad design with chromium ornament. A better technical knowledge would have enabled the consultant to persuade his client that his product should be completely redesigned.

A brief examination of American consultant designers will bring out certain points which may well be characteristic of consultants in this country ten years from now. The American consultant is a species of advertising agent who tells his client that good design will mean good business. Although I cannot quote from their books,

American consultants have said, in fact, that a product will sell better if they have designed it. Now there are, unfortunately, several reasons, which we have already discussed, for supposing that a high design standard has nothing to do with an increase in sales. What partly accounts for the success of American consultants is their facility for satisfying the manufacturer's craving for novelty. Faced with a buyer's market, British manufacturers may well demand similar service. Again, as the American consultant operates a large staff, personal responsibility for a design standard tends to be obscured by the need for making money. But, on the other hand, he is exceedingly thorough in his work. With his own staff of technical specialists he tackles every aspect of a job and their combined intelligence will inevitably locate one or two points where cost can be saved. Their knowledge of mass-production methods is very wide and they provide one of the links which enable one industry to benefit by adopting the practices of another. Here, then, is the main reason why sales often increase when consultants are employed, as it also helps to explain the high status enjoyed by American consultants: a status which *can* enable them to propose a high design standard and get the manufacturer to accept it. And if difficulties do arise the consultant is always prepared, like Loewy and his motor-car, to prove his point by building the design himself.

THE RETAIL TRADE

This comparatively short chapter on the retail furnishing and domestic equipment trade can do no more than suggest some ways in which department stores, chain-stores, large and small shops can help to improve design standards. To tackle the problem effectively, each different type of domestic product would have to be considered in relation to stores all over the country in much the same way as the manufacturers we have just dealt with. The task would be a large one and would require a separate book. Instead, out of the fifty-six stores I have visited in different towns while collecting material from manufacturers and designers, I have selected thirteen which I think are representative of the retail trade. In these stores I asked questions about their methods of buying, display and selling. I also asked about their expected rate of stock turnover, their opinion of public taste and their attitude towards it.

Before dealing with the stores in detail there are some general points to be made. It may seem to be a fruitless supposition, but if there were no retailers, no middlemen, between the manufacturer and the public, the design standard of industrial products would undoubtedly improve. While I am well aware that such a situation would mean the end of quantity production and inexpensive goods, the supposition does emphasize the point that a direct collaboration between producer and consumer would lead to a closer understanding of each other's wants, and consequently to an improvement in design standards. Above all, it would mean that a manufacturer would no longer have to gauge public taste by his intuition of guesswork, and by what the retailer tells him, before putting a new design into production. Let us face it. The retailer *is* an obstacle in the way of improved design standards. We have only to compare the designs of all types of goods, especially carpets, textiles and light fittings, for sale in stores with those produced under the direction of an architect or designer for special contracts, for shops, offices, aeroplane interiors and private houses and occasionally also for hotels, passenger liners and theatres.[1] The designs of the first type are on the whole mediocre and tend to be sentimental, vulgar or fashionable, while the latter have a vital and individual character. The truth of the assumption that all contract work is controlled by gifted and competent designers does not matter; the point is that a great many retailers are bad conductors of public taste. It follows that at least some sections of the public are not being allowed to buy the goods they intuitively want. Although in many respects this will always be the case, it remains the responsibility of the retailer to widen the range of his stock by including some pieces of modern design.

The firms I visited may be divided into three groups: those which stock no modern designs at all, those which stock a few, and those whose stock is almost entirely devoted to modern designs. In the first group, which represents more than 80 per cent of the retail trade, I saw four firms. One is a general retailer of furniture and soft furnishings in Scotland, another is a chain store in the North of England selling similar, though less expensive goods. The other two firms are in the Midlands. One deals with hardware, fireplaces and fires, while the other stocks all types of electrical products, including light fittings. Amongst the retailers who have made an effort to sell modern designs I saw four firms. The first, in a Scottish city, sells furniture, furnishings, glass, china, etc., to the professional and upper classes. The second, in a north country city, is similar, though its furnishing departments comprise only a small part of its total business. The manager of the third firm, a furniture store in a Midlands town, said that his clients came from the upper working class and middle class. The fourth firm I saw is a very large department store in the Midlands selling all types of domestic furnishing and equipment.

As the types of store already mentioned can be found in any town or city, I felt that it would be better to see five stores which are mainly concerned with modern designs. One of them, a small but high-class furniture shop with some textiles, is in the West Country; the remainder are in or near London. The largest store deals with all types

[1] See the account of the interior furnishing design in The Time and Life Building, *Design*, March 1953.

of furnishings, two others of medium size stock mostly furniture, while the last is very small and is concerned mainly with hand-made furniture and *objets d'art*.

Speaking at a conference in 1946 on design, H.A.Holmes, then Chairman and Managing Director of Selfridge's, Ltd., said '...when we get to the goods in which both fitness for purpose and decorative function are combined, my feeling is that the public not only does not fully appreciate good design, but is often in an extraordinary muddle about the whole thing. These, of course, are the items bought very infrequently by the average customers, and about which they know least when they buy.'[1] If that is a fair generalization of the public's response to domestic furnishings, what does the retailer do about it? In the case of the first group of stores I found that stocks were limited to those items which had always had a steady sale. This meant that the design of furniture, textiles, light fittings, fires, etc., on show was not different in kind from the best-sellers of 1930 and earlier. For the public, accustomed to buying in these stores, there were no alternatives available and consequently no modern designs were shown.

What are the reasons for this? First of all it was explained to me that the public did not want modern designs. The public's response to war-time utility furniture and the designs which immediately followed it was usually the first point made.[2] Then the financial risk of trying anything new was emphasized. It is usual for a retailer to invest a large amount of his capital in stock. With the present-day margin of profit, which is lower than before the war, a furniture retailer should sell his entire stock every ten to twelve weeks. This means a rate of four to five turnovers a year. As that is the average for all designs, those which sell slowly must be balanced by quick-selling designs. It is clear that even if a modern design were stocked, its trial period of three months might well be too short for it to be recognized and sold. In a small furniture shop the problem was put another way. I was told that if a bedroom suite remained in the shop for a year it would cost the owner £12. With that type of risk involved all the retailers in this group preferred to stock only designs similar to those which had sold before. In addition, the buyers who obtained the designs from manufacturers and the salesmen who passed them on to the public were conservative in their choice in order to maintain a steady commission on their work.

Much has been said about the retail buyer and modern design.[3] Most of it has been

[1] H.A.Holmes, 'Design and the Retailer', a paper read to the Conference on Industrial Design, September 1946, arranged by the Council of Industrial Design and the Federation of British Industries. Reprinted in the *Journal of the Royal Society of Arts*, 25 October 1946.

[2] See pp. 5–6.

[3] See, for example, J.W.Waterer, *Leather in Life, Art and Industry*, Faber & Faber, 1946, p. 226: '...the trade buyer whose ethics visualise nothing nobler than buying down to a price rather than up to a standard', etc.

derogatory and there is little to add. Designs, good or bad, meant little to the buyers I met in this group of stores. Their one responsibility was to buy the goods they knew could be sold at a profit. After them comes the salesman in the store, whose job is to please the customers. In furniture shops, for example, few salesmen have a knowledge of design, timbers, construction and function. 'Salesmanship consist[s] in the art of concentrating the customers' attention on some superficial point of novelty provided for this purpose by the manufacturer.'[1] The more decoration, bright filigree knobs and contrasting veneers there are, the more the salesman will have to talk about. A modern design of wardrobe or dressing table provides different subject-matter for conversation, which can only be brought out by training.

The standard of design in the hire-purchase chain stores is invariably the worst to be found. There are several reasons for this. The method of sale on credit (for example, two bedroom suites worth £150, for 30*s*. a week) appeals to the lower end of the trade. Yet, why cannot furniture worth 30*s*. a week be of modern design? The commercial spirit is, of course, very pronounced in these stores. The question which the salesman asks is not what would you like, but how much can you afford to put down as a first instalment? Once the figure is fixed he shows you its equivalent value in a limited selection of furniture, and you make your choice. The design of all of it is atrocious anyway, so no critical comparisons are possible. Few managers in chain stores will accept the responsibility for the designs stocked. One I met said that his was only one store in a large distributing company and that his stock was part of bulk purchases made at his head office. Again the procedure is the same. The buyer bases all his purchases on the sales figures of his company's stores. What sold last time is bound to sell again.

The two Midlands stores, one selling hardware, etc., and the other selling electrical products, catered for upper-class tastes. The design standard of their goods, apart from the plain utilitarian items, was appallingly showy and vulgar. In both shops special displays presented various flickering electric imitations of coal fires. Although as a type they were the most expensive fires on show, they represented in one store 75 per cent of the sales of all fires. The other store, where I was informed that flicker-coal fires appeal to a 'cultivated taste', sold nearly as many. Also in this store 90 per cent of the fireplaces sold had insensitive shapes and were faced with bright glazed tiles in beige and brown. The remaining 10 per cent were of bricks, stone and marble, a little better in design, but more expensive. Some of the ceiling light fittings I saw in a very overcrowded collection were of the chandelier type, antique in design, and

[1] Dennis Chapman, 'Families, Their Needs and Preferences in the Home', a paper read to the Furniture Design Conference, July 1949, arranged by the Council of Industrial Design, which also issued a report.

made of dark stained oak or thin 'wrought' iron. Others, in what is called the 'Adam' style, consisted of plain brass chromium-plated tubes with parchment shades and crystal glass drops suspended underneath. The managing director of the store commented: 'I would hate to be involved in the selling of some of the hideous light fittings in stock.' In defence of this he pointed out that the manufacturers and the public were responsible. He merely passed the goods from one to the other. When I asked why his ceiling was confusingly littered with light fittings, a typical sight in any electrical showroom, I was told that the maintenance of a very large stock was inevitable. New models, in small quantities, were always being ordered from the manufacturers, not because they had any particular merits, but because they were new and different. His salesmen could not talk about the forms and functions of the designs, except to say: 'Look, madam, this is a new model, just come in!'[1]

The first two of the four firms which make an effort to sell modern designs can be dealt with together. They are the only stores in two large cities where a variety of modern furnishings can be seen. Both stores began selling expensive and exclusive designs of furniture and some textiles in the 1930's. I found that they still had a similar trade but the bulk of their goods was comparatively inexpensive. The remainder of their stock consisted of imitation antiques and poor designs of no period. Though both stores were owned by large companies, the managers in charge were wholly responsible for the design of all that was bought and sold. Moreover, they were both interested in modern design and well aware of the market it commanded. This market, I was told, had been extending gradually, especially since the war. The salesmen were equally interested in the modern as in the other types of design. They had visited furniture factories (one had had some experience of machine weaving) and were better educated and trained than the salesmen we have discussed so far. The display methods used were also different. Instead of the tedious, badly-lit rows of wardrobes, dressing tables, sideboards, etc., to be seen in most furniture emporia, I found suggestive schemes showing how designs of furniture, upholstery, carpets and light fittings could be arranged together. In the case of one store some of the furnishing schemes contrasted modern with period designs, a method which, I was told, brought a more active, critical response from the customer. Instead of limiting them to the shop window spaces, these schemes were used over most of the floor area.

Such ideas are elementary, and yet it is surprising to find how little they are used. Another concerns the buyers. The buyers, particularly of furniture, carpets and textiles, should attempt to correlate their opinions on design and price before any stock purchases are made. Only in the smallest stores is all the buying carried out by one

[1] See p. 65.

man. For the others it is in the hands of several people of differing taste and experience. It is the same for the salesmen, who also work in separate departments, although they frequently have the responsibility of dealing in turn with customers who wish to buy all the furnishings for one or more rooms in the same store.

As few firms have made any deliberate effort to test the public's reaction to modern designs, it is worth describing in detail the enterprise of the third firm in this group, Hopewell's, Ltd., of Nottingham. The firm wished to introduce to the public modern designs in tax-free furniture, and in June 1950, arranged its own exhibition. There were fifteen groups of modern furniture, including dining-room, bedroom and unit furniture, upholstery and incidentals such as small tables and chairs, etc. Eleven furniture manufacturers sent pieces. The exhibition, which was the first to be staged by the firm, lasted seventeen days. At the same time a free open competition was held asking visitors to place, on a special card, each group of furniture in order of preference. When doing this they were asked to consider three points: 'general appeal, attractiveness of design and fitness for purpose.' A first prize of £20, second prize of £10 and third prize of £5, were awarded to the three persons who placed the groups nearest to an order determined by the popular choice. In this democratic way the order of merit was determined by all the votes cast.

The highest points were scored by a bedroom suite, in second place came upholstered furniture, and in third place a natural oak dining-room suite together with upholstered armchairs. The competition card also included a questionnaire which invited visitors to state their preferences on design, wood, finish and shade. The results, in terms of percentages, were as follows:

		Per cent
Design	Reproduction	15
	Semi-traditional	26
	Modern	59
Wood	Oak	50
	Walnut	33⅓
	Mahogany	16⅔
Finish	High gloss	11
	Semi-bright	10
	Matt	39
Shade	Dark	15
	Medium	55
	Light	30

It was estimated that no less than 10,000 people visited the exhibition. Of them, 856 entered for the competition and 710 filled in the questionnaire. The managing director of this store, which caters for the upper working and middle classes, said the competition aroused considerable interest among young people. The object of the competition was to get the public to state which modern furniture designs it preferred. The results, together with the answers to the questionnaire, gave clear evidence of the public's interest. Apart from this it would be unwise to attach too much significance to the questionnaire results, as they do not represent actual purchases. Only 1 in 14 visitors took part and it is often the young rather than the older people who answer such questionnaires.

The very large department store I visited is part of a combine controlling other similar stores. Most of its stock is obtained from bulk purchases made by the combine. For instance, once a year about 200 furniture manufacturers are invited to submit easy chairs ranging in price from about £4 to £19. From this collection the sales managers, directors and buyers select roughly thirty designs, and place large orders for them. The store caters for three types of customer which were described as follows: the working class, with very small incomes and 'no coherent idea of what it wants'; the middle class, conservative in taste and looking for something with 'character'; and the professional class, which wants a definite, usually good, type of design at the cheapest price possible. Only in the furniture department had the store deliberately attracted attention to modern design. As far as textiles, carpets and floor coverings, hardware and light-fittings are concerned, the inclusion of modern pieces did not appear to be the result of a definite policy. However, I collected evidence from all departments in an attempt to assess the taste of the public.

The range of carpets, as can only be expected, offered little choice. The salesman explained that carpets were usually the first items to be chosen when furnishing a house. He said he had difficulty in selling the more garish floral designs. Linoleum, enjoying a strong sellers' market, was atrocious in design, but felt base, a cheaper and more plentiful material, had much better patterns. Mourzouk, a type of matting made in India with the worst Cubist and jazz patterns imaginable, should be mentioned as a cautionary example from another country. It had a popular sale for kitchen use. In upholstery and curtain materials flowers and leaves in browns and greens made up the bulk of the designs. While these inexpensive weaves and prints were neither good nor modern, they showed an improvement on similar qualities of cloth fifteen years ago. Quilted bedspreads in diamond stitching and pastel colours satisfy the desire of many people for richness and opulence. I was told that people save up to buy them rather than the cheaper bedspreads with coarse printed patterns. In the crystal glass department rejected exports in criss-cross cut patterns were popular, but not more so than a group of plain, glass bowls and decanters in satisfactory designs. The really nasty enamel transfers of ships and huntsmen, etc., were selling badly. The salesman for aluminium hollow-ware told me that, from his experience of public taste, 'no amount of "art" makes up for bad utilitarian qualities'. Apart from some decoratively embossed teapots, most of his stock attained a satisfactory design standard. In the department for light fittings I learned that customers were sceptical of the very fussiest lamp shades and wanted to try them out to see if they worked. Light greens and all shades of red were the most popular colours. However, for ceiling fittings in the £2 to £9 price range, designs with imitation candles were not as popular as the straightforward fittings made of wood and metal combined with glass bowls. In the same department I was told that an expensive reflector fire of modern design sold better than an imitation coal fire of similar price.

In general the situation was not inspiring. Even with the limited number of first-rate modern designs available it was clear that a more stimulating display could have been arranged. In the same store, the furniture department showed how this could be done. The department occupied a very large area and the different types of furniture were portioned off in island groups. All the easy chairs were clustered together. So were the beds, the three-piece suites and the sideboards. But, at one end of the area and near the windows there was a display of modern designs arranged in a series of furnished schemes. This section began in 1949 with unit furniture. As sales more than justified the floor space occupied, the section was allowed to expand, and to include most types of furniture, and several of the best modern designs. The salesman told me that although the process was slow, the public was becoming increasingly interested. If this were maintained he would be allowed to expand his section steadily. The salesman insisted that modern designs must be given prominence. If they were distributed amongst the other groups on the floor they would hardly sell at all. The segregation of the modern from the indifferent designs, and the placing of one man in charge of them, appeared to be wise because of the commission he got on each item sold. It is interesting to note that the store failed in an attempt to start a similar section for reproduction antiques.

With the retailers who are mainly concerned with modern designs, the chief point to emphasize is their uncompromising faith in the best work of contemporary designers. Indeed, apart from describing some of their methods and accounting for their success, there is little more to be said. The picture of the enlightened factory owner or managing director will now be familiar to the reader. Men of the same quality are to be found in the retail trade, so there is no need to deal with them in detail. From my experience of the five stores in this group I was able to view critically the remainder of the trade, including the firms already described. Apart from the small shop selling hand-made furniture the stores in this group share several characteristics. They are managed by their owners. Their stock is a reflection of the owners' taste for the modern designs available. Quality of workmanship also influences their choice and before the war most of their goods were expensive and often made by hand. I found that since then medium-price furniture, textiles, etc., are in the majority. Three of the stores have their own small factory workshops producing special pieces of furniture for contract work and a limited number for sale, sometimes in the inexpensive ranges. The owners of these stores are also responsible for most of the designs made by them. Furniture of this kind tends to be of better quality than furniture brought in from another factory and sold for the same price. This is because only one set of profits is entailed.

Another characteristic I found amongst owners of these four stores was their willingness to collaborate with a manufacturer who wished to bring out a new design, usually in furniture or textiles. It was customary for the owners, after seeing and perhaps advising on a suggested design, to place, in advance, orders which would make the manufacturer's first production run worth while.[1] In return, the store could expect to have the exclusive right to sell the design within its area of distribution.* The larger articles in these stores are usually displayed as parts of complete furnishing schemes. The salesmen are carefully chosen and trained to take a critical interest in modern designs. And finally, the buyers attempt always to correlate the design and prices of the various articles in stock.

The small shop selling hand-made furniture, hand-woven rugs, textiles, pottery, glass and *objets d'art*, belongs to a special category. It is perhaps of only marginal interest in this context. The exclusive and precious qualities of its stock reflect the difficulty of trying to interest some wealthy sections of the community in modern design. Although most of the designs were extremely good there was a tendency in this shop to favour the novel and the unusual, not as incidental elements in a design, but for their own sakes.

The position of the retailer, under modern conditions of production and distribution, as an indispensable link between the manufacturer and the consumer has been emphasized. If the standard of industrial design is to improve, the retailer must be an active intermediary. His task has several aspects. It is often supposed that a retailer places himself in a precarious position when he tries out a modern design. This might be true for the small shopkeeper, but it does not apply to the larger

[1] See the Good Furnishing Group, p. 17.

stores.[1] The risk they run is far less than that which is faced by an enterprising manufacturer who has to meet all the production costs of a new and untried design. Again, retailers might do much more to collaborate with manufacturers who are willing to produce modern designs. As H. A. Holmes pointed out '. . . retailers in the past have been too prone to play safe all the time. How many stores are willing to take a share of the reasonable risks run by enterprising manufacturers in laying down a line of production? And yet their interests are common.'[2] Moreover, this collaboration could be extended to several industries at once, with the retailer as the central adviser.

It is, of course, the retailer's responsibility to stock only those articles of which he can be proud. If he has no faith in his own judgement (one retailer I met confessed that after twenty years in the trade he had lost all his artistic standards) then he should engage someone to advise him. In Copenhagen the co-operative stores have all their stock checked by Professor Steen Eiler Rasmussen to ensure that it reaches the required design standard.[3] The question of the design, both interior and exterior, of retail stores can only be touched upon here. As a direct attraction for the customer, as well as being an indirect influence on his taste, the design of retail stores is very important. Dunn's of Bromley is an obvious example of how modern architecture and modern furnishing design can be interrelated.

Another aspect of the retailer's task is the training of his staff in design appreciation. 'It should always be remembered that while the power of any particular individual to influence industrial art may be strictly limited, the aggregate influence resulting from the improved training of distributors will eventually be very great, and that such influence is cumulative.'[4] The report from which this passage has been taken surveyed the whole question of training the staffs of retail firms to appreciate design. Today there are numerous schools and classes for the retail trade where design is included with other subjects, but probably the most effective training is embodied in the short but intensive courses arranged periodically by the Council of Industrial Design.[5] Visits to factories can also do much to stimulate the interest of buyers and salesmen. By this means they can become personally concerned in the design of the articles they handle in the store. A salesman often deals with customers who are unaware of modern design and yet willing to learn about it. One of his chief objectives, therefore, should be to teach people about design; for, to quote H. A. Holmes again: '. . . It is

[1] See, for example, the encouraging response received from retailers by the Council of Industrial Design, pp. 215–16.

[2] H. A. Holmes, *op. cit.*

[3] See The Hon. Margaret Lambert's Report on her visit to Scandinavian countries in 1946, in the possession of the Council of Industrial Design.

[4] *The Art Training of the Middleman*, 1927 (the report of a Committee of the British Institute of Industrial Art).

[5] See p. 215.

easier to sell furniture to people who know a little about furniture, and who are interested in design, than it is to sell it to people who do not care.'[1] In connexion with this, the method of sale and advice used by the twelve Gas Boards in this country might well be followed in retail stores.[2]

Then there is the question of how to display modern designs. For stores that only stock a few articles of modern design it is usually sufficient to display them along with articles of no artistic distinction. Such forced contrast will tend to focus attention, or at least catch the eye. But, with stores that are seriously concerned with design it is better to segregate the modern pieces. This can be done in each department or, as in the case of Lewis's Ltd., by setting up a separate section altogether. This venture, called the 'Design for Living' department, was started by the Liverpool branch of Lewis's in 1932. Later a similar department existed at the Birmingham branch. Since the war neither has been restarted. To begin with, an interior designer, Arundel Clarke, was called in to choose the articles and arrange the display. The department was managed separately from the rest of the store and although its turnover was small compared with other departments it had several valuable results. Apart from locating a new public demand, it influenced the taste of buyers and salesmen and so affected the whole store. The department was most successful with furniture, which it displayed in furnished rooms, together with textiles, pottery, glass, etc. An interior decoration service was also available.

A similar venture, this time in Sweden, should also be mentioned.[3] Since the war Nordiska Kompaniet, a large department store in Stockholm, has started a separate department for furniture and furnishings of modern design. The idea came from one of the directors, Elias Svedberg, who is also an architect. Everything in the department, which is called 'N.K.Bo', is of moderate price and generally cheaper than articles in the parent store. To show customers how their homes could be furnished there are scale models of the furniture on sale. The N.K.Bo is run by an interior designer, Fru Lena Larsson, who also arranges courses on home-planning.

Ventures like these may seem to be too optimistic for many retailers, especially those who would not risk the slow turnover of stock which might occur for the first two years. But there is no reason why small separate displays could not be arranged, either centrally in the store or in each department. This separation should not imply a difference in price, which was the case before the war, for this at once creates opposition to modern design. Another method is to arrange a series of exhibitions,

[1] H.A.Holmes, *op. cit.*

[2] See p. 56.

[3] See Jean Stewart, 'Selling Design in Sweden', *Design*, February 1949, pp, 11–12.

together with questionnaires. The significance of Hopewells' experiment lay in not simply asking the question 'Do you like modern furniture design?', but, 'Which modern designs do you like most?'

The people who are developing an interest in modern design belong to no particular class, or category of wealth, though from the evidence I obtained it seems that younger people are the most enthusiastic. With support and guidance from a responsive public, the retailer could greatly assist the manufacturer and his designer. As the best designs depend upon the designer's knowledge of the apparent or intuitive wishes of the public, the retailer's duty is clear.

ART SCHOOLS

From the evidence presented in the foregoing pages it is possible to draw one general conclusion. If a manufactured article is to achieve a high standard of design, then a person with artistic ability and training must be concerned in its stages of development. The next step in this investigation must be to give an account of the existing training establishments in this country which have made it their task (or one of their tasks) to educate for any form of activity connected with industrial art. I have attempted to obtain a representative picture by visiting fifteen art schools. They may be divided into two groups. First, the regional art schools situated in large cities and serving, in part, the needs of several counties. Included in this group are the art colleges and schools at Leicester, Bradford, Birmingham, London, Glasgow, Edinburgh and Manchester. In the second group are the schools which are either representative of, or primarily concerned with, the training of students for one industry. In this group are the schools at Stoke-on-Trent, Stourbridge, Kidderminster, Walsall, High Wycombe and Beckenham. I have also included the L.C.C. Technical College for the Furnishing Trades, Shoreditch, and the School for Jewellers and Silversmiths, Vittoria Street, Birmingham. Many of these schools are much more than monotechnics, and I have tried to present a balanced account of their activities. Finally, and in a separate section, comes the Royal College of Art. To supplement the evidence from these sources I have interviewed two H.M. Inspectors of Art Schools and three of the Ministry of Education's examiners.

From the outset it must be stressed that I was concerned mainly with one aspect of art education, namely the training of skilled workers and of designers for industry. In each school I met the Principal and some of his staff, and the information I obtained from them concerns the following points: the school's connexion with local industries,

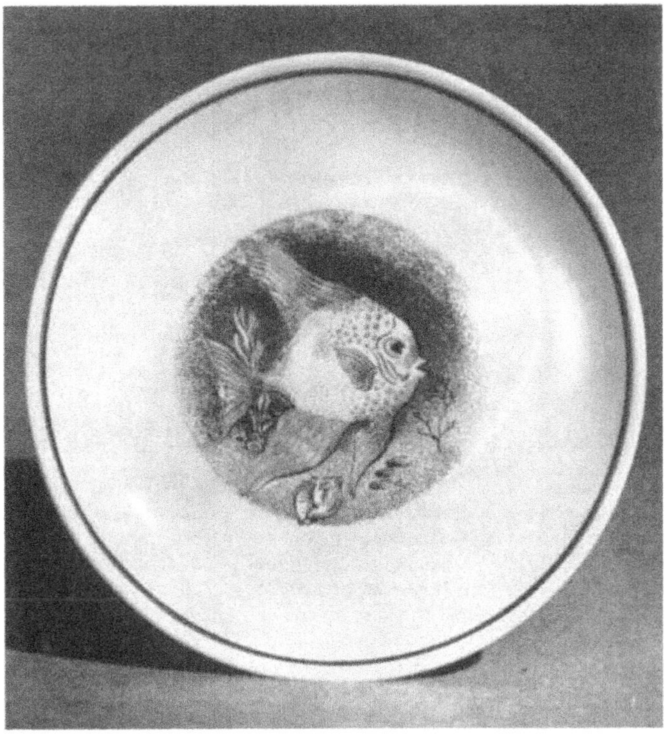

1. Instruction in pottery design is carried on in the heart of the industry. The coupe-shaped plate has a lithograph decoration by a fourth-year student. *Designed by Eric Elliott at Burslem School of Art.*

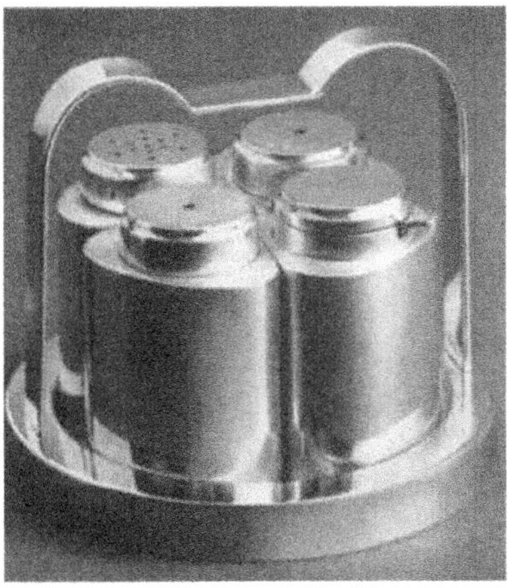

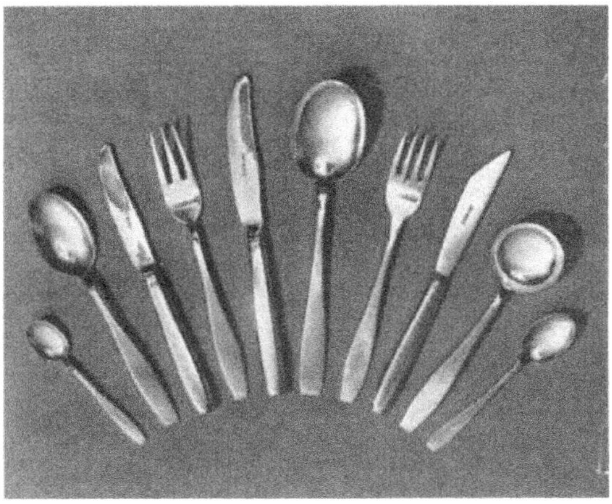

2 and 3. Staff at the Leicester College of Art are expected to carry on their own work as designers or craftsmen. The students share in this and so get a realistic impression of the designer's problems. Both these designs were developed at the College. Left, *a cruet set for B.O.A.C.*, right, *flatware for Mitchells and Butlers Ltd; designed by Kenneth Holmes and N. R. G. Poynton for Gladwin Ltd.*

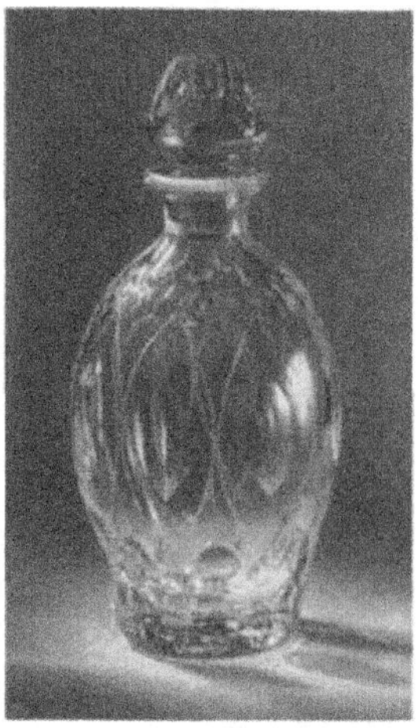

1. Full facilities for making and cutting glass are provided only at Stourbridge. Decanter by a senior student. *Stourbridge Art School.*

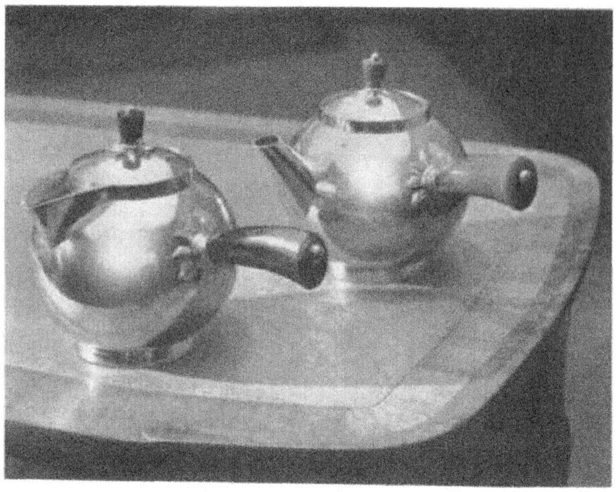

2. This is a refreshing exception to the unpromising standards of silver design in art schools. It is rich in detail, decorative in the use of horn for handles and knobs and yet boldly unfamiliar in conception. *Designed by Alexander Wilkie at Glasgow School of Art.*

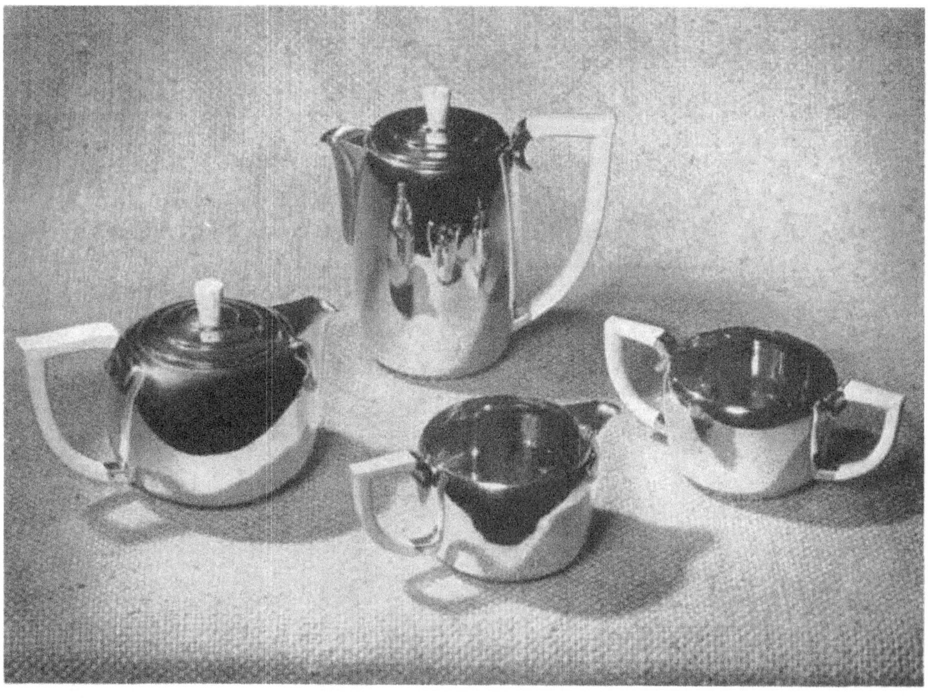

3. Industrial firms often have neither the time nor the inclination to experiment with designs for new materials and techniques. Art schools could take on the work and a beginning was made here with stainless steel at Leicester. *Designed by Kenneth Holmes and N. R. G. Poynton (assisted by students) for Andrew Brothers (Bristol) Ltd., at Leicester College of Art.*

and its ability to supply the designers needed. More especially, I looked for a sort of employment bureau which, with the aid of an advisory committee of manufacturers, would ensure that all graduates obtained posts on leaving the school. With an estimate of the school's reputation as a source of designers in mind, I then dealt with the design courses in terms of the facilities, equipment and the methods and standards of instruction available to students.

LEICESTER COLLEGE OF ART

It is advisable to begin with the College of Art at Leicester, for four reasons. First, it shares the same buildings, and is run in conjunction with, the city's Technical College. Secondly, it has a very close link with the City Council, through the Education Committee, and by its efforts civic design, especially lettering, has been improved. Thirdly, for a college of its size, it exerts an unusual influence on the 125 neighbouring primary, secondary and grammar schools. And fourthly, it can show a continuity of development which is shared by few other schools. This continuity may be traced back to the late B. J. Fletcher, who was Principal of the College from 1900 to 1920, and helped to initiate the change from craft to industry. The College was one of the first English schools to realize the necessities of the new century, and it has succeeded in impressing its ideas upon more than one of the local industries. Leicester printing and lettering has for a long time been famous all over the country. It originated in the College, as did the project of developing a local cane-chair industry, on the lines of that which flourished in Austria.[1]

The present Principal, Kenneth Holmes, was appointed in 1935. The tradition of industrial design training for the main local industries (hosiery, knitted fabrics, printing and commercial art) has been carried on and improved under his direction. He has also widened the activities of the College to include design training for the light metal trades, and he has built up the College's prestige so that it now advises the City Council on display-work and printing, and often carries out all the production work on the College's machines. Its prestige may be indicated in another way. The College was recently offered free a large stand at the local trades' exhibition, so that it could demonstrate its relationship with Leicester's industries. The influence of the College on neighbouring schools, as mentioned above, is no less advanced. Under the direction of the Principal, the staff selects the fabrics and pictures which are to be purchased by headmasters for equipping their schools. In addition, the College began

[1] B. J. Fletcher induced H. H. Peach to take this up in 1907, and by 1914 the factory was giving work to 200 people. H. H. Peach and B. J. Fletcher became founder members of the Design and Industries Association. See p. 193.

a circulating collection of objects of good design so that it could be used educationally in neighbouring schools. This service is now carried on by the Leicester Museum.

The total number of students at Leicester is about 2600; of these, 550 are full-time students, 800 part-time, and 1250 evening and occasional students. Each year the average number of students which passes out of the various schools is as follows: corsetry, 10; dress designing, 15; commercial art, 6; metal work, 4; furniture, 10; art school teachers, 24. Apart from those who will teach, all these students are found jobs in the industry suited to the design training which they have received at the College, but not all of them are taken on as designers straight away. It is significant that the College refuses to enrol more students than can be placed in jobs in industry or in teaching posts in other schools.

The two most important departments in the School of Industrial Design at Leicester are printing and metalwork. In the printing department there is extensive equipment and full-scale industrial machines, such as printing presses, type-setting machines, lithography and photo-lithography apparatus, all of them up to date. In the metalwork department, started in 1945, the equipment is less spectacular, and it is calculated to give the students an adequate knowledge of the production techniques which they will eventually use in industry. On the one hand the department is directly connected with the older silversmithing school and on the other its facilities extend to the cutting of tools for its own plastic moulding press. The course is designed to give the student an opportunity of becoming used to workshop practice and obtaining a knowledge of the limitations both of machine- and hand-tools. It also gives him experience in preparing working drawings and perspective sketches.

The furniture industry in Leicester is not sufficiently large to take many designers from the school. Nevertheless, the department possesses the complete equipment for a modern factory, and students intending to be skilled workers carry out all their work under the most modern conditions.

The College has carried out several commissions from local manufacturers, especially from those in the printing and metalwork trades. In most cases the students are partly responsible for the work from start to finish and a full professional fee is charged. A portion of this fee is handed to the student, the rest is put into a departmental fund for trips to London and other cultural centres. Where instructors in the department help in a design, they are only paid for the work done in their spare time.

The College insists that all members of the teaching staff show evidence of continuing their own work as designers or craftsmen. They are expected to use the College facilities when carrying out their own private commissions.

BRADFORD COLLEGE OF ART

The Bradford College of Art is a school which shows a continuity of development comparable with that of Leicester. The Principal took over the College in 1938 and put into practice an educational policy which he had experimented with in a smaller school. Briefly, he starts each student on a particular and confined line of study, i.e. a craft, and when sufficient knowledge of that is obtained, the student's horizon is broadened to take in several aspects of art for which the College is equipped.

Bradford is one of the few colleges which sets its own examinations in all subjects, but as a regional college it must expect to take in students from smaller local schools who have already started to work for the Ministry's diplomas. In the College, therefore, there are two kinds of student: those working for the internal examinations and those working for the Ministry's.

Because of its location, the most important subjects at Bradford are woven and printed textiles. In the first year, the incoming student's work done at the secondary school is assessed and he begins to study either weaving or printing. Both types of student receive similar training in their first year. The work is designed to encourage an imaginative experimental approach to form and colour. This free, 'loosening-up' process has the effect of removing all the bad ideas which the student may have assimilated.

In the second year the student has to pay more attention to detail, especially in plant and flower drawing. He is trained to record visual impressions quickly. At the end of the second year, the student makes a definite choice for weaving or printing. For instance, he will be required to spend three months in intensive practice of his craft at the College. At other times during the third year he will spend several days in local mills and in talking to trade designers. In weaving, cloth construction taught in the nearby technical college is stressed equally with the student's own creative work on hand-looms. Later the student designs on point paper and operates Jacquard looms.

Design students for printed textiles receive a similarly thorough course. They are given problems stating the price of production and the particular type of cloth to be used and are required to design within these limitations. Several students in each final year are expected to produce designs for sale to local manufacturers. Before they leave the College, students of both woven and printed textiles will have had jobs in industry arranged for them. The number of students on the four-year textile course is eighty; twenty of them leave each year for jobs in industry, the majority of them as designers.

BIRMINGHAM COLLEGE OF ART AND CRAFTS

The Birmingham College, situated in this country's chief centre of industry, occupies an important position. The College was founded in 1884 as the first municipal art school in England, and was also the first to be conceived under the influence of the Morris movement. Under Principals such as Catterson Smith that tradition was kept alive, and soon after the First World War training for industrial production began to take the place of that which was only applicable to the arts and crafts. This was due to B. J. Fletcher, who moved to Birmingham from Leicester. The College is the centre of an educational system comprising thirteen schools of art. For our purposes only one of these schools (Vittoria Street School for Jewellers and Silversmiths) is important and it will be dealt with later. Again, only two departments in the College were of interest to my inquiry: silversmithing and jewellery and industrial design.

A municipal art school in a large English city is a very complex unit. In order to judge rightly the results obtained, one has to be quite clear as to the various aims pursued. The Birmingham College trains artists, architects, commercial artists, craftsmen, skilled workers and designers. We are here concerned with the fifth and sixth categories, above all the sixth. What is done for the education of the industrial designer? So far as experimental craft-work goes—which is certainly necessary to the creative designer—most of the classes at the College would be of advantage; so far as development of truly modern requirements of design for machine-production is concerned, a good start was made when the College recently began to equip a nearby workshop with machines for basic processes. But, in order to obtain an insight into the complexities of machine-production, much more machinery ought to be available in the Birmingham College than has been hitherto acquired. In some classes the attitude of the artist-craftsman still rules unchecked. This is particularly true of the Silversmithing and Jewellery School where students are taught by practising free-lance designers. The stress is upon hand-work. The great demand in Birmingham for better designs of machine-pressed nickel-silver, which should fall within the scope of this school, appears to be totally unrealized. In other courses, such as cabinet-making, demand in the Midlands for special designers is too small to justify much stress on the designing side. The furniture school caters almost entirely for day-release students aged between fifteen and twenty. They come from about sixty Midland firms most of which are concerned with cheap mass-production furniture. Faced with this and a very small demand for furniture designers, the school has been wise to stress mastery in craftsmanship carried out inevitably in the tradition of Gimson and the Barnsleys.

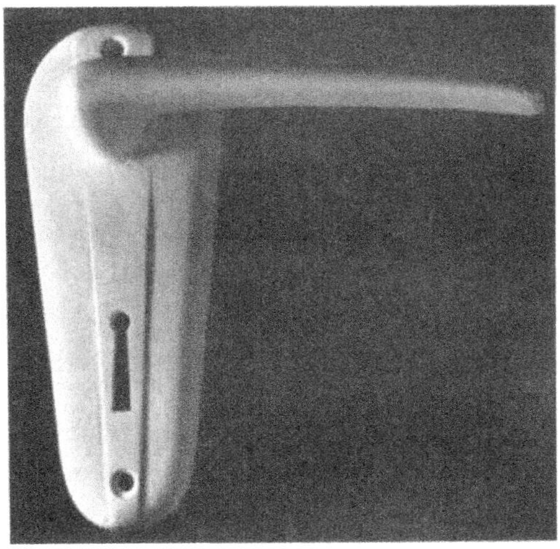

1. The making of prototypes and models for commonly used articles helps the student to blend creative ability with practical knowledge. Prototype door handle. *Birmingham College of Art and Crafts.*

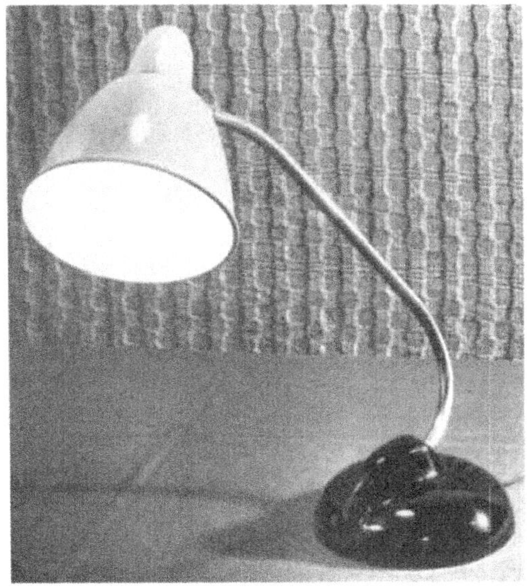

2. Much can be gained by persuading local firms to take on students full-time during vacations. This table lamp was made at Best & Lloyd Ltd., by a student. *Birmingham College of Art and Crafts.*

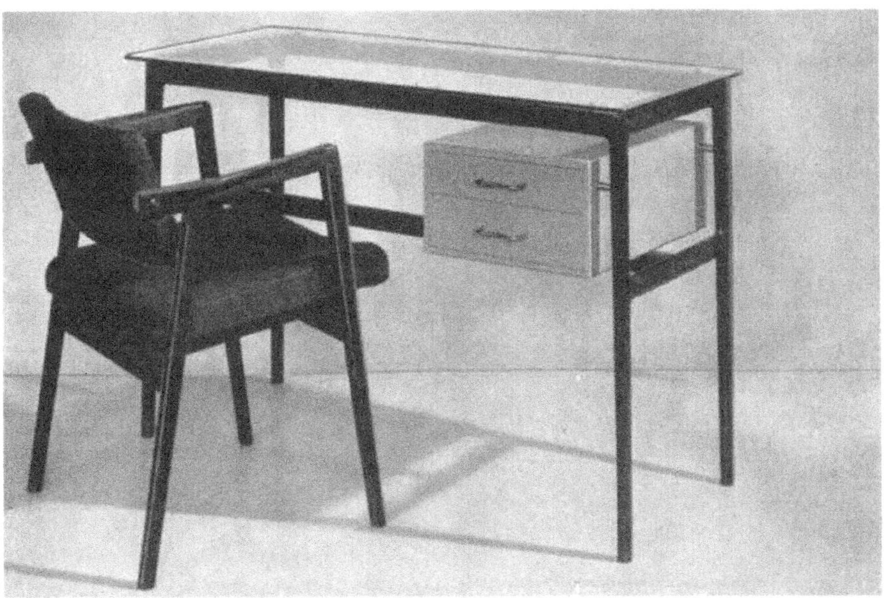

3. Through visits of study to European countries and a constant stream of magazine articles students can keep abreast of the newest developments. *Designed by J. Padiak at London County Council Central School of Arts and Crafts.*

PLATE LXXX

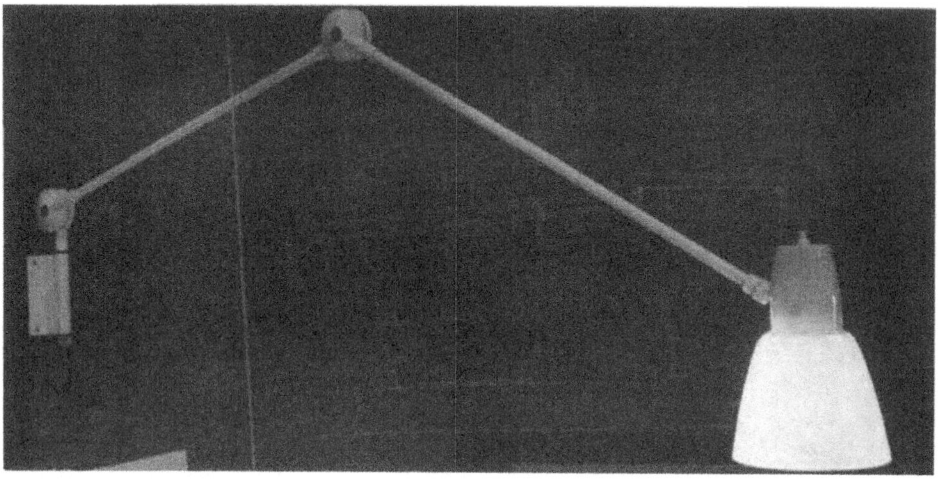

Experimentally, five third-year students working as a team designed and produced at school a hospital lighting fitting. First sketches, general arrangement and detailed part drawings, were followed by jig and tool making, production of a short run, assembly and testing. This is probably the first time in a British art school a light engineering product has been controlled throughout by students under realistic 'factory' conditions. Selling price estimated at under £5. *Designed* (1954) *by D. M. Pearson, W. Schejbaal P. J. Farrell, J. W. Cooper and C. R. Cheetham at School of Industrial Design, L.C.C. Central School of Arts and Crafts.*

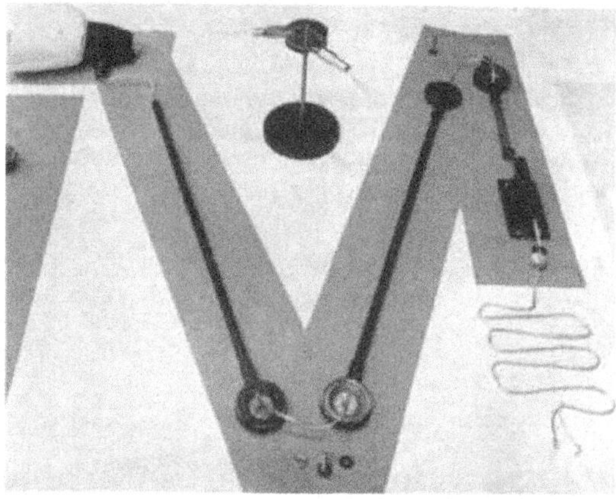

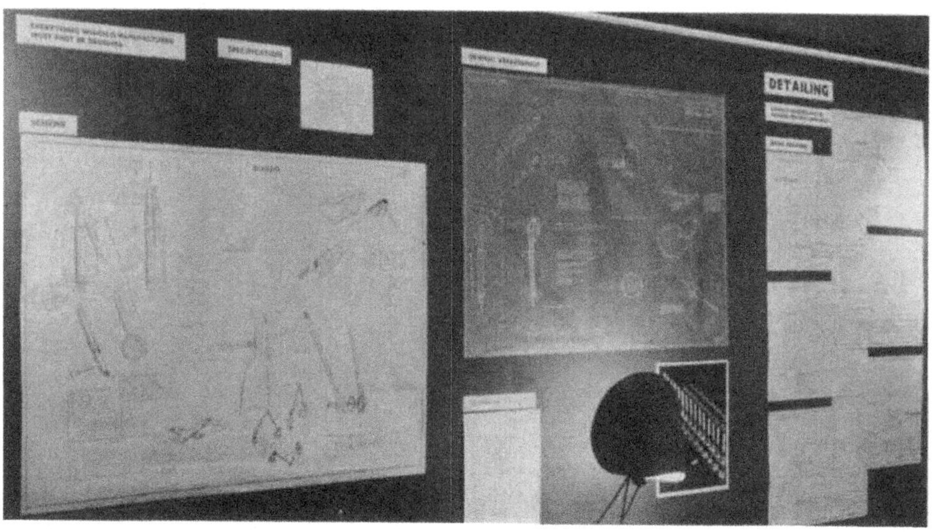

In order to provide education for the industrial designer without restricting the development of handicrafts, a separate class for industrial design was established in 1925 and is still directed by A.E. Harvey, a well-known free-lance designer in the Midland light metal trades. Since the war he has been assisted part-time by no less than three industrial designers coming from nearby firms. His students are drawn from the various craft classes and I got the impression that for many of them industrial design was not their first choice of subject. The curriculum compares very favourably with that of the Chicago Institute of Design, which will be dealt with later. Its philosophical approach is similar. In the first year experiments in form, abstract exercises and three-dimensional work in plaster and sheet-metal provide an excellent basis for product designing. In the second and third years, students are encouraged to work out design problems in domestic appliances and other light metal products. As an additional practical training, students are sent in their vacations to factories where they can study methods of production, and in some cases construct prototypes of their own designs. However, much has still to be done to convince Birmingham manufacturers that the best students in this school could, if given the opportunities, become first-class industrial designers.

LONDON COUNTY COUNCIL CENTRAL SCHOOL OF ARTS AND CRAFTS

The problem here is different from that in any other city, because the size of London does not permit the centralization found in Birmingham and Leicester. The L.C.C. has twenty-six schools giving instruction in arts and crafts. The Central School in Southampton Row corresponds, to a certain extent, to the Birmingham School, at least in so far as the advanced courses are concerned. The other twenty-five schools either resemble Birmingham branch schools or are organized with a bias towards one special trade, or they work in conjunction with the regional polytechnics or technical schools. The Central School is closely connected with them, especially by a system of scholarships, which gives students in them a chance to take advanced courses. Like Bradford, the School holds its own diploma examinations and encourages students to select one craft at the beginning of their training for special study. Such an approach is traditional in the Central School, whose foundation in 1896 was due to the influence of the Arts and Crafts Movement of which W.R. Lethaby, the first Principal, was one of the leaders.

At the Central School there are four departments which are of importance here: interior design (furniture, pottery and stained glass); textiles; silversmithing and jewellery; and industrial design. The interior design department co-ordinates the furniture, textile and other sections of the school which concern interior designers. Chiefly noticeable is the lively attitude towards furniture design which is stimulated by instruction from practising designers. The care and precision of workmanship, which was so much a part of the Arts and Crafts Movement, is the basis of the work, and has been creatively used. Besides wood, metal and plastics are used experimentally, though hand-work rather than machine techniques characterizes most of the designs. It should be remembered that the biggest furniture factories are in and near Tottenham. How can they send draughtsmen and designers for afternoon classes, or even full-time instruction, to the Central School?

Hand-loom weaving, both Dobby and Jacquard, is taught in the textile department. Industrial methods are not stressed, chiefly because of the small demand for designers of woven textiles in the London area. Textile printing, on the other hand, is carried to a point where some of the work of the best students is immediately accepted by the trade for commercial production. Block- and screen-printing of dress and furnishing fabrics are studied and the standard of work is probably the highest in any British art school. The silversmithing and jewellery department is one of the best attended in the school and enjoys a wide reputation for skill in the crafts. This is recognized by the trade in London, which supports the department by sending students part-time and also employing some of those who complete the full-time course. In each year there are about thirty full-time students, roughly divided as follows: eleven engravers; ten jewellers; nine silversmiths. There is little machinery apart from spinning lathes for silver, waste-wax and centrifugal casting apparatus for jewellery. What equipment there is reflects the London trade, which unlike Birmingham is fundamentally craft-based.

Lastly, there is the School of Industrial Design which, under the principalship of H. G. Murphy, began in 1937. After the war it was revived, in 1946, and owes much to an experiment which the present Principal, William Johnstone, carried out at a smaller London art school just before the war. In each year there are about ten students, some of them from engineering colleges. As in the class at Birmingham, three-dimensional exercises are given initial importance. The work is suited to the light metal trades and students use lathes, drills, small presses and plastic moulding presses. For more detailed mechanical instruction they spend one day a week at a technical college. At least two product designers and one mechanical engineer give part-time instruction in the School. An experiment which began in January 1954 should be outlined for it is, as far as I know, unrivalled in any other British art school. Five third-year students formed a team to design a light fitting through all stages of planning, making working drawings and moulds, to the construction of a short run of finished models, as if the School were a factory.* See Pl. LXXX.

GLASGOW SCHOOL OF ART

The School is unique in that it occupies the most architecturally interesting art school building in Britain. The Glasgow School of Art was designed by C. R. Mackintosh between 1899 and 1906. It is the best known and one of the most impressive of his works. In contrast with the initial promise which this building provokes, I found that in all departments the emphasis is placed on types of training suitable for teachers of art. For instance, instead of interesting students at once in a particular craft, there are two years preliminary work in which a large number of general subjects are taught, none of which is made to lead towards industrial design. Indeed, industrial design has only been studied since the war, while in general the School's work is still very close to that of the Arts and Crafts. Again, the School's connexion with the Glasgow Royal Technical College is too slight to matter much. It has practically no links with local industries, such as furniture and carpets. Although the staff did not impress me with its willingness to co-operate with industry, the Glasgow manufacturers whom I interviewed were not prepared to give the graduate students a trial or to patronize the School in any way.

Under the heading 'Design' in the syllabus, the following subjects are listed: embroidery and weaving; textile design and fabric printing; mural decoration; stained glass; silversmithing and jewellery; bookbinding; industrial design; interior design; lettering, writing and illumination. Altogether there are about ninety students in the Design School. Textiles and industrial design are the most progressive departments. In the textile department simple weaves are carried out on hand-looms but, for three weeks of the year, serious design students attend the Woollen Technical College at Galashiels, where they can study machine processes, though they are not able to weave their own designs. During the war, when the department was under the direction of Agnes McCredie, a practising designer of tweeds, there were no training facilities and no raw materials. Her students collected and learned to spin their own woollen yarns and they made and used vegetable dyes.

The industrial design department at Glasgow is, as yet, not very spectacular. It was started in 1946, takes six students a year, and gives them an introduction to commercial metalwork and the theory of plastics design. Students make working drawings and mock-ups of small domestic articles. In contrast to the other departments in the School, I found that students here were able to study techniques in a number of local light metal factories.

The need, which has recently been fulfilled in London by the Royal College of Art, for a general training in interior design is met in Glasgow by a most interesting department.

Students learn to use basic materials such as wood, metal, glass, fabrics, etc., in combination with plastics. The training makes the student begin with abstract designs for articles which may be functional, or at least useful. Most of the students leave to become art school teachers, although there are openings for the most promising amongst them in shop decoration, or even as interior designers in private practice.[1]

EDINBURGH COLLEGE OF ART

The tendency to concentrate on training art teachers, which I found at Glasgow, is even more apparent at the Edinburgh College of Art. This is understandable. Edinburgh is a city with, apart from printing, less industry than is to be found at Glasgow. However, there is also a flair and chic about some of the work in the School of Design and Crafts (where these qualities are called for) which is only attainable in a cultural centre. For instance, fashion drawing at Edinburgh has a quality of draughtsmanship that would be of immediate interest to the best stores and publicity firms in London. Dress materials also show elegance and lightness of touch, a quality which I found in very few schools. Nevertheless, Edinburgh and Glasgow are very much alike in attitude and method, and there will be no need to deal with Edinburgh in great detail. As in Glasgow, students have two preliminary years of general art work. At the beginning of the third year they select a craft in which, two years later, they will be examined for the College's diploma. In the Design School there are about a hundred students. In 1949, out of forty graduates, twelve went into the following industries as designers: printing, 4; textiles, 4; furniture, 2; pottery, 2.

In the fabric weaving department the students can work hand-operated Dobby looms, but for commercial experience they visit the Woollen Technical College at Galashiels. For fabric printing there is both block and screen equipment, and in their third year students visit printing works in Manchester and Carlisle. The standard of work in the glass department has great promise, but at the present time there is no furnace and only cutting, intaglio and engraving can be taught. To this, sandblasting equipment may soon be added. The printing department is well-equipped and leads the student from elementary hand-setting up to fully automatic equipment.

MANCHESTER REGIONAL COLLEGE OF ART

The Manchester College is in the centre of the Lancashire textile industry and is devoting an increasing amount of time and energy to the training of furnishing and dress print designers. Although out of sixty-two students who took the Ministry's

[1] Two graduates from this department were concerned in the re-designing of the London showrooms of the Standard Motor Co. See *Design*, July 1953, pp. 8–13.

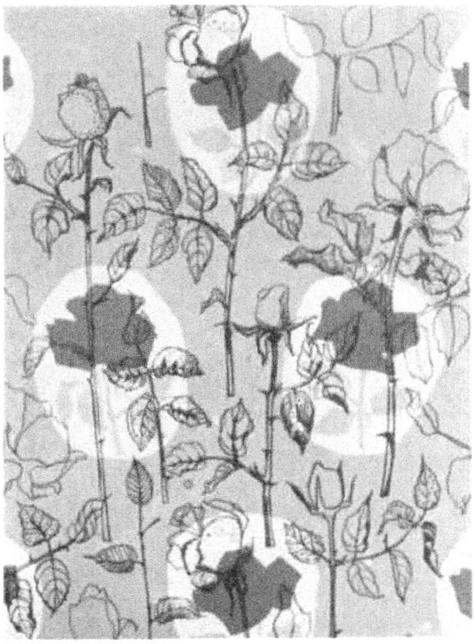

1. Wallpaper design could be closely linked with the popular and usually large painting classes in art schools. *Designed by Margaret M. Stewart at Glasgow School of Art.*

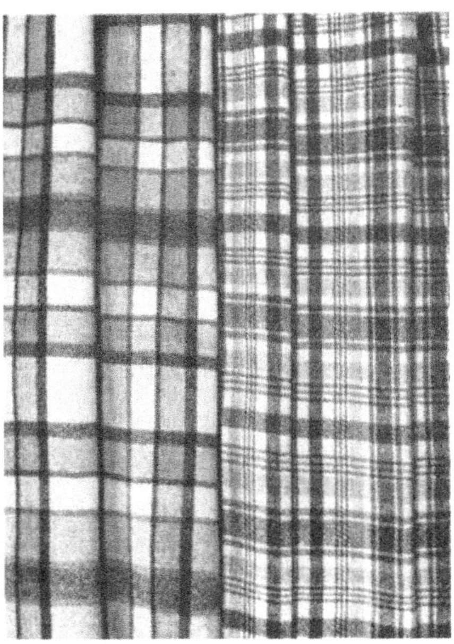

2. Close connexion with design in the local industry should be the primary function of an art school. Two woven fabrics. *Designed* (left) *by Margaret Broadbent*, (right) *by Jean Falding at Bradford College of Art.*

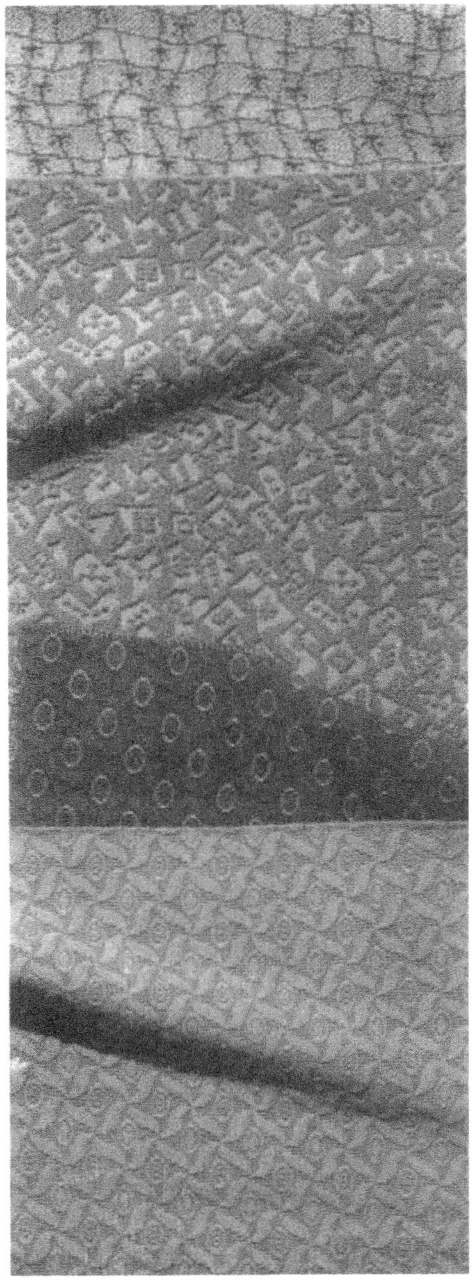

3. Realistic equipment, time for mistakes and a stimulating atmosphere are essential requirements in a school sending designers for staff jobs in industry. Two uncut moquettes (top and centre) and a rayon and cotton tapestry weave (below). *Designed by Constance Mary Holt at Leeds College of Art.*

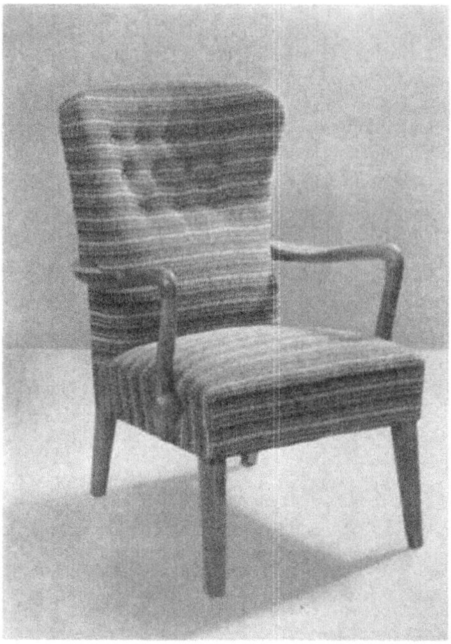

1. In most schools part of a student's work would be carried out under examination conditions. This chair was for the National Diploma in Design. *Designed by L. Dandy at High Wycombe School of Art.*

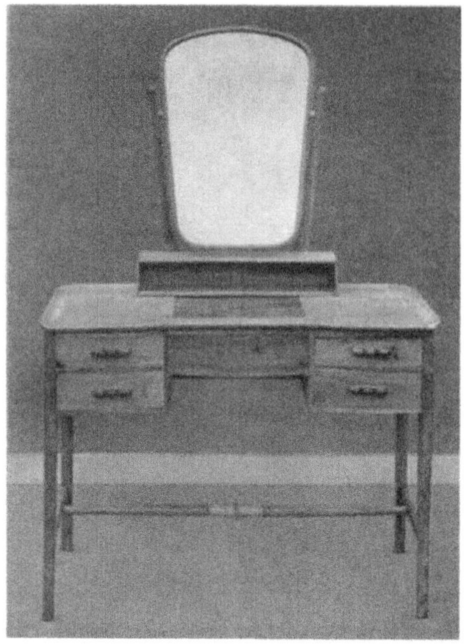

2. The potentialities of laminated plywood for furniture have been carried further at Beckenham than at other schools. The mirror is held by a continuous laminated form. *Designed by D. Wellstead at Beckenham School of Art and Crafts.*

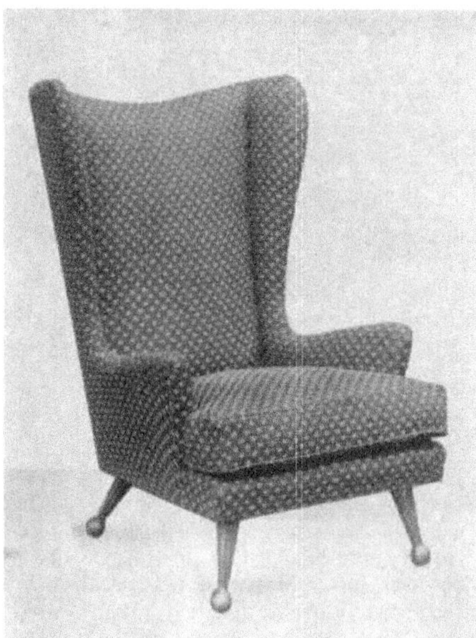

3. The factors involved in teamwork can be introduced at the art school. *Chair designed by senior students, upholstered by part-time day-release students, at L.C.C. Technical College for the Furnishing Trades, Shoreditch.*

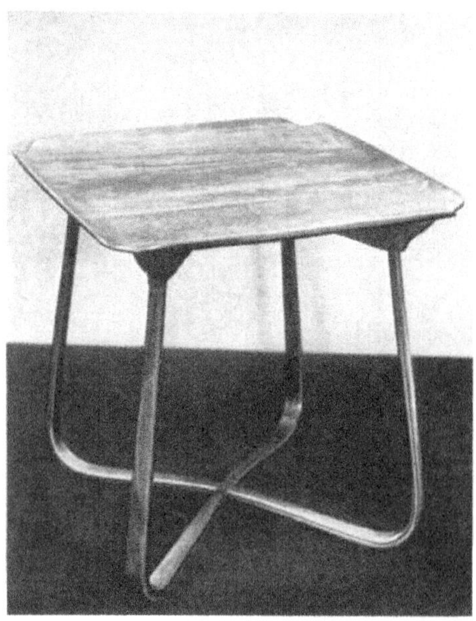

4. A further exercise in plywood structure. The laminated supports are decorative: small circles and spots of aluminium and brass are recessed in the table top. *Designed by students at Beckenham School of Art and Crafts.*

diploma in design in 1950 only three went into the industry, the numbers have since improved considerably and now all student textile designers can find jobs, mainly in local firms.

The need for the Lancashire cotton-weaving and cotton-printing trade to have a *Fachschule*, i.e. a school preparing for one industry exclusively, has at last been recognised. The College has recently acquired another building which will be fully equipped to provide a comprehensive training for designers. The School of Textile Design, now inadequately housed in the main building of the College, will shortly be able to take about 20 students in each year, several of them recruited from abroad. This centre for textile design should make more manufacturers aware of the importance of having a local source of talent adapted to the needs of the industry.

The salaries of junior designers in firms has been disappointingly low and in the past many students have preferred to set up as free-lance designers or begin teaching straight away in another school. The College hopes that in future out of those students who enter the industry fewer will become mere studio hacks and more will have the opportunity of producing original work of their own. At present students' work for screen-printing is promising and derives much of its strength from plant drawing.

Worth mentioning also is another source of designers in Manchester at the Municipal College of Technology with its specialist department of textile technology. Here scientific instruction of university standard is given in spinning and weaving, with the aid of a remarkable plant of looms, etc. It is chiefly concerned with students who take a full degree course and who intend to become producers and mill managers. They may specialize in design during their final year. Since the war there has been an average of about two design students each year who have already had an art school training, and their work on the looms showed great promise. The instructor was once on the staff of a well-known weaving mill and approaches the problem of weaving design as one of experiment with yarns in the loom.

STOKE-ON-TRENT SCHOOLS OF ART

There are four art schools in Stoke-on-Trent, at Burslem, Hanley, Stoke and Longton. Each school has its own master-in-charge, but they all come under the direction of the Principal of the Burslem School. This is the chief School in the group, and it is exceptionally well equipped to cater for the pottery industry. There is equipment for hand-throwing, turning, jolleying and mould-making. There are also three kilns for firing students' work. For decoration, hand-painting, transfer printing and lithography are taught. Instruction received from the permanent staff is supplemented by designers and craftsmen from the local industry who attend part-time. Since the war, and largely because of the pioneering work of a recent principal, R. W. Baker, the industry has taken an increasing interest in the Stoke-on-Trent schools. As already mentioned in the chapter on pottery, the British Pottery Manufacturers' Federation makes an annual award of five scholarships to full- or part-time student designers employed in the industry. At Burslem more and more attention is being paid to the training of a few first-class designers for full-time work in local firms. But at the same time about fifty are released by manufacturers for one day a week, while a further 450 attend during the evenings.

Another interesting fact is that Burslem acts as an intermediary between Royal College of Art students and the pottery industry. These students spend half of their time in the Potteries, either at the School or in the design department of a local firm. While some firms are willing to co-operate in this, the majority are as yet unwilling to devote the time and trouble entailed. However, almost without exception, the industry looks to the Burslem School to supply it with designers. Another point is that the school in Stoke-on-Trent, under the direction of Frank Dearden, has been the industry's 'research centre' for lithography. Its pioneer work in developing the process of photo-lithography has been especially important. The standard of pottery design in all the schools is encouraging, and it is particularly good at Burslem, which caters more for the full-time student. As the work is carried out in the centre of the industry, it has, apart from artistic merit, a sound commercial basis.

STOURBRIDGE ART SCHOOL

Stourbridge is the only school in this country which offers full facilities for glass design. In addition to cutting, intaglio and engraving lathes, a glass furnace and annealing plant were installed in 1934. The instruction in making and decorating glass shapes is by craftsmen who are closely associated with the local industry. However, the demand for glass designers is very small and there are only about three or four full-time design students at the School at any one time. After taking the Ministry's Diploma in glass design, the best of them aim to complete their training at the Royal College of Art. In this respect the School has been very successful. From 1937 to 1950, fifteen students went to the Royal College of Art and more than half of them have afterwards entered the industry as full- or part-time designers.

In spite of this local manufacturers have not stated clearly what type of training they would like the School to provide and their support is only half-hearted. The situation is as deplorable now as it was in 1935. In the year 1950–1 there were fifty-four part-time students who were not released during working hours but attended during the evening. Their main object was to improve their skill, thirty-two of them as glass-blowers, the remainder as decorators. Though the design instruction they received was incidental, it could be effectively used by the firms in which they were employed. However, I investigated one case where the young designer of a nearby glassworks was attending the School for one full day and two evenings per week. He hoped that in four years he would be able to gain the Ministry's diploma.

The standard of the best work in the School is promising. The Principal is interested and influenced by Swedish engraved work, and is encouraging students to make the best possible use of the traditional Stourbridge methods of cutting glass as a means of producing modern designs.

KIDDERMINSTER SCHOOL OF SCIENCE AND ART

The school is in the largest area of carpet-weaving in this country. Before the war, although manufacturers were quite willing to give their youngsters an afternoon off and pay their fees, there was little value in the carpet-design training available. This was due partly to inadequate equipment—there were no hand- or machine-looms—and partly to the low standards of design amongst the instructors. Since 1949, when a new Principal was appointed, the outlook has been much improved. Manufacturers still continue their support and 24 students from their design studios attend for one day and one evening per week. The course lasts for four years, and the instruction given in carpet design is backed-up by practical experience of hand- and machine-looms. It is hoped that Kidderminster will soon be able to start a school of carpet technology which would also include a full-time course for students who wish to enter the industry as designers.

Such a design course at Kidderminster would be of great importance. At the moment there is no other school, including the Royal College of Art, which can offer full facilities for carpet designing. But, unless the staff instructors will be prepared to make an entirely fresh approach to the problem of carpet design, the appallingly low standards of the local firms will dominate the students' work. The risk of this is, of course, specially great in the case of part-time students who come from neighbouring factories.

SCHOOL FOR JEWELLERS AND SILVERSMITHS, BIRMINGHAM

The School for Jewellers and Silversmiths in Vittoria Street was founded in 1890. It was first organized by the Jewellers' and Silversmiths' Association, but was later taken over by the city's Municipal Education Committee. It is all the more worthy of special attention, as it is one of the very few *Fachschulen* or monotechnics in the country, though it differs in many respects from the Central European monotechnics.

Above all, the training is mostly done in evening classes on a voluntary basis. There are no full-time students. The aim of the courses is to give workers in the trade an opportunity of completing their knowledge and skill, which in most factories are used only for routine work in one special process.

Out of approximately 300 trade students, only 6 per cent are apprentices who intend to make their livings in the two crafts. Less than 50 per cent of them are allowed by their employers to spend one day a week at the school. These students are expected in addition to devote three evenings a week to practical work. The remainder are voluntary students and attend for evenings only.

There is a tendency in the school to work at, and become proficient in, only one craft. Scant attention is paid to the problem of design for the E.P.N.S. trade. The aim of most students is to acquire the skill needed for higher wages, and there is a very small chance that any of them will become designers with high artistic standards. The absence of initiative and imagination in design which characterizes the Birmingham silver and jewellery trades tends also to be the hall-mark of this School's work. As the School is a branch of the Birmingham College of Art and Crafts, and for its design course has been under the direction of the master in charge of silversmithing at the College since 1952, it is probable that its standards will improve.

WALSALL SCHOOL OF ARTS AND CRAFTS

Most of the fancy leather goods trade is carried on in Walsall. Because of this, the art school holds a trade class equipped with a skiving machine and two up-to-date sewing machines for persons engaged in this work. Twenty-five men and women attend for two evenings a week for a two-year course. None of them is released from daily work. They learn the theory of leather and its practical uses, especially the cutting of simple patterns and stitching. Their object is to become proficient pattern-cutters in their firms, a pattern-cutter being the closest approximation to a designer in Walsall. About one-fifth of these students attend for one extra evening a week for a class in drawing and design. The most promising of them could expect to win scholarships to the Cordwainers' Technical College, London, for a course which lasts two years. It has already been said that the prospects for a proficient designer in the Walsall trade are very bad, and the best students at the art school are inclined to go to London in search of better-paid jobs.

The School is noticeably more alive in subjects other than leather—for example, painting and decorating, furniture, and even plastics. This is not due to any fault in the teaching or even the equipment provided at the School, but rather to the attitude of the local leather industry.

PLATE LXXXIII

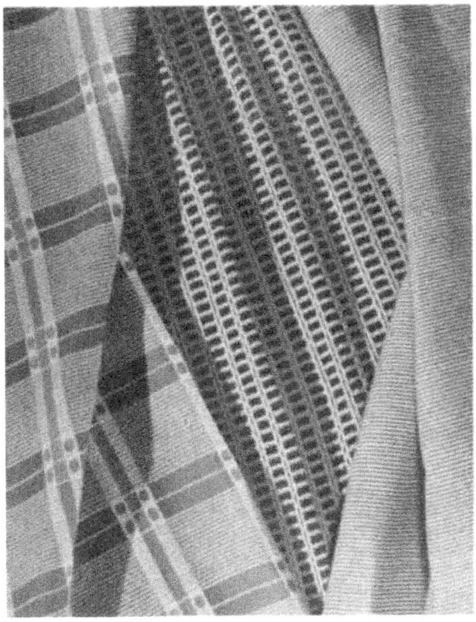

1. Dobby weaves. *Designed* (left and centre) *with fancy yarns by Ann Jennifer Bell,* (right) *two-beam plain weave by W. A. Brook at School of Textile Design, Royal College of Art.*

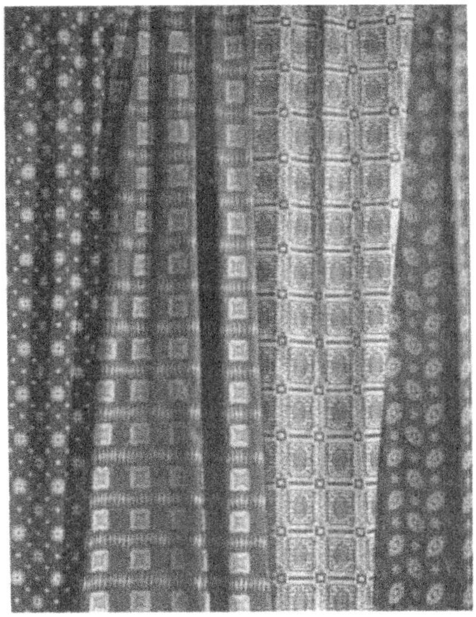

2. Jacquard weaves good enougn to rival any commercial range. *Designed* (left to right) *by Ann Jennifer Bell, Margaret Foster, F. Hoswell, Sheila Hankinson at School of Textile Design, Royal College of Art.*

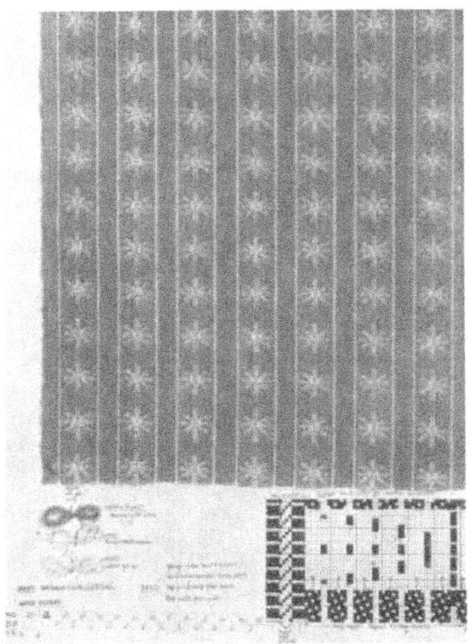

3. Method of laying out a sample Dobby weave. *Designed by Edna Penniall at School of Textile Design, Royal College of Art.*

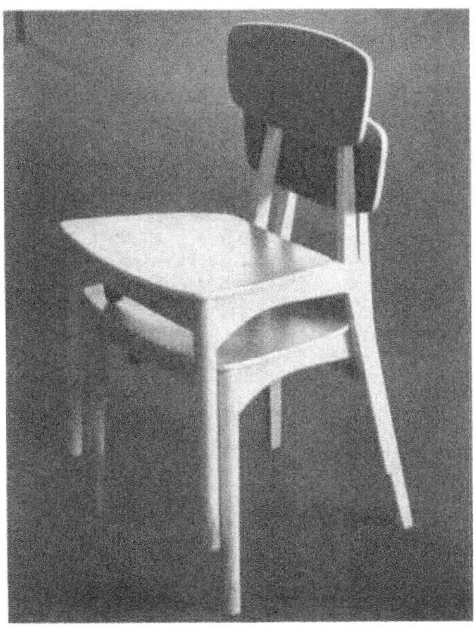

4. Stacking chairs with plywood seats. *Designed by V. A. Hindley at School of Wood, Metals and Plastics, Royal College of Art.*

PLATE LXXXIV

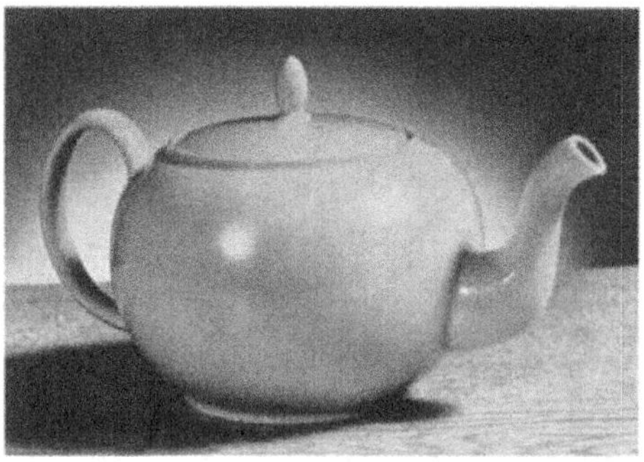

1. A refined form for traditional manufacturing. *Designed by James Rushton at School of Ceramics, Royal College of Art.*

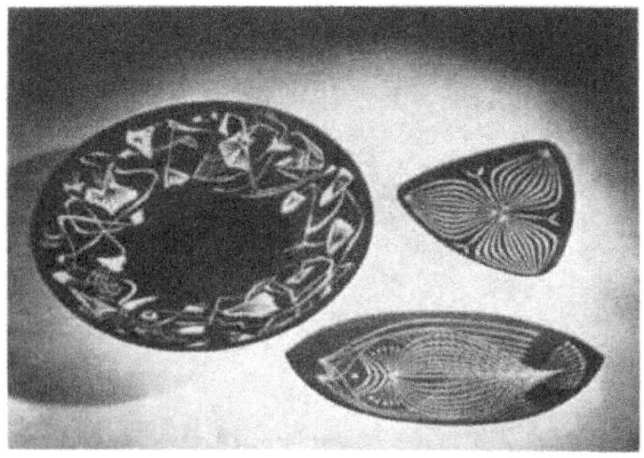

2. Glass dishes with intaglio and sandblast decoration. *Designed by G. P. Baxter at School of Industrial Glass, Royal College of Art.*

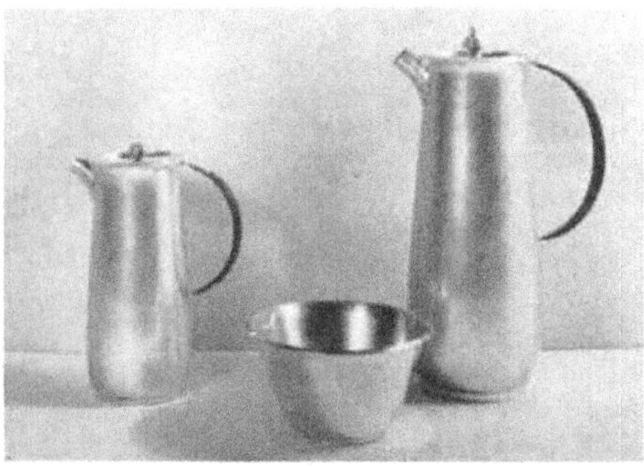

3. Silver coffee-set. Winning entry in the National Design Competition, Design and Research Centre for Gold, Silver and Jewellery, 1950. *Designed by David Mellor at Sheffield College of Art.*

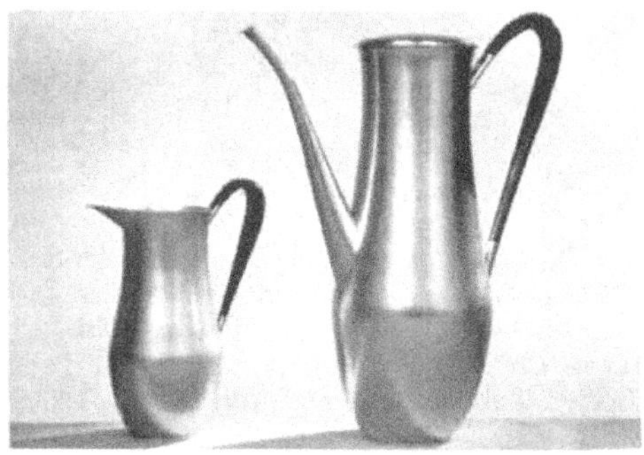

4. Silver coffee-set, 1951, showing the same student's development. *Designed by David Mellor at School of Silversmithing and Jewellery, Royal College of Art.*

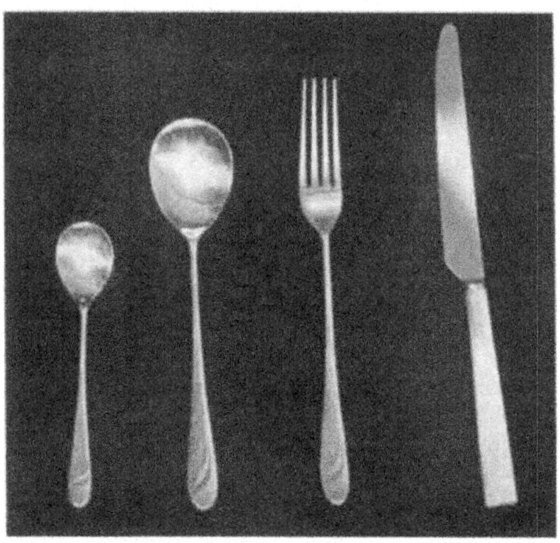

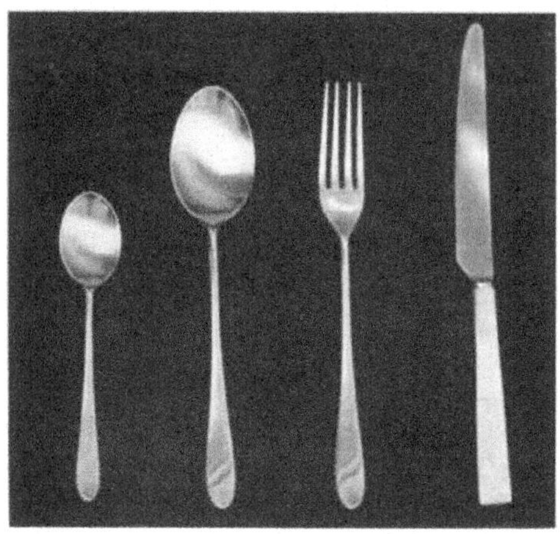

5 and 6. Few students are able to co-operate fully with a manufacturer during their training. David Mellor's first design (*left*) was taken up by a Sheffield firm. The pieces in this set are the first prototypes in silver. *Right*, the production versions in stainless steel and E.P.N.S. *Designed* (1951) *at School of Silversmithing and Jewellery, Royal College of Art.*

L.C.C. TECHNICAL COLLEGE FOR THE FURNISHING TRADES, SHOREDITCH

There are about forty students in this College on a full-time course of two years' duration. Their aim is to become competent craftsmen, designers and draughtsmen in furniture firms. A thorough craftwork and technical grounding is provided, though the machinery available, which can only be used by the instructors, is inadequate. This deficiency is partly made up by visits to local factories. No emphasis was laid on modern design; in fact, the finished pieces at the College's annual exhibition in 1950 suggested to me that there was a confusion of ideas: a confusion between the courtly styles of the eighteenth century and the low standards of design current in most firms during the 1930's. But, since 1952, the new Principal, W. J. Kape, has inspired students with better design standards; a factor of importance when one remembers the lamentable standards of the East London trade. I was told that a large proportion of the students became salesmen in furnishing stores. Those who do become designers or design foremen are often engaged by the many small local firms.

There is also a part-time course for juniors who are released from their work for one day a week. There are about 300 of them, and their chief concern is to gain skill as furniture craftsmen. Only the functional properties of design are considered. Other courses at the College can be taken in upholstery and French polishing.

HIGH WYCOMBE SCHOOL OF ART

This School is situated in one of the important centres of furniture manufacture, and one of its main tasks is to train students to become furniture designers. On a course varying from three to four years' duration full-time students spend a good deal of their early studies on a general art training. Their practical experience of furniture construction is limited mainly to hand-work; design training is of a satisfactory standard. The chief instructor is a well-known free-lance designer who works for several local firms. The students are generally interested in modern design standards and their work tends to draw inspiration from the eighteenth-century styles of hand-made furniture which are traditionally associated with High Wycombe. Attention to working details, specification and costing are all parts of the course. Particular emphasis is placed upon commercial methods and business management, for it is realized that amongst local firms, especially the smaller ones, a full-time creative designer would be a liability unless he were able to undertake other duties. About ten students in each year find jobs in the local trade.

BECKENHAM SCHOOL OF ART AND CRAFTS

Though this School is not situated in a furniture manufacturing area, the students in its furniture department achieve a high standard of design and practice. The system for training craftsmen and designers is kept fluid, and each student gets individual attention. The main object of their instructor is to get his students accepted as designers by furniture firms in or near London. As at Stourbridge, it is quite common for students, when they have finished their course, to be accepted for more advanced training at the Royal College of Art. In the course of three to four years' duration attention is paid to model making, craft work and manufacturing experience on the main types of machine used in the industry. Beckenham is one of the few schools I know which is experimenting in laminated-wood construction. An attempt is being made to use creatively the strong Scandinavian influence which is now being felt in this country. In addition, the problem of designing for cheap mass-production furniture of modern design is stressed throughout the course.

EVALUATION

In this survey I have been mainly concerned with those classes in art schools which are directly connected with the industries using industrial designers. I realize that it would be a narrow view if one insisted that these aspects of their work were the only ones which merited attention, but to be of the best use in training designers, an art school should provide a range of equipment similar to that in commercial use, and a staff of teachers with industrial design experience. No one art school can tackle the problem of design in all industries, and it is usual for each of the main schools to connect itself with one or two industries, preferably those which dominate in the area served by the school.[1] This is the situation at Leicester, Bradford and Birmingham. One further point needs emphasis at this stage. To be effective as an improver of design standards, an art school should not only make students aware of the present production methods (and design standards) in the industries it serves, but also carry out experimental work, based on present commercial practice, to prove that better standards of design are practicable and economically possible. As a research centre in this way an art school would give a positive direction to its students' work, maintain their enthusiasm and most important of all, influence the manufacturers.

An art school should be the meeting ground for the employer and the designer. If

[1] 'The practical utility of the art school in an industrial area depends on the extent to which it advances the work of those industries in the district where design and colour are important considerations' (*Board of Trade Working Party Report on Carpets*, H.M.S.O., 1947, p. 23).

design is to be improved this association is vital, though in many art schools its importance is not stressed enough. It is usual for a school to have an advisory committee of manufacturers. This should have five aims: first, to link the industry with the school. Second, to provide designers and craftsmen from industry to teach at the school part-time. Third, to supervise a type of employment bureau for graduate students who wish to commence work in industry. Fourth, to encourage local firms to take in full-time students during the vacations for one to three months' practical training and works experience. And fifth, to obtain small design commissions from the industry for the staff and students to carry out, not as trial projects, but as work which will be paid for. As has already been said, all schools should be confident of taking on such work for one or two industries. From my experience of the working of these committees in art schools I found that there is general agreement on the duties which they should undertake, but not nearly enough concern that these duties should be carried out wholeheartedly. For instance, most committees have made no attempt to assess the number of art school graduates which an industry could be expected to absorb each year. And very few committees have attempted to keep their schools in touch with current industrial practices by finding commissions for them to undertake.

In connexion with the training of designers, four more points should be emphasized. Less than 10 per cent of the schools I visited had taken full advantage of the practical and commercial training which local technical schools could offer. Lack of understanding on the part of principals and staffs, and consequently lack of co-operation, are recurring traits, which in far too many cases have been accepted as inevitable. The accurate costing and timing of work on a design should be, as in most Continental art schools, part of a student's training. Nearly all the schools I saw were not equipped with sufficient machines and tools to teach industrial design to students. The fact is recognized yet little is done about it. In the schools which possess architectural departments, it is not realized that a cross-current of ideas, if exchanged between the architecture and industrial design students, would be of enormous benefit. Here, the practice of the Bauhaus, about which something will be said later, has never been properly appreciated in British provincial art schools.

Apart from training designers, art schools provide classes for sculptors, painters, illustrators and commercial artists. I found that the general standard of work in the last two groups was satisfactory, and in many cases of a very high order. The reason for this is outside the scope of my inquiries, but it is clear that much of it is due to the constant demand from commercial firms, advertising agents and the printing industry for high standards of work. The situation is healthy, and competent students in this field can expect to find employment with relatively little difficulty. The work done by students of sculpture and painting is characteristically different. The techniques used lack discipline and there seems to be no vital *raison d'être* in the choice of subjects. The life class is the centre of the trouble, for in every school I visited there were students sitting, hour after hour, patiently copying the contours of the model on to canvas. The finished work may contain some subtleties of line and form, but I know of nothing more antithetical to Life than their paintings of nudes. The lack of direction towards some goal is the chief trouble, for there is no employment for these students, except in other schools, as teachers. There are some exceptions, of course, but it must be agreed that the situation of artists in contemporary society is so precarious and that so few of them succeed in their chosen work, that the amount of attention which art schools pay to training for fine art seems to be unwarranted. The painting and sculpture classes in all schools I visited attract many more students than the craft and design classes. 'Art for art's sake' is a disease from which few art schools are free and it is only to be expected that departments other than sculpture and painting are partly contaminated by it.

At this point something should be said about the role of the Ministry of Education. To obtain my information I interviewed two of H.M. Inspectors of Art Schools, one in London and one in Birmingham, and three of the Ministry's examiners. Briefly, the connexion between the Ministry and the art schools is as follows: it has no power to influence the running of art schools or to dictate the type of courses which should be held. Although an art school remains the responsibility of the local education authority, the Ministry may help by granting sums of money. Three main examinations are set by the Ministry: the Intermediate Certificate in Art and Crafts, the National Diploma in Design and the Art Teachers' Diploma. We are concerned with the first two of these. Until 1951 it was obligatory for students who wished to take the N.D.D. to gain the Intermediate Certificate. This meant that for the first two years' work at school they were required to disperse their energies by studying between six and eight separate art and craft subjects. For their third and final year they concentrated on one particular subject for the N.D.D. The changes which have now taken place are important.[1]

In the first place, they throw upon the art schools themselves much of the responsibility of assessing the candidate's qualifications. Only those who have completed an approved course at a recognized teaching institution are now eligible to enter the examinations and 30 per cent of the total marks are awarded by the schools themselves on the work of the students during the course. The schools are now able to make up any sort of a course they want and not be tied down as hitherto to the specific

[1] See *Rules governing the Award of the Ministry's Diplomas and Certificates in Art, 1951*, H.M.S.O., 1950.

requirements of the Ministry's examinations. For students taking the N.D.D., the Intermediate examination is obligatory no longer, but for those who wish to go on to take the A.T.D., the Intermediate may be taken afterwards. It is now fully recognized that the Intermediate is more applicable to the training of teachers than designers. It is important to note that an assessment by the school of a student's work should take into account his varying attainments during the whole course as well as the results of the school's own examinations.

The Ministry of Education will also give advice upon the appointment of a principal for an art school, but the local education authority has the final choice. It is interesting to find that the principal of almost every school is a painter. There seems to be no valid reason for this, except that designers or architects who are established in their professions may be unwilling to take on the task of governing an art school. It may be argued that the Ministry's control over art schools should be strengthened if a better design standard is to be attained. In addition, certain duties, such as the assessment of the rate at which students will be accepted in industry, could best be carried out by one body. It is surprising that at the present time the Ministry has no figures to show how many students each year become designers, or how many become teachers, when they leave their schools.

In connexion with this survey of art schools, it would be useful to refer to the report made by the Training Committee of the Council of Industrial Design.[1] The report stresses one important point which, so far, has only been mentioned. Once it is known which branch of design a student will take up, his training should depart from general art work to allow him to concentrate on one particular subject.[2] This would counter industry's chief criticism that art schools 'have a detached and impractical outlook'. The report points out the dangers of allowing fine art to dominate all other activities in art schools and thus render the whole conception of art education distasteful to industry. Once a definite course of study has been established it would be the duty of the school to encourage, gradually, the pursuit of wider interests. This would ensure that students establish their own discipline, and interest themselves in only those subjects which are relevant to their main study. We shall see later, when dealing with the Royal College of Art, how this approach works out in practice. It is also important, as the report indicates, that art schools and technical schools should contribute to a

[1] *Report on the Training of the Industrial Designer in England and Wales*, C.o.I.D., July 1946 (unpublished).

[2] 'It is the same with art and design as it is with general education. Only in a fairly narrow and concentrated field will a student's interests be sufficiently aroused to evoke his deeper creative instincts and only then...will he come up against the essential difficulties which must be faced. It is far better to learn through one subject the values which will be found later on to apply to many others' (Robin Darwin, 'The Training of the Industrial Designer', a paper read at the Royal Society of Arts, February 1949).

joint course of instruction for student designers. In the case of the draughtsman in technical schools, he should have his course 'impregnated throughout with considerations for aesthetic values, as in the case of architectural training'. A technical school trains persons for all branches of industry from distribution to management. And for these, to quote the report again, progress would be made 'less by the addition of a set subject than by a general shift of emphasis'. In order to make the best use of both art and technical schools, the needs of industry, and the rate at which it can absorb designers and other persons likely to influence the course of design in the factory, should for each district be accurately and regularly assessed.

THE ROYAL COLLEGE OF ART

The Royal College of Art was founded in 1837 in order to extend 'a knowledge of the arts and the principles of design among the people', using 'not theoretical instruction only, but also the direct practical application of the arts to manufacture'.[1] In spite of this precisely defined programme, the College in the nineteenth century became more and more a school for the training of drawing-masters. In the same year, 1837, a School of Design was set up to train students in what was called 'ornamental art' to distinguish it from the 'fine arts'. Nevertheless, the relation of crafts to design was forgotten and even the Morris movement could not, for several decades, change this attitude. It was not until 1889, when Walter Crane became Principal for a short time, that new regulations were introduced which led to the establishment of a design department with its attention focused on industrial needs. W. R. Lethaby, first Principal of the London Central School, was made Professor of Design and remained as such from 1900 until 1918. In 1901 the College was reorganized by being divided into the four schools of architecture, painting, sculpture and design, as well as a few craft classes, and in this form as far as design was concerned it remained unchanged until 1948.

But the new ideals cannot have carried very far, for in 1911 the Government Committee pleaded for the total dissolution of the College and later the Design and Industries Association, as well as the specially appointed Hambleden Committee on Advanced Art Education in London, 1936, still had good reasons for proposing a complete reorganization in order to restore to the College its *raison d'être*.[2] The suggestions of the Hambleden Committee that a body of governors on which industry should be strongly represented be appointed, and that the ultimate responsibility for

[1] See *Report of the Departmental Committee on the Royal College of Art*, London, 1911, pp. 27 ff.

[2] The proposals of the D.I.A. are contained in a memorandum, written in 1933, and published in Herbert Read's book *Art and Industry*, Faber & Faber, 1944, pp. 175–82.

the policy of the College be in the hands of the Board of Education, were adopted. But the really important proposals were not put into effect until 1948. The impetus came from the report, in 1946, of the Training Committee of the Council of Industrial Design, already mentioned. In that year Robin Darwin became Principal. A new Advisory Council was formed to make all staff appointments, and to deal with the administration which had previously been in the hands of the Ministry of Education. Seven *ad hoc* committees were set up to assess the number of designers which the various industries could be expected to absorb, and on their recommendation six separate schools were started for industrial design training. In scope they cover textiles, silversmithing, jewellery, glass, ceramics, graphic design, fashion and work in wood, metals and plastics. Practising designers were appointed as professors, specialized teachers were recruited from industry, and each school was provided with up to date workshop and factory equipment. This meant that six professors now took the place of the one professor of design.

The methods used in these six new design schools follow a similar pattern, and it will not be necessary to discuss each one in detail. To explain these methods I have chosen one of them, the School of Silversmithing and Jewellery. The course is adapted equally to those who will practise full-time in industry, as free-lance designers, and as independent artist-craftsmen. The different requirements of large and small scale production and of high- and low-price markets are considered concurrently, to emphasize the unity of process in designing work of all kinds. Throughout the course design is studied in relation to materials, tools, manufacturing processes and the function of an article. Training in craftsmanship is secondary to training in design, but the student who completes his course will have sufficient practical ability to enable him to try out his designs in the material and make prototypes. The course lasts three years and the student, while he has his one main subject always in mind, is encouraged to relate it to a progressively wider field of cultural interest. As the number of students on the course is between fifteen and twenty a personal relationship between each student and his professor is possible. For the examination the student submits a number of examples of finished and unexecuted designs selected by himself. This involves a test of both the student's proficiency and his power of discrimination. Successful students then join a firm for nine months to gain practical industrial experience. If this period is completed satisfactorily the diploma of the College is awarded. This diploma, Des.R.C.A., awarded for the first time in 1951, is a qualification which has been created to distinguish graduates of the Faculty of Industrial Design from those fine arts and graphic design students who receive an associateship, or A.R.C.A. For a student who would derive particular benefit from further study in the school a fourth-year course can be arranged. For designers already working in industry a short two-year course is available, with a scope which is reduced from the full three-year course so that most of the practical work is excluded.

The mechanical equipment of the College is a striking feature. It is best shown in the School of Wood, Metals and Plastics and in the School of Textile Design. The first amounts to a basic factory equipped with all except speed machines. Students learn to use them as tools to make a design accurately. Once they have understood each process the work, if repetitive, is finished off by one of the instructor's assistants. The same is true of the textile design school, where there are treadle, Dobby and hand-operated Jacquard looms, the latter run in conjunction with a card-cutting machine. In one respect the department of industrial glass is different from the others. It is only equipped for engraving, and has to send its students to the art school and glass-works at Stourbridge for practical instruction in making and cutting.

The Royal College of Art is no doubt providing the most promising and progressive industrial design training experienced in this country so far. While each school severely limits the number of students in each year, the quality of the training and the College's connexions with industry are good enough to ensure that most graduates can find employment as designers.* Part of this may be accounted for by the College's position in London which is a centre for the most important events in every artistic field. The type of education which this affords is the closest approximation to that of a university which industrial designers have so far enjoyed. Added to this, students work under professors and instructors who are practising designers with their studios and work-shops at the College. When a new commission comes in from industry they can watch and even participate in its development from a series of written instructions to a finished design. The design work of staff and students for The Lion and Unicorn pavilion on the Festival of Britain South Bank site should be noted here. Finally, the co-ordination of ideas between different schools at the College is one of the main tasks of the Department of Interior Design set up in 1951. In it students consider their work in relation to the other design elements in domestic interiors.

COMPARISON: THE BAUHAUS AND THE INSTITUTE OF DESIGN

In the summing up of my experience of design training in this country, certain characteristic features were emphasized. Their intrinsic and relative values, can best be assessed by comparing them with the most progressive practices in foreign schools. I shall be concerned mainly with two schools—the Staatliches Bauhaus, Germany, and the Institute of Design, U.S.A.

To begin with, the Bauhaus was the most original and influential school of design to be founded in this century. It was created by Walter Gropius, opened at Weimar in 1919, moved to Dessau in 1925, and closed finally in 1933.[1] All students were bound apprentices and had, during the early years of the Bauhaus, to pass the municipal journeyman's examination before they were admitted to the highest classes and to actual participation in the school's building tasks, such as the new building at Dessau. The Bauhaus combined craft training with art training—artists such as Kandinsky, Klee and Feininger were on the staff—but the central features of the curriculum were architecture and the building crafts. Manual craft-work was insisted on for two reasons. First, as an exercise in training hand and eye to appreciate the possibilities of basic materials. Second, to prepare the student for experimental design and model-making for standardized mass-production. The Bauhaus workshops were used as laboratories for evolving new designs. The new building at Dessau was partly paid for by royalties from licences granted to various firms for the mass-production of Bauhaus models. The items included furniture, carpets, textiles, porcelain, and light fittings. How many British art schools have adopted this approach and have a similar record of achievement?[2]

Instructors at the Bauhaus, architects, designers, artists and craftsmen were all practitioners. Students with some trade experience began with a preparatory course of six months to experiment freely and individually in the workshops. In this country few full-time students come to art schools with any trade experience. In Scandinavian schools initial trade experience is the general rule. The students saw and experienced all aspects of work at the Bauhaus and at the end of the period were helped in the choice of a particular subject which would form the core of their studies for three years. It must be remembered that there was a strong community spirit at the Bauhaus amongst students and staff, who were all working towards a common end. 'What we preached in practice was the common citizenship of all forms of creative work, and their logical interdependence one on another in the modern world.'[3] Close attention to one form of design did not imply a narrow intellectual outlook. Students in the earliest years worked all the time under two masters, an artist and a technical

[1] See W. Gropius, *The New Architecture and the Bauhaus*, Faber & Faber, 1935; W. Gropius, in *The Year Book of Education*, 1936, pp. 493 ff.; N. B. L. Pevsner, *Pioneers of the Modern Movement*, Faber & Faber, 1936, p. 42 (revised, *Pioneers of Modern Design*, Museum of Modern Art, New York, 1949, pp. 18–19; *Bauhaus 1919–1928*, Ed. H. Bayer, W. Gropius, I. Gropius, Museum of Modern Art, 1938; Charles T. Branford Co., Boston, 1952.

[2] When Professor Wagenfeld—who has been mentioned in connexion with glass—was teaching at the Bauhochschule, Weimar, in 1928–30, he designed several light fittings and articles in silver and brass. They were produced in quantity by industrial firms and sold through a sales department of the school. Out of the profits new machinery was purchased. A door-handle he designed gave the school for one year a monthly royalty of 300 R.M., which was more than his salary as a teacher.

[3] W. Gropius, *The New Architecture and the Bauhaus*, p. 57.

instructor. The system proved difficult to operate, and was later given up. In British art schools, after the preliminary trial period there is an insulation course between painters and sculptors on one hand and craftsmen and designers on the other. But at the Bauhaus there was a cross-fertilization of ideas throughout the training period. Students began with the simplest tools and gradually worked up to the applications of machinery. Unlike a factory-worker the student actively took part in all stages of production. As the Bauhaus itself did not possess industrial machinery, students were sent out to work in factories. This took place towards the end of the course, when questions such as costing and timing were considered. Few design students in this country get similar opportunities. At the end of the course, in a special research department, a student had to execute a design of his own before a panel of master-craftsmen, in order to obtain his Journeyman's Certificate of Craftsmanship. In addition, the Bauhaus set its own examination which demanded a high standard of individual aptitude for design.

The Bauhaus was a comparatively small school. At the time of the move from Weimar to Dessau there were twenty-four teachers and just under 200 students. Professor Gropius, its Principal from 1919 to 1928, is an architect, and architecture was the dominant subject which correlated all other work. Art, craft and design were not pursued as ends in themselves—as is the case in most British schools—but related to the wider aspect of designing a civilization fit to live in. The Bauhaus was looking for a unity of architecture, painting, music, film, theatre and industrial design. This makes it understandable that its characteristic design standard embodied a certain narrow formalism. But that was a by-product of its main object which was 'to exert a revitalizing influence on design'.[1]

The curriculum and methods of the Bauhaus were adopted by several art schools and technical colleges in Germany. They provided the guiding pattern for the recent changes at the Royal College of Art. The latest example of its influence is worth noting. Arrangements have now been made at Ulm, in Western Germany, for the foundation of the Hochschule für Gestaltung, an international school of industrial design, together with a research institute for product design. The United States Government, representing the occupying power for that part of Germany, has promised one million German marks provided that an equivalent sum is raised in Germany. Teaching at the Hochschule für Gestaltung will be based on the experience of the Bauhaus, while Professor Gropius, when at Harvard University, was included on an American sponsoring committee. Max Bill, the Swiss architect, is the Director of the school and designer of the building at Ulm.

[1] *The New Architecture and the Bauhaus*, p. 62.

The standard of design training might be very different in our art schools, if, after leaving Nazi Germany, more instructors and former students of the Bauhaus had been induced to stay in this country. But the opportunity, which came soon after the Bauhaus closed down, was missed by government and local education authorities. The United States held out fairer prospects and we lost some of the best teachers, including Walter Gropius to Harvard University and Lazlo Moholy-Nagy to Chicago.

The Institute of Design in Chicago, U.S.A., was started originally in 1937 as the New Bauhaus, but changed one year later to a non-profit making institution with the unprepossessing title 'School of Design'. The reason for this and the school's lack of support and sympathetic interest from American industry in its early years need not concern us here. The school was founded by Lazlo Moholy-Nagy, who was Director until his death in 1946, when Serge Chermayeff took over. Mr Chermayeff resigned in 1951, and since that time Crombie Taylor has been acting Principal. From December 1949 onwards the Institute has been a department of the Illinois Institute of Technology and associated with the School of Architecture under the direction of Miës van der Rohe.

To begin with there were twenty students, today there are about 250 on full-time courses and 500 on evening courses. Worth noting is its first aim: 'To establish standards of design commensurate with the social responsibilities of designers employing industrial means of production.' More specifically, the Institute trains all types of designers and teachers, and acts as a general design centre for research and propaganda material. Few British art schools have yet realized the significance of this. Experimental work on photo-lithography at the Stoke-on-Trent Art School is perhaps the most remarkable example.

The workshops at the Institute are equipped with hand- and machine-tools for experimental and simple work in wood, metal and plastics. The preliminary course for all students is particularly important. Uninhibited freedom is allowed for experiments with form (sculpture and modelling), colour and basic materials, so that students can discover their individual talents. In Birmingham there is a similar course. At the Institute students can then choose between 'shelter design', product design, visual design (including display, commercial art, typography), photography and film. Although craft-work plays a less important part than at the Bauhaus, a student will be expected to concentrate on one main subject throughout the four-year course. On the other hand, the cultural experience for all students is wider. There is a co-ordinated series of courses in mathematics, physics and engineering related to the development of space perception, light, colour and form. Where there are design courses in British art schools, a similar broadening of the intellect is not allowed for. Towards the end of the course students partake in projects related to research work, sometimes carried out for particular manufacturers. Product design students are encouraged to work out new uses for mass-production equipment. Time and cost estimation and industrial and business practices form part of the analysis made by students for each practical design problem. On leaving the Institute, graduates become teachers in the art and design departments of universities, or designers working for manufacturers on a full- or part-time basis.

Both schools have attempted to put the designer in his social context by first insisting on mastery of one subject and then gradually relating to it other cultural interests. This is now the policy at the Royal College of Art. The following points (which are absent in most British schools) were also stressed: art and craft as not ends in themselves, but means to enrich the sensibility and facilitate experimental work for industry; principals, teachers and instructors as practitioners, freed from routine and administration work (the same is true of Scandinavian schools); students belonging to a community working towards a common object; the school as an active centre of design, initiating new techniques; and students aware of the commercial requirements of costing, etc. But, more important than anything else, these schools were specifically set up to train architects and designers. They were not isolated classes attached to schools of art and crafts.

PART II

DESIGN ORGANIZATIONS: PUBLICITY AND PROPAGANDA

THE ROYAL SOCIETY OF ARTS

The Royal Society of Arts was founded in 1754 by William Shipley, a drawing-master from Northampton. In accordance with the ideas of the Age of Reason and of Mercantilism, its aim was to be 'the Encouragement of Arts, Manufactures and Commerce'. The importance of industrial design, although the term was then unknown, was recognized by the first members of the Society. At a meeting in 1754 when the first list of prizes was drawn up it was stated that 'the Art of Drawing is absolutely necessary in many employments, trades and manufactures'. Four years later there was another offer of prizes for designs suitable for weavers, calico-printers, cabinet-makers and coach-builders, as well as for manufacturers of iron, brass, earthenware, or 'any other mechanic Trade that requires taste'.

Towards the end of the century, when the effect of the Industrial Revolution was beginning to be noticed, the Society, out of harmony with the ideas of its time, fostered both drawing, which involved designing for manufacture, and industrial research at the same time. It was the conviction of the leading theorists of the neo-Classic movement that industrial art could be fostered only by promoting academic figure drawing, painting and modelling. Only thus, it was thought, could any reaction on the lower strata of industry and commerce be conceived.[1] In disagreement with these ideas, the Society annually presented prizes for draughtsmanship, including 'the best compositions of drawings or ornaments, being original designs fit for embroiderers, calico-printers, paper-stainers or any other art or manufacture'. Soon after 1840 they were discontinued but, beginning in 1846, a small number of annual prizes was awarded to students of design in art schools.

The Society was surprisingly unaffected by the movement started by William Morris. Since Morris had preached the importance of art in everyday life, a new outlook on the problem of industrial art became possible. The Society might have been remodelled according to the new standards, and have helped to establish them. But this chance was missed, and the Arts and Crafts Movement spread independently of the Society. Instead, the necessary step from hand-craft to machine-art was taken first by a few progressive individuals, and then by the circle of artists and manufacturers united in the Design and Industries Association. Not until 1924 did the Society renew its interest in the promotion of industrial design. The initiative was

[1] See N. Pevsner, *Academies of Art Past and Present*, Cambridge University Press, 1940, p. 186.

principally that of Sir Frank Warner, whose textile firm has already been mentioned. Funds were raised, and from 1924 until 1933 competitions were held for textile, furniture, pottery, glass and book design. They were open to all British subjects, not to art students only. Altogether £11,550 was spent on prizes. The competitions were given up in 1933 because each manufacturer who contributed to the fund stipulated different requirements and aims.

A similar scheme, called the Industrial Art Bursaries Competition, was started in 1938, but had to close down in 1941 because of the war. In 1946 it was restarted, and is now a regular feature of the Society's work. The object of the bursaries is to enable students from art schools between the ages of seventeen and thirty to travel abroad and study foreign design, or in certain cases to obtain art training and industrial experience in this country. An average of twelve bursaries of £150 each are awarded every year out of a fund partly contributed by interested manufacturers. The number of industries varies annually and the designs for each one are judged by a panel of four persons, including a manufacturer and a retailer. In addition, the Sir Frank Warner Memorial Medal is offered for the best design submitted in the carpet, dress textiles or furnishing textiles sections of the competition. Through its connexion with manufacturers, the Society has found jobs for many of the successful students. As the winning designs are exhibited each year in the Society's library, it is possible to see a representative selection of the best work done in art schools. Although the difference in taste between the panels of judges is often striking, it appears that the most promising designs are entered for printed textiles, light fittings, leather goods and domestic solid-fuel-burning appliances.

The Society has been concerned with a large number of original and important exhibitions. It began in 1760 by helping a group of artists to organize the first public exhibition to be held in England. It displayed some of the work of these artists who, in 1768, founded the Royal Academy. In 1761 the Society organized an exhibition of machinery—the first industrial exhibition ever held—but it was not until 1847 that an exhibition of industrial art, or 'select specimens of British manufactures and decorative art', was held. It was seen by 20,000 people, and followed by similar exhibitions in 1848 and 1849. In that year the Society—mainly through the agency of Sir Henry Cole—took the lead in helping Prince Albert to prepare the first International Exhibition. The Great Exhibition of 1851 was in the end not in any explicit way connected with the Society, but in spite of that its share in the preparatory work remains noteworthy, although the grafting of decoration on to industrial products, which characterizes most of the exhibits of 1851, is by no means what we should today call a satisfactory 'Encouragement of Arts, Manufactures and Commerce'. The most important exhibition of industrial art which the Society has helped to originate was the Exhibition of British Art in Industry held at Burlington House in 1935; this will be dealt with later. In 1951 the Society was mainly responsible for commemorating the Great Exhibition of a century earlier. In its lecture theatre it held an Exhibition of Exhibitions, which described its own work in this field.

A large number of lectures on industrial art have been given by members and guests of the Society. In recent years its lecture hall has become a valuable platform in London for providing information on design methods for new materials and techniques, as well as for giving philosophical and historical lectures on design. Eight of the weekly lectures in the 1948–9 session were on aspects of industrial design. All lectures are reprinted in the Society's *Journal*, which is published fortnightly.

The Royal Society of Arts has carried out its work without government aid and without endowment. It is a private body dependent almost entirely upon the annual subscriptions of its Members or Fellows. There is no precise qualification for membership: the applicant only needs to be supported by three Fellows. Although the use of the letters F.R.S.A. is not now the coveted honour it may once have been, the extent of the Society's membership has enabled it to maintain a valuable independence. On the other hand, the exclusive Faculty of Royal Designers for Industry, a body created by the Society, more than regains the prestige which is lost by having a large membership.* A separate section must be devoted to this body.

THE FACULTY OF ROYAL DESIGNERS FOR INDUSTRY

One aspect of the Exhibition of British Art in Industry in 1935 must be dealt with here. At that time it was unusual for a designer's name to be mentioned when his work was displayed.[1] But at the 1935 exhibition it was arranged that for each exhibit the designer, as well as the manufacturer, should be given credit. In fact, one of the aims of the exhibition was to bring industrial designers into prominence. At the same time it was realized that industrial designers had no status. Unlike architects, the designers had no body similar to the Royal Institute of British Architects, to which they could be elected. Although the Society of Industrial Artists had been in existence since 1930, it did not enjoy a prestige similar to the R.I.B.A., or demand an equivalent standard of professional ability from its members.[2]

[1] It should be made clear that to give full credit for work displayed had been a principle of the Arts and Crafts Exhibition Society ever since its first exhibition of 1888.

[2] See p. 229.

On the initiative of the Royal Society of Arts the position was altered. Instead of establishing a new body with unlimited membership the Society in 1936 created the exclusive distinction of Designer for Industry. It was intended to form a select panel of designers, not a professional body. Only the most distinguished designers could be elected and not more than thirty could hold the honour at any one time. Later, the number was increased to forty and the title Designer for Industry prefixed by 'Royal', after the King's approval had been received. Members are entitled to use the letters R.D.I. after their names.

The Faculty began with twelve designers: Douglas Cockerell, bookbinding; Eric Gill, sculpture, typography; James Hogan, glass; J.H. Mason, typography; H.G. Murphy, silver; Harold Stabler, silver, pottery; Fred Taylor, posters; George Sheringham, decoration; C.F.A. Voysey, interior decoration; Keith Murray, glass, pottery; Tom Purvis, posters. An honorary award was made to E. McKnight Kauffer, a pioneer in poster design, now an American. Excluding honorary members, there are now thirty-eight designers in the Faculty. Of these only sixteen are staff designers with full-time employment in factories. Again, out of the total number only thirteen are concerned with the industries described in this book. In both respects the Faculty is unbalanced, on the one hand with too many designers outside industry and on the other with a disproportionate number of commercial artists.

The members meet periodically for private discussions or to entertain foreign designers visiting this country. In addition they have carried out two ventures of their own. The first was the Design at Work Exhibition, at Burlington House in 1948, which was designed by members to display their work. The exhibition was disappointing because it clearly showed that only a very small number of new, first-rate designs was going into production. Almost everywhere the emphasis was laid on the display fixtures and not the case-histories, industrial prototypes and products. For their second venture the members were commissioned to design the interior furnishings of the Royal Pavilion on the South Bank site for the Festival of Britain. Altogether twelve members supplied special designs and arranged for their production by firms which held the Royal Warrant.

THE DESIGN AND INDUSTRIES ASSOCIATION

In 1915 a private association was founded for the promotion of art in British industries —the Design and Industries Association—to whose persevering propaganda probably more progress is due than to the activity of any other institution or individual. In spite of this, the Association is still far from holding the position it deserves with

PLATE LXXXV

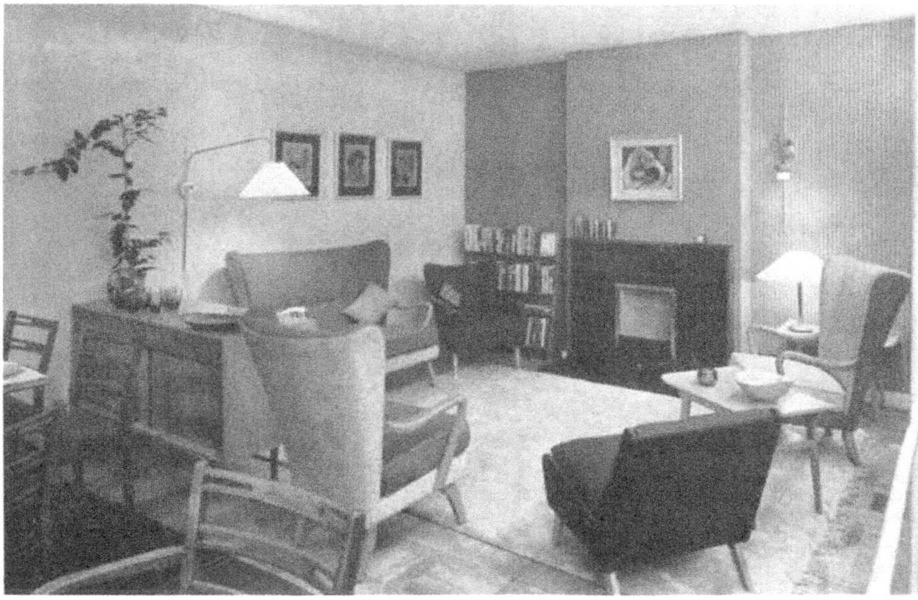

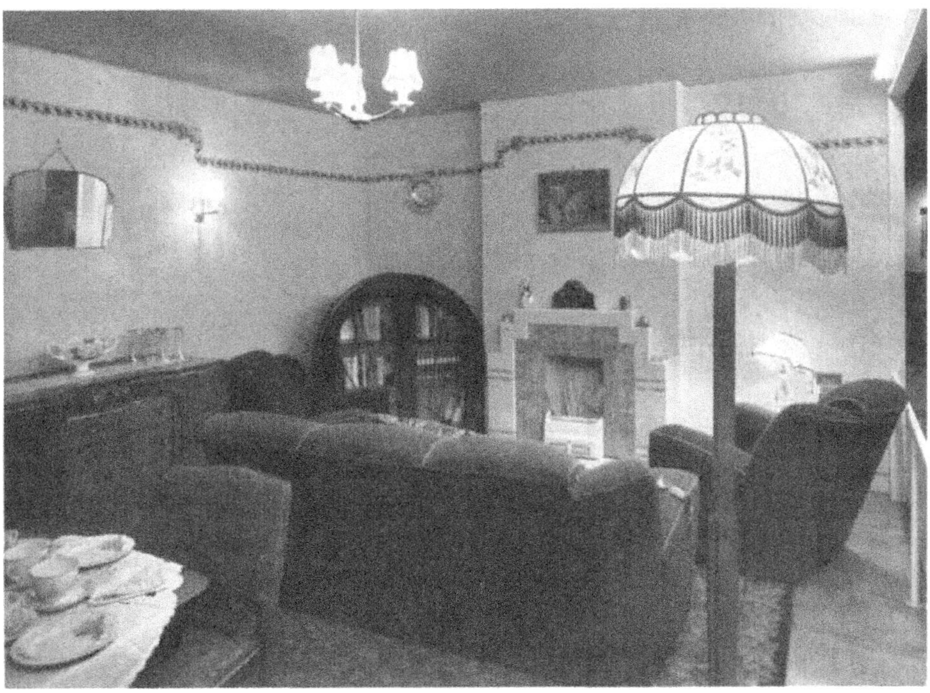

A bold and controversial exhibition was staged by the Design and Industries Association in 1953. Two rooms of equal size and cost were contrasted, one (*top*) was modern, the other (*bottom*) was equipped with furnishings claimed by the trade to be best-sellers. Without charge the visitors, 30,334 in three weeks, were asked to vote in secret, and the results showed that 60 per cent. preferred the modern room. The exhibition was afterwards shown in the Provinces. The D.I.A., founded in 1915, has continuously encouraged private individuals as well as manufacturers and retailers to insist on better standards of design in industry.

THE COUNCIL OF INDUSTRIAL DESIGN:

See p. 216

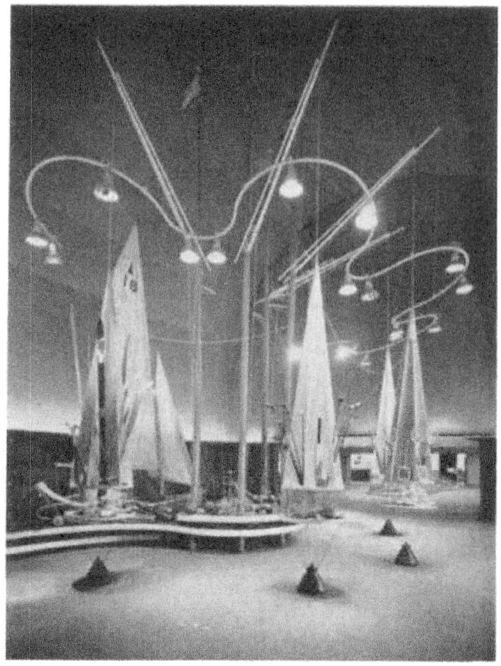

1. The Britain Can Make It exhibition, 1946: a view of the 'Sport and Leisure' section. The exhibition, the first of its kind to be government sponsored, was organized by the Council of Industrial Design.

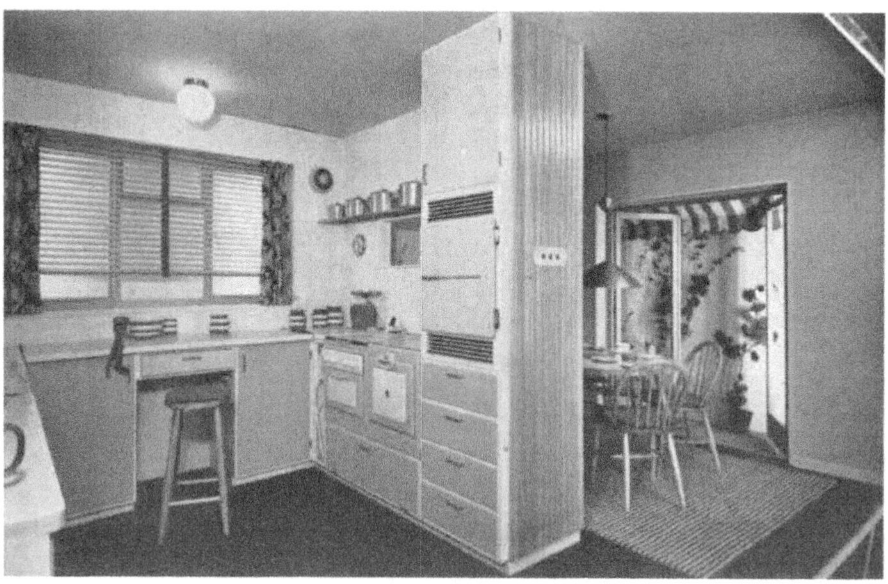

2. Kitchen and dining-room for a modern house at the Britain Can Make It exhibition, which was held in London and visited by 1,432,546 people during 14 weeks. Inquiries were encouraged from trade buyers interested in this first showing of post-war British design.

industrialists, retailers and the public; a position which the Deutscher Werkbund had attained in Germany shortly before and immediately after the First World War. Yet the effect of its incessant work in creating a more favourable atmosphere can hardly be overrated.

The avowed model for the D.I.A. was the Deutscher Werkbund, created in 1907. It is a remarkable instance of British level-headedness that just one year after the outbreak of the First World War this dependence on a German model was frankly admitted. In one way the D.I.A. was created by the war. This country was dependent on many domestic articles 'made in Germany', and when imports ceased the Board of Trade promoted exhibitions of 'enemy products', so that British manufacturers could take over and capture a new market. Most of the exhibits were cheap; many of them were cheap and nasty. A group of English artists and craftsmen who knew of the Deutscher Werkbund and had seen its exhibition at Cologne in 1914, seized the opportunity to persuade the Board of Trade to hold a further exhibition of German goods, but this time selected for their excellence of design and workmanship. The exhibition was held, and with it came the pamphlet *Design and Industry: a proposal for the foundation of a Design and Industries Association.* Two months later the inaugural meeting of the D.I.A. was held, with an original membership of 199. On the first Council there were men like Sir James Morton, H. H. Peach, Frank Pick and Sir Frank Warner from industry and commerce, C. G. Brewer, B. J. Fletcher, Ambrose Heal, C. H. St John Hornby, F. E. Jackson, W. R. Lethaby and H. Stabler from art and architecture. The first pamphlet published by the D.I.A. in 1915 already contained the famous thesis, so much discussed since then: 'The first necessity of sound design is fitness for use.' One year later another equally significant tenet was formulated. One of the paramount aims of the Association should be 'to harmonize right design and manufacturing efficiency, accepting the machine in its proper place, as a device to be guided and controlled, not merely boycotted'. This was evidently meant as a challenge to the anti-machinery attitude of Morris and his followers, and as such it still sounded revolutionary in this country at the time when it was put forward.

After this brilliant start, the Association developed slowly. The Deutscher Werkbund seven years after its foundation was strong enough to organize the most important exhibition of industrial art held in this century, the Cologne Exhibition of 1914, a token of victory for the Modern Movement in Germany. The D.I.A. remained an obscure, though by no means ineffective, body for more than ten years. At the Arts and Crafts Exhibition at Burlington House in 1916, the D.I.A. was allotted a section to display articles of good design selected from retail stores. Membership steadily grew from 292 in 1916 to 602 in 1928. A small journal began to appear in

1916, and the articles in it were remarkable for their vigorous and direct attacks on poor design standards and shoddy workmanship. There was a touch of humour too: 'We want less futile and more beautiful things in less futile and more beautiful houses and cities...[a demand] which has nothing to do with sandals or a diet of rigidly selected proteids, as is too commonly supposed.' From 1922 onwards there was published a series of excellently written and well-illustrated *Year-Books* plainly reminiscent of the pre-war *Year-Books* of the Werkbund. In 1927 the very satisfactory British Section of the International Exhibition of Industrial Art at Leipzig was organized by the D.I.A. under the direction of H. H. Peach.

Since then many activities have been undertaken. The D.I.A. made a point of going out of London to the Provinces with exhibitions and lectures to stir up public interest in design. Branches and local groups were founded in Birmingham, Manchester, Bristol; branches were also started in Northampton, Sheffield and Hull, but have now closed down. Once encouraged to begin, the branches supported themselves and carried out schemes for design improvement most suited to their localities. Meanwhile the London headquarters of the D.I.A. developed new types of propaganda. It published pamphlets and books, such as the extremely bracing and useful *Cautionary Guides* to several English towns, or the manifesto on the village pump. With these publications the D.I.A., under the leadership of Sir Laurence Weaver and Clough Williams-Ellis, went beyond the industrial into the social field of activity, and linked up with bodies which aimed at better planning in town and countryside. The year 1932 saw the first adequately produced periodical of the Association. It was called *Design in Industry*. After a year it was replaced by a new periodical *Design for Today*, which was never the Society's own paper but contained its official notes and remarks and was recommended and supplied by the Association to its members. At the beginning of 1936 another magazine was started, *Trend in Design of Everyday Things*, which, although it was more interesting in text and layout, had to be discontinued after only two issues had appeared—a depressing sign of the lack of public interest in the aims of the Association.

The slump of 1929, like the First World War, gave the D.I.A. a splendid chance to impress its ideas on manufacturers looking for something new to recapture lost markets. For most manufacturers this meant modernism, first seen at Paris in 1925, but for some who were already enlightened and for others who were influenced by the D.I.A., the slump period was the beginning of a rational policy of obtaining, making and selling new and good designs. The D.I.A. did all it could to introduce the best designs from abroad to this country. From 1925 onwards members went on travel tours to France, Sweden, Denmark, Germany and Austria, Holland, and Russia.

One of the most remarkable actions of the D.I.A. before the war is connected with the Dorland Hall Exhibition of 1933, which was promoted by *Country Life*. Though none of the official work was done through the office of the Association, and though its name is not to be found anywhere in the catalogue, it must be given the credit for the initiative as well as a great deal of the necessary spadework. The exhibition, the best of its kind hitherto held in England, was an outcome of the suggestions made in the Gorell Report.[1] Part of the exhibition was afterwards shown at Manchester, Halifax, Leeds, Edinburgh, Bristol and Stoke-on-Trent, admittedly under the auspices of the D.I.A. The Birmingham branch arranged an independent exhibition of Midland Industrial Art in the summer of 1934. The standard of this exhibition, which was held in the City Art Gallery, was by no means equal to that of the Dorland Hall Exhibition. Here, one of the first attempts was made in this country to assess public taste by asking visitors to fill in a questionnaire. Since then several exhibitions have been held, based on another principle. Beginning in 1935, the D.I.A. induced big shops or department stores to hold exhibitions on their premises consisting entirely of articles that these shops had in stock and selected by the D.I.A. Such exhibitions, the educational value of which appears beyond doubt, have been held at Bowman's, Harrod's, Peter Jones's and Whiteley's in London, Furlong's at Woolwich, Kendal Milne's at Manchester, Dunn's at Bromley, Jones's in Bristol and Rowntree's at Scarborough.

Another means of propaganda has been the exhibition of model houses completely furnished by the D.I.A. at a given cost. Such exhibitions took place at Manchester and at Welwyn in 1933, and at Birmingham in 1934. Here, of course, as well as in the exhibitions in stores, the aim was to impress upon the public that everyday objects of good design are obtainable at reasonable prices, if one seriously tries to find them. One of the Association's aims was to promote well-designed products for the 'positively or comparatively poor'.[2] At Welwyn and Manchester the complete contents of the house did not cost more than £200. In 1936 a D.I.A. member, Mrs C. G. Tomrley, made an inquiry to see how far well-designed, well-made furniture and furnishings were obtainable at prices a working-class family could afford.[3] The family income was assumed to be £4 per week. On the basis of Mrs Tomrley's report an exhibition of furnished rooms was held at the Building Centre in London in 1937. Of chief interest were the contents of a three-room flat which cost £50, for they showed plainly that £50 was not enough to buy a sufficient quantity of well-designed articles.

[1] See p. 201.

[2] 'Unless furniture, pottery and printing answer the common needs of a world in which most of us are positively or comparatively poor they can make no claim to fitness' (*The Journal of the Design and Industries Association*, no. 8, July 1918).

[3] *The Working Class Home: its Furnishing and Equipment*, Council for Art and Industry, 1937.

The year 1937 was the last fully active one for the D.I.A. before the war. A new experiment to bring members together was the week-end conference held at Bexhill. The programme included nine lectures, several discussions, and a film show. It was attended by 140 members and friends. Another new periodical was started, called the *D.I.A. News*, which ran regularly for a time but became more and more infrequent during the war. The most outstanding success of the year was the talks given by a D.I.A. member, Anthony Bertram, on the B.B.C., and the way in which the D.I.A. followed them up.[1] By planning in advance—about nine months before the talks were given—the D.I.A. provided sixteen identical portable exhibitions to illustrate Anthony Bertram's talks. Manufactured articles or, in the case of architecture, etc., photographs, were displayed. When set out each exhibition measured 16 feet wide and 5 feet high. Within twelve weeks, 111 exhibitions were held in retail stores and forty-six in art schools all over the country. The effect of a B.B.C. talks series on a little-known subject being brought home visually in the listeners' own towns was noticeable in shops and newspapers for several years. It was the Association's greatest success: achieved, it must be remembered, through the voluntary help of manufacturers, retailers and its own members.

In 1938 the threat of war and a trade depression—especially amongst furnishing firms—made progress for the D.I.A. difficult. Apart from arranging its own section of well-designed articles at the Women's Fair, Olympia, no major scheme of propaganda was undertaken. When war came the D.I.A. lost the services of many active members. By 1942 it was planning for peace. With the Army Bureau of Current Affairs, it designed small portable exhibitions dealing with house and domestic design problems, and sent them to Army camps, art galleries and schools. It held an essay competition asking schoolchildren to describe the houses and designs they would like to have after the war. There were 450 entries: 99 per cent of them wanted houses of modern design. In 1943 the first of the now very successful monthly lunch meetings was held in London. In 1944, the year in which the Council of Industrial Design was set up by the Government, the D.I.A. was incorporated.

Since the war the D.I.A. has not fully captured its former energy. It is in danger of becoming a social club where any expression of horror at the very many appalling designs still to be seen in shops, is not taken seriously. The week-end courses at Ashridge and the monthly meetings in London and the Provinces do much to keep members in touch with each other, but do little to stimulate campaigns of action. Where are the provocative booklets like those which typified the D.I.A. before the war? Surely the situation described in the first 186 pages of this book provides

[1] See p. 244.

scope for another stormy onslaught. As one of the founder members of the Association, Hamilton T. Smith, has recently pointed out: 'The heathen no longer stone us out of their cities: they are more likely to offer us a cocktail. And this frightens me. Most certainly they are not all converted: can it be that we have grown less exacting?'[1]

Some people may have thought that the Council of Industrial Design could very well carry on the work begun voluntarily by members of the D.I.A. It would be a great mistake to believe them. The D.I.A. has a prestige derived from years of useful and successful work. It is also an independent body, free to comment openly on all aspects of design in industry. These are assets not to be thrown away. At the present time the D.I.A. is short of money and has had to restrict its activities. But a greater number of seriously interested members would mean more money, more money would mean a more intense and widespread activity, and more activity in publications and exhibitions would mean a higher standing for the Association in popular and official recognition, and consequently a higher standard of design in industry.[2]

The most notable work of the D.I.A. since the war was in the organization of its Register Your Choice Exhibition at the Charing Cross underground station in 1953. The intention was to contrast two rooms of the same size, one furnished in a contemporary manner and the other stocked with furnishings that were not period reproductions but which the trade claimed to be best-sellers. The cost of each room was to be the same. An appeal amongst D.I.A. members had a generous response and just over £500 was raised to meet the costs of the exhibition. As the public was to be asked to vote for either room, it was essential to be impartial and not overstress the comparison. Mrs Phoebe de Syllas was commissioned to design both rooms with furnishings lent by retailers.

The exhibition, carrying the words 'These two identical rooms cost the same to furnish. Which do you prefer?', ran from 25 February to 21 March. It received very wide publicity in the daily and weekly press and 30,334 votes were cast. In the voting a distinction was made between men and women and between those over and under thirty-five. 18,188 people (or 60 per cent) voted for the room with contemporary designs, as did 59 per cent of the men, 61 per cent of the women, 53 per cent of the over thirty-fives and 63 per cent of the under thirty-fives. On the voting slips people were asked to state if they had spent £10 or more on furniture during 1952. A test

[1] *D.I.A. Year-Book, 1951*, p. 36.

[2] The following membership figures show signs of relaxing interest. In 1935 the membership was 819 (654 individuals and 165 firms); in 1952 it amounted only to 605 (468 individuals and 137 firms). The figure for 1952 includes 316 members in London, 85 in Manchester, 50 in Birmingham, 41 in Bristol, 10 in Northampton.

sample was taken and it is perhaps significant that in this case the 'contemporary' room obtained a smaller majority (55 against 45 per cent). But the main results showed that of the people who voted (most of them Londoners) the majority were against the type of furnishing thought by the trade to be most popular. And this surely means that if more retailers included in their stocks articles equal to the design standard expressed in the contemporary room, they would not lose business. Later, similar exhibitions were held in Manchester and Glasgow, under the supervision of the D.I.A. and in other provincial cities and towns on the initiative of retailers.[1]

THE BRITISH INSTITUTE OF INDUSTRIAL ART

The British Government realized as early as 1914 the necessity for some organized effort in order to obtain a better standard of industrial art. Immediately before the First World War the Board of Trade and the Board of Education decided to found an institute of a semi-official character. Its aim was to promote the collaboration of craftsmen, designers, manufacturers, retailers and the public. Its main activity was to be the arrangement of exhibitions. The First World War prevented the realization of these plans, but the work was taken up again in 1919, and in February 1920 the British Institute of Industrial Art was incorporated. Sir Hubert Llewellyn Smith was designated as chairman. A Treasury grant enabled the Institute, otherwise independent of official control, to start in a conspicuous way almost at once. In May 1920 a gallery for exhibitions of industrial art and handicraft was opened in Knightsbridge. Its principal aim was to find a market for the work of designers and craftsmen worth official support. During the year 1920 three exhibitions were held, from which the nucleus of a future permanent collection was purchased and for many years shown to the public in the North Court of the Victoria and Albert Museum. Owing to the first post-war slump, which became so menacing in 1921, the government grant was not renewed, and the Institute, started with so much energy and impetus, became entirely dependent upon private donations and subscriptions, though it maintained its semi-official status.

Considering these economic difficulties, its achievements are worthy of the highest praise. The gallery in Knightsbridge had to be given up, but various other exhibitions were arranged in the Victoria and Albert Museum and in provincial towns. Amongst

[1] See also the *D.I.A. Year Book, 1953*; *Design*, April 1953, pp. 8–11 and May, pp. 7 and 36).

the most important of these was the large travelling exhibition shown in Bradford, Birmingham, Huddersfield and Leicester in 1922, and the exhibition in the North Court of the Victoria and Albert Museum which was held in 1929 under the remarkable title 'Industrial Art for the Slender Purse'. It was perhaps the first public recognition of the fact that the most urgent need is to improve the artistic quality of inexpensive objects. The other main activity of the Institute was research. Its pamphlets on *Industrial Art and British Manufacture*, on *The University Teaching of Art in relation to Commerce*, on *The Art Training of the Middleman*, on *Public Departments and Industrial Art*, and on *The Training of the Silversmith*, were very well done and were the foundations of some more recent investigations.

Since the Government had withdrawn from the struggle for the improvement of industrial art in 1921, private institutions became indispensable. The Design and Industries Association, already dealt with, was founded as early as 1915, the Industrial Art Committee of the Federation of British Industries in 1921. The Board of Trade remained inactive until 1931, when it appointed the Committee on Art and Industry.

THE INDUSTRIAL ART COMMITTEE OF THE FEDERATION OF BRITISH INDUSTRIES

The year 1921, when the Federation of British Industries decided to set up an Industrial Art Committee, was the same year in which the British Institute of Industrial Art had to restrict its activities. The Committee is composed entirely of manufacturers representing almost every industry concerned with design. It was under the chairmanship of Charles Tennyson from its foundation until 1949, when Ernest Goodale took his place. As the authoritative spokesman for industry on design questions, the Committee has done very useful work.

In 1924 a Register and Employment Bureau for industrial designers was started. By the end of 1937 it had been used by more than 700 firms, and by 1946 over 1000 artists and students had been provided with industrial design work. Jobs and commissions were found for fifty to eighty designers a year, half of them commercial artists. There was also a card-index of firms that might be interested in designers. The Bureau's lists were never completely comprehensive. In 1936, therefore, it helped to establish the National Register of Industrial Art Designers, proposed by the Council for Art and Industry.[1] The Bureau then mainly devoted its attention to advising

[1] See p. 205.

students and helping them to find jobs in industry. Finally it was allowed to lapse in 1948, and subsequent inquiries were referred to the Council of Industrial Design.

Before the war the Committee arranged refresher courses for industrial designers at the Royal College of Art, promoted regular visits of design students to factories, and distributed instructional notes on various industries to art schools. In 1930 the Committee issued a memorandum on State-Aided Art Education. It proposed that regional colleges be established and that the Royal College of Art be reorganized for the benefit of design training. It followed this in 1935 with a detailed memorandum on industrial art training for the management and staff of firms—both on a full-time and a part-time basis—and with a similar training for foremen, travellers and salesmen. Courses were suggested for entrants into a trade as well as for those already employed. On account of its influential position with industry and its interest in design training, the Committee helped to formulate the Hambleden Report, 1936, already mentioned. In 1944 it again tackled the question of the Royal College of Art, and sent a memorandum to the Board of Education. In addition, each year the Committee awards a prize to a promising student at the Royal College of Art.

The Committee has shown active interest in the major exhibitions of industrial art. It brought out a very sound report on the Burlington House Exhibition in 1935. In 1946, at the time of the Britain Can Make It Exhibition, the Committee and the Council of Industrial Design, held a design conference attended by 900 delegates for industry. From 1949 to 1950, a series of discussion meetings on design policy and management was held with the help of the Council. Also in 1950 the Committee aided the Council's selection of exhibits for the Festival of Britain exhibitions.

In 1936 the Committee suggested that a co-ordinating architect be appointed for the layout of the British Industries Fair. This was not accepted at the time, but when the first post-war B.I.F. took place in 1947 a panel of architects was appointed 'to secure the greatest possible measure of architectural unity and harmony'. However, the greatest need at the Fair is not for better design of the layout, but better design of the products exhibited. It is a great pity that the Committee has never taken a stand to show well-designed products at the B.I.F. Its influence there would be considerable, because the majority of manufacturers would be prepared to listen to advice from their own Federation.

Many of the early duties of the Committee have now been assumed by other bodies, chiefly the Council of Industrial Design.[1] Although it meets infrequently, the Committee still has the duty of officially representing industry on questions of design.

[1] In 1944 the Committee issued a report on *Proposals for a Central Design Council and Industrial Design Centres*, to follow up the Weir Report which came out in 1943. See p. 208.

THE COMMITTEE ON ART AND INDUSTRY

Our survey has brought to light the fact that in most industries decisive improvements, where they have occurred at all, date back no further than 1930. What the Royal Society of Arts, the Design and Industries Association, the British Institute of Industrial Art, the Federation of British Industries, and private individuals had done before that time may be of the greatest historical value as pioneer work, but these efforts had remained isolated and sporadic. For the twelve years immediately following the First World War one can hardly speak of a live modern movement in British architecture and industrial art. But by 1930 animated discussions had begun in many quarters, and various activities were started under the auspices of numerous institutions—whose work, by the way, frequently overlapped and yet remained carefully independent. When things had gone so far, the British Government decided to return to the fray.

In 1931 the Board of Trade appointed a Committee on Art and Industry, under the chairmanship of Lord Gorell. Its task was to inquire into the production and exhibition of articles of good design in everyday use. Among its members were personalities such as Roger Fry, art critic; A. E. Gray, potter; C. H. St John Hornby, typographer; Sir Eric Maclagan, Director, Victoria and Albert Museum; H. Robertson, architect; Sir Hubert Llewellyn Smith, Chairman, The British Institute of Industrial Art; H. Trethowan, pottery designer; Professor E. W. Tristram, Painting School, Royal College of Art; Clough Williams-Ellis, architect and architectural critic. This first official body did not concern itself with the whole intricate complex of problems. Some palpable results had to be brought before the nation as soon as possible. Hence, with wise self-limitation, the Committee confined its activity to examining the one question, that of industrial art exhibitions, though this could naturally not be done without some side-glances at related questions.

The Report of the Committee was published in 1932. It begins with a succinct survey of the history of collaboration between industry, commerce and art from the foundation of the Royal Society of Arts in 1754 until the First World War. It then sketches out some requirements for art education, rightly stressing the point that nothing conclusive can be done unless the citizen is brought up to be far more aware of art and design than hitherto. Passing on to the actual object of the Report, preliminary investigations into the conditions of art in various industries are recommended. In conclusion, the Committee strongly supports the idea of a permanent exhibition of industrial art in London, and of travelling exhibitions in the Provinces.

It is not necessary here to go into details as to the ways recommended by the Committee for carrying this out.

The most significant fact about the Gorell Report is that it was an official document emphasizing the vital importance of improvements in British industrial art and confirming the urgent necessity of immediate action. And yet, in spite of this, there is (1954) still no permanent exhibition of industrial art in this country.*

THE COUNCIL FOR ART AND INDUSTRY

Following the suggestions which were made by the Gorell Committee, the Board of Trade appointed a Council for Art and Industry. The President of the D.I.A. at that time, Frank Pick, became Chairman. Among its members were industrialists, merchants, architects, artists and art critics. The Council started its activity at the beginning of 1934. During the first year a comprehensive programme was worked out. The various objectives are mentioned in the *Second General Report* of the Council, which was published in 1936.[1] They are enumerated under three headings: education of the consumer in the appreciation of designs, training in design for those engaged in dealing with it throughout industry and commerce, and encouragement of good design especially in relation to manufacture. Besides the English Council, there was a special Scottish Committee at work. The activity of both bodies was not limited to meetings, lectures and discussions; in contrast with many councils and committees, energetic research work began almost immediately.

For the education of the consumer, special research work was carried out in 1934 and 1935. There were separate reports for England and Scotland, entitled *Education for the Consumer*, and published in 1935. The reports dealt in a very concentrated manner with art education in elementary and secondary schools. In 1937 the Council held, under the auspices of the London County Council, an exhibition of educational material available for elementary schools, including industrial products which could be used as object lessons in good design. The exhibition was seen by 10,600 people in eleven days. Furthermore, the Council was invited by the same body to plan and equip a new elementary school built at Peckham. This school was completed in 1938. Also in 1937 the Scottish Committee held an exhibition of Scottish Everyday Art.

For information on the existing conditions of training for designers and craftsmen, evidence was collected from twelve trades. The first full report, *Design in the Jewellery*,

[1] See *The Board of Trade Journal*, 23 April 1936.

Silversmithing and Allied Trades, was published in 1937. Special committees to co-operate with local education authorities in the replanning of industrial art education were set up at Manchester for the cotton and rayon industries, at Stoke-on-Trent for the ceramics industry, and in Scotland for the woollen textile industry. The reports of all three committees were published in 1937.

Like the Gorell Committee, the Council had from the outset emphasized the importance of exhibitions for the education of the designer as well as for that of the consumer and the manufacturer. Here, again, action was taken almost immediately. A small exhibition of silverware was held at the Victoria and Albert Museum in 1934, showing objects of the past side-by-side with modern examples. A pottery exhibition seen by about 20,000 visitors followed in 1935. In the summer of 1936 an instructive collection of domestic metalwork was on show. In 1937 the Council was responsible for the British pavilion at the Paris International Exhibition, but owing to an inadequate grant of money this was not completely successful. Following its report on *The Working Class Home: its Furnishing and Equipment*, the Council in 1937 held an exhibition of model low-cost rooms at the Building Centre. This has already been referred to in connexion with the D.I.A.

The most influential report brought out by the Council for Art and Industry is *Design and the Designer in Industry*, published in 1937. It is based on the results of thirty-five meetings of the Council and on evidence received from seventy witnesses. The suggestions contained in the report are summed up on p. 7 as follows: 'The Report suggests, on the one hand, that manufacturers should help to secure better art education for young designers and for all ranks up to the heads of firms themselves, should employ in the design rooms more fully-trained designers and give them better pay and prospects, and should secure the part-time services of skilled artist-craftsmen, and of selected architects, painters and sculptors, and on the other hand that all experts in art education should seek to understand the industrial point of view more fully, that the courses of training in our art schools should be modified and extended for industrial purposes and the teachers be secured a wider training and practical experience in industry, and that, by the closest co-operation between all concerned, the local art schools and colleges, or their industrial departments, should in future become, in effect, an organic part of industry working enthusiastically for industry, albeit remaining centres of liberal art education.' The suggestion was backed up by an account of the training and status of designers in industrial centres abroad. Later, a committee of the Council in the Midlands produced the excellent report already referred to, *Design and the Designer in the Light Metal Trades*, published in 1944.

One of the most interesting aspects of the Council's work is outlined in its 'Proposal for an Industrial Art Centre'. This is the title of a report dated 3 December 1937, which is unpublished. On the first page the Council states that '...the time has come to institute an active exhibitions policy in this country, in the hope that if regular and comprehensive arrangements were made to show the public (including manufacturers and distributors) the best specimens of current industrial production, a healthy stimulus would be given to industry, as well as to art and technical schools, designers and the whole community, and sound standards of design would gradually be established'. Again, on p. 4: 'The judgement of individuals is mainly formed by visual discrimination, and it is therefore necessary that systematic methods should be adopted for bringing regularly to the notice of as many people as possible, throughout the country, the best modern goods which are made by craftsmen and manufacturers.'

The report outlines the proposals made by the Gorell Committee for permanent and travelling exhibitions, and considers that they are in general satisfactory. Before making its report the Council had taken evidence which concerned similar exhibitions abroad. Amongst those briefly mentioned are the regular exhibitions held by the Svenska Slöjdföreningen; a permanent collection of modern designs held in Munich by the Abteilung für Gewerbekunst; temporary exhibitions at the Stuttgart Museum; a case of patronage by a manufacturer in Hanover who endowed, for a short period, a collection of exemplary machine products and a travelling exhibition; and, finally, the Grassi Museum, Leipzig, which held regularly an exhibition of selected articles of contemporary design during the annual Leipzig Fair. Also in the evidence presented is the suggestion by the Museums Association that a central bureau of industrial art should select, assemble and circulate to local museums exhibitions of contemporary industrial products. It was thought that local museums might also form small collections themselves.

The Council's own proposals are as follows: An exhibitions organization should be established in this country, it should be housed in a permanent exhibition building in London, and hold a constantly changing series of exhibitions, each presenting a reasonably complete picture of some section of industry. Apart from British goods, some examples should be obtained from abroad for comparative purposes. It is also suggested that makers' and designers' names and retail prices be given on the exhibits. To protect the manufacturer and designer against plagiarism the Council thought that it might be necessary to impose a time-lag of, say, six months before new products were shown. The Council considered that it should be possible to obtain most of the British exhibits on loan from manufacturers without payment. Finally, it proposed that the Government should provide the accommodation and the staff, while the City

Guilds, trade associations, educational and philanthropic bodies would, it hoped, provide the necessary money.

As one of the original, though now apparently forgotten, objects of the Victoria and Albert Museum 'was to influence and display current production' the Council thought that the Museum's collections should be extended to form the exhibition of contemporary design rather than set up an entirely distinct organization. Instead of the term 'museum' the Council suggested 'Industrial Art Centre'. The Centre should have separate accommodation of its own, so that 'industry and craftsmen may realize that the institution has a character distinct from the Victoria and Albert Museum itself'. The officer in charge should have 'a definite measure of independence', subject only to the general supervision of a special advisory committee.

Amongst its proposals the Council thought that a limited number of the more distinguished examples of industrial art should be added to the permanent collections of the Victoria and Albert Museum. In the concluding paragraphs of the report it is stressed that the Centre 'should become a place of entertainment and education for the general public. It would seek to establish a current standard of design by developing a sanely critical attitude towards things of everyday use and service'. It is strange that after enlarging upon all its proposals the Council fails to raise the one question of central importance—namely, how should the exhibits be selected?

Another important aspect of the Council's work was the setting up of the National Register of Industrial Art Designers. This will be described in the next section. The work done by the Pick Council, as it has become to be known, is a notable and valuable achievement. It is greatly to be regretted that neither the Council nor a similar body exists to continue research into the problem of industrial art today. After carrying out the most comprehensive and thorough field-work industrial art has ever received in this country or abroad, the Council had to suspend its work on the outbreak of war. It was never called on to resume its task.

THE NATIONAL REGISTER OF INDUSTRIAL ART DESIGNERS

Following a recommendation by the Council for Art and Industry, a semi-official body known as the National Register of Industrial Art Designers was founded by the Board of Trade in 1936. It was endowed with a grant from the Treasury and provided with offices in London for the registrar and secretary. The Register, which began work early in 1937, was controlled by a board of fourteen governors under the chairmanship of Charles Tennyson. All designers working in this country could apply for registration as members. They were selected by a panel of governors who could call upon 128 honorary advisers representing approximately forty industries to judge the applicants' work. The first aim of the Register was to improve the status of designers by allowing its members to use the letters N.R.D. To begin with there was resistance from some manufacturers who thought that if their own full-time designers were put on the Register, other manufacturers might be tempted to offer these designers more favourable terms of employment! To avoid this it was decided to publish only an alphabetical list of registered designers which did not state classifications or addresses. As we shall see later, this deplorably weak attitude was characteristic of the policy of the National Register.

The other aim of the Register was to bring manufacturers into contact with the designers whose work was represented by drawings and photographs at the London office. In this way it became an employment bureau under an obligation to try to find work for its members. By the end of 1938 there were 425 registered names. Of this number approximately 200 were free-lance designers, fifteen architects, twenty-one designer-producers and the remainder full-time staff designers. In 1938 forty-nine firms approached the Register and the names of ninety-seven designers were put forward. In addition to this the Register gave its members many hundreds of letters of introduction to manufacturers who might be interested in their services. The list of members grew steadily. In 1942 there were 672 names and 780 in 1945.

In the *D.I.A. News* for August 1938, the chairman of the National Register pointed out that its object 'is to compile a complete register; selection cannot be confined to the exponents of any particular school or style, and the governing body has wisely adopted the policy of admitting all those who are technically qualified, and are engaged in the production of original design for British industry, whether as whole-time employees or free-lance'. Such a policy all but vitiated the hopes of the Pick Council and the aims of the Register itself. No doubt its dependence on a Government department, which measures success numerically, made it difficult for the Register not to show an increase in membership every year. The problem of selection was also difficult. The question before the governors was this: if a designer is technically proficient while his work has no artistic distinction, should he be admitted to the Register? In far too many cases the wrong answer was given, in spite of protests from designers already on the Register who worked to keep the standard undiluted. After all, national registration can only help the designer's status if it carries with it a guarantee of quality. By its inability to maintain a high standard when selecting members the National Register did much harm to the profession.

The Treasury grant was severely cut on the outbreak of war and finally withdrawn. At the same time the National Register was taken over and maintained by the newly-formed Central Institute of Art and Design. During this period the Register produced a very good report, with criticism and suggestions, on *Design and Development in the Leather Industries*. It was carried out in the manner of the Pick Council, but published by the Leather Footwear and Allied Industries Export Corporation, Ltd., in 1945. In 1945 the Register was disbanded and its list of 780 names handed over to the Council of Industrial Design. The use of the letters N.R.D. by existing members was allowed to continue, but no further appointments were made. The Register's list is now a part of the Council's own record of designers, to which manufacturers may apply for help. Unlike the National Register, the Council is under no obligation to try to find employment or commissions for designers on its record.

THE CENTRAL INSTITUTE OF ART AND DESIGN

At the outbreak of war in 1939, when the Pick Council had to suspend its work, a new body called the Central Institute of Art and Design was formed. It was a voluntary organization supported by its members. Its object was to protect the interests of about forty artistic movements and organizations in the country while the war lasted, mainly by providing an information centre in London.

Industrial design was only one of its activities, although its chairman, Charles Tennyson, and director, T.A. Fennemore, both came from the National Register. Amongst its activities the Central Institute organized several courses of two weeks each to refresh the minds of designers already working in industry. From 1946 until 1948 it was helped in this work by the Council of Industrial Design. Beginning in 1942, the Council Institute published a monthly bulletin, *Art and Design*, which is now a useful source of reference on the work of artists and designers during the war. As already mentioned, the Central Institute maintained the National Register of Industrial Art Designers, and carried on the work of finding jobs for designers in industry. In 1945 an Exhibition of Historical and British Wallpapers was staged at the National Gallery, which was also the Institute's headquarters. Finally, in 1948, it held the Inn Crafts Exhibition to demonstrate the work of designers and craftsmen for brewers and innkeepers. The Central Institute was dissolved in 1948.

THE COUNCIL OF INDUSTRIAL DESIGN

The Council of Industrial Design, the first organization of its kind in the world, was founded in 1944. The formation of such a body had its immediate origin in the Report[1] of the Sub-Committee on Industrial Design and Art in Industry set up by the Post-War Export Trade Committee of the Department of Overseas Trade, under the chairmanship of Sir Cecil Weir, in 1943. The Report recommended that a Central Design Council be appointed by the Lord President of the Council to set up a permanent centre for changing exhibitions of industrial design. It was stressed that this new body should be independent, have executive powers and not be controlled by a government department. However, an interdepartmental report was submitted to the Presidents of the Boards of Trade and Education, and the recommendations of this, the Meynell-Hoskin Report of 1944, were the chief reasons why the Central Design Council was established as a grant-aided body financed on the Board of Trade vote. It was set up on 19 December 1944, by the President of the Board of Trade in the Coalition Government, Dr Hugh Dalton, and called the Council of Industrial Design.[2] In 1945 premises were found for it in Westminster and at the same time the Scottish Committee of the Council was established in Glasgow.

The Council's terms of reference follow closely on the recommendations of the Weir Report. Briefly, its task is 'to promote by all practicable means the improvement of design in the products of British industry'. Its main functions are to hold exhibitions—the organization of the Britain Can Make It Exhibition was its first major task—and to give publicity to good design; to co-operate with the education authorities in matters affecting the training of designers; to advise government departments when asked to do so on the design of articles to be purchased by them in this country or exhibited by them overseas; to encourage the establishment of design centres, and to be a centre of design information itself for the public and for industry. To a large extent it had to work out its own activities. This was the job of the Chairman, Sir Thomas Barlow, and the Director, S. C. Leslie, until 1947, when Dr R. S. Edwards as Chairman, and the present Director, Gordon Russell, took over. In 1953 W. J. Warboys was appointed to succeed Dr Edwards.

Acting in the capacity of a board of governors the members of the Council are appointed by the President of the Board of Trade to advise on policy. Amongst the

[1] Unpublished.

[2] The Council began with a grant of £55,000 from the Board of Trade. From 1947 to 1950 it received £164,000 per year, and for its work on the Festival of Britain it was paid £318,000. For the year 1952–3 its grant was reduced to £100,000. The Council's permanent staff now amounts to about 100 persons. When big exhibitions take place temporary helpers are engaged.

PLATE LXXXVII

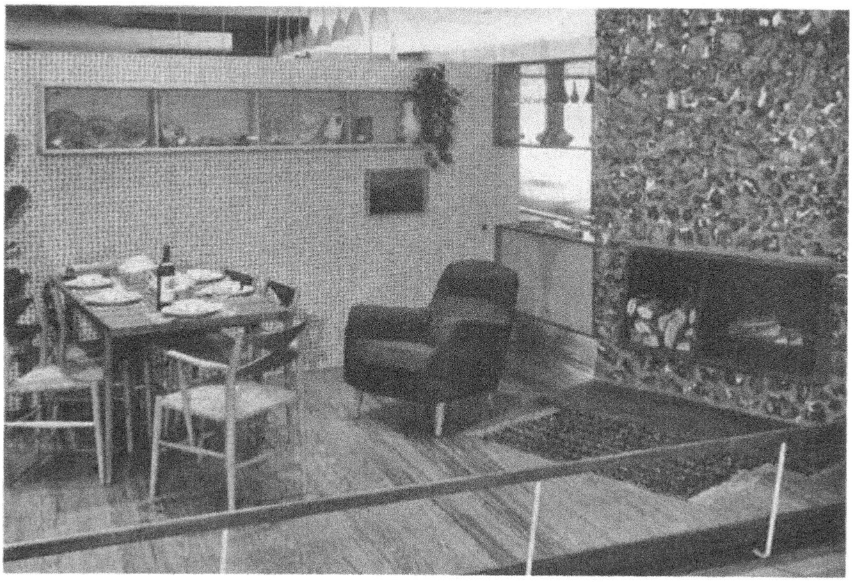

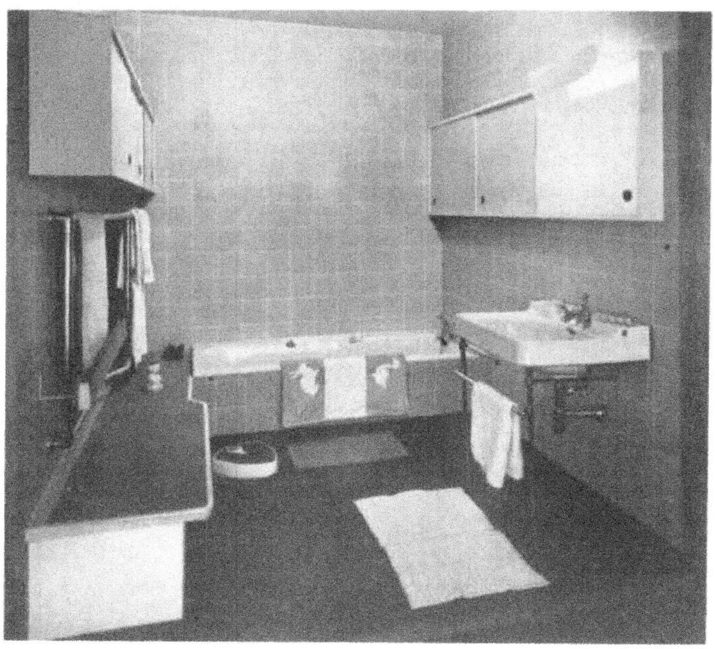

The part of the Council of Industrial Design in the Festival of Britain exhibitions, 1951, is very briefly indicated in these two views of the Homes and Gardens pavilion on the South Bank. The Council was responsible for selecting suitable British articles in current production; the exhibitions in London and the Provinces gave the Council its biggest chance to stimulate improvements in design. Where possible the exhibits were arranged as integral parts of a sequence of rooms.

PLATE LXXXVIII

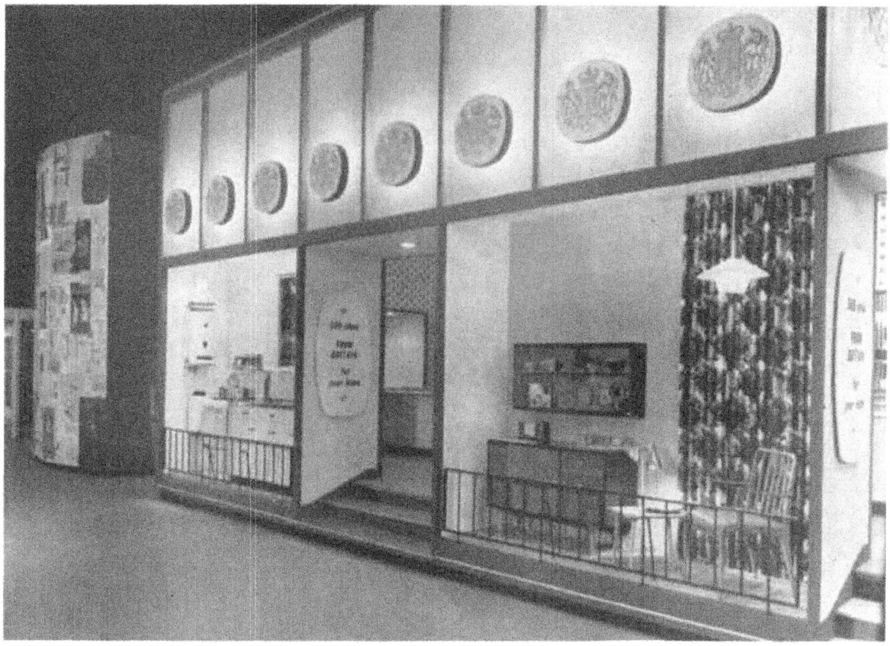

1. To focus attention on modern designs for export the Council of Industrial Design has organized a pioneering series of exhibitions abroad. This stand was seen at the Canadian National Exhibition, Toronto, 1953. With new designs the experiment was repeated in 1954.

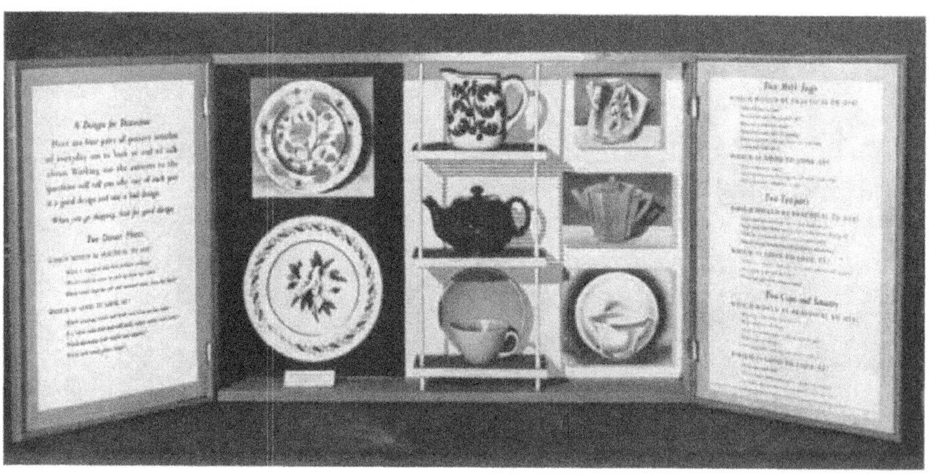

2. Helping people to discriminate between good and bad design is one of the tasks of the Council. This portable showcase belongs to a series on pottery, lent to schools, art galleries and retail stores.

members in 1952 were Sir Colin Anderson, director of Anderson, Green & Co. Ltd. (the Orient Line); Sir Leigh Ashton, Director, Victoria and Albert Museum; Robin Darwin, Principal, Royal College of Art; Geoffrey Dunn, managing director, H.G. Dunn & Sons; John Gloag, writer on industrial design; Sir Kenneth Lee, Chairman, Tootal Broadhurst Lee Co. Ltd.; Sir Charles Tennyson.

In its first annual report the Council gave this definition: 'Industrial design is not simply the plan of a particular product. It is a unity in the industrial process, a governing idea that owes something to creative design, something to the machine, something to the consumer, and links them all together.'[1] To meet its various duties, the Council in 1947 regrouped itself into two divisions, industrial and information, and it has retained this arrangement ever since. The Industrial Division is concerned with manufacturers and designers and the setting up of design centres. It keeps a record of designers available to industry, advises on design training and maintains a library and photographic library. The Information Division is responsible for propaganda, for retailers, schools and the public. It organizes exhibitions, publishes pamphlets and handbooks for public use and *Design*, a monthly journal intended for manufacturers, retailers and designers. The purpose of the two Divisions is plain enough. 'Without an effective demand for better design there can be no adequate incentive for the supply of it and without examples of good design being put before the public there can be no standard of comparison on which to base the demand. The twin drive for better design must therefore be pursued at the same time—the Industrial Division, to encourage a supply of well-designed products from, and the flow of trained design talent to, industry, and the Information Division to stimulate public demand for higher standards.'[2]

The task of the Industrial Division has not been easy. As we have already seen, the attitude of many manufacturers towards advice from persons outside their industries is likely to be cold and disdainful. The Council was aware of this from the start, as the following passage shows: 'Sometimes, however, the Council and its representatives have found themselves the victims of lingering suspicion borne of past encounters between industry and doctrinaire critics....Difficulties arise sometimes from the natural sensitiveness of some manufacturers to 'outside' criticism. When a body—especially one with the Government crest on its notepaper—approaches them with a programme for the improvement of design, they assume that it comes on an assumption of superior knowledge, which the manufacturer has no reason to concede. The truth is that the Council has its own difficulties. It cannot possess expert knowledge

[1] *First Annual Report, 1945–46*, Council of Industrial Design, H.M.S.O., 1946, p. 11.

[2] *Fourth Annual Report, 1948–49*, H.M.S.O., 1950, p. 1.

of design in every one of fifty or sixty consumer goods industries. It is therefore in the position of placing a programme of action before industries whose members are far more expert than itself in the details of their own processes and problems. Here the Council can only build upon the goodwill and commonsense of industrialists themselves....It is impossible to argue that improved design immediately and invariably means improved sales.'[1] It is particularly significant that this statement was made in the Council's first report. Instead of unguarded optimism we find a sober awareness of the immense size and difficulty of its task.

Before going on to deal with the Industrial Division's permanent programme, its work for the Festival of Britain exhibitions should be outlined. This began late in 1947, when it was announced that the Council was to be responsible for the selection of all industrially produced exhibits and certain handicrafts which were shown in the exhibitions. The Council was repesented on the Festival Council and the Executive Committee. Work started in 1948 on an industrial survey to find the most interesting design developments. It was intended that the 1951 exhibitions should tell the story of British achievements in the arts, sciences, technology and industrial design, and the industrial survey provided material to illustrate the theme. Along with this, the Festival of Britain authorities formed a Presentation Panel, and on this the Council was also represented. The co-operation of manufacturers was enlisted in a series of discussion groups held for each industry. To supplement this work, a pictorial card index of approved designs was started in June 1948.[2] It was called the 1951 Stock List, and it provided the material references on which the theme and industrial content of the exhibitions were based. Manufacturers were invited to contribute photographs of their best products. By the end of March 1949 the Stock List had about 1400 photographs, a year later the figure was 3790, and by the time the Festival began it had reached approximately 20,000, representing the work of about 5000 firms. The designs taken from the Stock List and the way in which they were selected and presented will be dealt with in a separate section later.

[1] *First Annual Report, 1945–46*, p. 14.

[2] The index was similar to the *Deutsche Warenkunde*; an illustrated record of good designs in German industry. The *Warenkunde* was started by the Deutscher Werkbund in 1921, and was originally confined to designs of the highest standard. It was based on the *Werkbund-Dürerbund-Warenbuch*, 1914. The *Warenkunde* was first shown to the public at the Die Form exhibition at Stuttgart in 1923; later, in 1928, it was included at the Der Stuhl exhibition, also at Stuttgart. In 1934, the Werkbund was dissolved by the Nazis, whose own organization, the Reichskulturkammer, took over the *Warenkunde*. The R.K.K. made considerable additions to the record, and published the first volume in 1939. In this country the first and subsequent volumes can be seen in the library of the Council of Industrial Design. Since the war, the *Warenkunde* has returned to the re-formed Werkbund. The *Warenkunde* should not be confused with the *Schweizer Warenkatalog*, published 1949, compiled by the Swiss Werkbund.

After the Festival, certain sections of the Stock List were maintained and given the title 'Design Review'. As a permanent free service to trade buyers, writers and lecturers, etc., from home and abroad, Design Review was reopened in July 1952 for inspection at the Council's London headquarters. At the end of 1952 it contained photographs of about 2600 items catalogued under sections which include furniture and furnishings, tableware, leather and travel goods, jewellery, domestic appliances, hardware and building fittings. In addition a section on street furniture is maintained for the benefit of local authorities and other public purchasing officers.

One more aspect of the Division's work for the Festival should be mentioned here. In 1949 the Division formed the Festival Pattern Group, consisting of twenty-six manufacturers of different products. The aim was to use the scientist's diagrams of the structure of crystals as theme for pattern designs. A Cambridge scientist, Dr Helen Megaw, agreed to supply crystal structure diagrams to the Group.[1] The manufacturers, who contributed to the expenses, then selected suitable designs which were afterwards interpreted by their designers. The samples produced were criticized by members of the Group and the best of them were used to decorate the Regatta Restaurant at the South Bank Exhibition. It was also hoped that the designs would be incorporated in the firms' ranges of products. The 'crystal' designs were suited mainly to textiles, both woven and printed, carpets, wallpaper, laminated plastic sheet and, in one case, pressed-glass sheet. Of chief importance, perhaps, was the fact that the formation of the Festival Pattern Group showed how the Division could enlist the help of manufacturers in a design project.

To return to the Council's 'long-term work of advice and persuasion'.[2] The effect of the Industrial Division's work on manufacturers has, as can only be expected at this stage, not been spectacular. In a file of case-histories of products submitted by about fifty firms for the Stock List, but of an insufficiently high standard of design, most of the entries concern the re-design of existing models and often the alterations are slight. The Division's aim is to encourage manufacturers to make better use of their staff designers. But where outside designers have been recommended to the manufacturers the results have usually been successful. It should be especially noted that each case appears to have originated when the manufacturer's first design had been rejected as unsuitable for the Stock List. The implications of this are obvious and we shall have to come back to it later. Here are some of the entries: A firm sent in a light fitting for a bed-head with a weak and clumsy switch; upon receiving the Council's criticism, a further prototype to the firm's own design was produced. In

[1] See *The Architectural Review*, April 1951, pp. 236–9.
[2] *Sixth Annual Report, 1950–51*, H.M.S.O., 1951, p. 5.

the same way, a radio cabinet was submitted for inclusion in the Stock List, but when the decoration on the dial was criticized the manufacturer agreed to remove it. Again, after a range of street lighting columns was rejected, the Division's advice was accepted and the whole range completely redesigned by the firm. In all three cases the improved designs were included in the Stock List.

The Industrial Division has helped to establish three design centres: the Design and Research Centre for Gold, Silver and Jewellery, 1946, the Rayon Industry Design Centre, 1948, and the Crafts Centre, 1948. The first design centre, the Cotton Board's Colour, Design and Style Centre, which had been set up in 1940, provided a valuable working example to be followed by the others. The first two, and the Colour, Design and Style Centre, will be described later in a separate section. In 1945 the Council worked out the scope, functions and organization of the centres it wished to propose, together with the degree of industrial support they should receive. Discussions between the Council and thirteen industries began at once. By 1946 one centre had been established and arrangements had been made for the other two mentioned above. Early in 1947, after the Britain Can Make It Exhibition had closed, the Industrial Division made further attempts to convince manufacturers that design centres for their industries were needed. A wider number of industries were covered for, apart from those already dealt with in this book, they included boots and shoes, clothing, dress and fashion accessories, hosiery, lace and toys. Composite centres, similar to the Interior Decoration Design Centre, which includes carpets, furnishing textiles and wallpapers, were proposed.[1] It was found, however, that much more preparatory work was needed before the new centres could be launched successfully. An aspect of the work included the nationalized industries and the Division's influence on design matters began to be felt by the British Electricity Authority, the G.P.O. and London Airport. In addition, the Division was represented on several advisory committees set up by the ministries.

Two further activities of the Industrial Division should be mentioned here. The first is the maintenance of a record of qualified designers suited to serve most types of industry. This record is based on the National Register of Industrial Art Designers, already dealt with, which the Council took over in 1945. The list of members of the Society of Industrial Artists was also referred to. As new designers, and those already practising who were not on the National Register, applied to the Council, their work was examined and, if it showed competence, their names were added to the record: 255 designers were added in 1946, 281 in 1947, and so on. Requests from manufacturers showed a similar progressive increase, from forty-five requests in 1945 to

[1] See pp. 227–8.

127 in 1946 and over 350 in 1953 (of which 50 per cent concerned industries in this book). The number of designers on the record increased as follows: there were 626 names in 1947, 1360 in 1949, 1894 in 1951 and over 2000 in 1953. It is usual for each request for a designer to be treated as a special case and for both parties to be interviewed by a representative of the Division. The designers' record is for the benefit of manufacturers and not an employment bureau for designers.

The other activity is advice on the training of designers and, to an increasing extent, advice on the training of management in design methods and appreciation. In 1946, the Council produced the *Report on the Training of the Industrial Designer in England and Wales* already mentioned. Although the Report was not published, it was discussed with representatives of the Board of Trade and the Ministry of Education.[1] During the same year two refresher courses were held for junior draughtsmen and designers from engineering, metal and electrical firms. Until 1948 the Division co-operated with the Central Institute of Art and Design, in conducting similar courses over a wider range of industry. And in the same year a course was held on colour and lighting in factories and on machines. It was attended by 180 architects, designers, paint manufacturers and others. A similar type of course was held on furniture design in 1949, when 168 persons attended.

In 1951, during the Festival of Britain, the Division organized an international design congress for manufacturers. The subject of the congress, which was attended by more than 250 manufacturers, was 'Design Policy as a Responsibility of High Level Management'. Papers were read by twenty-three manufacturers from Britain, Europe and the United States, who had shown a consistent interest in the design of their products and services. As a preliminary precaution the speakers were asked to draft their papers to a similar plan, dealing first with the character of their design policy, how it was devised and carried out and its commercial importance. Several of the papers were supported by small displays of products made by the speakers' firms. The congress ran for two days and was divided into three groups which met concurrently. The first group dealt with furniture, textiles, pottery and glass; the second, transport services and motor-cars; and the third, domestic appliances.[2] It is hoped to hold a similar congress in 1956.

The concern of the Information Division is propaganda for a higher design standard, directed mainly at retailers and the buying public. The method or approach adopted to carry it out has been based on the experience of other bodies especially

[1] Several of the underlying ideas in the Ministry of Education's *Rules Governing the Award of the Ministry's Diplomas and Certificates in Art, 1951*, H.M.S.O., 1950, were put forward in the Report.

[2] A synopsis of the congress appeared in 1952 under the title *Design Policy in Industry*.

the Design and Industries Association. The importance of this aspect of the Council's work was recognized from the start. The Council aimed to rouse the public to a state of alert sensibility and, after studying its circumstances and needs, to explain the principles of good design by means of comparative examples. For its purposes the Council divided the public into three groups: children of school age; progressive adults anxious to be enlightened; and, by far the largest group, the uninterested people 'who believe that they know what they like'?[1] It was clear that the Council's difficulties lay with people in the last group. 'There is no point in exaggerating the amount of interest they at present feel in design or minimizing the difficulty of rousing their critical awareness. The way to do it is to find the points of common interest and work from them....'[2]

The work of the Information Division is divided under the following headings: education, exhibitions, and retail trade. Added to these are the services of the research and information department (now closed), the press office and *Design* magazine. The Council's educational work began in 1945 with general lectures to mixed audiences. Between 1946 and 1948 a group of inexpensive books on design, in a series called 'The Things We See', were written for publication by Penguin Books Ltd. Altogether six books were produced, costing 2*s.* 6*d.* each, and dealing with a range of well-designed articles in common use. Added to these were two 'Puffin Books' for children. Between 1946 and 1947 experiments were made with various visual methods of teaching design appreciation. A short film called 'Designing Women' was produced, and had a fairly wide circulation. The first of the Council's film strips, 'Our Homes', a general introduction to the social background of design, was also made.

After these early experiments the Division, when it was formed in 1947, decided to develop those aids to design education which were permanently visible. The first of these were the wall cards in sets of from twelve to twenty, costing between 10*s.* and 20*s.* each. They were especially suitable for children at school, and their reception encouraged the Division to begin a new series of aids called Design Folios. The first one dealt with teapots and was published in October 1948. It was followed by others based on a similar pattern of presentation which had an introductory essay and twelve large photographs describing the historical design changes for each article. A series of six folios cost 25*s.* The idea was successful from the start. In March 1949 more than 1400 schools were subscribing; in January 1952 the figure was 2108. Altogether eighteen folios had been published.

In 1947 a new venture was started. This comprised a small travelling exhibition set of six boxes which, when opened, would illustrate the manufacture of furnishing

[1] *First Annual Report*, p. 20. [2] *Ibid.*

fabrics. The set proved to be popular in schools, art galleries, and retail stores where it was used for staff training. Later it was followed by another one on pottery design. The sets could be hired free of charge except for carriage costs.

On a larger scale were the Design Weeks, which also started in 1947. With them the object was to bring the persons and organizations most interested in design in one chosen town together, in a corporate effort to stimulate public interest in the subject. The Design Weeks varied in each of the nine towns in which they were held, but in all of them, except the first, there was a small central exhibition called 'Design Fair', for which the Division was responsible. Total attendance was 239,438. It was encouraging to find that very many retailers were willing to stage small displays of good designs in their particular trade, whether they stocked them or not. The main activities during a Design Week were conferences for manufacturers and retailers and students, and for housewives, lectures and film shows. Local schools were visited by the Division's education officer in order to demonstrate the Council's visual aids.

The response of retailers during the Design Weeks encouraged the Division to set up in 1948 a special retail section to persuade retailers to pay more attention to good design, especially as an aid to increase sales. As a result of the Design Week in Birmingham, a conference was held between the Division's staff and twenty-two Midland buyers and managers. It was found that the best way of co-operating with retailers was to start a scheme for staff-training. In May 1949 a special three-day course was held for retail staff-trainers. Practical discussion-group methods of teaching design were demonstrated by staff of the Bureau of Current Affairs. An interesting outcome of the course was the formation of the Midland Retailers' Design Discussion Group, which will be described later. In addition, a well-known London store asked the Division to arrange a course of training for its own sales staff. In the following twelve months four courses were held at Attingham Park near Shrewsbury. Three of them were for furniture salesmen, and one for salesmen of soft furnishings. They were all fully booked and shops from all parts of the country sent representatives. The courses are now a regular activity of the Division.

A more direct contact between the Division and the public, through the help of retail stores, came in 1949. The contents of several rooms of a house furnished by the Division at the *Daily Mail* Ideal Homes Exhibition, were sent on loan to a retailer in Chester. At the Exhibition the following year the Division had a similar display which it afterwards sent out as a touring collection of good designs to be staged in retail stores. This time seven stores took the collection for a period of three weeks each, and paid all the expenses. The main object was to give stores the opportunity of showing modern furniture designs to their clients without having to risk buying them for stock until they were sure of the demand. By 1950 several stores had, on their own initiative, staged their own exhibitions. Many of them asked the Council to recommend a designer who could select the objects, often using the Stock List, and arrange the exhibition.

One further activity which concerns retailers should be noted. In April 1951 the Hatfield New Town Development Corporation asked the Division to furnish one of the first of its new houses. Four local shops were invited to stock and supply the furnishings. This provided a unique opportunity for future tenants to revalue their judgements of design before beginning to furnish their own houses. Five thousand local people visited the house in three weeks.[1] Also in 1951, the Division furnished two houses in the new town at Peterlee, County Durham. The first house was furnished with modern tax-free designs, while the second showed how heirlooms and old furniture which tenants might have already possessed, could be combined with new designs. In this case the cost of furnishing the houses was borne by four nearby retailers in exchange for suitable advertising on the site. Again, at the Festival of Britain Architecture Exhibition at Lansbury, Poplar, the Council furnished a typical house and flat. Since 1951 there have been many more instances of this work.

Before dealing with the exhibitions organized by the Division, it will be necessary to describe the Council's first big venture in this field, the Britain Can Make It Exhibition held in 1946.[2] It was the first Government-sponsored exhibition in which particular stress was laid on design. The Council received a special grant of £234,757 for its work. For nearly a year the Council devoted most of its energies to organizing the exhibition and some of its work in the wider field of design propaganda was delayed. However, one of the immediate effects of the Council's work in arranging this exhibition was to bring its staff into direct contact with manufacturers. The designing and planning of the exhibition, which covered 90,000 square feet in the Victoria and Albert Museum, were carried out by James Gardner, with Basil Spence as consulting architect. Between them they directed the work of the seventy-five designers who produced the individual displays. British manufacturers were asked to submit their best designs, and the Council appointed its own selection committees for each group of products. 5259 articles were selected from 1297 firms out of a total of 3385 who submitted their goods. No space in the exhibition was sold to manufacturers, but each exhibit was credited to the maker and designer.

[1] There was a precedent for this type of propaganda on the part of the Council in 1946, when the National House-Building Industries' Joint Standing Committee invited the Council to furnish twenty-two demonstration houses in all parts of the country. The cost of all furnishings was limited to £300 for each house. Questionnaires were available to give visitors an opportunity to express their opinions. The Standing Committee afterwards issued a fully illustrated report, *Demonstration Permanent Houses, 1946*.

[2] The wider aspects of this exhibition, including its historical position amongst the larger industrial design exhibitions in this country, are discussed on pp. 235–7.

Domestic articles were not shown separately, but included in display features such as those called 'Shopwindow Street', 'What Industrial Design Means', and 'Great British Designers'. There were several furnished rooms, a section on packaging and one on raw materials. The exhibition ran for fourteen weeks and was seen by 1,432,546 people, three-fifths of them under forty. It was possible for trade buyers from this country and abroad to get in touch with the manufacturers of products which interested them.

Now to return to the general work of exhibitions. Between 1945 and 1953 the Council took a stand at the chief annual exhibitions of domestic equipment organized by daily newspapers. The first of these was the *Daily Herald* Post-War Homes Exhibition, 1945, where the Council was given space and financial help to erect its stand. A design quiz competition was held and 37,000 entries were received. Later, in 1946, at a similar type of exhibition, the Council furnished a one-room flat. The relatively low cost of well-designed articles was emphasized at the *Daily Mail* Ideal Home Exhibition in the same year. For this the Council appointed a designer to furnish the Ministry of Supply's aluminium house for not more than £250. The prices of all items were displayed in each room. A similar venture at the same exhibition in 1949 has already been mentioned. Amongst the trade exhibitions at which the Council has had a stand were the British Industries Fair in 1947 and 1954, the Building Exhibition in 1947 and the British Plastics Exhibition in 1951.[1]

In 1947 the Information Division decided to provide an exhibition hall at the Council's headquarters, and this was then used for displays covering most of the industries which concern the Council. Chief amongst them were packaging, textiles, printing and five pre-Festival of Britain exhibitions of material in the Stock List. In addition, the Tea Centre in Lower Regent Street, London, has provided space for five special exhibitions. One of the Division's more recent ventures, its collaboration with the B.B.C.'s Television Service, may become very important in the future. For some studio broadcasts where no artificial scenery is needed, the B.B.C. has followed the Division's advice by providing modern furnishings.

The Division's first opportunity to show its work abroad came in 1949 when, in collaboration with the British Council, it presented the British section at the International Exhibition of Industrial Design in Amsterdam. In 1950, twelve screens showing photographs from the Stock List were erected in the British Pavilion at the German Industries Fair in Berlin. From 1953 onwards this has intensified with many exhibitions in foreign markets, notably the two held in Toronto in 1953 and 1954.

[1] The Council had a stand devoted entirely to Coronation souvenirs at the B.I.F. 1953.

A suitable building to house a permanently changing exhibition of industrial designs is one of the Council's chief requirements. As already mentioned this idea was recommended in the Weir Report, 1943, but since then the difficulties of finding a building have been great, and the Council had to rely on its own small exhibition hall attached to its offices. In 1952 the hall was partly taken over by Design Review and exhibitions can no longer be held there.

The Council's illustrated monthly magazine, *Design*, is also the responsibility of the Information Division, although it is in fact addressed to manufacturers and designers as well as retailers. It was first published in January 1949 and costs 2*s*. 6*d*. It reviews recent design developments in this country, and maintains a series of short notes on foreign trends. Among its regular features are the case-history reports of satisfactory solutions to design and production problems. It has dealt with design in the engineering industries, exhibition design, competitions, the work of art school students, and interior decoration, as well as design matters from all the industries dealt with in this book. It has also had features of a reflective nature under such titles as 'Patrons of Design', and several articles on traditional design in particular trades. From September 1949 to March 1951, the magazine included, in shortened form, *Design Abstracts*, which had, since 1947, been published separately by the Council. Information on design from about 400 British and foreign periodicals was presented briefly in note form.

Another aspect of the Division's work should be mentioned. This was a department for research and information which began in 1945 with a wide programme. An attempt was made to find out what people abroad wanted from British industry. In this the Council received little encouragement from firms in this country, and the problem was left unsolved. Since then, however, the Division has co-operated with the British Export Trade Research Organization on design inquiries about the needs of foreign markets. In 1946 four working parties were sent to Germany by the British Intelligence Objectives Sub-Committee. They gathered information on design in the British, American and French zones of Germany, which was afterwards collated by the Council's staff for publication. The information is contained in the four B.I.O.S. Reports. In the same year another party visited Scandinavia for a similar purpose. This section of the Division's work was not allowed to expand as much as had been hoped when it was first set up. In 1952 it was closed down.

Part of this research work was done by the staff of the Council's library, which also answered more than 10,000 inquiries a year. Apart from a large stock of periodicals the library has a collection of about 2500 books connected with design, which can be consulted without charge. The Council has issued since 1945 a fortnightly calendar of lectures, exhibitions, etc., in this country connected with design. In addition it has since October 1948 sent out a regular newsletter about recent events intended mainly for schools and other educational bodies, but also for industry and the Press.

The Council's photographic library was started early in 1946 with 700 photographs. A year later it had 5648 photographs, and by 1952 more than 35,000. Its main task is to collect and classify photographs of all new products from this country and abroad which reach a reasonably high design standard. Apart from the Council's own staff, the library is open for use by manufacturers, designers and the Press. It also contains a lantern slide collection from which slides can be borrowed. For students of design the material and services provided by the library are extremely valuable.

Finally, there is the Division's press office. Its task is to persuade the Press to publish details of the Council's activities and the work of any body concerned with good standards of design. To supplement this it supplies special articles on design and issues photographs which illustrate the Council's work. Several of the articles have been published in the house magazines of commercial firms while its news notes and photographs are often used by periodicals. Part of the press office's work has always been directed to foreign publications and, in 1951, it was responsible for trade press publicity, both at home and abroad, for the Festival of Britain.

For the sake of clarity I have not included the work of the Scottish Committee of the Council in the foregoing account. As nearly all its activities are essentially the same as those of the Council itself, there will be no need for me to give full details of them. The Chairman of the Committee was Sir Steven Bilsland (now Lord Bilsland) until 1948, when R. A. Maclean took over. The chief officer was A. P. Middleton until 1948 and Wyndham Goodden until 1951, when Alister Maynard took over. Its permanent staff numbers about fourteen, and it receives a portion of the Council's annual grant.

Chief amongst its early activities was the exhibition, Enterprise Scotland, held in Edinburgh in 1947. Like the Britain Can Make It Exhibition it was set out thematically, with industrial designs made in Scotland being shown together in domestic interiors. As the Committee had always interested itself in domestic architecture, Scottish housing design and methods were included as a separate section. The Exhibition coincided with the season of the Edinburgh Festival and was seen by 456,000 people. At the same time fourteen shops in Princes Street held complementary displays. In the following year a travelling version of the Exhibition visited some of the remoter country districts of Scotland.

In 1948 a furniture competition was organized to stimulate Scottish manufacturers to improve their design standards in time for the Festival of Britain. Forty firms took part, some of them submitting as many as twelve pieces to the panel of judges which consisted of designers and retailers. The judge's written criticisms were attached to each piece of furniture and the manufacturers were invited to meet the judges to discuss their choice. Later, the winning designs were shown to the public. In 1949 the Committee included craft-work amongst its permanent activities. As in the industrial field its work has been advisory, though it was partly responsible for setting up the permanent Crafts Centre in Edinburgh in 1949. In 1950 an exhibition of light fittings was held in Glasgow; the majority of the fittings were brought from the Continent and they had the effect of showing manufacturers in this country how far behind they are in this field. For its part in the Festival of Britain the Committee was responsible for organizing the Living Traditions of Scotland Exhibition, which will be dealt with later. In 1954 it organized a design congress for 200 business men and others.

I have presented a full account of the Council's work for two main reasons. First, it is the first organization of its kind in the world where the care of design is the concern of government, and distinct from private organizations.[1] Secondly, it is the logical successor to the Gorell Committee and the Pick Council, and a remarkable instance of sustained interest in design on the part of successive governments. The change of attitude since the years before the war should be noted. Whereas the Pick Council occupied itself mainly with research, the Council of Industrial Design concentrates, almost exclusively, on active propaganda. The Weir Report, from which the Council originated, is a brief document. It recalls the main recommendations of the Pick Council, sums up the need for a central authority on design, and makes its own proposals. Although the Weir Committee suggested that an independent body should be set up, it should not, I think, be regretted that, in the end, the Board of Trade decided to resume its former interest and take on the task of forming the new Council. There are obvious reasons why an independent unofficial body could make

[1] There are several design organizations which are supported *partly* by grants from their respective governments: Canada, National Industrial Design Committee, Ottawa, founded 1948; Czechoslovakia, Ustředi Lidové a Umělecké Výroby, Prague, founded 1945; Denmark, Landsforeningen Dansk Kunsthaandvaerk, Copenhagen, founded 1907; Finland, Taideteollisuusyhdistys, Helsinki, founded 1875; Western Germany, Rat für Formgebung, Darmstadt, founded 1952; Holland, Instituut Industriële Vormgeving, The Hague, founded 1950; Japan, Kogei Shidosho, Kawasaki, founded 1928; Norway, Landsforeningen Norsk Brukskunst, Oslo, founded 1918; Sweden, Svenska Slöjdföreningen, Stockholm, founded 1845; Switzerland, Schweizerischer Werkbund Geschäftsstelle, Zürich, founded 1913 (part support by city of Zürich and not Swiss Government). For further details on these organizations and an account of those which receive only private support, see the *D.I.A. Year-Book 1951*. In Warsaw the Instytute Wzornictwa Przemyslowego, founded 1950, is run by the Polish Government.

a more direct and more rewarding attack on bad design, but there are equally obvious reasons why such a body, with the scope of the present Council, could not afford to exist. The suggestion in the Weir Report to establish 'a Pavilion of British Industrial Art as a permanent building, with changing exhibitions of the best modern design' has been held in abeyance for ten years. There is a pressing need for a permanent exhibition centre of industrial art, especially for the benefit of the public, as I have attempted to show in Part IV.*

As to the activities of the Council of Industrial Design, it seems that its most promising work is being carried out by the Information Division. The Design Weeks were very successful, and, if money could be found, they should be reintroduced for the benefit of all provincial towns. This would help the Division to increase its already noticeable influence on retailers. The courses for retail salesmen are a valuable aspect of this work. The furnishing of show houses in New Towns and housing estates has been another rewarding task. On the other hand, the Division might concentrate more interest on the education of children. Its design folios were helpful, but only when they were supported by an enlightened teacher. They could be re-formed to comprise contrasting examples of good and bad design so that children can learn to discriminate for themselves. To further this aim the Division might work out a simple method of critical analysis to guide the teacher. Experiments on these lines might be carried out by its own staff in one or two London schools.[1] With the same underlying object, the Division should take every opportunity to contrast its design standards with those exhibited by manufacturers at trade fairs. The Weir Report was clear in its views on this: 'We strongly recommend that an exhibition of this kind [the B.I.F.] should not be held under Government auspices without an opportunity being taken to stage an exhibit sponsored by the Central Design Council.' The B.I.F. is held annually by the Board of Trade, and the Council has exhibited three times; notably the B.I.F. Design Centre in 1954. Finally, one would like to see a magazine, complementary to *Design*, produced for the benefit of the public.

The task of the Industrial Division is much more difficult, as the reader of this book will appreciate. Its most effective agent for persuading manufacturers to improve design standards was undoubtedly the Stock List. In this the Division used a legitimate incentive by making it clear that only the best products would be included in the Festival of Britain exhibitions. It now does the same thing for smaller exhibitions which draw on material selected for Design Review. On the other hand, the enormous

[1] A series of small exhibitions of basic materials and well-designed goods made from them has been organized by the L.C.C. with help from the Council, for showing in London schools (see Sydney Foott, 'Design Appreciation in London schools', *Design*, January 1953, pp. 26–9).

number of names on the designers' record (in 1953 over 2000) appears to overstress the amount of talent available. If a manufacturer takes trouble to consult the record, the Division has to ensure that only the best names are put forward. Again, one aspect of its work concerns the training for designers, but in all the art schools I visited, with the exception of the Royal College of Art, I could detect no influence from the Division. On the other hand, its large-scale conferences for manufacturers and designers have been very successful. The most recent one, held in 1951, showed how the Division could assume European leadership in this type of work.

If enough money could be found the Council, as an agent for design propaganda, could spread its activities more widely throughout the country. I do not deny that London should remain the centre of its work, but if manufacturers, retailers and the public in the Provinces are to take a responsive interest the Council must constantly go out of its way to meet them. Design Weeks involve local collaborative interest, but they are short-lived in each town. Because of this it seems reasonable, if the cost is not too great, to suggest that the Council should appoint members of its staff to the position of regional officer in each of the main centres of population and industry. The officer could establish a two-way traffic of information. To do this he would need first to appreciate the situation in his area, and then enlist the aid of local loyalties. This would do much to encourage the inhabitants of provincial towns to adopt a responsible attitude to the Council's aims.

These aims should not, I think, be confined to propaganda for the purpose of conversion. The word propaganda has a sinister sound. Much more would be achieved ultimately if the Council deliberately tried to provoke critical response and discrimination. In order to do this a high design standard would have to be set up and maintained. The Council is the only organization in the country which could perform this service convincingly and on a large scale.

DESIGN CENTRES

The 'Design Centre' is a British phenomenon. It is an organization set up by an industry, or a group of industries, to provide the following services: technical research information and the way in which it can be related to industrial production with the help of a designer; a library with British and foreign publications; market research at home and in the export markets; exhibitions for the information of the industry or to attract the attention of foreign buyers; a meeting place for manufacturers and designers; and, finally, a link between the art schools and industry. The broad aim of a design centre is to improve the design and (if the distinction must be made) quality of an industry's products. Before the war the idea of a design centre with these duties was not put forward. The bodies we have already dealt with, from the Design and Industries Association onwards, had got as far as saying that either a permanent or a temporary exhibition of designs of a high standard was necessary. For a brief period in the 1920's there was a permanent exhibition in London, and since the advent of the D.I.A. there have been many temporary exhibitions. But I do not know of one exhibition where an industry, or its representative organization, was responsible for the selection and display of designs of a high standard. There are several large firms and many trade associations which carry out most of the other duties now ascribed to a design centre, but in no case is the emphasis placed on design throughout. The Building Centre, founded in 1932, though unique in this country and of great importance to architects and building contractors as well as to the public, has never limited its permanent or changing exhibitions to design of a high artistic standard.

THE COLOUR, DESIGN AND STYLE CENTRE

The first design centre, the Colour, Design and Style Centre, was set up in Manchester by the Cotton Board in 1940. The idea came from Sir Raymond Streat, chairman of the Cotton Board, who believed that a design centre would be able to improve the design of cotton goods and increase sales for manufacturers, especially in foreign markets. Accordingly, the Centre was established and supported by the Cotton Board out of the levy imposed on raw cotton. At first there was fierce opposition from a large number of manufacturers, particularly from directors of the biggest firms. These manufacturers were already conducting for themselves many of the activities which the Centre was to undertake and they resented the fact that their competitors, particularly the smaller firms, would be able to enjoy equal benefits.[1] However, this incredibly short-sighted attitude did not prevent the Centre from making a bold start. After two years the tension had disappeared and the Centre had established itself with the general approval of the industry. Much of the credit for this belongs to J. Cleveland Belle, the Centre's Director in the early years, who staged several startling and original exhibitions which attracted fresh attention to the cotton industry, particularly from the couturiers of London fashion houses.

The Centre's policy has been to raise the status of cotton, especially as a fashion fabric, by demonstrating its qualities and proving that it is worthy of carrying the best standard of printed designs. Although the question of fashion design is outside

[1] See J. Cleveland Belle and D. H. Tomlinson, 'The First Design Centre', *Design*, September 1949, pp. 7–11.

the scope of this book the Centre's work should be described. As we have already seen, many Lancashire firms weave and print cotton for dress goods and for furnishing fabrics. Often the same designers are used for both. The Centre has formed its own register of textile designers, carefully graded to distinguish between experienced free-lance designers and those who have only recently finished art school training. The register is used mainly by the smaller firms. More rarely, the Centre is asked to nominate a designer for a permanent staff position. The Centre's policy of holding small exhibitions on its own premises has been a notable success. It began in 1940 by displaying designs by leading painters and sculptors to provoke interest. By 1951 it had held over fifty exhibitions; some of the most useful dealt with fabrics from foreign markets. The aim of some of the exhibitions has been to correlate the design standard of the best products in several different industries—furniture, pottery, glass, wall-paper, etc., and textiles. Although intended for manufacturers, designers and art school students, it was always possible for the public to see the exhibitions. Much of the success in recent years has been due to the Centre's consistent use of two brilliant exhibition designers, Robert and Roger Nicholson. In addition, a full stock of current textile periodicals, a library, and a collection of foreign fabrics and designs are maintained at the Centre. And yet, with all this activity in the centre of the industry, manufacturers are still producing hundreds of furnishing fabrics of poor, enfeebled designs. One wonders indeed just how much propaganda is needed before a general improvement in the design of cheap and expensive fabrics can be noticed.

Between the setting up of the Colour, Design and Style Centre and the next design centre there was an interval of six years.[1] During that period the Council of Industrial Design had started work and, as we have seen, one of its objects was to encourage industries to set up design centres. Under the terms of the Industrial Organization and Development Act, 1947, the Board of Trade was able to make grants, equal to the amount subscribed by each industry, towards the foundation and maintenance of design centres for a trial period of three years. To do this the Board has acted on the

[1] The work of the Leather, Footwear and Allied Industries Export Corporation, Ltd., active from 1940 to 1953, should be noted here. While its interests were primarily centred on the overseas market, it did exert some influence on leather design in this country. The Corporation initiated an investigation into design and development in the trade, which was carried out by the National Register of Industrial Art Designers. See *Design and Development in the Leather Industries*, published by the Corporation in 1945. The Corporation was not a 'design centre', although it carried out market research for the trade and held 'Leather Weeks' in retail stores abroad. It also issued twice a year an engaging and well-illustrated magazine called *Envoy* and supervised the presentation of the leather industry's exhibit at the B.I.F. In these and in other ways the Corporation had a salutary influence on design in the trade.

advice of the Council of Industrial Design. The problem of convincing manufacturers that design centres for their industries were needed was tackled by the Council, in the way already described. In addition, the Council took the opportunity of advocating design centres in the evidence it gave to the Board of Trade's industrial working parties which were active between 1945 and 1948.[1] The working parties of chief importance here were concerned with the cotton, wool, linoleum and felt base, carpet, furniture, jewellery and silverware, pottery and domestic glass industries. They were concerned with all aspects of these industries, but the question of design was given special prominence in the last five mentioned. Except in the pottery industry, the working parties endorsed the Council's proposals for design centres.

THE DESIGN AND RESEARCH CENTRE FOR THE GOLD, SILVER AND JEWELLERY INDUSTRIES

This was the first design centre to be set up with the help of the Council of Industrial Design. Because of the pressing need for research work in these industries, the Department of Scientific and Industrial Research proposed that research as well as design should be included in the Centre's activities. Manufacturers in the gold, silver and jewellery industries were generally agreed that a centre of this type was necessary, but the chief credit for carrying the proposals into practice belongs to the Worshipful Company of Goldsmiths. The Company had been mainly responsible for design improvements in the past by means of competitions, scholarships, commissions and exhibitions. In 1946, when the Centre was set up, the Company provided it with offices at Goldsmiths' Hall. As will be seen later the Centre had to restrict its activities early in 1953.

Apart from research, the Centre's main duties were to keep a register of designers, hold competitions and exhibitions and award scholarships. The register provided details and photographs of the work of designers, many of whom were teaching in art schools. There was very little evidence to show that manufacturers, after using designers on the register, had improved the artistic standard of their products. The Centre held a national design competition annually on the basis of one year for speculative designs on paper, to give students and free-lance designers an opportunity to send in their work, and the next year for designs that have been made up in factory or workshop.[2] The sum of £600 was given away in prizes for each competition and

[1] For terms of reference the President of the Board of Trade asked each working party: 'To examine and inquire into the various schemes and suggestions put forward for improvements of organization, production and distribution methods and processes in the industry, and to report as to the steps which should be taken in the national interest to strengthen the industry and render it more stable and more capable of meeting competition in the home and foreign markets.'

[2] The competitions succeeded those held by the British Jewellers' Association until 1945.

the winning designs were exhibited at Goldsmiths' Hall, and in Birmingham and Sheffield. The competitions covered all sections of the industries, though the attention paid to electro-plated articles was meagre. For example, the competition in 1951 asked for fourteen articles in silver, four in gold, but only five in E.P.N.S. When it is realized that more than three-quarters of the industries' total output is in E.P.N.S. ware, the inclusion of only five articles appears to be quite insufficient. Moreover, it is the manufacturers of electro-plate who most urgently need to improve their designs. The Centre also awarded annual scholarships which amounted to about £800. The scholarships were intended mainly for art school students and the winners were advised to travel abroad, often in parties supervised by an art school instructor.

The Director of the Centre and his staff spent much of their time visiting manufacturers, designers and art schools in Birmingham and Sheffield. At the British Industries Fair the Centre usually had its own stand. Also, it has organized several foreign tours for manufacturers and designers. For example, a small group made a successful visit to factories, art schools and a trade fair in Switzerland in 1949. Later in the same year another party went to Paris. Finally, the Centre administers the purchase-tax relief scheme for hand-made silver articles of approved design; this has already been described.[1]

After seeing the order in which the results of competitions are placed, after interviewing some of the designers on the register and, after seeing the articles of 'approved design' relieved from purchase-tax, I have no doubt that the Centre was upholding a very mediocre standard. A glance at its *Bulletin*, which was published periodically, will immediately confirm this. As the industries it serves are notoriously unenterprising in design, it is particularly sad to find that the unexpected presence of a design centre in their midst has done practically nothing to improve design standards. Early in 1953 the Centre had to restrict its activities. For a large part of its funds, which came from the trades concerned, the Centre was dependent on the Jewellery and Silverware Council which closed down in December 1952. An informal committee of manufacturers has now been set up to find out what design and research facilities are wanted. In the meantime a nucleus staff is maintained to deal with all inquiries, run the library and administer the purchase-tax relief scheme. The other activities of the Centre already mentioned have been suspended.

THE RAYON INDUSTRY DESIGN CENTRE

Acting again on behalf of the Board of Trade, the Council of Industrial Design shared equally with the British Rayon Federation the responsibility of setting up a

[1] See pp. 34–5.

design centre for the industry in 1948. Financial support also came from member firms. Until it was abandoned in favour of a meagre design section on the Federation's own premises in Piccadilly, the Rayon Centre occupied an eighteenth-century town house near Grosvenor Square, furnished throughout with modern designs of a high standard. The Centre was fortunate in having as its first Director Dennis Lennon, who is a talented architect and designer of furniture and display-work. Until it closed down in 1952 the Rayon Centre was the first place to take a foreign visitor who wanted to see some of the best examples of British industrial art in actual use. The Centre carried out no scientific research, but through its connexion with the British Rayon Research Association it helped to relate design to new techniques. An up-to-date record of British fabrics in production was kept at the Centre and used by foreign buyers who sought advice. For manufacturers in this country there was a reference collection of fabrics from abroad. A register of available designers was also maintained. In some cases the Centre purchased designs to sell to member firms.

The Centre held several small but enticing exhibitions in its own display room. In 1950 it exhibited textile designs by members of the Society of Industrial Artists; twenty-six of the designs on paper were offered for sale. This original idea was followed later in an exhibition held for textile design students at the L.C.C. Central School of Arts and Crafts, where some very successful sales were made. The Centre increased the prestige of British rayon, mainly through its exhibitions and its sumptuous magazine *Rayon and Design* which, unfortunately, only lasted for three numbers. Nevertheless, a lot remains to be done to raise the general design standard of rayon furnishing fabrics for, like the Cotton Centre, the Rayon Centre found that its most spectacular successes were mainly concerned with fashion goods.

THE BRITISH COLOUR COUNCIL AND THE INTERIOR DECORATION DESIGN CENTRE

The British Colour Council was conceived by a group of manufacturers in the late 1920's. It was founded in 1930 and started work the following year. The Council is wholly supported by its members and the fees it receives for special services. In 1949 the Division of the Council dealing with interior decoration was enlarged to become a Design Centre. This venture was sponsored by two trade associations and two large firms, together representing carpets, linoleum and wallpaper. The Government is in no way responsible for the Centre. For some time the need for 'a permanent meeting place for all sections of the furnishing trades' had been felt.[1] And, as we have seen, no trade needed it more than the carpet and linoleum trades.

[1] *Board of Trade Working Party Report on Carpets*, H.M.S.O., 1947, p. 24.

The Centre is divided between two objectives. On the one hand, it has to attempt to demonstrate how the work of the British Colour Council can be helped to improve design standards and on the other it has to provide a meeting-place for manufacturers and designers. To link together an intelligent use of colour and the maintenance of a high design standard is a useful objective only when carpet manufacturers and others are prepared to employ better designers and to risk experiments. There is no necessity here to discuss the relationship between colour and linear or three-dimensional design. The two are interdependent in most manufactured articles, particularly in carpets, textiles, and wallpaper. But the point to stress is that, when beginning to suggest improvements to firms in these trades, it is essential to make a distinction between colour and design, otherwise there will be immediate confusion.

The Design Centre is, unfortunately, divided in its aims. First, it is required by the parent body to suggest colour schemes for houses, hotels, factories and showrooms. See, for example, the booklets entitled *Colour in the Home*, *Colour and Lighting in Factories and Offices*, *Colours for Machinery*, etc. The production of a *Dictionary of Colours for Interior Decoration* was also undertaken. These activities represent the Centre's contribution to colour and *design*. Secondly, the Centre has attempted to improve design in all industries concerned with domestic interiors. This has been done, not by working to a design standard, but by emphasizing features which various domestic articles have in common. Inevitably, the only common visual feature which is bound to turn up regularly is colour. So colour becomes the criterion of excellence, and the point that a good design standard springs from a common feeling amongst contemporary designers in different industries, and not from superficial resemblances in their work, has been entirely missed.[1] It is significant that for an exhibition at the Centre in 1949 based on a theme from the *Dictionary of Colours for Interior Decoration*, the products of three participating manufacturers came into hopeless conflict.

Because of this, it is doubtful if the very well attended conferences for staff designers held periodically at the Centre and in the Provinces can do much good. It is true that the much needed 'meeting-place' has been provided, and provided in London. Some of the lectures have been profitable: a discussion in 1950 under the title 'Some recent designs and why I produced them' indicates the approach which *ought* to be followed. Nevertheless, few if any significant improvements in design will be made by means of the Centre while it remains in the shadow of another organization with entirely different aims.

[1] 'While materials may conform to a predetermined colour scheme, the design, pattern or character of the materials is at the discretion of the manufacturers, who can proceed confidently knowing that their products will be "fashion right", if kept within the selected colour range. The whole output of widely different products is thus related in co-ordinated tonal effect for the ultimate consumer.' Cf. the *Prospectus* for the British Colour Council.

The general idea behind a design centre is sound. If a centre succeeds in becoming the focal point in its industry for design, manufacturers will compete with each other for the highest honours. And consequently the industry's prestige will increase. At the same time the centre should enlist the interest of the public, especially through its exhibitions. It should provoke discussion so that people who are otherwise visually inarticulate, will become critical and express their opinions. A design centre, unlike the small retail store, could offer this service to manufacturers without fear of losing money. Throughout, the most promising design centre has always been the one concerned with the cotton industry. Unlike the other three, it has a clear idea of what it wishes to accomplish and is guided by definite standards, even if these are in part influenced by fashion. Other design centres should follow, but only after the mistakes made already have been properly understood.

THE SOCIETY OF INDUSTRIAL ARTISTS

The Society of Industrial Artists is an independent body founded in 1930 'to secure a recognized status for artists working for industry and commerce, and to provide a controlling authority to advance and protect their interests'. According to this programme, the Society might have developed into something like a trade union; some of its activities in fact tend in that direction. Nevertheless, from the beginning the Society considered its task to be the promotion of improvements in commercial and industrial art education, the organization of exhibitions and the fostering of a high standard of industrial design. The Society participated in the Dorland Hall Exhibition of 1933 and acted as a witness for reports made by the Council for Art and Industry. Also before the war, the Society attempted to develop local branches in provincial centres, and for a time the branch at Stoke-on-Trent worked very effectively. The Society has grown slowly. In 1936 there were 250 members, 400 in 1946, and 700 in 1951. During the war there was a lapse in its activities, but in 1944 the Society produced a well-reasoned report on *The Training of Designers for Industry*.

Important changes came in 1945. In that year the Society was reconstituted to form two groups, one for commercial artists and the other for industrial designers. Each group was to be equally represented on the Society's governing Council. Members' activities were subdivided still further into sections concerned with posters, illustration, typography, packaging, exhibitions, fashion, textiles and general industrial design. At the same time the principle of selective admission was enforced, membership being open only to practising designers whose work had been approved by a selection committee. Members of the Society are entitled to use the letters M.S.I.A.; there are also a number of fellowships (F.S.I.A.). Students taking a *bona-fide* course in design at recognized art schools can be registered at the Society to become Licentiates. Also in 1945 it was ruled that all members should observe the Society's Code of Professional Conduct. By adhering to this Code members can claim professional status as commercial artists or industrial designers. In addition, the Society has set out in precise terms the regulations governing competitions, conditions of contract, briefing forms and schedules of average fees, although none of these is binding on a member or his client.[1] These codes are especially useful as guides for the young designer entering the profession. The Society will also obtain legal and technical advice for its members. Finally, there is a Register Advice Committee which submits short lists of designers to manufacturers who apply to the Society.

Among the activities of the Society are social meetings, design conferences, discussions and exhibitions. In 1949 a Summer School was held at Ashridge for members to hear lectures on colour, crystallography, geometry and dynamic form. A journal of design news is brought out six times a year, and in 1947 the Society began to publish a biennial review called *Designers in Britain*. The four volumes already published illustrate the work of designers, not all of them members of the Society.

The Society's presence is felt more strongly in London than anywhere else.[2] Since the war no attempt has been made to set up branches of the Society in the regional centres of industry. I interviewed many designers in Bradford, Manchester, the Potteries, Birmingham, etc., who were not members of the Society simply because London is too remote. As we shall see in the next section, a group of designers in Birmingham has felt the need to form a local association on its own initiative. At the present time the Society is most anxious to increase its membership, so that it can exert a wider influence for the profession. Between 1950 and 1957 the Society hopes to enrol 1300 designers, and so nearly treble its present membership. On paper this may seem possible; in 1953 the Council of Industrial Design had a record of 2000 names which it used when manufacturers applied for suitable designers. But in practice it means inevitably that the Society's selection standards will have to be lowered. Although the problem is difficult it would be a pity to sacrifice the Society's most important aim—'to secure a recognized status' for designers—merely because it cannot afford to carry out its work on a large scale.

[1] When formulating these codes the Society received advice from the Industrial Art Committee of the Federation of British Industries.

[2] There is one remarkable exception to this. In 1949 a new body, the Association of Canadian Industrial Designers, was formed with aims similar to those of the Society. In fact, the Association has based its practice on the Society's codes and schedules, etc.

But the chief criticism to be levelled at the Society is its *apparent* preoccupation with commercial art. The four volumes of *Designers in Britain* are striking evidence of this. While the first actually dealt with commercial art before industrial design, the second devoted more than two-thirds of its space to commercial art. The third and fourth volumes were similarly unbalanced. In this country there is a most pressing need for an association which will benefit staff designers, actively engaged in mills and factories. That is why the plan launched in 1954 to form a new Product Design (Light Engineering) Group, should be attractive to many of these designers. At the same time the new Consultant Designers' Group, formed late in 1953, helps to set a standard for the outside designer capable of working for several different industries.[1]

THE MIDLAND INDUSTRIAL DESIGNERS' ASSOCIATION

A Design Week held in Birmingham by the Council of Industrial Design in 1948 succeeded in bringing together a small number of Midland staff designers. At the end of the year they had formed the Midland Industrial Designers' Association with the object of establishing a focal point in the Midlands for all matters concerned with design. In 1951 the Association had just over seventy members, drawn from factories, retail stores and art schools. Meetings are held during the winter months for lectures and discussions of members' work. A collection of photographs of their work has been formed and is exhibited at the Association's meetings. In 1950 the collection was given a three-day exhibition in the Birmingham City Art Gallery.

Two of the Association's ventures are particularly interesting. Beginning in the summer of 1950, a large department store in Birmingham, Lewis's Ltd., allowed the Association to select one article of good design every two weeks and display it on a specially designed stand in the appropriate department. The reasons for the Association's choice were attached to each article. The first item chosen was a leather portmanteau, and while it was on view there was a noticeable increase in sales. Later in the same year, at the time of Lewis's annual Home-Makers' Exhibition, a popular event for its customers, the Association staged a display by contrasting

[1] The Society of Industral Designers in the United States of America should be briefly mentioned here. It was founded in 1944. Eight years later it had less than 150 members, all of whom were capable, according to the Society's rigid specifications, of designing not less than three types of industrial product. The Society does not include commercial artists.

domestic articles of good and bad design. Visitors to the Home-Makers' Exhibition were forced to see this display because it was staged in a small department which connected the two parts of the exhibition. Again, on each article was a written statement giving the reasons why the design was good or bad. In most cases, the Association concentrated on functional defects or advantages. The degree of co-operation which the Association received from Lewis's may be gauged by the fact that two pieces of furniture from its own stock were displayed as 'bad' designs. Apart from attracting public interest in design, it is evident that the Association has already given many Midland staff designers a fresh impetus to impress upon their employers the need for higher design standards.

SELECTIVE EXHIBITIONS OF INDUSTRIAL ART

At the present time hardly a month passes without one selective exhibition of good designs being held somewhere in this country. Much of the initiative for this comes from the Council of Industrial Design working on its own or with retailers. London gets most of the exhibitions, but in the Provinces, especially in the smaller county towns, there appears to be a steadily increasing number of retailers who are willing to hold small exhibitions of designs selected by interior designers often not on their staffs.

The situation before the war was very different. Exhibitions of designs chosen simply for their artistic merits were rare and of sufficient importance in the history of industrial design to make a brief reference to the best of them worthwhile. Most of them have already been mentioned when dealing with the organizations and institutions concerned with design. The first relevant exhibition was staged by the Board of Trade at Goldsmiths' Hall in March 1915. Its subject was 'enemy products', or German designs revealing the early influence of the Deutscher Werkbund, selected by the group of enthusiasts which soon afterwards founded the Design and Industries Association. Later in the same year the D.I.A. held its own first exhibition under the title Design and Workmanship in Printing. From then until 1938 the Association held many exhibitions and was always trying out new methods of influencing manufacturers or the public. Its most important exhibition, held at Dorland Hall in 1933, was the best survey of modern design to be seen in the 1930's in this country. During the previous decade the British Institute of Industrial Art had, for a short time, its own gallery in Knightsbridge. When that was closed the Institute held several temporary exhibitions in London and in a few of the larger industrial cities. But in none of them did the exhibits show a consistently high standard of design which could, at the same time, be connected with the Modern Movement. The reasons for this were made clear at the exhibition held at Waring and Gillow's in 1928. Part of the exhibition was designed and selected by Serge Chermayeff, who had been given a completely free hand and had travelled much on the Continent in order to obtain exemplary modern articles such as pieces of furniture, textiles, domestic fittings, glass, china, metalwork, etc. The exhibition was arranged as a series of rooms of different types. It was one of the decisive ventures which helped to establish the Modern Movement in this country.

In 1934 London saw an excellent display of photographs and models of International Modern Architecture at the newly built headquarters of the Royal Institute of British Architects. In the same year there was another exhibition at Dorland Hall which was less successful because the designs were made exclusive and, though modern, too orthodox. At the beginning of 1936 the R.I.B.A. held an Exhibition of Everyday Things. Domestic equipment and articles for the building trade were interesting, but the other sections were uneven in quality. Later in the year Heal's showed an exceedingly stimulating exhibition of modern rooms, designed by Maxwell Fry, Jack Howe, Raymond McGrath, Marcel Breuer, Christopher Nicholson, Hugh Casson, Christopher Heal and Brian O'Rorke. From 1937 until the war the D.I.A. and the Council for Art and Industry were separately responsible for several small exhibitions.

THE ROYAL ACADEMY EXHIBITION OF BRITISH ART IN INDUSTRY

The exhibition at Dorland Hall in 1933 was a most impressive venture and it introduced the principle of reserving for the promoters complete control over the selection of exhibits. By the time the Exhibition of British Art in Industry, held in 1935 at Burlington House, had closed, the principle was almost wholly discredited. As this exhibition received so much adverse, and in some cases malicious, criticism, it is necessary to report those facts which one needs to know in order to judge its organization and results. First of all, it was a revolutionary act on the part of the Royal Academy to open its rooms to an exhibition of industrial art, and it must be recognized as such. It must also be recognized that the Royal Society of Arts by being, with the Royal Academy, jointly responsible for the exhibition, had shown its earnest desire to help industrial art where help was urgently needed. This should not be forgotten, whatever one may criticize in the exhibition itself.

The following were given as its chief objectives: 'To impress upon the British as well as the foreign public both the importance of beauty in the articles they purchase, and the fact that British Manufacturers in co-operation with British Artists, are capable of supplying such articles in all branches of industry.' The other aim was to enable manufacturers to study to what extent British artists could supply 'original, attractive and technically suitable designs for the production of articles by mechanical means'. The general conditions stated that all exhibits must be of original, contemporary British design, created by designers of British nationality and working in Great Britain, manufactured in Great Britain by a British individual or firm, if possible from British materials. The honorary Organizing Secretary was John de La Valette. The preparation took more than a year, and the amount of work done during that time was great.

The exhibition consisted of the following departments: dress materials, furnishing fabrics, carpets, ceramics, glassware, gold, silver and jewellery, leather goods, book-production, posters and commercial printing, domestic equipment, furniture, plastic mouldings. A special room was devoted to an exhibition of shop-window display, which was changed fortnightly. The decoration of the galleries was entrusted to the following designers and architects: Maxwell Fry, H. S. Goodhart Rendel, John Grey, Sir Ambrose Heal, Oliver Hill, Charles Holden, L. Irving, E. Maufe, O. P. Milne, P. A. Steynes, G. Grey Wornum, and designers from the Rowley Galleries. To almost all the objects shown the name of the designer was attached and—as often as possible—the retail price. This, however, could not be made a hard-and-fast rule, owing to the marked reluctance of many manufacturers.

So much for the aims and the arrangement of the exhibition. By far the most serious criticisms were that too much in the exhibition was expensive and too little within the reach of the majority of visitors (to say nothing of the majority of the population); that too much was modish and eccentric and too little of sound and plain workaday character; that there was too much exclusive craftsmanship and too little mass-production. Nevertheless, at this exhibition several designs of a high standard (many of them are still influential in industry) were shown to the public for the first time. Amongst the designers of these were Susie Cooper, Harry Trethowan and Keith Murray, pottery; James Hogan, Raymond McGrath and Eric Ravilious, glass; Harold Stabler, Bernard Cuzner, Omar Ramsden, A. E. Harvey and H. G. Murphy, silver; Sir Ambrose Heal, Gordon Russell and Betty Joel, furniture; R. D. Russell, radio cabinets; Allan Walton and Alec Hunter, textiles.

Not all the articles shown were expensive. There was cheap pottery, there were some cheap dress materials and articles made from synthetic resins (plastics became known at this exhibition), and there were other inexpensive articles as well. But the proportion between these and costly articles was unbalanced. Since the exhibition was called British Art in Industry, and since its President was the then Prince of Wales, who more than once pleaded for 'fitness for purpose', 'simplicity of line, form and decoration' and 'working for the majority instead of for the needs of a minority',[1] the great quantity of expensive goods was a slap in the face not only to those seriously concerned with the matters in question, but also to those ignorant and well-intentioned visitors who came to find out whether the Modern Movement could after all mean something to them.

This is certainly a fair and a weighty argument against the Burlington House Exhibition. In comparison, it is less important that some of the best British designers complained of not having had their share in the building up of the exhibition, and that manufacturers complained of not having known exactly who was in the end responsible for the selection. But there is one more objection which is so obviously dependent on the exclusive character of much of the exhibition that it cannot be omitted. Some of the articles had been made for the sake of publicity only, and not for future commercial production. This was rare amongst pottery, dress and furnishing textiles. But the very existence of such cases, partly accounted for by the basic condition of the exhibition not to show anything which had been on the market before, is a serious matter, because it is apt to invalidate some of the most important results of any such exhibition. If the visitor cannot be sure of finding the objects which he especially likes in the ordinary, or at any rate in available, shops, and if the expert cannot be sure whether the show represents individual or industrial production, the practical utility of such an enterprise is gravely affected.

THE BRITAIN CAN MAKE IT EXHIBITION

Between 1935 and 1946, the year when the Britain Can Make It Exhibition was held, there were no national exhibitions devoted exclusively to industrial design. The principle of selection according to excellence in design returned to its originators in this country, the D.I.A., to be used only on a small and local scale. When the principle was revived for the Britain Can Make It Exhibition the response, especially from manufacturers, showed that it was vested with suspicion and distrust—plainly a legacy from the Burlington House Exhibition. Although the exhibition has already been dealt with in connexion with its organizers, The Council of Industrial Design, there are several points which properly belong to this context.

The Britain Can Make It Exhibition had three aims: 'To intensify the interest of manufacturers and distributors in industrial design and their awareness of the

[1] Address delivered at the opening of the Burlington House Exhibition (John de La Valette, *The Conquest of Ugliness*, p. 7) and Message to the D.I.A., 4 November 1931 (*D.I.A. Quarterly Journal*, January 1932, no. 17).

desirability of rapid progress; to arouse greater interest in design in the minds of the general public, as consumers; and to stage a prestige advertisement before the world for British industry, industrial design, and standards of display.'[1] The fact that the exhibition was held at all, so soon after the war and by a newly-formed body with a permanent staff of less than thirty, was due almost entirely to the driving enthusiasm of the late Sir Stafford Cripps, then President of the Board of Trade. In some ways the exhibition was staged at an inopportune moment. In most industries manufacturers were busy with the change-over from war-work to peace-time production. Many new designs were not in production, or even in prototype form and manufacturers were reluctant to disclose details of them. There were shortages of some of the most suitable materials, and utility schemes, particularly in the furniture trade, hampered individual initiative. Finally, in nearly every group of domestic articles, there was a sellers' market and therefore little incentive to bother with new designs. But on the other hand it must be said that the exhibition provided an excellent opportunity to show manufacturers the need for better design standards before they began full-scale production. A useful ally was the public, tired of war-time austere designs in domestic equipment and ready to be attracted to new designs in a colourful exhibition. At the same time, as part of the policy for exporting more British goods, the promoters wanted to put on a show which would appeal to foreign trade buyers.

The exhibition was arranged in the Victoria and Albert Museum by seventy-five designers working under the direction of James Gardner, with Basil Spence as consulting architect. The selection of the exhibits was carried out by separate committees for each industry, or group of industries. They were appointed by the Council of Industrial Design, while assistance was available from 'technical assessors', nominated by the industries. It has been already stated that the method of selection aroused the suspicions of many manufacturers, especially in the Midlands. For instance, it was thought that the selection was in the hands of a metropolitan clique who had no right to tell manufacturers what should or should not be made. Some complained of being allowed no opportunity to discuss the selectors' choice, which was said to be arbitrary and could only result in a manufacturer's refusal to co-operate or his unthinking acceptance of the standard set. It was also said that the selectors' taste was unallied to the known demand in some foreign markets.

What then was accomplished by the Britain Can Make It Exhibition? First of all it was seen by nearly one and a half million people in fourteen weeks. Of the trade buyers who saw it, 43,000 were British and over 7000 came from abroad. The articles exhibited showed that the selection committees had maintained a design standard

[1] *Second Annual Report, 1946–47*, Council of Industrial Design, H.M.S.O., 1947, p. 4.

which was much more satisfactory than its counterpart at Burlington House in 1935. A few of the articles were exemplary, notably amongst radio cabinets, woven furnishing textiles, leather goods, glass and domestic equipment. The worst designs were found amongst wallpapers, carpets, clocks and jewellery. The exhibition was presented in various themes and many of the best articles were to be seen three or four times. A great many of them, of course, were originally produced before the war. Others had not then been tried on the market. As the promoters did not state that all articles should be in, or about to go into production, there were a few which were never intended for public sale. Although, in general, the cost of articles was not high, this was certainly not an exhibition of industrial art for the 'slender purse'. The exhibition has been severely criticized for its preoccupation with new forms of display technique. Although the paper, bunting and strings did sometimes deflect attention from the exhibits, I feel that their gaiety and freshness did exactly what was wanted, which was to attract people towards an exhibition of industrial art. (See Pl. LXXXVI.)

EXHIBITIONS FOR THE FESTIVAL OF BRITAIN

In 1951 the Festival of Britain was held to mark the centenary of the Great Exhibition. The Festival provided this country with its most outstanding opportunity to improve, by means of propaganda, the standard of industrial design—one of the four subjects chosen for special attention in the official exhibitions. The other three subjects were the arts, science and technology. Apart from the exhibition of central importance in London on the south bank of the Thames, there were nine smaller exhibitions in London and the Provinces in which industrial design was illustrated. For our purposes the most important were the Exhibition of Industrial Power at Glasgow, the Living Traditions Exhibition at Edinburgh, and the two exhibitions which toured the country, one by sea and one on land. The Festival exhibitions were designed to tell the story of the material achievements up to 1951 which were due to the initiative of the British people. The ways in which this theme was presented, together with the long list of architects and designers responsible, need not detain us here.[1]

In these exhibitions the Council of Industrial Design was responsible for the selection of manufactured articles, all of which had to be in current production, designed and made in Great Britain. The Council was represented on the Festival of Britain Council and the Executive Committee, while its Chief Industrial Officer, Mark Hartland Thomas, was a member of the Presentation Panel. The Council's task was to procure suitable articles to illustrate the theme. Its preliminary work for this has

[1] For the South Bank Exhibition, see *The Architectural Review*, August 1951, and for the other exhibitions see *The Architectural Review*, September 1951, pp. 191–202.

already been described.[1] The initial selection of exhibits was made by the Council when it compiled its 1951 Stock List. The work began in 1948, and the Council was assisted by small industrial panels appointed by each industry to approve the chosen articles or reject them because of faulty manufacture or functional inefficiency. Final selection was made by the 'theme conveners' and designers employed for the exhibition, who had to choose all their exhibits from the Stock List. By limiting selection to one central source, the Council was able to prevent an article being shown more than once in the same exhibition. Naturally, this procedure meant that more than the best was drawn out of the Stock List, and that an unusually large number of firms in each industry was represented. The effect of this on the general standard of design of the exhibits will be discussed later. Although free exhibition space was provided, the manufacturers of articles selected had to lend them free, at their own risk, for the duration of the Festival and pay all transport costs. As there were altogether more than 10,000 exhibits a great deal of credit is due to the Council for successfully arousing the interest of manufacturers in the project. There were, in fact, very few cases of stiff opposition to the principle of selective exhibitions. Although the Festival exhibitions were not trade fairs, it was possible for all visitors to obtain addresses of firms and the prices of their products on show.

At the South Bank Exhibition industrial design was most prominent in the Homes and Gardens pavilion. The building was designed on an irregular plan with intriguing changes of level. In a series of five sections new ways of making the best use of a small living space were suggested. Each room in the house was dealt with in turn, beginning with the nursery. Most of the furniture shown was in the tax-free range; the textiles were comparatively more expensive and so were the light fittings. The decorated pottery and glass were, of course, unobtainable on the home market. All these things, as well as domestic electric and gas equipment, aluminium hollow-ware, plastic-ware, cutlery, radio cabinets, wallpaper and carpets (only in plain colours) were arranged as integral parts of a sequence of rooms. Jewellery was to be found in an introductory pavilion called the Pantechnicon, and in the Power and Production pavilion, where there was also more pottery, glass, silver and E.P.N.S., furniture and textiles. As a final sequence in the Homes and Gardens pavilion there were five parlours, arranged by different designers, who attempted to combine modern designs with Victorian and older pieces. Motor-cars were shown in the Transport Pavilion. Leather and domestic metalwork were to be seen in various places. One special feature at the South Bank Exhibition was 'Design Review', which exhibited all the photographs and samples in the 1951 Stock List.

[1] See p. 210.

At Kelvin Hall in Glasgow, the subject presented was industrial power, and relatively few opportunities were provided for showing domestic articles. This must have been disappointing for the people of Glasgow because, except for the spasmodic inclusion of some furniture, textiles and metalwork, and a small section devoted to the effect of hydro-electricity on a crofter's cottage, there was no indication of how industrial power has influenced the design of domestic articles. The exhibition at Edinburgh, for which the Scottish Committee of the Council of Industrial Design was responsible, was more interesting for the general public. Most of the domestic articles shown were arranged in furnished rooms. The exhibition was stimulating, chiefly because modern designs were placed in juxtaposition to traditional Scottish examples. The exhibition which travelled by sea to visit ten coastal towns, was accommodated aboard the *Campania*, a converted aircraft-carrier. Space was severely limited, but all the smaller domestic articles, including some furniture, were displayed, though not in furnished schemes. The land-travelling exhibition, which visited four provincial cities, dealt mainly with industrial design and production. Unlike the one at Kelvin Hall, this exhibition succeeded in illustrating the relationship between design and production technique, as well as displaying the finished articles. The process was then completed by suggesting solutions to specific domestic problems—making the most of the small living-room, arranging for television, keeping warm, etc.

When all the articles in the exhibitions were considered there was an alarming disparity in design standards. In making its selection for the Stock List it seems that the Council of Industrial Design did not adopt a sufficiently high standard and stick to it. Instead of rejecting all but the very best examples, the Council preferred to make its choice in many industries as widely representative as possible. When describing this method, the Director of the Council, Gordon Russell, said: 'Our aim was to find something worth showing from the greatest number of firms possible, otherwise we should run the risk of all the exhibits in certain trades coming from three or four firms, which could hardly be tolerated in an operation financed by public funds on this scale.'[1] In certain groups, such as jewellery, silver and pottery, the design of the selected articles was far below the general standard set by furniture, textiles and glass. Would not the exclusion of such articles be tolerated when the alternative meant that the general design standard would be lowered? After all, the Council is supported by public funds to improve 'by all practicable means' the standard of industrial design. This cannot be achieved if the Council gives its official sanction to

[1] See *The Architectural Review*, February 1952, p. 74. Cf. *The Architectural Review*, December 1951, pp. 349–57.

designs that do not meet with its whole-hearted approval. If it had rejected such articles manufacturers—even whole industries—might have thought again about design and employed better designers. The distinction of having a design exhibited in the Festival exhibitions was a legitimate incentive to manufacturers which could have been exploited more. Such opportunities are rare.

PERIODICALS—BOOKS—THE B.B.C.

PERIODICALS

In periodicals dealing with industrial art a distinction must be made between trade magazines and those which deal exclusively with the fine and industrial arts. In circulation extent, quantity and influence, trade magazines are infinitely superior. In the design standard of products illustrated they are incomparably inferior. Although that is the general situation, there are trade papers which devote articles periodically to industrial design, chiefly as the result of exhibitions. As most of them are published weekly, they are concerned with items which have a 'news interest'. It is exceedingly rare to find a trade paper deliberately and on its own initiative creating an intelligent interest in high design standards. The general policy of a trade paper is to help manufacturers to sell more goods to the public. But it is fair to say that most trade papers have shown a sympathetic interest in the Council of Industrial Design, and have printed long extracts from its Press 'handouts'. The same is true of daily newspapers, except the popular Press, where design propaganda is so overloaded with sentimental irrelevancies that it loses all value.

Among periodicals of artistic standing, the most widely diffused is certainly *The Studio*. It appears monthly, but besides this magazine the same publishers in 1926 started a periodical called *Commercial Art and Industry*, since changed to *Art and Industry*. It remains the only surviving private publication which deals with industrial design, although its main interest has always been with commercial art. The Studio, Ltd., also publishes a *Year-Book of Decorative Art*, and other books on art and design. As they lack a reliable policy towards design standards, the tendency in all these publications is not to be too dogmatic. Much of the material discussed and illustrated belongs to what is known as the 'Mayfair Style'. But their large circulation makes *The Studio* and its additional publications important instruments of design propaganda. The most comprehensive magazine dealing with our subject is *Design*, the official monthly organ of the Council of Industrial Design. It is intended mainly for manufacturers, designers and retailers, and has already been described. Directed at a sophisticated public, sympathetically inclined towards antiques, is *House and Garden*. When new designs are reviewed the most fastidious standards are insisted on.

When *The Architectural Review* devotes a whole number, or even a few pages, to industrial art, it is an event of some importance. The *Review's* irreproachably strict standard, its outspokenness and its admirable thoroughness have made it the most valuable journal of modern architecture in this country. It has discussed and illustrated all questions which concern industrial design. Two special numbers of the *Review*, December 1935 and October 1946, were compendia of good design in articles of everyday use, with names of manufacturers, designers, and retail prices—necessary adjuncts to the national exhibitions of design they were concerned with. *The Architects' Journal*, similarly insistent on high standards, is the most outspoken of the weekly periodicals. Industrial design is dealt with from the architect's point of view.

The most important foreign magazines are: *Form* and *Kontur*, both produced by the Svenska Slöjdföreningen; *Domus* and *Stile Industria*, Italy; *Architektur und Wohnform*, Germany; *Werk*, Switzerland; *Dansk Kunsthaandvaerk*, Denmark; *Art et Décoration* and *Art et Industrie*, France; *Interiors*, *Industrial Design*, *Everyday Art Quarterly* and *Furniture Forum*, U.S.A.

BOOKS

There are at least fifty books on industrial art, or closely connected subjects, which are worth reading. The following is a brief description of those which seem to me to be especially interesting. For an account of the forces and events which first made a problem out of industrial art, and how a solution was proposed and acted on, the only book is *Pioneers of the Modern Movement*, by N. B. L. Pevsner.[1] Dr Pevsner begins with the mid-nineteenth century and traces the changing relationships between art and industry from the confusion of the Great Exhibition, through William Morris and England at the turn of the century to Germany, the Deutscher Werkbund, Walter Gropius and the Bauhaus. It is a forcefully written direct statement of the origins of modern architecture and modern industrial art, essential for an understanding of the situation today. Dr Pevsner's account ends chronologically, with the foundation of the Staatliches Bauhaus at Weimar in 1919.

The development of this school and its methods, which several English art schools rightly attempt to emulate, are described by its originator, Walter Gropius, in *The New Architecture and the Bauhaus*.[2] Professor Gropius's teachings codified the

[1] Faber & Faber, 1936; revised version, *Pioneers of Modern Design*, The Museum of Modern Art, New York, 1949.

[2] Translated by P. Morton Shand, Faber & Faber, 1935.

pioneering work done by architects and designers first in this country and afterwards in Germany. Among the wholesome influences on industrial art in this country his has been the most outstanding and the most pervasive, for it is mainly because of Gropius that we accept the machine as a necessary adjunct to man's creative energy.

Soon after the influence of the Modern Movement was felt in this country, it became necessary to reinterpret design and explain the new principles. In 1934 a number of books were published. The most influential of them has been *Art and Industry*, by Sir Herbert Read,[1] whose aim is not to suggest direct improvements of existing conditions but rather to shape the frame of mind necessary for future improvements. He does not believe in a practicable reform, unless a new philosophical (aesthetic) creed is accepted. Hence he deals in the first part of the book with formal values of machine art. The author discusses the possibility of establishing a standard by which a work of art may be judged. He instances the 'Golden Cut', and says: '...if we are conscious of a law of proportion, and then slightly deviate from it, to avoid its precision, we shall produce a more beautiful effect'.[2] To support this idea of deviation from a numerical law, Sir Herbert Read says: 'The same argument is used of the metrical irregularities of verse, where it certainly has some application.'[3] But surely this is unacceptable. It presupposes a fixed code of metrical rules from which poets can deviate to gain effect. While the rules undeniably exist, it cannot be supposed that poets begin by inventing deviations. On the contrary, they begin by finding words to suit the way they think and feel an experience should be expressed. The fact that the product of their work forms itself metrically or unmetrically is, to this extent, irrelevant. In the same way one cannot suppose that a designer, who sets out to plan a useful household article, has a fixed, quasi-Platonic notion, of what it *ought* to look like.

Sir Herbert instances Mondrian, Hélion and Ben Nicholson and says that such 'pure', abstract artists could 'keep untainted...the formal essence of all art'. Their art will 'occupy, in the future, a relationship to industrial design very similar to the relationship pure mathematics bears to the practical sciences.'[4] I am not able to say if this comparison is justified, but where fine art and industrial art are concerned his belief in an intellectually apprehended standard of excellence is highly suspect. Abstract art, which according to him inspires the industrial designer, is, he admits, distinct from humanistic art and therefore segregated from 'human ideals and emotions.'[5] Yet six pages later he quotes approvingly a remark by Walter Gropius,

[1] Faber & Faber, 1934, revised 1953.
[2] *Art and Industry*, p. 28.
[3] *Ibid.*
[4] *Op. cit.* p. 53.
[5] *Op. cit.* p. 49.

describing the Bauhaus: 'Our guiding principle was that artistic design is neither an intellectual nor a material affair, but simply an integral part of the very stuff of life.'[1]

The second part of Sir Herbert Read's book has short and concise paragraphs on the conditioning factors of material and manufacturing processes in pottery, glass, metal, wood and textiles. An exceedingly interesting chapter shows that in modern mass-production the real task of the creative designer lies less in the inventing of details than in the assembling of standardized parts to create new shapes of a functional as well as an intuitive order, a task which is similar to that of the architect who builds with standardized bricks, windows, etc. The third part discusses colour and ornament, the fourth, art education.

One of the first books to assess the situation in this country in the early 1930's was *Industrial Design and the Future*, by Geoffrey Holme.[2] It consists of a manifesto of some ten pages, pointing to existing troubles and desirable improvements, and a questionnaire which Geoffrey Holme had sent to manufacturers, business men, commercial artists and architects in England and the United States. The answers received are published *in extenso*, and constitute the most important feature of the book. Although, naturally, they vary in accuracy, they nevertheless give a remarkable picture of the prevailing situation. All kinds of deficiencies are mentioned and remedies suggested, but no consistent solution of the problem results either from any individual answer, or from the whole of the answers taken together.

A book calculated to interest the general reader is *Design and Civilization*, by Noel Carrington.[3] The writer believes in the dependence of art and design on the social conditions of a period. Thus one-third of his book gives short descriptions of what design was, and of how it was related to society, during four periods selected from English history: the Middle Ages, the Tudor period, the 'Golden Age' of the eighteenth century, and the Victorian age. On the same lines Noel Carrington sketches in the social background of art today. He then deals with William Morris, the growth of the Modern Movement and the position of the designer and manufacturer as it is and as it ought to be. The spheres of State activity and of individual activity are defined, and in conclusion the necessity of the artist-craftsman is asserted.

Undoubtedly the most prolific writer on design in this country is John Gloag. Since 1921 he has written more than twenty-five books on architecture, design, craftsmanship, interior-decoration and furniture. His chief work on design is *Industrial Art Explained*.[4] The most interesting of his other books are *The Missing*

[1] Cf. *The New Architecture and the Bauhaus*, p. 59.
[2] *The Studio*, 1934.
[3] John Lane, 1934; revised 1947.
[4] George Allen & Unwin, 1934; revised 1946.

Technician in Industrial Production, 1944, *Plastics and Industrial Design*, 1945, and *Self Training for Industrial Designers*, 1947.[1] In *Industrial Art Explained*, he deals with the immediate antecedents of British industrial art since the middle of the nineteenth century. The stress is laid on present conditions, less from the manufacturer's than from the art lover's point of view. He begins with an account of architectural tradition and industrial architecture in the nineteenth century, then deals with design in relation to the manufacturer and engineer. This is followed by a provocative chapter on the situation today, after which he discusses the influence of materials on design, and finally a short but useful chapter on the training and status of the designer.

In John Gloag's writings two points are stressed repeatedly: that 'design is a normal business operation' and that 'good design may be a potent sales factor'. The first point would receive wide agreement, but the second, if pushed too far, would confuse the salesman with the designer and commercial with artistic standards. For instance: 'Although a few progressive British manufacturers are beginning to realize that the industrial designer is the missing technician in industry, they have still to realize that he is a secret salesman.'[2] The point is amplified by Raymond Loewy in the Appendix to *Industrial Art Explained.*

At this point some mention should be made of *An Enquiry into Industrial Art in England*, by Nikolaus Pevsner.[3] For the last sixteen years it has remained the only book which analyses the whole field of industrial art from a critical standpoint and makes suggestions for improvements. Its approach to the problem was new, and it has not been followed since in this country, on the Continent, or in the United States. In its pattern of presentation, survey of trades, organizations concerned with design, conclusions and suggestions for improvements, it has been the model for this book.

For the public, interested only in the finished works of industrial art, there is a very excellent guide to flat dwelling and home-planning called *The Flat Book*, by J.L. Martin and S. Speight.[4] Unfortunately, it is now out of date, for all the prices given have been superseded, and most of the articles described are unobtainable. The book is divided into contemporary furniture, furnishing, room arrangement, colour and texture. The emphasis throughout is placed on modern design, where to buy it and how to use it. In the absence of a new edition, one can refer to several brief, illustrated guides to furnishing issued by the Council of Industrial Design. Anthony Bertram's talks on the B.B.C. in 1937 have been described in connexion with the Design and Industries Association; in 1938, with help from the D.I.A., he brought out an inexpensive popular book based on his talks.[5] It provides an excellent primer of industrial

[1] George Allen & Unwin.
[2] *Industrial Art Explained*, 1946, p. 134.
[3] Cambridge University Press, 1937.
[4] Heinemann, 1939.
[5] *Design*, Penguin Books.

art. Also amongst the popular books on design is the *The Things We See* series.[1] Six books on everyday articles were produced. They are well illustrated, brief but to the point.

Books on design written by practitioners are valuable. They are also rare, especially in this country,* but in the United States three have been published recently and should be included here. The first is *Industrial Design (A Practical Guide)*, by Harold Van Doren.[2] Mr Van Doren is a consultant designer, and from his experience in American industry he describes the designer's approach and methods for the benefit of students. The book deals with domestic articles as well as commercial equipment and machinery and is mainly concerned with three-dimensional articles in metal, plastics and glass which are mass-produced. At the end three case-histories are fully documented. Mr Van Doren stresses that the object of the designer is to increase the sales of his client's products. While he gives a unique study of American industrial practice, he appears to lack a reliable, or even a satisfactory, design standard. Amongst the preliminary chapters there is a thorough and concise account of the origin and practice of streamlining.

The second book is *Design this Day (The Technique of Order in the Machine Age)*, by Walter Dorwin Teague.[3] Mr Teague has also written with his eye on students, but his book has a wider scope to interest other people who are already sympathetic towards design. He puts industrial design in its social context and relates it to architecture, civil engineering and town-planning. Mr Teague's criterion of excellence in design is 'Order', which is achieved by doing things in the right way. He also discusses 'Ultimate Form', derived from the perfect understanding of the purpose involved, and implies that it is both recognizable and not limited to the age for which it was created. There is no space here to question and discuss these very doubtful assertions. From the point of view of the contemporary designer in a fully mechanized production system, he deals with the problem of maintaining order in our civilization. From a very large assortment of accomplished and projected works, both now and in the past, he describes the philosophical approach and the industrial techniques necessary for solving the problem. The core of the book is made up of the following chapters: sources of form, fitness to function, materials and techniques, unity, simplicity, rhythm, dominance, accent and scale, balance and symmetry, style. As might be expected Walter Teague thinks that the 'Golden Cut' is a necessary standard for the designer. An appendix to his book sets out the proportional scheme of the Parthenon. The book's chief value lies in its definition of the designer's position in

[1] Penguin Books. [2] McGraw-Hill, 1940.
[3] Harcourt Brace & Co., 1940; *The Studio*, London 1947.

some sections of American industry, although one's attention is too often distracted by the author's high-powered optimism in the capabilities of men and machines.

Lastly there is Raymond Loewy's book on his work as a consultant designer which has already been mentioned.[1] Mr Loewy's narrative is light-hearted and engaging, but the precise development of his career is difficult to follow. Verbatim accounts of interviews with business executives together with descriptions of his own practices are helpful when trying to understand the American view of art in industry. His few references to this country show that an alert and penetrating mind has been at work.

Other American books on the wider aspects of design should be briefly mentioned here. *Technics and Civilization* and *The Culture of Cities*, by Lewis Mumford,[2] are too well known in this country to need description, but his more recent work *Art and Technics*[3] provides the opportunity to indicate Mumford's characteristic approach to art, industrial techniques and civilization. In spite of his professed willingness to accept machinery and presumably mass-production as necessary to civilization, Mr Mumford states that 'the greatest developments to be expected of technics in future, if once the philosophy I have been advocating becomes generally accepted...will consist in using machines on a human scale, directly under human control...'.[4] His romantic attitude is fairly explicit: 'art is that part of technics which bears the fullest imprint of human personality; technics is that manifestation of art from which a large part of the human personality has been excluded in order to further the mechanical process'.[5] He goes on to deal with society's unsympathetic approach to art and the effect that this is having on art. Rightly he stresses that a living art must have a responsive audience and that the present schism between art and technics will lead to nihilism. But his remedy seems to be none other than a return to the practices of William Morris. At their best machines and machine products can only 'express order and subserve power'.[6]

The same problem is approached by another American writer, Siegfried Giedion, in a different way. His *Space, Time and Architecture*[7] has been widely read, but the more recent book *Mechanization Takes Command*[8] is directly relevant to this context and should be discussed. This very large book, there are over 700 pages, deals mainly with the development of industrial techniques in the nineteenth and twentieth centuries and their effect upon Western civilization. Dr Giedion begins by making it clear that

[1] *Never Leave Well Enough Alone*, Simon & Schuster, New York, 1951. See p. 148.
[2] Harcourt Brace & Co., 1934 and 1938. [3] Oxford University Press, 1952.
[4] *Op cit.* p. 78. [5] *Op. cit.* p. 21. [6] *Op. cit.* p. 81.
[7] Harvard University Press, 1941; New edition Oxford University Press, 1951.
[8] Oxford University Press, New York, 1948.

mechanization has brought society to a chaotic condition because it is the chief agent of an unbalanced civilization where thought and feeling are dissociated. Dr Giedion takes pains to trace recent techniques to their source—the second section of the book deals with the beginnings of mechanization—and he gives long, profusely documented case-histories to show how mechanism has superseded the hand. He is scarcely interested in the appearance of new products and criticism of artistic standards forms no part of his thesis. The facts in each case are given a philosophical interpretation which leads him on to the hopeless conclusion that basic human values are incompatible with the mechanized life of today. The reader can enjoy his catharsis, for Dr Giedion's tale is always fascinating, but he is offered no encouragement to face the tragedy which still continues within and around him.

THE B.B.C.

The B.B.C. is one of the most powerful agents available for design propaganda. Its interest in the subject began in the early 1930's. From April to June 1933, that is, before the opening of the Dorland Hall Exhibition, a series of talks on 'Design in Daily Life' was given by such men as Wells Coates, Maxwell Fry, John Gloag, Frank Pick, A. B. Read. The lectures were published with a number of excellent illustrations and an introduction by Noel Carrington. In view of its date, this piece of initiative is greatly to the credit of the B.B.C. Since then the B.B.C. has included in its programmes many talks and discussions on design. Mr Bertram's series of talks has already been mentioned. They were given in 1937, and repeated in the following year on television with the aid of photographs and examples. For the first series there were 300 organized listening groups throughout the country, apart from people who listened privately. It was found that teachers formed by far the largest group of any class of listener. Since the war the Corporation's interest has been sustained, although it appears to lack policy and direction. The talks given have been too spasmodic and unrelated to make much impression. Nevertheless, a very good programme series for school children, called 'Looking at Things', was held first in 1949 and since continued. To back it up the B.B.C. issued a small booklet with more than 100 illustrations.

Although the main task of the B.B.C. is to entertain, it gives each week a large number of talks and discussions on farming, education, domestic affairs, politics, etc., as well as on music, drama, literature and art criticism. Industrial design, in spite of its great importance to all types and classes of persons, receives comparatively little attention. In defence of this, the B.B.C. puts forward two arguments: first, that design propaganda is often too didactic; and secondly, that it is primarily a visual matter and unsuited to its General Advisory Council. Already the Council has forty-nine members representing most forms of human activity. A policy for industrial design programmes could perhaps be worked out on these lines. To begin with, there should be a comprehensive explanation of the background of design, say over the last sixty years, followed by a survey of the present situation. Then, to awaken or to sustain interest, feature programmes on materials and factory methods should be given, preferably with visual illustration. This, and what follows, could best be done by the Television Service. As television is essentially a family form of entertainment coming into the home, the articles dealt with should be limited to those which are of chief interest in domestic furnishing. There are about sixteen groups altogether. The background explanation should be followed by studio discussions amongst manufacturers, designers, retailers and customers, around actual examples in current use. Controversy should be encouraged here. Once established as regular features, these programmes could devote part of their time to the reporting of current trends in design, exhibitions and trade fairs. Particular aspects of the design policy could be developed further for broadcasts to schools. For a more sophisticated audience the Third Programme could be used, for example, for talks or discussions on the evolution of a contemporary design standard, design in its social context, the training of designers, and reports from foreign countries. Altogether the policy would achieve two objects. It would make the public more critically aware of its environment and it would provide an incentive for manufacturers to produce better designs, worthy of broadcast comment.

PART III

CONCLUSIONS

EXPLANATORY NOTE

It will now be clear that design standards depend upon a variety of forces. Chief amongst them are manufacturers, designers, materials and techniques. The scope of my inquiry has been limited to domestic articles and I have dealt with the effect of these forces on furniture, carpets, linoleum, metal windows, ornamental ironwork and door furniture, hollow-ware in silver, electro-plate and aluminium, jewellery, gas stoves and fires, refrigerators, electric fires, irons, small kitchen appliances, clocks, light fittings, radio and television cabinets, woven and printed furnishing fabrics, wallpapers, hollow-ware in china, earthenware and glass, leather suitcases, handbags and fancy leather goods, articles in plastics and motor-cars. In addition I have dealt with consultant, free-lance and staff designers, architects, craftsmen, the retail trade, art schools and a number of organizations, institutions, exhibitions and other media for design propaganda. Before any suggestions for improvement can be made a large number of facts are needed. Outspoken disapproval is also necessary, for one must know exactly how bad things are before steps can be taken to alter them. My object has been to present factual evidence and, although much of it has been accompanied by critical comment, the result is bound to appear confused.

The purpose of the third part of this book is to attempt to clarify the conflicting evidence by analysis under different headings. These headings have, in turn, been chosen for two reasons which must be dealt with beforehand. First, we must relate the evidence to a sequence of industries placed in an approximate order of merit according to design standards. Secondly, this must be compared with the sequence of industries which was compiled in 1935 from a similar point of view. These two reasons provide critical standpoints from which the evidence can best be assessed.

SEQUENCE OF INDUSTRIES ACCORDING TO DESIGN STANDARDS, 1953

It is evident that the products of the industries discussed differ greatly in standards of design. Moreover, each of the industries has its own varying standards. So my approximate order of merit is determined by the number of firms in each industry concerned with high or good design standards for the majority or minority of their products.

First. Representing industries where most firms have high standards in most of their products: leather suitcases, metal windows, gas and electric stoves, watch-cases.

Second. Representing industries where a few firms have high standards in most of their products: furniture, radio and television cabinets, pottery, woven furnishing fabrics.

Third. Representing industries where most firms have some products of a good standard: printed textiles, gas and electric fires, refrigerators, irons, crystal glass, fancy leather goods, aluminium kitchen utensils, articles in plastics, motor-cars.

Fourth. Representing industries where only a few firms have some products of a good standard: light fittings, wallpapers, clocks, door furniture, pressed glass, carpets, silver and aluminium tableware.

Fifth. Representing industries where nearly all firms maintain an unsatisfactory standard in most of their products: jewellery, linoleum, ornamental ironwork, electro-plate.

SEQUENCE OF INDUSTRIES ACCORDING TO DESIGN STANDARDS, 1935

The following sequence was compiled from a similar point of view in 1935.

First. Representing industries where most firms had high standards in most of their products: leather suitcases, metal windows, watch-cases, motor-cars.

Second. Representing industries where a few firms had high standards in most of their products: furniture, radio cabinets, pottery, woven and printed furnishing fabrics, gas and electric stoves, gas and electric fires.

Third. Representing industries where most firms had some products of a good standard: fancy leather goods, aluminium table- and kitchen-ware, clocks.

Fourth. Representing industries where only a few firms had some products of a good standard: light fittings, wallpapers, silver, articles in plastics, crystal glass.

Fifth. Representing industries where nearly all firms maintained an unsatisfactory standard in most of their products: jewellery, linoleum, ornamental ironwork, electro-plate, door furniture, carpets, pressed glass.

The comparison between the situation in 1953 and as it was in 1935 is very revealing. The significant result is that six products have been added to the group where most firms have some products of a good standard. To account for this four products (motor-cars, printed textiles, gas and electric fires) have come down, while two (crystal glass and articles in plastic) have come up. In addition, gas and electric stoves have gone up to join the products in the first group. In this way there has been an improvement more of quantity than of quality. Industries responsible for products in the first and third groups manage to put a larger number of acceptable designs on the market now than in 1935. Good design standards have spread in the crystal glass and plastics industries and to a much lesser degree amongst firms making carpets, pressed glass and door furniture. On the other hand, one cannot now say that most motor-car firms maintain a high standard. Instead, they, in company with most firms making printed textiles, gas and electric fires, achieve a generally good standard in some of their products. A more serious decline can be seen in the clock industry, for there are now only some products in a few firms which maintain a good standard. But the most important result of all is that there are only two products, gas and electric stoves, where most firms now achieve a higher standard than in 1935.

The sequences do not, of course, take into account the very best or the very worst designs produced by isolated firms. I am well aware of this and the other limitations of these simplified sequences. What has been said already in the detailed analyses of the industries concerned will, I hope, make up for the shortcomings here. My object is to focus attention on the general situation, for it is with the general situation that we are most concerned.

DESIGN STANDARDS

REASONS FOR DIFFERENCES BETWEEN INDUSTRIES, REASONS FOR CHANGES SINCE 1935

The sequences are the outcome of explicable conditions which, with the aid of the evidence collected, we ought now to be able to explain. For instance, are the high design standards in some industries and the appallingly low standards in others the result of the producing, the distributing, or the consuming forces in the trade process? Or are they the result of all three? As far as they depend on the producer, does that mean on him personally, on the organization of his factory, on certain methods of manufacturing, or on something else? We shall have to deal with each of these and other questions separately to find out which of them may yield serviceable answers. A first group of questions concerns the factory and the designer; a second group, the retail trade; a third group, the taste of the public; a fourth group, various extraneous matters which affect design.

OLD AND NEW INDUSTRIES

THE MEANING OF TRADITION

There are several questions to be dealt with under this heading, though no conclusive answers can be given. The older industries are often to be found centred in particular localities for reasons of fuel, materials, skilled labour or the market. Glass is made in Stourbridge near coalfields and claypits. Pottery is produced in the Stoke-on-Trent area for similar reasons. Carpets are woven in Kidderminster because of a local craft of Flemish origin, now extinct. Brass is founded in Birmingham where there are coalfields and sand for making moulds. Walsall started a leather industry to serve a nearby agricultural market. And there are several more examples. Amongst the newer industries the grouping is far less marked, but where it exists it can be accounted for by the presence of skilled labour, created usually by one or two pioneer firms. The motor industry in Coventry is a striking example. The plastics, radio and aluminium hollow-ware industries are widely scattered. So is the furniture industry, although it is in origin one of the oldest in the country.

The location of an industry appears to bear some relation to the design standard of its products. This is chiefly a matter of environment. How can the manufacturer working in the jumbled jewellery quarter in Birmingham relate his products to a civilization in which any reasonable standards of town-planning, architecture and design have a value? What influence have the grimy streets and alleyways of Kidderminster on designers working and living in the town? And what is likely to happen to the artistic sensibility of men brought up in the appalling atmosphere of that shapeless agglomeration which is now called Stoke-on-Trent? Although no answers to these questions are available it is evident that a thorough research into them would produce fascinating results.

At least the situation has been recognized by manufacturers since the early years of the Industrial Revolution. Their reaction to it has been almost wholly negative. On the one hand, they have camouflaged the exteriors of their factories with all types of artistic styles to serve as a barrier against the squalor which lies outside. See, for example, many of the weaving mills in the West Riding of Yorkshire. And on the other hand, when the day's work was done, they withdrew to country houses surrounded by woodlands and grass. Exactly the same attitude exists today. A tour round London on the North Circular Road will reveal how self-conscious manufacturers are about their places of work. But, apart from this, in both old and new industries one also finds manufacturers who do not care what their factories look like and who consider that their only duty is to work and make money. Elsewhere one finds complete withdrawal of factory and workpeople from the industrial scene; a recent example is Wedgwood's at Barlaston. The design and layout of the factories themselves are another question, and something will be said later about the studios used by staff designers.

One can do no more than suggest that the location and environment, the design and layout of factories, exert some influence on the design standards of the articles produced. But it seems probable that a factory which is orderly and well-planned, has healthy working conditions, and has close cultural connexions with the surrounding neighbourhood will, by nature, produce articles of a high design standard. It is tempting to speculate on what would result if Cadbury Brothers made furniture or Lever Brothers wove carpets.

Likewise, it is not possible to make any conclusive statement on the relation between the age of an industry and the design standard of its products. The older industries, furniture, pottery, glass, woven and printed textiles, silver, jewellery, etc., tend to be conservative in taste and to resist a change of character in the design of their products. Each of them has a strong tradition which is fostered rather for the sake of the artistic styles which belong to it than for the *materials, techniques and skills* upon which it was established. Thus, few of the traditions are alive today in the sense that they afford fresh scope for creative work. The point has been elaborated in connexion with pottery and glass. In most old industries, tradition is an encumbrance because the techniques and skills, in some cases the materials, have changed irrevocably. One finds cases of mechanized craftwork (pottery and furniture), where the break with tradition has been hesitant and only partially accomplished. In leather, silver and jewellery there has been no real change and craftsmen are still employed at all stages of production. Textiles are woven by comparatively old methods. Nevertheless, the value of a tradition is conditioned first of all by the degree of change which has taken place in the methods and materials of manufacture. Where change has been slight, it is still possible to keep the tradition alive by creating new designs which embody an artistic standard high enough to make all former products of the tradition slightly less applicable to the contemporary demand. This is what is meant by revaluation; and each designer, before he sets out on a new design, must look back to see what, from the contemporary viewpoint, is lacking in the products of the past. By doing this he gains both knowledge and inspiration. There are, indeed, few cases of this happening, for it is almost universally assumed that tradition can only be kept alive by constant imitation of traditional designs, a process which means certain death to any tradition in any form of art.

New industries naturally have short traditions and to this extent are less encumbered. It is characteristic that design in new industries begins by copying the superficially similar character of other objects. This is mainly because most new industries using new techniques with unfamiliar materials have to establish themselves by producing substitutes for products already on the market. Motor-cars, wallpapers, electric light fittings, and articles in plastics, electro-plate and pressed glass all began in this way. After the early period the old forms are discarded and designs with a new character begin to set their own standards. For example, this was accomplished quickly in motor-cars and plastics; in electric light fittings and wallpapers the transition has been gradual and is not yet complete; in articles made of pressed glass and electro-plate the stage of transition has not even been reached.[1] Radio cabinets are a special case where early hesitancy caused most firms to model their work on the characteristic standards of furniture design. Watch-cases are exceptional; as they were not designed to replace another product they had to work out their own design standards from the start.

SIZE AND ORGANIZATION OF FACTORIES

There does not seem to be any tangible relation between the size of a factory, in terms of employees, and the design standard of its products. As a rule, the small factory is run by one person or a small number of persons; the big factory has a more complex and less personal form of organization. In small factories design is usually either very good or very bad. A factory may have remained small because the owner wants to limit his output to products of a particularly good quality, or because he is just lacking in enterprise. He may be an artist-craftsman or a man with exacting personal taste, who is not willing to embark upon volume production, or simply a narrow-minded businessman who has inherited the factory and wants to live comfortably on its profits instead of investing them in improvements of plant or design. He will muddle on in the same groove as his father and grandfather, as long as no serious crisis forces him into something new. The first type is to be found amongst the highest class furniture makers, weavers and printers of furnishing fabrics and is characteristic of one or two firms making pottery, light fittings and wallpapers. The second type is very frequent amongst the Birmingham silver, electro-plate, jewellery and light metal manufacturers, the Walsall leather firms, the High Wycombe furniture factories, and the Stoke potteries.

[1] These industries are 'new' only in the sense that their design standards have not yet developed. Glass was pressed in moulds on Tyneside in the eighteenth century and electro-plate was invented in 1838.

SIZE AND ORGANIZATION OF FACTORIES

Design in large factories is seldom very good and is often very bad. The business of a large firm and its design policy depend upon a board of directors which is in turn responsible to a company of shareholders. Some of the shareholders may be intimately concerned with the running of the factory, but in most firms the bulk of them are interested only in the market value of their investments. It follows that if the firm takes a risk—and design improvements often involve an initial risk—the profits for the next two or three years may drop a little. The public may greatly benefit from the improvements, but the shareholders will have their dividends reduced. In at least a dozen factories I was told that the directors' first duty is to retain the confidence of those who lend money to the firm. Their second duty is to keep the regular staff employed by maintaining a steady volume of production. Lastly, they have a duty to the public by way of the retailer. If the demand from the market weakens the advertising agent would be consulted long before the designer. This type of firm can be found in all industries, but its effect on design standards has been most disastrous in furniture, textiles, linoleum, carpets, radio cabinets and pressed glass.

In large factories one does not often find the personal and human element which can have such a favourable influence on the design standard of the goods turned out. If the managing director is interested in design he can, if he has the energy, influence his whole staff and see that his ideas are carried out. I found such cases in the radio, printed and woven textile, pottery, leather, plastics and motor-car industries and in two crystal glass firms and one carpet firm. But the managing director's influence is often weakened by the stratified organization of factory departments. This is most noticeable in the larger textile firms in Lancashire. There I found that valuable new ideas, coming as they often do from the design studio, would, if not checked in time, upset the carefully planned routine of the material, dyeing, production, sales and management departments. Each department is expected to carry out a set amount of work. Any alteration to the set amount will be resisted by the staff concerned. So, unless the designer's ideas are backed wholeheartedly by the managing director, few design improvements will take place. A similar situation is characteristic of the very large firms in all industries. I interviewed the managing directors of several but found, amongst those who were personally interested in design, that they could not spare enough time to bring about noticeable improvements. They had, of course, delegated some of their duties to managers of sales and production, but none of them had set up a director of design. Their designers did not have executive status and so there was no sympathetic, knowledgeable and reliable spokesman for design matters on the boards of directors.

PRODUCTION METHODS

Again dealing first with small firms I found that production is usually on a limited scale and dependent upon frequent design changes. This practice can be very good for design or it can be deplorable. It rests with the manufacturer to determine which result should be achieved. 'Where only a small quantity of a product is sold before the design is changed the machinery used in the trade will very largely be simple, adaptable and cheap.'[1] In the case of the Birmingham jewellery, electro-plate and light metal trades, the few designers employed are incapable of setting a high standard. This means that novelty is the only criterion which matters, a condition which is aggravated by the ceaseless demand for something different from the retail trade. Firms construct new designs from a variety of components bought from trade stampers, who periodically invent new shapes for ornaments, handles, brackets, etc. A special design for a component would cost the firm three times more than one which is generally available to the trade. I found a similar state of affairs in the smaller potteries; it is also true of the Walsall leather goods trade, but not so disastrous in its effect, for with all leather manufacturers there is an underlying sense of fitness. On the other hand, limited production and few machines controlled by an enlightened owner and gifted craftsmen can produce some of the best designs of furniture, light fittings, wallpapers, pottery and textiles on the market. Such methods are flexible and can, if necessary, produce experimental designs without losing money. There is a direct link between design and execution under the owner's personal supervision.

In large firms where great quantities of an identical design are turned out the situation is entirely different. To begin with, mass-production methods can result in the separation of design and execution. When this happens the designer loses control of the production process which, for reasons of speed, economy and efficiency, is inflexible and unsusceptible to creative experiment. Production on a large scale implies a reduction in the quantity of patterns or models run at the same time—the fewer the models, the fewer the different manufacturing operations—which means that the public's choice is limited. The situation is much more serious now than it was in 1935. Firms everywhere have reduced the quantity of patterns and models since the war. Mass-production can also make it difficult for the managing director to supervise all stages of manufacture including design. And moreover, as the production run gets longer the sales manager's interference with the design standard increases, and at the same time the designer's influence steadily decreases. Mass-production

[1] *Design and the Designer in the Light Metal Trades*, H.M.S.O., 1944, p. 4.

implies a mass market. To play safe, most firms adopt a low or mediocre design standard based, not on a fresh appreciation of the market, but on what has been sold in the past.

The worst results of the mass-production process can be seen in the furniture, carpet, linoleum, pottery, wallpaper, textile, radio, motor-car and pressed glass industries. The best results are to be found amongst firms who realize that each increase in the output must be accompanied by more intensive thinking and planning behind each new design. This in turn makes a designer and an experimental craftsman essential. Such firms can be found in the furniture, pottery, textile, radio, motor-car, plastics and domestic equipment industries. The inclusion here of furniture, textile and plastics firms represents a marked improvement since 1935.

Yet, if we consider the quantities turned out by the first group of mass-production firms and think of all the houses into which their goods go, and the human beings who will be surrounded by them, how depressing it is that more advantage is not taken of the mass-producer's great opportunities for increasing the artistic harmony of our environment. Here is the furniture firm selling 1000 dining-room suites a week, here the carpet firm where best-sellers are measured in terms of miles, here the wall-paper firm affecting 20,000 rooms with one design, the pottery selling a million pieces in one design every five years or the pressed glass firm producing between six and eight million of the same design. What an overwhelming responsibility! Again, in 1935 the quantities of identical designs were not so large. Firms with such a volume of trade ought to have sufficient money and understanding to counterbalance this concentration on a few models by an equal concentration on the best designers available as, in fact, is the case in the firms in the second group, for an increase in the quantity of each design by no means necessitates a reduction in its artistic standard.

MASS-PRODUCTION AND DESIGN STANDARDS

In this section two questions already raised will be discussed further: the relation between design and execution, and the relation between design standards and mass-production. In all industries which I examined, even those that depend on highly skilled labour, the designer does not carry out any manufacturing operations although he may be intimately concerned with them. He may exert some control over them, helped by an experimental craftsman and the production engineer, but in all cases he is distinct from the man who makes. The men who blow or cut glass, beat silver, make a wrought iron gate, throw an earthenware bowl, stitch leather, or construct a chair may possess admirable dexterity, but they are not artists. It is the designer, and not these men, who creates shape, proportion, colour and pattern. Whether he prepares a drawing for a skilled worker or for a 'hand' in a completely mechanized factory the designer's task is essentially the same. Great changes have taken place, and sad changes, no doubt, between the days when manufacturers still *manu faciebant*, and our century, when the 'hand' has become the slighting expression for an entirely unskilled operative. However, we cannot put the clock back. Had these changes not occurred everywhere, the problem of the designer, the problem of industrial art as opposed to *les arts décoratifs*, would not exist.

As to the artistic value of articles produced under present-day conditions, this is likewise bound to be different from that of past centuries. The charm of a piece of medieval handicraft lies to a large extent in its being the live expression of something human. No contrast of attitude existed between the goldsmith chasing a chalice and the painter painting an altar. The picture was not pure art, and the chalice not pure handicraft. Either represented an indivisible ('individual') unity of soul, mind and hand. The situation changed as soon as the artist began to think of repetition, no matter whether in pottery or the graphic arts. Even a fifteenth-century woodcut depended as a rule on two separate personalities, the artist who invented the composition and drew it out, and the cutter who transferred it to the wood-block. Only one of them needed to have artistic inspiration; the other's natural gift was skill, appreciation or sensibility, but not the genius of art. The final effect certainly always depends on both, in that deficiencies on either side can spoil the whole. But while no limit to artistic genius exists, the craftsman's limit is reached when he has rendered the designer's intention completely. Now in many industries the machine is infallible in doing that, as long as the designer knows how to design for the machine. Hence it is just pretending if one insists that a hand-cut leather bag must be artistically better than a machine-cut one, or hand-printed wallpaper better than machine-printed. This was true only so long as designer and craftsman were one. Incidentally, it is one of the reasons for the low artistic standard of some industries which use little machinery, such as jewellery and pottery. By relying entirely upon skilled craftsmanship and doing without proper designers, they mistake technical dexterity for creative imagination.

All these considerations show that neither machinery nor repetition-work in itself is detrimental to the artistic quality of a design. On the contrary, there is one great possibility in mass-production methods of which advantage is seldom taken. I explained earlier that a firm turning out very large quantities of goods will necessarily tend to reduce the number of different models, in order to simplify production processes. By concentrating upon a few designs more money can be spent on each of them, higher fees for designers should be available, and better men can be obtained. This has in fact been the policy of some radio factories, some motor-car works, and plastics firms. I say the policy, for it is a matter of deliberately chosen policy and not the necessary outcome of a certain method of production. It depends inevitably on the personal attitude of the manufacturer.

If I say that standardization does not in itself preclude the highest design standards, I do not deny that mechanical methods of production may sometimes restrict certain improvements of design. A designer who creates new models for a large furniture firm will, in the proportions which he chooses, often be controlled by the standard sizes of mechanically produced units, such as legs, drawers and backs. A designer for roller-printed fabrics must not exceed a certain size of repeat and a certain number of colours. A designer for cheap car-bodies has to consider the chassis and available steel pressings. A designer for plastics must take into account what shapes can be easily moulded. These are only a few instances of what one comes across in almost every industry. But whereas the designer in all such cases is undoubtedly less free than his colleague designing for hand-production, the consequence, if he is good at his job, is by no means a reduction in design standards. Every architect is just as dependent on by-laws and on the standard sizes of bricks, window frames, etc. His artistic genius comes out in reconciling awkward circumstances and creating out of them, and in spite of them, a new and personal unity.

While the conditions discussed so far limit only the range and not the value of the design, there is another limitation imposed on design by machinery, which is indeed a more serious obstacle to artistic improvements. I said in connexion with motor-cars that the cost of a new set of tools is astonishingly high. The same is true of the engraving of rollers for cotton printing, of the making of tools for articles in plastics, aluminium, etc., of the printing of a complete edition of lithographs for pottery. In all such cases, a considerable sum has to be invested before production can begin. It would be natural, therefore, for the manufacturer to shrink from experiments. As long as he 'gets away' with his goods, any change seems a foolish risk. To take such a risk, he must be conscious of his duties to the community. It is easier to embark upon a new design in hand-tufted rugs, hand-printed textiles, hand-mounted jewellery, hand-thrown pottery, hand-blown glass, hand-beaten silver, where a failure means no loss of capital, but just the loss of time and material spent on the one article or on a small range. In view of the figures given earlier—a pattern for lithographed pottery being a failure unless at least 50,000 are sold, a radio cabinet in plastic a failure unless at least 10,000 are sold—one cannot condemn an industrialist too severely because he does not jump at every suggested artistic improvement. It would be absurd to suggest to the manufacturer that he ought to ruin himself for the community, but what one is justified in asking is that he should consider carefully and conscientiously how he can best obtain a balance between his business and his public duties. And this can best be reached by experimenting on a small scale with designs of a better standard.

So, after considering the relation between a factory's age, size, organization and production methods and the design standard of its products, the only result is that the large firm can be helpful by using its abundant means and progressive spirit in favour of design, and the small firm can be helpful by using the personal taste and energy of one man controlling the production for experiments on a small scale. But in practice the ideal representatives of both these types are rare. Most mass-production factories rely on routine work for the sake of easier money-making, and most small manufacturers are too indolent or too much concerned with securing orders and keeping afloat to recognize their social duties. So it seems that it is not the age, size, organization or production methods which decide the matter, but once again the personal attitude of the manufacturer.

THE CREATOR OF DESIGNS

The designers available to manufacturers are of four types: the free-lance, the architect, the consultant and the staff designer. Before discussing the relative merits of articles designed by them, it is necessary to outline the basic requirements of an industrial designer. He should have complete knowledge of the limitations and possibilities of materials, techniques and machines in his industry, and he should be aware of technical developments in related industries; without this he can create nothing new. By his artistic training and his knowledge he should be able to set a high design standard which is economical to maintain, for wherever possible he should strive to reduce production costs, which in turn implies intimate knowledge of his firm's systems of costing and selling. He should have a status high enough for him to lead a team composed of those who undertake research and experiment, those who produce, and those who sell. To copy the work of other designers is fatal to his prestige. Instead, he should be capable of interpreting imaginatively the evidence from the retail trade and market research. The ultimate duty of a designer is to serve his public. If he can serve and uplift its standards at the same time, he is a good designer; but if he allows the centre of his interests to shift from service to education, his public will diminish and he will be in danger of becoming redundant.[1]

Free-lance designers are found chiefly in industries where designing means surface decoration and not the shaping of the article itself—e.g. printed textiles and wallpapers. It is very exceptional to find a successful free-lance designer of carpets and woven fabrics. The free-lance may be an art school graduate with or without craft or industrial experience. He may be a commercial designer with a fully staffed studio of his own, or he may be a painter (even a sculptor). In all cases the free-lance does his work outside the factory. Free-lance designers are sometimes used for jewellery, carpets, and woven textiles where their work is invariably used to bolster the imagination of the staff designer, and so becomes artistically ruined. This is also true of some printed textile and wallpaper firms. The free-lance, after he has been paid for his work, cannot object to this treatment, though he may help to avoid it by understanding thoroughly the techniques for which he is designing, and in pattern goods this is not difficult. In this respect the commercial designer, who is often a Continental, is in a strong position, for he knows exactly what is wanted, not generally by the whole industry, but particularly by individual firms. He is more business-like than the other types of free-lance designer, but his artistic standards are very much lower. As far as generalizations are permissible, I would say that the majority of the best designs on printed textiles and wallpapers are the work of art school graduates and professional painters working as free-lance designers. The enterprise of Allan Walton as a converter of very good printed fabrics is exceptional, but worth noting; throughout the 1930's he bought designs from painters, giving them first a free hand and then considering, with them, the technical limitations to be observed; he paid for each piece of work and afterwards provided a royalty on the profits.[2] The fees paid to free-lance designers for pattern printing vary: 5 to 8 guineas for a young designer's sketch; 15 to 25 guineas for the work of an experienced commercial or free-lance designer. Higher fees are paid to the exceptionally talented free-lance, and sometimes to painters.

Although much more goodwill is needed on the part of staff designers and manufacturers when dealing with free-lance designers, I am in sympathy with those who find outside artists difficult to deal with, unbusiness-like and touchy. This is so, I am

[1] 'They [i.e. the designers of the next generation] must resist the natural reaction to recoil in horror from the worst examples of popular taste and, instead, when confronted by some really terrible piece of furniture, understand that this grisly, depraved exhibit may contain the secret formula for which they are searching.' R.D. Russell, in his inaugural lecture as Professor of Wood, Metals and Plastics at the Royal College of Art, February 1950; reported in *The Architects' Journal*, 2 March 1950.

[2] The firm has since changed hands, and is now known as George Henderson Textiles Ltd.

afraid, partly because, being artists, they often have to be like that and partly because, having been brought up with the idea of the great message of free and independent art, they want to be like that. Since about 1800 (Schiller, Shelley, Beethoven, Turner) the artist has begun to consider himself a superior being to whom the ordinary person ought to look up. His task is to express his own infinitely important and interesting experiences and thoughts. We still suffer badly from this conception. The medieval painter was a craftsman. As such he would not have thought of objecting to the carrying out of orders. Holbein designed cups and ephemeral decorations, and he painted the sign for a schoolmaster as well as his portraits. We may have gained much by that liberation of the artist which started during the Renaissance, but in the end this irresistible development has proved fatal. Without it we should not have had Rembrandt, but neither should we have had the unbearable over-supply of artists whom nobody wants and whose private feelings are totally uninteresting in a century such as this. And we should also not suffer from the equally disastrous lack of good designers for industry.

This consideration determines the attitude which the socially conscious critic ought to adopt today. If the painter is not so great as to express with the utmost intensity and vitality the essence of his time, let him starve, unless he is willing to find the way back to serving the community. Take some of the best commercial artists in London. They know their job and design posters, packages and catalogues, the social value of which is higher than that of any good portrait or landscape which they might paint and exhibit, while their artistic value is certainly not lower.

In complete contrast to the free-lance designer is the architect. As designers, architects have had some notable successes in industry (Wells Coates and Serge Chermayeff, radio cabinets; Keith Murray, pottery and glass; Jack Howe, clocks; C.F.A.Voysey and R.Y.Goodden, wallpaper; Basil Spence, Eric Lyons, E.Brian O'Rorke and Dennis Lennon, furniture). Most of the pioneers of the Modern Movement in Germany were architects (Peter Behrens, Walter Gropius, Bruno Paul, Riemerschmid, Miës van der Rohe and Marcel Breuer). The reasons for this are simple. The architect is an artist with a professional status. He has been trained to produce work which is a satisfactory synthesis of his own artistic standards and the wishes of his client. And he is adept in creating a new unity out of standardized parts. The architect may not know much about industrial techniques but he has a working method which enables him to find out about them. And, at the same time, he can impress the manufacturer with his thoroughness. Architects frequently come into contact with industry because of the fittings required for their building contracts. Where nothing suitable is available they may produce their own designs. We have discussed in connexion with metal door furniture the influence which architects have had on trade design standards. Light fittings is another case. Brian O'Rorke's work as the co-ordinating designer for five ships of the Orient Line is an outstanding example of how the architect can obtain from several industries a very high standard of design.

Nevertheless, the suggestion, so often made in the past, that the architect is the most suitable person to design for industry provides no solution to the problem. After all, the architect has other duties to the community for which he has been specially prepared. He may, with help from the manufacturer, produce one or two ranges of startlingly original designs, but he cannot be expected to keep it up. Except in rare cases the architect does not retain his hold on industrial design. And, indeed, why should he? His own field of building is more lucrative and, in normal times, far more secure. The Bauhaus was primarily a school for industrial designers, in spite of the fact that architecture was the co-ordinating study. This and other considerations make it pointless to accept the architect as the most promising designer, for in order to do so we must forget the example of the Bauhaus and, in addition, we must set aside the efforts of the Royal College of Art and the regional colleges to train students to be full-time designers in industry.

As I have already dealt with the consultant designer in a separate section I need only reiterate the main points. The consultant is an independent expert with a staff, studio and, in some cases, a workshop of his own, who is called in by a manufacturer to take charge of design matters. By virtue of his appointment and the fees he charges, the consultant should be able to gain the respect of his client's sales, production and technical managers, form them into a team, and lead them. That is the ideal procedure which has, in several cases, worked in practice. I found instances of it in connexion with domestic appliances, radio cabinets, plastics, fluorescent light fittings, furniture, and motor-cars where only the bodywork is concerned. In this country, though not in the United States, consultants do not set up their own staff with studios in a client's factory. I know of at least one instance of this happening in Germany. Just after the war Professor Wagenfeld was engaged as consultant designer to the Württembergische Metallwarenfabrik for glass and metal products (cutlery). At the factory he has been able to build up a design laboratory and a staff of assistants. His contract with the firm entitles him to royalties on sales of $1\frac{1}{2}$ per cent, and a minimum advance of 1200 Westmarks a month; the money to be payable for fifteen years after his death. In this country royalties are not usually paid; the conventional form of payment is a retaining fee, perhaps with separate sums, calculated on an hourly basis, for each piece of work executed.

A consultant designer, if he is thoroughly conversant with the firm's methods and its market requirements, is probably the best type of designer for a firm which requires few new designs each year, and keeps them in production for a long time. Typical examples are radio cabinets, and gas and electric equipment. In addition, a consultant designer with sufficient experience could be engaged by several firms in allied industries serving similar markets to correlate their design standards. This might work in the light metal trades if co-operative and enthusiastic manufacturers could be found. However, consultant designers in this country have not been working long enough to accumulate sufficient prestige. Moreover, they depend too much on the skill and experience of their client's staff. Often their so-called 'all-round knowledge' of industrial practices is a myth. I know of several cases where consultants have taken on work in too many varied industries. In order to justify their existence they have resorted to 'styling' the client's products with superficial embellishments.

The question of the consultant's prestige as an expert on design is of great importance. It is, of course, preferable that he should be recognized as a successful artist and not as a successful trader in designs. In the United States artists and architects are looked upon as influential members of society, valued chiefly for the artistic standard of their work. But the American consultant designer receives only commercial recognition, with a prestige measured by his annual earnings.

Our discussion of the free-lance, the architect, and the consultant designer has revealed that the scope of these designers is limited because they cannot grasp the immense and growing variety of materials, techniques and machines used in industry, in addition to the demands of the wide range of markets served. In firms with long production runs and few products the consultant has a place if he makes it for himself; the free-lance is particularly suited to pattern design; the architect *can* apply himself to most industrial practices, but he does not usually do so for long. Where constant supervision and an intimate knowledge of complicated techniques are required some other type of designer must be found. Such a designer must be employed as a full-time member of the staff.

I have said earlier that for creative work the designer must know the limitations and possibilities of the media with which his firm is working. Where techniques are varied and complicated, as in pottery, there is no alternative to the staff designer. The same is true of plastics, where new materials and processes are being developed every year. For woven textiles constant experiment with yarns in the loom is imperative. In blown glass the best designs come from personal direction of the maker and the cutter. In carpets and furniture (in large firms) the inflexible production methods must be observed constantly if creative work is to result. This is even more true of mass-produced motor-cars where techniques are numerous and change often. In light fittings, new technical developments are less frequent but a large number of designs are required for short production runs. The same is true of jewellery, silver, electro-plate and light metal work. In all the firms where these things are made a staff designer is essential. And working with him, to interpret his ideas, must be an experimental craftsman.

Now most, perhaps 70 per cent, of the staff designers I met are incapable of reaching anything more than a mediocre design standard, either because of their own limited talents or because of the circumstances of their employment. Instead of repeating the reasons for this, as they apply to each industry, I want to outline the type of staff designer which I believe manufacturers *should* employ. By this I mean that he is the only type of designer who can achieve a high artistic standard in the articles mentioned above. My designer has been trained in one of the art colleges of Leicester, Bradford or Birmingham, or at the London Central School or the Royal College of Art. He has had at least five years' experience in his own industry, has studied the methods in two or three others catering for a related trade, and has travelled on the Continent, especially in Italy.[1] In his firm he has a studio of his own, perhaps a small staff, and a status equal to that of the chief engineer and the production and sales managers. He is immediately responsible to the managing director. He is paid between £800 and £1500 a year according to the size of the firm, the experience needed, and the volume of the work. He is not paid a royalty on sales, as this would create envy amongst other executives in the firm, who are all responsible for the way in which a product sells. His access to all departments in his firm is unlimited, so that he can study methods of production and costing. Although the consultant may be more aware of new material and technical developments, the staff designer is able to relate them immediately to his requirements. In addition he will have spent at least six months in the sales department. He will visit agents and retailers frequently in home and foreign markets. And at the firm's expense he will make at least one journey abroad each year to refresh his mind and widen his interests. Such a designer is not an impossibility. I have met him several times in firms making furniture, light fittings, pottery, woven textiles, glass and plastics.

Provided that the staff designer is a creative artist, the first requirements to stress are his immediate responsibility to the managing director and his executive status in the firm. I do not know of more than twenty designers in this position today. Payment for staff designers is universally too low. The figures vary, but £1000 a year in all

[1] I say Italy, because it is there that some of the best examples of fine and industrial art, past and present, can be seen almost side by side.

but the motor industry, and sometimes the furniture, carpet, textile and wallpaper industries, is difficult to attain no matter what artistic ability and experience the designer possesses. Again, I have met staff designers who have been told so many times what cannot be made and what cannot be sold that their imaginations have atrophied. This is the disease in the carpet and woven textile industries. In provincial industrial towns designers are cooped up year after year, often in ill-lit, inadequate studios, at the beck and call of the sales manager.

Continuous interference from the sales departments is a serious problem. Nobody would object to the interference of engineers who tell the designer what is technically necessary and advisable. Nor would anybody object if the sales manager were to tell the designer which of the year's new patterns or new models were successful and which were not. But when the sales expert tries to indicate what ought to be done in future, the result is as a rule artistically disastrous. There is nothing astonishing in that. A layman may well tell the painter that he does not like a certain combination of colours, but there would not be many laymen who would venture to suggest to the painter what he should do to get a better colour scheme. Accordingly the sales manager will usually be good at negative proposals and unqualified to give anything more than general positive suggestions. As a result the designer soon finds himself incapable of creating anything with original vitality, let alone a design of high artistic standard. Unlike the consultant he is in a position to obey and cannot afford to terminate his contract. Indeed, the badgering interference from others is so acute in some firms, that I found the designers spent their evenings at the local art school doing craftwork as a palliative to the sickening routine of the factories.

If the designer is to create new, significant works of industrial art, he must be a free and responsible agent within the framework of his firm, with a live understanding of the possibilities of his media, be they looms, presses, tangent benders or spinning lathes. To be an engineer is not enough, though at the moment it appears to be so when 'fitness for purpose' is the governing criterion of excellence. Some designers are owners or directors of their firms. Examples can be found in all the industries which cannot do without staff designers. I do not think it necessary for the designer's name to be attached to products for which he has been partly responsible. It is far better for the manufacturer, and indeed his whole staff, to receive the public's reaction to a new product and apportion the praise and blame amongst themselves. If the design is artistically unsatisfactory the designer should be turned out. Bad design cannot be abolished without abolishing bad designers. Surely nobody would suggest that worthless pictures should be bought in order to keep mediocre painters alive.

THE MIDDLEMAN

The success or failure of a designer's work depends upon its reception by the retailer. It has been already stressed that if quantity production of articles at reasonable prices is to be maintained the manufacturer is forced to rely upon the retailer's judgement of new designs. And this judgement can only be based on what the public is offered in the retailer's store. The retailer's responsibility as the link between maker and consumer is considerable and should increase in proportion to the size of his business, for it is at all times his moral duty to provide the manufacturer with an accurate account of public taste: accurate in the sense that it is derived from the sales of articles fairly selected by him or his staff to include designs of the highest standard in each price group.

In an earlier section I divided retailers into three groups: those with no modern designs, those with some, and those with many. I added that more than 80 per cent of the retail trade is represented by the first group, which includes chain and department stores and many private businesses of all sizes. It is with these stores that I am most concerned. When dealing with them I found that the initiative required for design improvements must come from the owner or manager of the store. In chain stores and some department stores a deliberate policy must have the authority of the head office of the company. Without it, no manager of a branch store, however enlightened, can achieve satisfactory results. This is because his goods are bought in bulk by the parent company which is, in fact, not in direct contact with his customers or any members of the buying public. The situation is difficult. If no incentive for improvements can come from the public, the only alternative is for the chairman, or one of his directors, to feel personally that the design standard of articles bought and sold by his company is unsatisfactory. But what chance is there that men in these positions will find time to bother about the taste of the public? Their full-time duty is to trade in articles which can be sold at a profit. The fact that these articles have an artistic value is irrelevant to the company's main interest. If sales fall prices will be reduced, advertising campaigns will be started, fewer quantities of each article will be stocked, novelty and up-to-date gadgets will be stressed in every window display. A higher standard of design will not be considered, for it is not one of the accepted devices for forcing the public to buy what it does not want.

Retailers of furniture, soft furnishings, and the more expensive articles of domestic equipment (silver, pottery, glass, radio sets, etc.) do not realize with sympathy that their public is inexperienced in buying such things and cannot afford to take risks.

For example, tables and carpets are bought only once, maybe twice, in the lifetime of most people. They are unfamiliar purchases and, unlike shoes, millinery, tinned food and magazines, which are bought frequently, are not likely to invoke a positive response. Faced with a hesitant public, the retailer has a choice of action. He can take advantage of it by limiting his stock to designs which have always sold in the past, or he can display a fair selection which includes modern designs. If he takes the latter course he must train his salesmen to appreciate modern designs. It is imperative that they should be able to demonstrate their qualities to many customers who will at first be sceptical and unsure. For it is, indeed, a social duty to make people aware of those domestic articles which have been specially designed to give pleasure over a long period of years.

Apart from salesmen, whose taste in each department should be correlated with a standard of which the retailer is proud, the buyers must also be taught to discriminate between good and bad designs. It is their duty to ensure that within each price group articles of comparable artistic quality are purchased. A recently established organization which attempts to educate the staffs in retail stores should be noted here. It is the Midland Retailers' Design Discussion Group, which was started in September 1949, as a result of a Design Week and a retailers' design conference, both held by the Council of Industrial Design. The group meets periodically to hear lectures on aspects of design, but it appears to lack the enthusiasm of its complementary body, the Midland Industrial Designers' Association. For instance, it was this Association, and not the retail Group, which conceived and carried out the idea of making special displays of good designs in the departments of Lewis's at Birmingham. The Association, together with this and other ventures, has already been dealt with.

One of the most effective ways of introducing better design standards to the public is to hold small exhibitions in retail stores. These should be staged at regular intervals and the articles selected by a design critic. Before the war the Design and Industries Association showed how exhibitions of this sort could be effectively carried out. Now, similar initiative is being shown by the Council of Industrial Design. The object of the exhibitions is to awaken interest in the buying public, which should be able to make purchases on the spot. This means a new demand can be located and its size estimated for the information of the manufacturer. The experiment at Hopewell's store in Nottingham is an excellent example of how to elicit public criticism. To obtain advice and active help when staging an exhibition, retailers in each town should encourage students at the local art school to take part. The retailers would benefit from this willing and youthful assistance, and the students would profit by having their studies directed towards a real project in which all could share. It is evident that such joint enterprise would be of value to both parties, and yet I do not know of one example. If a whole exhibition, however small, is too much trouble for the retailer, he could achieve similar benefit by allowing students to dress one of his windows. There was an example of this on a large scale in Regent Street, London, in 1950, when teams of students from twenty-one art schools in London and the Home Counties took part in a competition to arrange displays for the windows of thirty-five stores.[1] It had been agreed that each display should remain in its window for one week, yet only one day after being judged some of them had been removed. On the other hand, one store (Hamley Brothers) was so impressed by the competition that it awarded an annual scholarship of £200 to encourage display students in the competing art schools.

There would indeed be a remarkable improvement in the present situation if one retailer in each town would regard his store as a public design centre, a place where one could be sure to find the best designs from each industry which serves, say, the domestic furnishing market. Before the war Lewis's 'Design for Living' Department showed how a big store could accomplish very good results in this way, and at a time when the best designs were comparatively much more expensive than they are now. Nevertheless, there have been for many years a few medium and small size stores (Heal's, London; Dunn's, Bromley; Gane's, Bristol; Bowman's, Camden Town) which set an admirable example to the rest of the trade. In general, retailers do not realize that for them the risk of introducing an article in a modern design is far less than that incurred by the manufacturer who originated it. As a corollary, retailers might do more to help a small manufacturer, anxious to produce promising new designs, by placing orders in advance.* At the same time, they should attempt to check their ceaseless quest for novelty—for the 'something that is different'—which distracts manufacturers from making steady design improvements. But, above all, retailers should be aware of their social duty to improve design standards and relate it to their more obvious business of maintaining a turnover of trade.

THE TASTE OF THE PUBLIC

Throughout this book I have tried to lay stress on the importance of allowing the public in shops to see (and purchase from) a fair and representative selection of articles which must, in each price group, include some modern designs. For this there are two reasons. The first concerns the manufacturer's and retailer's estimate of public taste, the second the critical faculty of the public itself. It is not generally

[1] The competition was organized by the Council of Industrial Design on behalf of the Regent Street Association.

realized that a new design from any factory passes through the hands of six different people (designer, manufacturer, sales manager, traveller, retail buyer, retail salesman) before it reaches the public. In several cases one or more wholesale distributors would have to be added. Although the public's taste is originally estimated by the demand for what is available in the shops (an evident limitation) it may, by the time it has been passed back to the manufacturer, by way of at least four different people, contain several inaccuracies. The four intermediaries are usually paid on the basis of a commission on sales transacted and they are not anxious to risk their livelihood on unfamiliar designs. This is especially true when unfamiliarity is caused by a standard of design which is higher than usual. Every one of them has a right to say what he personally likes or dislikes, but none of them has a right to withhold a design from the public or a piece of information about taste from the manufacturer because he personally disapproves of it. As far as we are concerned, this is one of the most insidious forms of dictatorship. Its effect is cumulative and yet in isolated instances most difficult to detect and eradicate. The only satisfactory solution is for manufacturers and retailers to estimate public taste by the sales results of a representative selection of designs. Moreover, they should ensure that new and outstanding designs are allowed to remain within the public's view for a much longer time than the familiar designs that have a steady sale. For it often happens that such new designs need a lengthy trial period before being assimilated and judged critically by the public. We find exactly the same situation in the other arts. New poems, paintings and musical scores are not appreciated or even recognized for months, and maybe years, after publication.

This brings me to my second reason why all stores should provide a representative selection of designs. The public is unaware of exactly what it wants until it sees it. The public may have an intuitive conception of its requirements; this comes out most clearly when the functional aspect of design is of decisive importance (kitchen equipment, gardening tools, gadgets). Where artistic quality is involved in a design the public will be hesitant and uncertain until its critical faculty is aroused. In most people (I think more than 90 per cent of the total population of this country) the critical faculty is undeveloped. Often it is swamped by such factors as snobbery, nostalgia, sentiment, exchange value (of antiques, etc.), love of things hand-made, and many more, all of which are irrelevant as far as the artistic quality of a design is concerned.

The development of a critical faculty in people depends in the first place upon being able to compare good and bad designs. The extent of the development depends on the number of comparisons that can be made. People can be encouraged to do this by retailers; they can be taught to do it by demonstration in schools, universities and adult classes. Although the critical method is simple, it needs energetic application because, in all cases, analysis of varying complexity is a prerequisite to true judgement. For each man or woman the final result of the analysis will be personal, and therefore slightly different from another person's. Nevertheless, the aggregate result of analyses made by people with backgrounds, educations, habits, intelligences, etc., which are broadly similar, should provide the manufacturer with a working knowledge of public taste.

But what about the other 90 per cent, the people with prejudices and preconceptions, half-formed by themselves and half-formed by others, or inherited *en bloc* from the past? It would not be difficult to make a fairly reliable estimate of their taste. At least, that is what manufacturers and retailers tend to think. Often they share the same characteristics themselves and are readily sympathetic when they hear that other people (members of their public) crave for similar satisfaction. That is the reason for the seemingly insatiable demand for imitation of bygone styles (eighteenth-century furniture, Adam chandeliers, Oriental carpets, etc.). And yet the manufacturer only *hears* that there is a seemingly insatiable demand for these things. He does not *know* that he is satisfying the demand. And he certainly does not *know* that he is fulfilling the most profound wishes of his public. For even the 90 per cent would (and in fact do) manifest a response of varying degrees of intensity when confronted with a choice between good and bad design. The success of inexpensive furniture and textiles of modern design in competition with mediocre examples provides sufficient grounds for this assertion. The minority of manufacturers are aware of this; the majority have yet to realize that the taste of the public can only be ascertained by comparing the sales results of good and bad designs in a common price group. To achieve this the manufacturer must join with the retailer to ensure that comparisons can be made.

MARKET RESEARCH

I have emphasized the value of sales results as an indication of public taste (i.e. in cases where the public is given a choice) because most people when called upon to spend money are induced to think more carefully about design. In addition, manufacturers and retailers naturally attach great significance to sales results. After all, these results are expressed in figures, tangible and more easily understood than personal opinions. Attempts have been made to estimate public taste in other ways:

by questionnaires to be filled in at exhibitions (Hopewell's), or on the doorstep by specially recruited inquiry agents. If people are confronted with a range of articles of differing standards of design there is some point in giving them some time to deliberate and then asking them to state their choice. This usually is impossible in the case of the agent on the doorstep, and his results are consequently without value. But in both cases the people, when making up their minds, lack the mental discipline which the expenditure of their own money imposes. They are apt to forget what they can afford, and indulge their fancy to the extent of giving thoroughly unreliable answers.

Another method would be to carry out an inquiry on the lines of this book, among representative groups of consumers. The central question would not be 'why don't you buy this design or that design?' but 'why do you furnish your house in this way?' A recent inquiry somewhat on these lines was carried out by Dennis Chapman amongst retail furnishing salesmen in Liverpool.[1] His evidence concerns working-class and middle-class families who are able to furnish three or four rooms, either by paying cash or by hire-purchase. He relates 'the social, the emotional, the aesthetic and functional elements in domestic life', to the design standard of furnishings. It is perhaps significant that he finds that women are primarily responsible for choosing the designs. Mr Chapman supplements his evidence with a complete inventory of the contents of a random sample of fifty-one parlours in a working-class housing estate. Amongst their contents he found forty-five three-piece suites, thirty-three pianos, thirty carpets and forty occasional tables. Unfortunately, he does not describe their standards of design. Inquiries like this, broader in scope and based on detailed descriptions of private houses, would be useful additions to the manufacturer's estimate of public taste from sales reports.

But we cannot afford to ignore market research as far as it concerns the collection of answers to questions from groups of people. In this country there are at least two well established organizations for carrying out this work, and many more in the United States where the idea was originated by Dr Gallup in 1933. A growing number of American manufacturers of widely selling products are relying on market research workers to keep them informed of trends in public taste not only to help in the preparation of new designs but to estimate the value of different types of advertising. Manufacturers in this country compete on a smaller scale and comparatively few of them feel the need to support market research campaigns. Although less than 5 per cent of the manufacturers I interviewed had carried out market research, my estimate of the situation may not be accurate as the information, or even the fact that it had been obtained, is confidential.

[1] See p. 154.

It seems necessary therefore, to deal with this hypothetical situation by pointing to the chief disadvantage of market research. The important role taken by accountants in modern business practice fastens what seems to be undue significance to figures. Armed with figures which indicate by percentages the public's opinion of certain new ideas in design the manufacturer can face his designer with what he believes to be an accurate statement of public taste. The designer must comply with the results of research. In doing so he will set aside his imagination which will become atrophied progressively as the initiative for new ideas passes from him to the market research organization which, again working by percentages, will be able to present a convincing picture of what the public would like to buy in the future.

This very real danger in market research appears to be unanswerable. As to its advantages I can see none which could not be obtained more accurately and with less expense by the results of sales transacted where the public is given, in the same price group, a choice which includes good modern design. The manufacturer's understandable liking for figures must be recognized and in fact supported, for it is the only method by which he can be convinced that people prefer one standard of design to another. More experiments, run on the basis of the D.I.A. exhibition, Register Your Choice, but permanent and situated in retail stores where the exhibits can be bought, will be the best means of finding out what people really want.

DESIGN STANDARDS AND PRICE

A point which has been emphasized in this book is that the artistic quality of the inexpensive product is the question which really matters. It is therefore now necessary to re-examine the results of my inquiry in various industries with regard to the relation between price and design standards. I mentioned in connexion with public taste that in furniture and textiles (woven and printed) it is possible to find modern and mediocre designs in the medium price range. Indeed, the majority of modern furniture designs *are* in this range. Inexpensive textiles of modern design, especially weaves, are more difficult, and sometimes even impossible, to locate in a small provincial town. In 1935, and throughout the 1930's, the situation was entirely different. There was Plan furniture, and inexpensive pieces imported from Finland, but the handful of small British firms interested in modern design was concerned almost exclusively with hand-made furniture of a high price for an exclusive London market. In the same way, hardly any textiles of satisfactory design were available to the working-class customer.

In carpets nowadays a few modern designs of a reasonable standard can be found at a medium price; in 1935 this was not possible. Linoleum at all prices is as unsatisfactory as ever. There are some designs of a good standard to be found in inexpensive door furniture; in 1935, satisfactory designs were rare and often too costly for anyone but the architect with a special contract. Silver, jewellery, leather goods and crystal glass cannot be discussed here because they are either subject to heavy purchase-tax or partially restricted to the export market, two factors which will be dealt with later. In plain white pottery there has been no significant development in design since before the war; then, as now, there were a few inexpensive pieces of satisfactory design. The situation in decorated ware is different; a higher percentage of good designs is now on the market than before the war. Plain shapes in crystal glass have always been expensive, and the working-class is forced to rely on mass-produced pressed glass. Before the war there were no satisfactory designs in pressed glass, but now there are a few if one is prepared to hunt for them. Electro-plate is cheaper than silver, but it cannot be called inexpensive; in any case, design standards have not altered during the last fifteen years. Handbags, suitcases, etc., covered with plastic cloth, cheap, and of very good design, are more plentiful now than in 1935. Aluminium kitchen utensils are now as cheap and as good as they were in 1935, and aluminium tableware is still cheap but good designs are now less easy to find. Amongst the cheaper radio cabinets a few firms have high design standards; the situation was similar before the war. Good and inexpensive light fittings were less plentiful in 1935 than they are now; the best designs are rarely available to the working-class customer. Purchase-tax still affects some types of domestic appliance and comparisons cannot fairly be made; in general, it may be said that there are more and better designs available in the lower price groups. In 1935 there was no inexpensive wallpaper of satisfactory design; today there are several exceptions to this, but the best designs are mostly in the price ranges where one has to pay for hand-block printing. At present, motor-cars bear a heavy purchase-tax and every model is beyond the means of the working-class; before the war it was at least possible for upper-working-class people to save up to buy a 'Baby' Austin, which was, at that time, a classic example of good design at a reasonable price. Finally, articles in plastics have always been inexpensive; now there are many more satisfactory and very good designs available.

It is evident that there has been a general improvement in the design of inexpensive articles. It is most marked in furniture and plastics, but noticeable in textiles (especially printed fabrics), door furniture, domestic appliances, plastic handbags and suitcases, light fittings, and wallpaper, and to a lesser extent in pressed glass. The change is just noticeable in carpets. No change for the worse can be detected in radio cabinets and kitchen utensils. Linoleum is still unsatisfactory, while aluminium tableware of good design is now more rare.

As we have seen, mass-production means that articles will be *ipso facto* less expensive than articles produced by hand. At the same time they can achieve design standards which are as high, and even higher, than hand-made articles. But it must also be recognized that mass-production methods can enable a manufacturer to offer an article at a price which is low enough to entice people to buy it, regardless of its design standard. This can affect the sales result of any article, whether it is of good or bad design. Some plastic condiment sets of excellent design sell at Woolworth's partly because of this, but so do some atrocious designs in aluminium teapots. In these circumstances public taste is a factor of no significance unless a comparison between the excellent and the atrocious is possible amongst similar articles within the same price group. If one cannot make this comparison, the manufacturer must be held responsible for the design of his products, according to the extent to which he undersells his competitors.

On the other hand, when cheapness ceases to be necessary, the manufacturer immediately recognizes that the design standard of his product is of first importance. Here one cannot generalize; the best designs in woven textiles are expensive, partly because they require expensive materials, but mainly because they are produced on a small scale. But exactly the same is true of some of the worst designs in woven textiles. In most firms making light fittings, clocks, furniture, textiles, and electric fires, the higher the price the worse the design standard becomes. Painstaking imitation of antique styles is particularly characteristic of expensive articles. The manufacturer feels it to be his duty to offer an abundance of 'Art' to excuse the high price of his article.

The most significant change since 1935 is that, apart from light fittings, wallpapers and woven textiles, the highest design standards no longer depend upon the patronage of an enlightened and wealthy class. This does not mean that public taste has generally improved. (If one can judge from the circulation of some daily and Sunday papers, from radio programmes and films, one would be more inclined to say that public taste has got worse.) On the contrary, it simply means that in 1935 there *was* a market for designs of a satisfactory and high standard which manufacturers and retailers knew nothing about. What are the implications of this? The process which has already marked a wider acceptance of satisfactory design standards has only just started. It seems reasonable to assume that there are millions of people who are still not aware that these standards can be achieved. I am led to this conclusion by the relatively small output of satisfactory designs and the relatively small number of stores in which they can be found. Today the demand for satisfactory and high standards of design is not so much limited by price as by the manufacturers and retailers who refuse to recognize that it exists.

I have already outlined how the mass-producer can introduce, with the aid of a designer and experimental craftsman, new and better designs. At first they will be in small quantities but, if successful, they can be transferred to the appropriate mechanism for mass-production. The extra cost of producing in small quantities articles which should be inexpensive must not be passed on to the consumer, but borne by the manufacturer. This is essential if the working-class market is to be served. The risk and the cost entailed together constitute a remarkably small price to pay for the discovery of a new demand in the working-class market. The retailer must also recognize this, and allow a long enough trial period for each new design to be appreciated. It is indeed fortunate that in the industries where small-scale experiments are impossible (domestic appliances, plastics), there are a sufficient number of inexpensive but satisfactory (and often exemplary) designs to prove the existence of such a demand.

SNOB APPEAL

In connexion with public taste I said that there are several factors which swamp the critical faculty of some people when confronted with a choice of designs. The most pervasive and vicious of them is snobbery. In this context I am only concerned with its most common form, derived from the desire to associate oneself with an envied class. In most cases this means imitation of appearances only and can be detected in dress, in tone of voice and, most important of all, in one's possessions. As might be expected with our still sharply-defined social differences between the poor, the working-, middle- and upper-classes, snobbery is a characteristic trait in the British character. Snobbery leads people to forget the functional quality of design and to attach too much importance to its artistic element, not for itself alone, but for the way in which it can be associated with wealth and position. It so happens that antique furniture and silver, Oriental carpets and tapestries are expensive and are the characteristic possessions of wealthy people, living in spacious houses with a staff of servants. While this is, admittedly, a generalization, it does equate with the popular conception of the upper-class milieu. Furthermore, the spacious house and the staff of servants do represent the background and mode of life for which these things were originally designed.

When the manufacturer responds to, and for the sake of trade encourages, a person's desire to emulate his betters, the most absurd results can be expected. This absurdity increases in inverse proportion to the price which can be paid. One finds mass-produced furniture imitating a mixture of eighteenth-century styles. To fit into the front parlour of a semi-detached villa the original dimensions are reduced, or perhaps disregarded, in favour of the more obviously characteristic carving. The function of the chairs, table and sideboard as useful pieces of furniture is of secondary importance and suffers accordingly. The most ornate eighteenth-century silver articles were never very useful, so imitations of them begin with an insurmountable disadvantage. Oriental carpets, as I have explained, cannot be reproduced on mechanical looms. Tapestries become curtains and chair covers, but there again the machine cannot do the work of a hand-weaver.

Snobbery can entice people to accept ridiculous circumstances. The public 'will make any sacrifice physically and mentally painful, it will go to any lengths in making an exhibition of itself if it is assured that it is the thing to do'.[1] Snobbery is ministered to by manufacturers and advertising agents. How many illustrated advertisements does one see every day showing quasi-antique furnishings in a real antique house, a mass-produced motor-car beneath the portico of a country mansion, a brand of toothpaste used by a film star?

Snobbery is vicious because, more than all other irrelevant considerations when buying domestic goods, it effectively obscures the purchaser's critical faculty. Both artistic and functional qualities in a design cease to affect the judgement. It may be expedient to suggest that the upper-classes should be induced to furnish their houses with exemplary modern designs. But such an event, if it ever took place, would be just as pernicious in its effect as the latter-day *ancien régime* which is now in evidence, for it avoids the crucial issue that snob appeal does positive harm by preventing people from judging a design on its own merits for itself alone.

PUBLIC DEPARTMENTS AND PUBLIC TASTE

It is possible to set a design standard to which people will respond without appealing to snobbery. The essential point is that whatever is designed must be functionally and artistically related to the ethos for which it is created. Whether it reaches a high standard or a low one, it is bound to set an example which, though it cannot be followed or imitated by others, will awaken in every person some degree of critical

[1] F. V. Burridge in *The Journal of the Design and Industries Association*, No. 9, October 1918, p. 14.

response. We have already discussed the possible influences of environment on manufacturers and designers—how Stoke-on-Trent and Kidderminster, for instance, by their overwhelming proportion of congested living space and fossilized architecture, are not conducive to the development of a creative artistic sensibility. In such cases the bad influences in the environment so far outweigh the good that the critical response of the inhabitants is deadened, and acquiescence results.

Anything more than a brief outline of the situation is beyond the scope of this book, and consequently no definite conclusions can be reached. It is axiomatic that with so much mediocre and bad design about, the more the design standard of the environment is improved, the more alert and positive will be the critical response of the public, a response which will inevitably be turned towards works of industrial art. To begin with, modern theories of education lay great stress on the architectural standards of schools. Since the war there has been a remarkable increase in the number of schools in which space, light and colour are intimately related to the needs of children. Before the war practically all domestic architecture in which modern design standards were implicit was paid for privately. Now, with many more enlightened architects working for local authorities, the greater proportion of houses set a good, and in a few cases exemplary, design standard. But most significant of all, the New Towns represent a progressive attempt to revalue, on a large scale, all the influences which planning and architecture can exert on our environment.

The situation appears to be far less satisfactory when considering buildings designed specifically for the use of the public departments themselves. For the most part they are retrogressive in design. Public transport is another influential element in our environment. Since the war a progressive trend has been noticeable. The British Railways Executive, employed until 1954, a full-time staff of architects, who, amongst other duties, designed the interior furnishings for new passenger trains. The British Overseas Airways Corporation also has a staff for similar purposes.

Progressive design schemes to be carried out by public departments invariably need the initiative and energy of one man, or a small group of enthusiasts, in a powerful and responsible position. The only outstanding example here is the late Frank Pick, a founder member of the D.I.A. His life's work is too well known to need detailed description. Before the First World War, when still a young man, Pick had the opportunity of providing Underground Railway posters for the London Passenger Transport Board. He commissioned work from progressive artists and so began a series of posters now recognized as being mainly responsible for the remarkably high standard of British commercial art. Pick's influence was cumulative, for by the time of his death he had been responsible for all aspects of design in London's transport services. He had strict honesty and unbending faith in the value of harmonious, simple design. By his efforts he reformed the whole appearance of the transport commodities of the world's largest city. He set a standard by example which encouraged the London public to develop its critical faculty along progressive lines.

There are many public departments in London and the Provinces that could use their influence to set a comparable example. As with industrial firms, the responsibility rests with those in the highest positions to use architects and designers capable of creating a contemporary environment of the highest functional and artistic standard.[1]

CIVIC DESIGN STANDARDS

Responsibility for the public's environment is also shared by the council in each city and town. Here again, only an outline of the situation can be given. Amongst other duties councils are responsible for town-planning, the arrangement of open spaces, and public buildings such as civic centres, museums and art galleries. They operate on a small scale and their actions involve the loyalty, patronage and general interest of the inhabitants. Unlike the public departments (Home Office, Trade, Works, Labour, Health, Housing, etc.) the councils can be expected to have a sympathetic understanding of local conditions and local needs.

As regards the artistic aspect of the regional environment, few councils have recognized their responsibilities. Art is painstakingly collected by councils to be bottled up in art galleries and museums. The working city outside remains as dismal as ever and the inhabitants, who would welcome a centre for leisure where they could enjoy the environment, are forced to escape. Instead, art has to remain a permissible relaxation, and life a vale of tears. It is extremely characteristic that it was the big industrial towns such as Birmingham and Manchester that bought the best collections of pre-Raphaelite pictures, whose lasting success can be accounted for only as a means of escaping real life. To protest against this attitude in Birmingham, a private Civic Society was formed as long ago as 1918, 'to stimulate civic pride in the domestic and civic life of its citizens, by urging the adoption of the highest standards of architecture for domestic buildings, offices, warehouses, factories, etc.'[2] The Society relies on private patronage and its activities are now almost exclusively confined to the preservation of social amenities. It is evident to anyone familiar with Birmingham that the Society's protests against bad design have been ineffectual.

[1] See George Williams, 'Street Furniture', *Design*, September 1954, pp. 15-33. Also see the many articles by Gordon Cullen in *The Architectural Review* from December 1949 onwards.

[2] *The Work of the Birmingham Civic Society, 1918–1946*, published by the Society, 1946, p. 4.

Artistic reform in any town or city, if it is to be effective, must be enthusiastically and powerfully supported from within the council itself. This seems to be the case at Leicester with the Principal of the College of Art acting as the Council's artistic adviser. I have already described how the College carries out display work, lettering and printing for the Council and how it began a circulating collection of articles of good design to be used in local schools.[1] During the war the College obtained control over the city's emergency design requirements, air-raid notices, school and hospital equipment. The art school at Carlisle is also developing an interest along these lines, but few other schools have shown similar initiative. Their scope could be very wide: the furnishing of public buildings, design of municipal stationery, bus or tram tickets, street name plates, outdoor seats, etc. In most cities it is distressing to see how art students, like the art collections in the gallery, are confined physically and mentally within the four walls of the school. Not only is harm done to students who spend their energies exclusively on fictitious design schemes, but the citizens fail to benefit from the youthful design talent they have paid to educate. The situation is ironical, for cities cannot afford to do without guidance on artistic matters.

I have already suggested that one retailer in each city or town should make his store into a public design centre. Such a project might take shape more quickly if the incentive came from private enterprise. But it should be noted that most cities have in their art schools and art galleries the framework of a ready-made public design centre.[2] The art school provides the trained staff for exhibition display work, the art gallery has the space. The task of selecting the articles for display in this permanent but changing exhibition of industrial art, could be performed by a small committee of representatives from the art school, art gallery, industry, retail trade and the general public. It would be interesting to discover which collection in the art gallery attracted most attention over a period of years: the types of painting epitomized by those acquired by the Chantrey Bequest, or objects of modern industrial art.[3]

[1] In this connexion the Department of Circulation in the Victoria and Albert Museum should be mentioned as a special case. The Department makes loans to other museums and to approved educational institutions.

[2] Cf. Patrick Geddes's conception of a Civic Museum as a permanent centre for exhibiting fresh data and bringing up-to-date existing knowledge of city and region, in terms of town planning and social activities, *Cities in Evolution*, William and Norgate, 1950.

[3] 'Museums might well begin to have a really beneficial formative effect on popular taste, which is not a fixed and unalterable thing as many people seem to suppose.... How valuable it would be to show the best of both hand and machine work side by side with a critical appraisal of each exhibit. How else can we convince ordinary people of the tremendous importance of both in a country which claims to be civilized? What museum has ever tried to show the relation of the fine to the applied arts, the relation of, say, painting and sculpture to industrial design? How many museums realize what a first-rate discipline good lettering can provide? Or that on ticketing, display and lighting they share many problems with

MARKETS FOR BUYERS AND SELLERS

We must now return to the manufacturing aspect of industrial art to see how certain economic forces affect the choice of designs available to the public. As long as firms are able to keep their machines going steadily they will hardly be inclined to consider drastic changes of design. Consequently a trade slump or buyers' market seems favourable, a trade boom or sellers' market detrimental, to the maintenance of a vital contemporary standard of design. It is not for me to go into the basic premisses of these bewildering phenomena. Probably we are here faced with one more outcome of the uncontrolled development of private enterprise in the nineteenth century. For the effects of a trade slump a number of instances can be given. Germany became the foremost exponent of the Modern Movement in architecture and industrial art at the time when her industries were suffering most acutely from the consequences of the First World War and the Versailles Treaty. Today, after a much more serious defeat, many German manufacturers are again making use of high design standards in their attempts to recapture lost markets. British firms remained reactionary for many years after the First World War. The first noticeable improvement in design came between 1930 and 1934, when this country was involved in an economic crisis. The United States was also regarded as extremely reactionary in taste until her economic crisis of 1929–33 brought about a change. The design of American cars, for example, was far behind European standards up to that time; afterwards it took the lead, when the vogue of streamlining was created in order to induce people to acquire new cars.

The main reason for design changes during a slump appears to be this: The manufacturer sees that his business is decreasing. Machines have to be stopped, shed after shed lies idle. This is a moment for speculating on new methods and new incentives. One of them concerns the design of his products, which is then seriously examined. The manufacturer says to himself: 'Suppose I try something new on a couple of machines. They are idle in any case. Perhaps the retailer and the public will like the new line and buy large quantities. It may even be worth engaging the services of a first-rate designer whose ingenuity perhaps even his name, may be of some help.' And while for such reasons he tries to improve the design standard of his products, some other manufacturers in the same distressed industry will have had similar

industry? How many take pains to ensure as friendly a welcome as a first-class shop? The result of this sad neglect is that most manufacturers think that art has nothing to do with industry, to the very great detriment of both.' (Gordon Russell, 'The Development of Co-operation between Museums and Industry', a paper given before the International Council of Museums Conference, July 1950.)

ideas. Design competition arises, and will in its turn lead to further and intensified efforts.

However, a slump and a buyers' market can also have a very different effect on design standards. In an attempt to attract custom away from his competitors, a manufacturer may force arbitrary changes in what may be already satisfactory designs. His sales manager, advertising agent and the retailer may insist on a succession of designs which are new and different. To meet these demands the manufacturer may have to resort to creating what is known as 'artificial obsolescence'. Some American motor-cars show the effect of this practice, there are definite signs of it in British furniture and radio cabinets, while in light fittings it is rampant. Apart from satisfying the quest for novelty the manufacturer is often forced to reduce his prices. We have already dealt with the effect of underselling on the public's choice of designs. For its causes I must again quote from the excellent report on design in the light metal trades:[1] 'In attempting to escape from the pressure of his competitors each manufacturer attempts to build up his own little monopoly. The frequent introduction of new patterns is one of the principal means whereby he does this. The buyers who represent his market, however, attempt to smash down the monopolies and preserve the price-cutting forces of competition. The customary procedure adopted to achieve this is for them to buy one only of a particular product and hawk it round among all the firms in the trade, offering large orders at a price substantially below that of the original supplier. All too often the result of this is not, as the apologist of intense competition would suggest, lower prices to the consumer and a weeding out of inefficient firms in the trade, but products made shoddier by the substitution of a baser metal—brass for bronze and tin for brass.' When this sort of thing happens design standards cease to matter much.

If a buyer's market is only partly favourable, then a sellers' market need not be wholly detrimental, to the achievement of high design standards. There is no law forbidding improvements of design in times of flourishing trade. If the moral energy of a group of manufacturers is strong enough and there is a demand for satisfactory or high standards, good industrial art can be produced at any time. But, taking the normal human being and especially the normal manufacturer, the stimulus needed for serious consideration of the importance of design is not strong enough to lead anywhere as long as business is flourishing. It is obvious that exchanging new tools for old, new rollers for old, new stampings for old, means stopping production for a certain time, and slackening it for a still longer period, even if there is not also a serious loss of sales after the change. Against all such understandable hesitations

[1] *Design and the Designer in the Light Metal Trades*, H.M.S.O., 1944, p. 6.

stands the one moral argument that it is unpardonable consciously to offer for sale goods which are badly designed. I have already indicated how improvements *can* be made by small-scale experiments, but there is no doubt which course is generally preferred.

While enjoying the condition of a sellers' market manufacturers tend to reduce, rather than increase, the number of different designs in production. The public's choice is consequently limited to those designs which have hitherto had the largest sale. This was for almost ten years after the war the situation in carpets, linoleum, woven textiles, aluminium hollow-ware, domestic appliances and motor-cars. In 1935, faced with a buyers' market, all firms making these articles offered the public a much wider range of designs than they do today.

UTILITY SCHEMES—PURCHASE-TAX—EXPORT MARKETS

As we have seen, the public's choice between good and bad designs can be restricted in both a buyers' and a sellers' market to the point where all good designs are excluded. There are other causes of restricted choice and three of them, utility schemes, purchase-tax, and export markets, may be usefully discussed together.

During the war the shortages of labour and materials in firms serving the domestic market prompted the Government to introduce utility schemes. Although the schemes varied for each type of product, the main object was to provide a sufficient quantity of well-constructed goods at reasonable prices. Shoddy articles were not tolerated, and the competitive forces in industry worked to equal or surpass the standard of quality fixed by the Government. The utility schemes were continued for a long period after the war. In this context it is unnecessary to describe the effect of utility standards on each domestic product. The schemes were altered from time to time and have now been suspended altogether. Their general effect on the choice of goods before the public caused, inevitably, a reduction in the number of different designs.

Purchase-tax has been levied on different domestic products to raise money for the Exchequer and restrict their sale on the home market, so that manufacturers are forced to sell in foreign markets. Its effect has been most noticeable on the design of decorated glass, silver, jewellery and leather goods and, to a lesser extent, on the design of carpets, electro-plate, woven textiles and motor-cars. Manufacturers of these items have had to suit their design standards to the needs of markets abroad, with the result that the public in this country has had to be content with many designs created specifically to satisfy foreign buyers. Because of purchase-tax, manufacturers have had little opportunity to conduct trials of new and perhaps promising modern designs on the home market, before risking them in foreign markets.

From the evidence which I collected it appears to be a fair generalization to say that buyers from the Commonwealth, and in fact most foreign countries, do not demand modern designs. Their taste is reactionary and is usually satisfied if the design standard of these items is 'traditionally British'. There are three exceptions to this. In leather goods it seems that thoroughly satisfactory shapes are augmented by incidental details which are slick, cute or stylish. In motor-cars the superior manufacturing methods and standards of comfort in American cars are being adopted over here by several firms while others, also anxious to capture a market in North and South America, have resorted to using superficial style clichés known collectively as borax. In decorated pottery the 'Californian style' demanded in the U.S.A. has induced British firms to supply a more progressive and generally better design standard than was thought necessary two to three years ago.

As the Government has insisted that firms should attempt to increase their export trade, most manufacturers have preferred, for economic reasons, to concentrate on designs which sell abroad. Anything which remains after the foreign buyers have given their orders is passed to the home market where, except for decorated pottery, purchase-tax is charged on it. The choice before the public is therefore severely restricted. There is no alternative, except in the case of woven textiles where tax-free goods are available. But, as I have pointed out already, design in such fabrics is handicapped by the materials which must be used.

However, there are a few instances where the demand in export markets has induced firms to improve their design standards. In connexion with metal door furniture we saw that competitive trade with Holland before the war had forced manufacturers in this country to adopt higher design standards. Once the new designs were in production, some of them inevitably found their way on to the home market. Carpets designed since the war to suit Scandinavian buyers provide another example. In crystal glass, at least two Stourbridge firms have found satisfactory markets for cut glass of modern design, but only a few examples of their work have yet been placed on the home market.

In recent years the most notable instance of selling designs of a high standard in foreign markets was the enterprise of The Scottish Furniture Manufacturers Ltd. Following the lead of their trade association the directors of seven furniture firms in Scotland formed a new company to take over and increase their business with foreign markets. The company was concerned exclusively with modern design and much of the necessary inspiration came from its Chairman, the late William Johnstone. From the start the company's policy had been to rely on specially commissioned designs from architects and free-lance designers. Their work was first shown when the company made its memorable début at the British Industries Fair in 1950. The large stand consisted of six rooms based on the layout of a one-storey American house of modern design. In addition to the furniture, modern designs in Scottish textiles, carpets, glass and light fittings were shown.[1]

In spite of the fact that the company had to cease operation in 1953 for reasons too specialized for discussion here, the venture was remarkable for it showed that a belief in the necessity of modern design must be backed wholeheartedly by a firm's directors. The principle is similar to my earlier suggestion that firms should devote a small part of their production to experiments with higher standards of design. Whether in the home or export markets this is, from the manufacturer's point of view, the safest method of ensuring that the public gets a fair choice.

ENFORCING DESIGN STANDARDS

So far we have accounted for typical instances where the public is offered unsatisfactory designs without being able to compare them with designs of a better standard. As might be expected there are several cases where the procedure has been reversed. The most notable example was the limited range of utility furniture designs made compulsory for all furniture manufacturers from 1942 until 1948. Although I have already dealt with utility furniture, it is necessary to say again that any public body (in this case the Board of Trade) which compels an industry to produce all its goods in one unvarying design is bound to store up resistance on the part of manufacturers, retailers and the public.[2] The austere designs of utility furniture no doubt made the reaction to 'Freedom' furniture in 1948 more intense, but it was the principle of defying the personal initiative of the manufacturer which was so wrong.

Another method of ensuring that no designs of an unsatisfactory standard are offered to the public is now used by the nationalized gas industry.[3] But in this case manufacturers are free to create their own designs. For the distribution of appliances to the public, manufacturers have to rely on twelve Gas Boards which will only accept designs passed by a small committee at their London headquarters. It is not widely known that this 'beneficent' dictatorship exists, because the general design standard

[1] See *Design*, July 1950, pp. 16–19. [2] See pp. 4–6. [3] See pp. 55–6.

of gas appliances is satisfactory. From the manufacturer's point of view it has many advantages, and when interviewing some of them I heard no complaints. For the public there is still a wide choice between the exemplary and merely satisfactory designs. Nevertheless, this dictatorship, however lenient, is wrong in principle and should not be allowed to continue its overriding control of the whole market. Power of this sort can be abused too easily. Who can tell when a manufacturer's outstanding design might be refused by the committee because it would make unsaleable all other similar appliances in stock?

Another form of dictatorship is illustrated by the purchase-tax relief scheme administered by the Design and Research Centre for Gold, Silver and Jewellery.[1] In practice it has failed to establish a high design standard: in principle it must be condemned.

A far preferable method of establishing a design standard is by indirect control. This task is within the scope of the Government, and the Council of Industrial Design was set up to carry it out. The Council merely puts forward, in exhibitions and publications, those products which it has selected. They may be of a good or very high design standard: they may be thoroughly unsatisfactory, but each one of them can be compared with similar products in retail stores. The public may be guided by the Council's attempt to focus attention on good designs, but it is not forced to accept the Council's standard. In the same way one cannot object to retailers whose stock consists exclusively of modern designs. This cannot be called dictatorship for at the present time the public has a disproportionately wide range of choice in many other stores. The worst designs imaginable are readily available to anyone who wants them in all parts of the country.

GOOD DESIGN AND GOOD BUSINESS

This title is inspired by the often repeated maxim: 'good design is good business.' The assertion that better design leads by itself to increased sales is to be found in discussions and manifestos everywhere since the eighteenth century. It is most prevalent today. But, from my experience of industries serving the domestic market, I am convinced that it is scarcely a half-truth and never susceptible to generalization. The greatest mistake that propaganda can make is to lure manufacturers and retailers into improvements of design by promising them an increase in sales.

Unquestionably there are in nearly every industry manufacturers who devote a part or the whole of their production to articles of exemplary design. Most of them had

[1] See pp. 34–5.

started on a small scale to supply new markets which they have created. They have had setbacks, and progress to begin with has been slow. In all cases described in this book I have tried to emphasize this general condition which attends all persons who try out something new and unfamiliar, for only in exceptional cases did I hear of immediate success. The people who now have the most successful trade in modern design began with a strong belief in their personal judgement. Their motive for introducing modern designs was not altruistic, but based on the intuitive awareness that a sufficiently large section of the public would respond in their favour. The nearer their prices were to those of goods already on the market, the greater was the response. While a few firms have increased their sales in this manner, the majority of cases I have described prove beyond doubt that successive improvements in design *can* maintain sales at their normal level. And that, I suggest, is the balance between private business and social duty that every manufacturer and retailer should strive to attain.

PART IV

SUGGESTIONS FOR IMPROVEMENTS

UNIVERSITY RECOGNITION

We have been concerned up to now with the situation as it is, the relation which existing forces bear to each other and to the standard of British industrial art. I now want to suggest some improvements and deal with the situation as it might be.

Much has already been said about the generally inadequate status of designers. That is only one aspect of a much wider deficiency, the unrecognized importance of industrial art itself. The other arts, literature, music, painting, sculpture, are practised and discussed in private groups and public halls throughout the country. Approaches are made to all levels of taste, from the universities to news theatres. Newspapers, periodicals, books, sound and visual broadcasts, films, all maintain a steady supply of information. In the minds of more than half the population there is general receptive interest that can achieve various levels of discrimination. Industrial art has no such basis of popularity. To most people it is a high-sounding name given to everyday objects.

I have already suggested what newspapers, magazines, books, and the B.B.C. can do to cultivate a wider public interest in industrial art. Art schools, private and public bodies, the Design and Industries Association, and the Council of Industrial Design can do much to help. But the essential need is for leadership at the highest level. The present-day confusion which has arisen on account of rival interests, would be lessened considerably if a body with sufficient prestige could be persuaded to undertake authoritative research, and put forward suggested directions in which thought and experience could be developed profitably. The task recommends itself to the universities at Oxford and Cambridge to set up departments of industrial art.

At the present time neither university has recognized industrial art as a profitable subject for discussion within or without its walls. From time to time single lectures, perhaps a short series of lectures, have been given by visiting speakers. But, one asks, is this sufficient recognition of a subject which can so unremittingly affect, every three years, the lives of more than 15,000 undergraduates, their future and that of their future wives and families? For this alone it merits more attention, but one must also remember that many of these undergraduates will become teachers in colleges and schools, and many of them will become manufacturers or retailers. For all these people it is of paramount importance to provide an education in industrial art, for they will by their example in later life influence the lives of countless thousands.

What then can be done at Oxford and Cambridge? First, two Fellows should be elected to alternate between work on outside research and the giving of lectures. Their object would be to develop the critical faculty of undergraduates in terms of industrial art. The method should consist of demonstrating the difference between good and bad design by means of critical analysis. The background and the broad outline of the present situation could be provided in lectures with lantern slides. The work of critical analysis should be carried out in small study groups of not more than fifteen persons. Personal discrimination can best develop in an atmosphere where response is uninhibited, where the disclosure of one's uncertainties is not detrimental to pride and social prestige. The discussion of artistic values, though it is a personal exercise, can degenerate into irrelevances unless one is face to face with another and equally concerned intelligence. There is no need to enlarge on this. The essential function of the 'seminar' or the 'tutorial' has long been recognized at these and other universities.

Up-to-date information on the many aspects of art in industry is, as I have tried to demonstrate, a prerequisite to intelligent discussion of industrial art itself. The information available is but a fragment of what is needed. The situation in this country has been partially covered by public and private research. The situation abroad has not been touched. The immense and influential developments in the United States alone warrant a thorough survey, for the information obtained would be of the greatest use to manufacturers, designers, retailers and teachers in this country. But who could support, foster and give authority to such research work better than the universities of Oxford and Cambridge? This, I suggest, is the other duty to be performed by the holders of industrial art fellowships.

A MUSEUM OF MODERN INDUSTRIAL ART

The idea of a permanent exhibition of industrial art is not new to this country. In 1852 a Museum of Ornamental Art was established at Marlborough House by the Royal Commissioners for the Great Exhibition. Its object was not to interest the general public but to afford inspiration and instruction for designers and craftsmen. In 1862 it was moved to a new building in South Kensington, now a part of the Victoria and Albert Museum. Recent changes at this Museum have allowed different examples of artistic work (in metal, leather, wood, glass, etc.) from historic periods to be shown together, but there is still no space provided for a permanent up-to-date collection of industrial art.[1]

I have already mentioned the Gorell Report, 1932, and the Weir Report, 1943. They made specific proposals for setting up permanent exhibitions, but in both cases they resulted instead in the establishment of public bodies with terms of reference embracing all aspects of art and industry. The first of these bodies, the Council for Art and Industry, held a variety of temporary exhibitions. In 1937, in an unpublished report, it outlined its proposals for an Industrial Art Centre for changing exhibitions to be set up in association with the Victoria and Albert Museum.[2] Since the war the Council of Industrial Design has carried out similar work as far as temporary exhibitions are concerned. Much valuable experience has been gained, and it seems possible that the present Council will soon be making preparations for a permanent exhibition.* The most recent proposals for an exhibition of this kind were made in the correspondence columns of *The Times*.[3] As they do not appear to have provoked response from the Government or any other body, there seems to be some excuse for outlining my own suggestions.

To begin with, a permanent exhibition should be set up in London *to benefit the public*. There ought not to be any confusion about this. Manufacturers and designers already have sufficient opportunities to exhibit their work, even if most of them have no permanent centre. The exhibition should be housed in a specially designed modern building which could be enlarged when necessary. This Museum of Modern Industrial Art, as it might be called, would be divided into five sections. The main section would be devoted to articles of the highest design standard arranged to represent about six furnished rooms. Changes should only be made when an article of equal standard but greater contemporary vitality is produced. The second section would be smaller and hold temporary exhibitions, each devoted to the best products of one industry. This would enable developments in design over a period of, say, ten years to be summarized. The exhibits in both sections would be confined to articles on the market at a reasonable price. Each would be briefly described with the names of the manufacturer and designer, the materials used, the date of design, the retail price and, if practicable, a list of retailers who stock it. These exhibits would be borrowed from

[1] At the City Art Gallery, Manchester, in 1933, a collection of examples of modern industrial art was started. It included furnishing and dress textiles, pottery, glass, rugs and furniture. Many of the examples were presented by manufacturers on the understanding that they would not be publicly exhibited until three years had elapsed. In fact, no display of this collection was made until the summer of 1939. Since the war the collection has been represented only in isolated showcases in branch galleries.

[2] See pp. 204–5.

[3] Robin Darwin, 28 July, Sir Leigh Ashton, 3 August, and John W. Waterer, 5 August 1951.

manufacturers who would be required to pay for the space occupied. I shall come back to this point later.

The third section would consist of articles purchased by the museum to illustrate the highest design standards of ten, twenty, thirty and forty years ago. Where possible they too should be shown as furnished rooms. The fourth section would illustrate the best current work in British art schools. The fifth section would contain modern industrial art from foreign countries. As most of it would have to be purchased outright, the section would be comparatively small. The museum would be linked with provincial museums and art galleries for the purpose of circulating small collections of industrial art. Otherwise it would have no outside duties such as propaganda, lectures, advice to manufacturers, designers, public departments, etc. But it would contain a public advice bureau on house planning and interior design.

The museum would be comparatively small, with a floor area of approximately 4000 square feet. It would inevitably start in London, but branches could follow in other parts of the country, possibly with the help of art schools, as I have already suggested. A parent body would be needed and here, surely, is a suitable opportunity for the Design and Industries Association, a wholly independent body, to exploit its goodwill amongst manufacturers. That it is well suited to the task cannot be doubted by anyone who remembers its pioneering achievements in organizing exhibitions before the war.

The suggestion that the D.I.A. is suited to the task is supported by the existence in Copenhagen of a similar museum, Den Permanente. It was founded in 1931 by members of the Landsforeningen Dansk Kunsthaandvaerk ('Arts and Crafts Society', the Danish equivalent of the D.I.A.). The venture began on a small scale and is supported by charging exhibiting manufacturers a rent for stands and a commission on sales transacted by the museum. Articles are selected by the Director, Dr Åsge Fischer, and two architects. For Danish manufacturers and craftsmen admission to the exhibition has become a coveted honour. They value the jury system because it adds to the prestige of those selected.[1]

[1] A related example, this time from the United States, should be mentioned here. In 1950, with the financial support and co-operation of the Merchandise Mart of Chicago, the largest commercial building in the world, the Industrial Design Department of the Museum of Modern Art, New York, organized the selection and exhibition of a special Good Design section, consisting of domestic furnishings. It was held in the Merchandise Mart again in 1951, 1952 and 1953. There are a total of about 1000 designs in the exhibition, which is changed twice a year to admit each time about 500 new items. The designs are selected from those sent in by manufacturers or specially requested by the exhibition's staff. Designs from any country are eligible if they can be bought in the United States, are new since the last exhibition and do not imitate the past. The selection committee, which also changes twice a year, is appointed by the Museum of Modern Art and consists of one business man actively interested in art or design and one designer. The permanent chairman, and Director of the exhibition, is Edgar Kaufmann Jr. The ex-

In this country, it seems possible that the museum I have outlined could derive sufficient support by only charging fees for space. The industries in this country are larger, wealthier and consequently more competitive than those in Denmark. Once the prestige of the museum had been established, the manufacturers of selected articles would enjoy valuable, but inexpensive, publicity. This in itself would provide the incentive for other manufacturers to compete. Their goal would not be lower prices and shoddier goods, but higher standards of design. This process, which would be cumulative, depends on two fundamental points, initial success and a highly selective standard. Although the museum would begin work on a very small scale, it would need financial help for at least five years. This is, of course, the chief drawback to any venture of this sort. Several suggestions can be made, but a direct appeal to the public or to an educational trust would probably be the most effective means of obtaining support.

The question of selection is of central importance. It would be unreasonable to suggest that one man, however enlightened, should be responsible for selecting all exhibits. But I do believe that one conscientious and distinguished person should be the leader and chairman of a small committee. His position should be permanent for a period of at least six years. The committee would consist of three members of the public, one manufacturer, one designer and one retailer. They would be replaced annually, and their names drawn more from the telephone directory than the memory of the chairman. While it is important not to give visitors a muddled impression, owing to a committee with hopelessly divergent views, it is equally necessary to avoid expressing the precious taste of a sheltered clique. If the chairman's taste appeared to be dominant, one should not mind. After all, the object of the museum's collection would be to provoke in its visitors a lively interest and a critical but also sympathetic response.

hibition is widely advertised and also supported by retail stores which make use of the 'Good Design' labels which manufacturers can attach to selected designs. As Rene d'Harnoncourt, Director of the Museum and Wallace O. Ollman, General Manager of the Mart pointed out in a joint statement in January 1950: 'It is the first time an art museum and wholesale merchandising centre have co-operated to present the best examples of modern design in home furnishings.'

A similar case in Canada might also be reported. In 1947 the National Gallery of Canada set up a design information office in Ottawa, and one of its functions was to provide a selective photographic record of all items of original Canadian design (The Canadian Design Index). In 1948 a group of manufacturers, retailers, designers and educationists was appointed to form the National Industrial Design Committee. Its promotion of Canadian design through exhibitions, competitions and booklets prepared the way for setting up a permanent Design Centre in 1953. It occupies 2000 square feet of floor space for exhibitions, contains a library and reading room and is open to the public. Amongst other exhibitions it holds annually a display of products which have received the Design Merit Awards to Industry, similar to the Good Design labels of the Museum of Modern Art. The Awards are made by the N.I.D.C., by means of a special jury which reviews each year all new products selected for the Design Index. The Centre and all its activities are paid for by the Canadian Government through the National Gallery of Canada, although some financial help is received direct from industrial firms.

INDUSTRIAL ART TRAINING

A NEW SCHOOL FOR DESIGNERS

Industrial art education should mean that people can be trained to discriminate between good and bad design. Its object should be to offer a method of analysis that enables people to formulate a personal standard. Put even more simply, it should cultivate a critical awareness of industrial art. At the present time there is no school or class with this specific object. Instead, one finds 'schools of thought', some for one acknowledged standard, some for another. Children, adults, designers, manufacturers are expected to agree or to disagree, never to question, to analyse, to revalue. Opinions are presented ready-made for acceptance or rejection, though in all groups where they are put forward the possibility of rejection is not seriously considered. In other words, people are told what is good design, and not why it is so. The same is true of most teaching of English literature, music and (where it exists) fine art. The presence of well-known names in the syllabus is never questioned. Their works are presented and one learns what to think about them, in much the same way as one learns what to think about hair cream and toothpaste while waiting for a train.

As I said in the Introduction the specific aim of this book is limited by its attempt to describe and evaluate the social, economic, material, and, in general, extra-artistic aspects of the present state of industrial art in Britain. An understanding of these conditioning factors is necessary before the result, the industrial art itself, can be criticized relevantly. It is not, therefore, my purpose to discuss contemporary methods of teaching, or even to mention those which seem to me most likely to awaken critical interest. The question is of pressing importance, but in order to deal adequately with it specialized research work would have to be undertaken.

I am concerned here with some improvements which I think are necessary in provincial art schools and regional colleges for the training of industrial designers. The Royal College of Art is a separate question to be dealt with afterwards. In conclusion a new school for training designers is proposed. Few students entering an art school are capable of knowing to which profession (architect, artist, craftsman, designer, teacher) they are potentially suited. It should therefore be the policy of the school to open all its departments to first-year students and allow them to study and experiment in each one for short periods. This is preferable to the usual practice of confining first-year (and in some cases even second-year) students to drawing and painting. Industrial design departments, however elementary, should be started in all art schools. In the prospectus and in the school itself, design should be equal in importance to painting and sculpture. Only in exceptional cases will the department of design be similar in size to the other departments in the school. While industry cannot absorb many new designers each year, it does require students to be thoroughly trained. And thoroughness is one of the characteristic advantages to be enjoyed in a small class.

After having experienced all aspects of the school's work, those students who intend to take up design should choose one particular branch of study and give it constant attention. In most schools it is possible to choose between textile printing, weaving, silversmithing and woodwork. Light metal work and plastic moulding can be studied in a small number of schools; machine and product engineering practically nowhere. Serious design work for glass can be found only at Stourbridge, pottery at Burslem, leather at Walsall and the Cordwainers' Technical College. It would be unreasonable to suggest that all schools should provide equal facilities in all branches of design. Where a promising student cannot, in his own school, follow a particular course of study, arrangements could be made for him to transfer to another school. This is a matter for the Ministry of Education, which should not wait until the student has completed three years at one art school before he can be transferred to another. As most students cannot afford to spend more than three years on a course, they should be encouraged to specialize in a specific branch of design at the end of the first year. In the second year there should be practical application of design to craftwork, gradually moving on to the use of elementary machines. Where sufficient mastery can be achieved in one year, the specialized work should be gradually related to wider cultural interests—architecture, interior decoration, music, literature and fine art. Art schools should provide students with abundant opportunities to gain a critical insight into the work of other creative artists. For most students this will be their only chance for free thought and discussion. Full-time work in a factory could never offer similar scope. Towards the end of the third year, students from all senior departments in the school should work jointly on real or fictitious projects. The help of retailers and the civic authorities should be enlisted here. In all but the smallest provincial schools, provision should be made, and support found, for one or two of the most promising students to stay on for fourth-year post-graduate work. It is at this stage that sufficient time could be found for carrying out commissions for manufacturers. Working with their instructors, students should undertake the necessary research on production and costing methods and construct mock-ups, and possibly prototypes, of a new design. Standard fees should be charged and divided between students and the school funds. The repetitive work of manufacture, even if it could be undertaken on the school's machines, would not greatly benefit the students' training and would certainly occupy much valuable time.

It seems desirable that wherever possible art schools should ally themselves closely with the predominant local industry. If the manufacturers in most towns are distrustful and unwilling to take an interest in design training, the fault is chiefly theirs. Although they object that art schools are 'arty-crafty', they make practically no effort to offer constructive criticism. In connexion with the same schools, the teaching profession has made definite demands and is consequently well-served. Moreover, students know exactly what salaries they will receive if they decide to become teachers. But manufacturers, instead of giving art schools a clear indication of the type of training required, prefer to wait and see what turns up. No design student can be sure of the amount of his starting salary in industry and thus, in the case of several industries, a valuable incentive is ignored. Again, the number of manufacturers who are prepared to take students into their factories for short periods is inadequate for most schools. There seems to be no doubt that the advisory committees of manufacturers set up with the specific task of keeping schools in touch with industry should be much more active.

There is an obvious advantage in employing practising designers to teach in art schools. In industrial areas it should not be difficult to find suitable men, particularly amongst free-lance designers and the assistant designers to a consultant. To gain the services of the best men, art schools should provide them with studios. In return the students would be able to watch, and perhaps help, the development of new designs. To supplement this, staff designers could teach at evening classes. In order to co-ordinate the work and administer the department a full-time teacher is needed, preferably with some practical industrial design experience.

The high cost of up-to-date machinery and the lack of sufficient space in which to house it together constitute an almost insuperable problem for most schools. Free loans of equipment from manufacturers might provide a temporary solution, but the only satisfactory alternative is for students to carry out their practical work at the technical colleges. And yet this, it seems, can seldom be achieved because the art school suspects the technical college of being 'tradey' while the technical college thinks the art school is 'arty'. This ridiculous situation is, in most towns, aggravated because the principal of the art school is a painter and fundamentally unsympathetic to the needs of industry. While one would much prefer to see architects or designers in charge of art schools, the problem is now urgent and a temporary solution should be found. Interference from the Ministry of Education might not be welcome, but it would awaken the interest of local education authorities.

The question of classes in painting and sculpture need not detain us here. In nearly all schools they have a strong following of students, who spend three years being artists and then go off to teach other students. Without a close link with the civic authorities or other public bodies willing to provide commissions for murals, bas-reliefs, etc., students inevitably remain out of touch with reality. This can well be the fate of design and craft students as long as art schools rest in retirement from public life. It is too little realized that art schools have a duty to bring their students into contact with the communities which support them. There never was a greater need in our towns and cities for a cultural centre where artists, designers and the public could meet. Given a fresh and energetic spirit to inform its work each art school could provide such a centre, ready-made. One might suggest a dozen useful activities. I have already outlined how local design centres for the public could be set up with the help of the art galleries, and how students could enliven the display in retail stores. Another activity could be concerned with improving the visual aspect of the town. But the ultimate object in every case would be to eradicate the lingering prejudice, first in the minds of students who believe the public to be reactionary and unsympathetic towards art, and secondly in the mind of the public who thinks art students arrogant and otherworldly. These are, I know, but two facets of a widespread and deep-rooted misconception which will take many years to remove. The problem is already being tackled nationally. The Council for the Encouragement of Music and the Arts, the British Council and, in a more conditional way, the B.B.C., have made tangible progress. Nevertheless, on a local scale and backed by local loyalties, improvements can be more definite, more striking, and therefore more influential. It is clearly the duty of each art school to make itself responsible for bringing the vitality of life into the work of its students and art into the lives of the community.

The position of the Royal College of Art is different. As the result of recent changes it now houses the most promising schools of industrial design in the country. This is chiefly because it ensures that each student is able to make a creative contribution to design in one particular industry. There is, one hopes, no place for the dilettante in the industrial design schools at the Royal College. One of its chief tasks is to restrict the number of students to those who have reached a high standard in provincial schools. This can only be achieved when the Royal College has established its position as the foremost school—perhaps university—of design, to which, in their methods of training, regional colleges will direct their most promising students.

It is because of the success of the Royal College and because of its interest only in certain fields of design, that I want to suggest a way of bringing a greater number of designers, particularly for powered appliances and product engineering in metals and plastics into industry. My suggestion means that a new school of design should be started. Bearing in mind the disposition of the metal and plastic industries, with which the school would be most concerned, and the need for it to be closely associated with them, it seems that it should be set up either in London or Birmingham. The former has the advantage that a wide range of cultural experiences are within easy reach, but the latter gains from its central position in the country. The new school should be independent of the established education authorities and not directly associated with an existing art school or technical college. For its support it would depend wholly upon those industries it is designed to serve.

This is not so revolutionary as it may sound. Several industries have already set up, especially since the war, research and scientific development organizations in order to benefit each subscribing firm. No industry has yet set up a design school, although the arguments for doing so, especially in this case, appear to be convincing. Manufacturers are the first to admit that they need young and well-trained designers as they are also the first to protest that the art schools do not provide them. As we have seen, the majority of art schools are governed by the idea of fulfilling other needs, so the irresistible conclusion is that those manufacturers who need designers and recognize that a special training is necessary in order to provide them, must set up their own school. By doing so they would ensure, for their own good, that the technical training and mechanical equipment be up-to-date.

That prospect may be sufficiently attractive in itself, but the question of staffing the school does not yield a ready answer. It is in the nature of things that any body supported collectively can be directed by the best brains. One reason why most art schools are directed by painters is that the post and its duties would not be attractive enough to an industrial designer or an architect already in practice, or its remuneration sufficient. It is a fact that only a handful of art schools in the country offer members of the staff facilities and freedom to practise as designers. This new school could be radically different and in being so it would invite the services of the best practitioners we have.

The school would begin on a small scale with perhaps twenty students. These would be young, trained engineers selected from those put forward by manufacturers from their own firms. With a basic technical training, supplemented by workshop experience already acquired, the school's chief task with its students would be to bring out and redirect their inherent artistic ability. In some cases the students' sensibility would not make the training worthwhile, but in others the problem of freely reassessing conventional practices would have rewarding results. As we have already seen in the case of electric fires, cookers, light fittings, etc., the engineer, when unhindered by pseudo-artistic considerations, often produces exemplary designs. The students would need to work in the school for at least six months. They could bring with them one or more projects which their firms were developing, for in this way their work would be both realistic and commercially useful. The interests of rival companies would have to be safeguarded and this could best be done by avoiding, in each session, collective work on projects belonging to competitive firms.

The school would not only train engineers to become industrial designers, but produce solutions to specific design problems of immediate use to the supporting firms. While it is probable that a firm could find solutions to its own problems, it is certain that the combined experience of staff and students working collectively would produce a better design in terms of technical economy and artistic standard in a shorter time.

This school could widen the outlook of engineers and help to relate their work to current practices in other industries. For manufacturers it would operate in a way associated with consultant designers but on a much larger scale and with greater prestige. For the students it would offer a scheme of training which is unobtainable today and yet urgently needed.

THE RETAILER

The manufacturer's information on public taste is supplied by the retailer. Amongst all the cases I investigated there were no significant exceptions to this. We have already seen that the retailer's position as distributor is necessary if the present system of quantity production is to be continued, and if the public is to be offered manufactured articles at reasonable prices. His responsibility as intermediary cannot be overstressed, for it rests with him to place before the public a fair selection of articles. And this means that the best modern designs available in each price group must be included. Moreover, these designs must be presented with as much emphasis as the articles of less satisfactory design. In this way the public can be helped to form a coherent taste. A simple analysis of recorded sales will become a truthful indication of what the public wants and provide a guide to the manufacturer. Every retailer who makes a living by trading in industrial art must therefore be held responsible for the selection in his store.

THE MANUFACTURER

It may be thought that the manufacturer's position as the original and foremost arbiter of taste is untenable. But this is not so, even under the most competitive conditions of a buyers' market. No matter how much the manufacturer protests that he follows the wishes of his public, it must be remembered that his public is unable to express its taste until he has put his product up for sale.

The manufacturer forms his own market. Though he may often try, he cannot escape the responsibility which this entails or ignore his power over the lives of others, varying as it does with the size of his output. The more a factory mechanizes its means of production and reduces cost, the more important becomes the need for intensive thinking and planning of new designs. If a design is to begin a long and inflexible run of production, then it should be the best design which the manufacturer can obtain. We have recognized the social consequences of mass-production, particularly in carpets, wallpapers, pressed glass, and plastic articles. We have also seen how successive designs of a bad standard cannot fail to sell if the public has no choice. Where there is competition between manufacturers of similar products one finds that the merits of the designs are seldom considered. Sales are increased by improving the efficiency of articles, but also by cutting prices and by advertising. Here again it is the producer with the largest output who captures the largest share of the market. The majority of manufacturers are astonishingly short-sighted. Instead of improving the one thing which really matters, the design of their products, they will go to endless lengths to force the public to accept what they have decided to produce.

Once more it must be repeated that the public cannot originate design standards or be held responsible for them, if it has no choice. Sales results do not indicate the taste of the public where there is no choice. No one knows, least of all the manufacturer, if the public has satisfied its deepest need by making its purchases in a store where there is no choice. It is useless to tell customers in hundreds of provincial towns that Londoners in the West End are allowed to choose between good and bad designs of comparable price. Retailers have become indignant when it is pointed out that their stock is not representative of current design standards. And they are partly justified, for there is not at the present time a sufficient quantity of good designs on the market. Here then is the duty of the manufacturer, particularly the mass-producer. He must introduce, gradually but progressively, a better standard of design, first in one range of products and then in others. In order to achieve this, one can only hope that the day will come when commercial competition will be governed by design merit, not by advertising and price-cutting. And when this sort of competition grows, design standards will be forced to rise.

The public must be encouraged to discriminate, first between good and bad designs, then between good and better designs. The wider the choice, the greater becomes the responsibility of the public. Relevant criticism may only come from a small number of people, but it will represent the wish of the majority who begin to realize that their worthwhile intuitive desires can, after all, be satisfied. Anyone who doubts this should remember the universally acknowledged aim of education.

THE DESIGNER

By insisting that manufacturers should offer the public a choice, which implies a larger number of good designs, one assumes that there is a stock of design talent not being used. This is, in fact, true. My interviews with staff designers in factories, free-lance designers, consultants and architects outside industry, and the best students at art schools, leave me with no doubt that a considerable addition could be made to the number of products of a generally satisfactory design standard now on the market. Amongst all types of designer, but particularly amongst those already employed in factories, there are brilliant men and women whose talents have been ignored or unrealized, and often misused. There is nothing sentimental about these Cinderellas of industry. It is a plain fact, however regrettable, that far too many manufacturers employ designers, not for creative purposes, but to fulfil their own preconceived ideas.

Sales figures, even if they are the result of a fair selection of designs, are useful only in so far as they sum up the present public demand. If viewed in continuity over a period of years they may indicate, but they will not state, what future demands will be. Though useless to the manufacturer, this information is vital to the designer. His imagination takes over, and if allowed to assert itself a creative design is bound to result. In theory that is the situation: in practice it is often very different.

My last remarks are therefore concerned with the designer—the creative force in industry—without whom there can be no significant developments in industrial art. For each new design the designer must return to study with sympathetic insight the function of the article and the materials and techniques which are to be used. Each new design is a new problem to be assessed from the beginning. It is not the application of an idea to something which is supposedly incapable of change, but a creative reinterpretation of the media of industrial art. Variations on the appearance of existing designs, whether the designer's own or someone else's can have no artistic significance. The so-called 'tradition' in many of the older industries can often be used in an exercise of this sort, while in the new industries the practice can degenerate until there is nothing but novelties and freaks. The valid traditions are rare, as I have already stressed, because they depend for their life upon functions, materials and techniques which are static, or have changed only slightly; if one of them is replaced by a substitute, the tradition ceases to exist.

The elements which condition the artistic significance of a design are creative use of the media and contemporary relevance, or vitality. It is for the designer to contribute these elements, for only he can provide them. To produce works of industrial art of the highest standard great concentration of thought and feeling (which implies sensuous contact with the material), technical knowledge, and a critical awareness of the best which is being created in other industries, are all needed. It is no small accomplishment to design well for one industry. One should remember this, and not regret that it is fast becoming impossible to design successfully for more than one. All modern artists produce their best work in the one medium where their creative energies coalesce to serve, to enliven and to enhance that one medium. It is by the work of men like these in British factories that the highest standards of industrial art will be set.

THE AIM

Industrial art forms that part of our environment which is of most immediate importance. It is also that part which is generally shoddy, vulgar, meretricious and otherwise devoid of artistic significance. I am not exaggerating when I say that great improvements can be made in the designs of between 80 and 90 per cent of British domestic articles. We simply cannot afford to neglect and misuse an element in our society capable of enriching and enhancing the lives of each one of us.

The manufacturers and retailers whose industry and administration makes possible the intercourse between the designer and the public should not interrupt the natural flow of thought and feeling. They must appreciate that industrial art is the joint responsibility of creative designer and responsive public; that it is the product of a two-way process in which one makes the effort to create and the other makes the effort to achieve. Only in this way can there be an industrial art in which our pleasure increases with a value that remains.

ADDENDA

FURNITURE (P. 16)

Since 1951 there has been a significant change in the trade's attitude to unit furniture. The advantages of standardized dimensions for economy in production were obvious, but what had hitherto been unrealized was the 'sales appeal' of allowing the purchaser to accumulate individual units one by one until his house was complete. While some of the most advertised unit furniture today is a vulgar travesty of the pre-war Plan range, there are several firms with sensitive modern designs meticulously thought out to afford the purchaser a wide range of uses.

THE GOOD FURNISHING GROUP (P. 17)

In 1953 eight retailers were members of the Good Furnishing Group: H.G.Dunn & Sons Ltd., David Elder Ltd., Findlater Smith Ltd., P.E.Gane Ltd., Hemmings Bros. (Nottingham) Ltd., Holmes The Furnishers Ltd., Lee, Longland & Co. Ltd. and York Tenn. Under the chairmanship of Geoffrey Dunn the members collaborate in choosing or commissioning designs for furniture, textiles, carpets, etc. Manufacturers reserve the designs exclusively to the Group because sufficient quantities are ordered. First-rate modern designs are brought forward in this way for sale at a reasonable price.

CARPETS (P. 22)

I made my investigation in 1950. Since then I understand that a general increase in production has enabled manufacturers to satisfy, in terms of quantity, the needs of the home market. Any changes in the public's taste since 1939 are, however, unknown, for the industry in general still (1954) refuses to offer a selection of distinctive modern designs.

WALLPAPERS (P. 91)

At the close of 1953 The Wall Paper Manufacturers Ltd., took a wise risk and introduced a new range of roller-printed patterns from 4*s*. to 8*s*. a piece. Several of these patterns achieve a very good modern design standard and serve as pointers to future developments. This experiment with imaginative designs for the cheaper papers has had a success which leaves no doubt that more and better patterns should soon be on the market to make the choice between inexpensive modern designs full and varied. See *Design*, January 1954, pp. 7–10; June, pp. 28–9.

ART SCHOOLS (P. 168)

Students in the School of Industrial Design, L.C.C. Central School of Arts and Crafts, who gain the National Diploma in Design from 1954 onwards, qualify also as graduates of the Institution of Engineering Designers. This further certificate of ability as designers should do much to enhance the prestige of young designers entering industry.

THE ROYAL COLLEGE OF ART (P. 182)

Out of 50 students from the industrial design schools of the College in 1952, 35 obtained posts in industry; the remainder either became free-lance designers or teachers. In 1953, out of 59 students, 44 went to industry. These figures take into account the period of nine months which industrial design students, on leaving the College, are obliged to spend working in a firm before they can obtain the Des.R.C.A. diploma.

THE ROYAL SOCIETY OF ARTS (P. 191)

In its bi-centenary year (1954) the Royal Society of Arts decided to inaugurate an award to manufacturers, writers and others who have rendered 'eminent service to the cause of good design in some capacity other than that of designer'. Known as the 'Bicentenary Medal', it complements the distinction 'Royal Designer for Industry', which the Society has awarded to designers since 1936. The first person to receive the Medal is Sir Colin Anderson, a director of the Orient Line, whose insistence on a high standard of design in the interiors of his company's passenger liners has been remarkable for twenty years.

THE COMMITTEE ON ART AND INDUSTRY (P. 202); THE COUNCIL OF INDUSTRIAL DESIGN (P. 221); A MUSEUM OF MODERN INDUSTRIAL ART (P. 295)

It is hoped that a permanent Design Centre in London will be set up by the Council of Industrial Design in 1955.

Since 1952, *Design Review*, the Council's illustrated index of current good designs of British manufacture, has been steadily enlarged. It has formed the basis of exhibitions organized by the Council in this country: from 1953 onwards, with support from the Board of Trade, items from the index have been shown in an important series of exhibitions in foreign markets. For 1953 and 1954 these selective displays have culminated in the British stand at the annual Canadian National Exhibition, Toronto. Also in 1954, and for the first time, a selective exhibition of consumer goods was staged by the Council at the British Industries Fair. All these events pointed to the need for a permanent centre in London for changing exhibitions of good designs, and the Council, following the original suggestion of the Weir Report, began in 1953 to plan the Design Centre. Its recommendations were made to the Government which, in 1954, approved the setting up of a Design Centre for British Industry. It is hoped that this Centre will provide a home for small selective exhibitions of well-designed goods of high quality.

BOOKS (P. 245)

The first British book to be written on product designing from the practitioner's viewpoint is *Form in Engineering Design*, by J. Beresford-Evans, Oxford University Press, 1954. The sub-title of Mr Beresford-Evans' book is 'The Study of Appearance During Design and Development'. It is a brief but stimulating book, written expressly for the engineering draughtsman: the man already in the factory working as a designer, but not recognized as such. Mr Beresford-Evans realizes that these draughtsmen are responsible for the shapes of most things we see around us, although practically none of them has had an artistic training. Accordingly, his book sets out some of the essential principles, not rules, and backs them up with diagrams. Artistic standards are in nearly every case developed from logical, if not generally agreed, functional solutions to common problems in product engineering.

THE RETAIL TRADE (P. 159); THE MIDDLEMAN (P. 271)

An unusual method of improving the design standard of goods is employed by La Rinascente, the leading Italian department store with headquarters in Milan. Articles in current stock are considered by a development office and, if it is thought that one of them might be improved, the store's design office proposes modifications, or prepares drawings for completely new designs. If the manufacturer co-operates he would be assured of a sufficiently large order while reserving the design to La Rinascente. The system has been operating since January 1954 and the first results, with greatly improved appearance, include packages, tablecloths and a pencil case.

POSTSCRIPT

BY NIKOLAUS PEVSNER

Who would not agree with Mr Farr's conclusions that 80 to 90 per cent of British products are 'shoddy, vulgar, meretricious, and otherwise devoid of artistic significance', that good designers exist in Britain in sufficient numbers to alter this state of affairs, that the public would be ready to accept higher standards of design, if it were given a fair choice, that it is the responsibility of the distributors to give the public that choice by insisting on a selection of merchandise in each department including some of good modern design, and that it is the responsibility of the manufacturer to make good modern design available at all price levels?

But Mr Farr leaves us in the dark on some other problems of equal importance, and the first edition of this book[1] (which I wrote in 1935) was no more illuminating on them. It may therefore be worth trying to formulate them and attempt solutions. These problems are the following. We speak of the manufacturer's and the distributor's responsibility. Mr Farr is very free with his 'the manufacturer must' and 'the retailer must'. He (and I) would say that it is the manufacturer's and the retailer's duty to help in providing better designed, that is, aesthetically better goods. Why should it be?

This is primarily not a matter of aesthetics but of social ethics; private gain is not the social justification of making and selling goods. The manufacturer's and retailer's duty is to satisfy the material needs of a neighbourhood, a town, a country, a society. If bread is needed, the baker provides it. However, some like their bread white and some like it brown. The baker must satisfy both tastes. But bread may be—American bread for instance—dead white and taste of paper. Is the baker therefore justified in refusing to bake it? I would say No. But if it is at the same time bad for people's teeth and digestion, then an honest baker who thinks in terms of a responsibility to the community ought to keep away from it. If one finds he does not, then he must be either an irresponsible man or he must have reasons to deny the bad effect on the health of consumers. Substitute the publican for the baker and you have a much more often debated case, not the case whether there is good or bad wine—only idiots say in this case that it is all a matter of taste and that one man's taste is as good as another's—but whether wine is good or bad for people. Some say Yes, some say No,

[1] On the relation of Mr Farr's book to mine, see the Foreword, p. xxvii.

and the British with their genius for compromise have invented licensing hours to say Yes and No at the same time. Does the same argument apply to traffic in hashish? There are clearly boundaries of salubrity and morality which manufacturer and trader must respect. He may be expected to respect them voluntarily or he may be forced by law to respect them. Can all this be applied to design? Mr Farr's and my answer is Yes, but difficulties surround any effort to defend such an answer. Four questions chiefly must be tackled, and I am only too well aware that there is not a straightforward answer to any of them. They are:

What is good design?

Why should one plead for good design?

In addition to these we ask two more questions:

What is modern design?

Why should one plead for modern design?

And finally, in combining the first and the second set of questions, a definition of good modern design must be attempted and a campaign for its promotion defended.

On the first of my questions Mr Farr gives an answer which does not satisfy me. That a teapot should 'represent a creative revaluation of the possibilities of the materials and techniques at the designer's disposal' takes us nowhere. But Mr Farr has more specific arguments. The first is the functional one: does the teapot work? On the validity of that there will be no two voices, even if a few, for the sake of aesthetic autonomy, may uphold their stubborn faith in, say, a brilliantly proportioned but uncomfortable chair designed by Charles Rennie Mackintosh. The functional answer was the answer of the Design and Industries Association in its early years, and it is still the best answer in dealing with beginners.

However, Mr Farr uses further criteria: Does decoration enhance form? he asks. This as a principle is only partially applicable. It would not help much in the case of a scarf, for instance. Two more of Mr Farr's criteria are, I think, open to criticism. Could the same effect, he says, be achieved in another way? And: does it 'make all former teapots slightly less relevant to contemporary artistic needs'? Both these questions are ambiguous. What is 'the same effect'? No two teapots can ever have the same effect, strictly speaking, whether well, badly, or indifferently designed. And 'relevant to contemporary artistic needs' invalidates the other question. It cannot be answered without a definition of what contemporary aesthetic needs are.

Much closer to the truth, even if an unprovable truth, are Mr Farr's remaining criteria: 'Is it something of which we have been intuitively aware, but yet never managed to express?' That, in my opinion, is a sensitive way of trying to analyse what causes our pleasure in, and approval of, certain harmonies and forms. And finally

Mr Farr introduces the criterion of vitality—and of that I shall have to say more later.

When I wrote my book in 1935, I tried to make my job a good deal easier by steering clear of aesthetic arguments. What I said was this—and it may be useful for the defeatist or indeed the man of plain commonsense, to hear it repeated: 'Anyone who has ever been present at the meetings of the jury for a big art exhibition of really live importance will have noticed that, if out of 5000 objects sent in only 1000 can be shown, there will be 500 which, from the first examination, are accepted by all experts, at least so far as they belong to the war and post-war generations. There will be 3000 at least, say 3500, which hardly one of the jurors finds himself able to take seriously. The remaining 1000 will be the border cases, the cases where arguments arise and convictions clash. In a discussion on 3000 objects of industrial art (e.g. in the British Industries Fair) the procedure would certainly be the same. There would be, from the first, unanimity as to two-thirds at least being out of the question, unanimity as to one-tenth being definitely good or acceptable—provided again the jury consisted of experts, that is, men who are keenly interested in living art and industrial design, have seriously attempted for some time to evolve a power of discrimination, and are prepared to judge disinterestedly, objectively and conscientiously.'

I have quoted these lines for what they are worth. I still think that they form a sufficiently safe basis for a book such as this. The layman may have a suspicion that I am exaggerating the relative unanimity which would exist on quality in design. But he must not forget one thing: among the thousands of people who walk down Oxford Street, or Princes Street, Edinburgh, every hour of the day, there are only a few who would say that they like the architecture of the buildings put up along these streets; all the others simply do not see them. The same is true of design. Most people never really take in a teapot until they have to buy one, and then, of course, they are easily influenced by a salesman, or by the most glaring effects, or by associations.

However, it is perhaps cowardice to leave it at that and not to make any effort at a definition of good design. Some *matériaux à servir* can at least be offered. One is this. It is extremely improbable that good design will happen by chance. For good design to be achieved, it must have been taken seriously by someone. This, it will at once be seen, rules out much that we deplore in trade shows. That the someone must have talent, that is, aesthetic sensitivity combined with a certain knack of the hand, goes without saying, and talent remains an undefinable quality. But his mind and hand must also be trained, that is, he must have experience of the problems of design.

If talent is the undefinable quality in the designer, the corresponding quality in the design is beauty—a word out of fashion, but one which cannot be dispensed with.

Beauty, I would describe as the controlled consonance and dissonance of forms and colours, controlled because control brings back at once the human element. Discordance must not be left out. That was the discovery of the English authors on the Picturesque in the late eighteenth century, and especially of Sir Uvedale Price. He wrote: 'I would rather wish that some improprieties should be risked for the sake of effect...than that improvements should be confined to the present timid monotony' (*Essays on the Picturesque*, 1810, vol. II, p. 157), and 'Although smoothness be the groundwork of beauty, yet... roughness is its fringe and ornament, and that which preserves it from insipidity' (*ibid.* vol. I, p. 115). But impropriety must be controlled. The impropriety of Oxford Street cannot be compared with, say, jazz in music. People may like or dislike jazz, but it cannot be denied that it is music. If the same instruments were left to do their loudest, or if you like, their subtlest, the result would only by the remotest chance of luck be music. Similarly no aesthetic value can be expected in design without one man being in control. The design must be the consistent expression of one mind. That does not eliminate team-work; but it keeps it in place.

So much or so little on our first question. Now to the second. Is not all this much too much trouble taken over a teapot or a knife and fork? Is good design all that precious? I used the comparison with the baker and various kinds of bread. They are considered one as good as the other, although in my description of the American bread I betrayed my bias. It is a bread without character, and it is bad for teeth and digestion. The first part of this statement is a matter of taste, the second—if it is correct—of health. Only the second can call for action. Hashish is entirely a matter of health, and so is accepted as calling for action.

Does bad design at any stage call for action? Is there anything in bad design that can be called bad for people? The answer must, I think, be Yes. It is physiologically known that red is an aggressive colour and that green is soothing. Red means a sudden Halt, green means an even Go. Similarly rising parallel diagonals are aggressive, a circle is static and rests in itself. One step further, and it can hardly be denied that patterns can be restless or restful, simple or intricate, or merely confusing, and so on. In fine art, that is, art not appearing in objects for practical use, restlessness, a jog or a jar, may be just what is intended. You go to a painting ready to receive whatever its impact, then retire from it, let the waves of its effects subside and then return to something else, until on a later occasion you are prepared to undergo the same process. But a carpet, a plate, a sideboard, must be good to live with. Their effect is permanent and so it must not be violent or confusing. Now that is exactly what bad design is. It is true that few people are aware of the effects of their surroundings and that to the trained sensibility they are stronger than to that which is affected only subconsciously. It is also true that sensibility may be overdeveloped and too easily affected. But that has nothing to do with our problem. The fact remains that the riotous wallpaper will have some unsettling effect on everybody, and that the riotous wallpaper—as distinct, for instance, from the riotous wireless—can never be turned off.

That is the negative side of things. The positive is as important. Again the case of the ear is more widely accepted than the case of the eye. A man whose enjoyment develops from Edward German to Verdi and so to Mozart and Bach does not deprive himself of anything. He simply widens and deepens the range of his emotions. The same is true of painting and sculpture, and also of architecture and design. The ashtray can, needless to say, not do to you what Piero della Francesca does, but the smoothness of its form, the coolness of its colour can give a sensation at least of the same order as the frescoes of Arezzo. That is, provided the cigarette does not fall in or out, in which case the ashtray is lost for aesthetic consideration.

So the answer to the question why bad design ought to be fought is that it blunts one's sense, to the question why good design ought to be promoted that it widens one's range of valuable pleasures. And as the concert and the visit to the gallery are only occasional events, but curtains and crockery are with us all the time, their effect on our lives, if less deep at any one moment, is equally deep by their sustained presence.

One *caveat* is necessary, however. The design of a textile or an ashtray can never be too perfect, the design of a room or a whole house can. It is the immortal little story of the Poor Rich Man told as early as 1900 by Adolf Loos, one of the pioneers of modern architecture. The Poor Rich Man had his whole house designed by an architect, and after that it was impossible to shift or add anything. The catastrophe in the story is some presents given to the poor man by his nearest and dearest. The architect cannot approve of them in his scheme, and in the exasperation at the client's aesthetic obtuseness exclaims: 'How dare you have presents given you? Have I not designed everything? Have I not considered everything? You need no more, you are complete.' The conflicts here are between beauty and life, or between planning and accident, or between aesthetic values and values of affection.

Now, when the first edition of this book came out in 1937, there was in the best design in Britain something of that dictatorial quality. Ornament was taboo—and, curiously enough, Adolf Loos had been one of the first to write that 'the lower the cultural status of a people, the more lavish will it be with its ornaments'. Modern was rectangular, smooth, even, puritanical. It was easy at that time to define what we meant by modern design.

So we have reached the third of our four questions: What is modern design? Mr Farr's definition is not mine. He calls modern only what is 'in a contemporary spirit and of positive value'. What I query is the second half of this definition. I prefer to say that all contemporary design is modern design—the word contemporary used not in the sense of every design made in 1953, simply because it is made in 1953, but every design of the last sixty years which is not in imitation of a style of the past. It was about sixty years ago that the great pioneers—Sullivan, the English designers and book illustrators of the Arts and Crafts, the creators of Art Nouveau on the Continent, men like Gaudí and Horta, abandoned period imitation. Then came a first brief phase of riot, and then, already by Sullivan and after him by Loos and Perret and a number of others, the canons of style were laid down which are valid to this day, and imitation of the past as it had been the rule in the nineteenth century became an atavism. As such it is still rampant everywhere. So much to define modern design.

Or so much at least to define what we would have called modern design in the 1930's. Now we are less puritanical, less exclusive, and therefore can be historically more accurate. The style of the straight line and the annihilated ornament is not the whole modern style of design. The twentieth century is not a century without ornament, although we shall see soon that not all is well with contemporary ornament. The revolt against period imitation started with the creation of a completely new set of ornamental conventions—what we call the Arts and Crafts and Art Nouveau. The maturity of the new style in architecture in the 1920's, was again accompanied by the creation of new ornament, only its creators were not the architects. This new ornament of the 1920's which still survives here and there appeared in three forms—as the German *Kunstgewerbe* style with dainty, spiky, playful forms, as the Parisian style of the 1925 exhibition with opulent scrolls and fluted surfaces, and as Jazz. The German style was complete at a Munich exhibition in 1922 and soon after became universally popular in its Swedish version.[1] The Paris style is familar to all of us. Jazz is derived from both, and in addition from cubism in painting and its application to design in the early textiles of the Bauhaus. It was at its height when the first edition of this book came out. It is now, so Mr Farr assures us, dying out. Its clientele was the masses, whereas the Parisian style belonged to Mayfair and the cinema, and the Swedish to the enlightened who did not want to abandon all decorative pleasure. The successor of Jazz is bogus streamlining or borax.

We are now ready to tackle the fourth question: Why modern design? The best way of attempting an answer is the negative way. Why should design not be in imita-

[1] See, for example, A. Koch, *Das neue Kunsthandwerk*, Darmstadt, 1923; *Kunstgewerbe*, edited by R. Bosselt, H. Busch and H. Muthesius, Berlin (Wasmuth), 1922.

tion of styles of the past or at least inspired by them? The answer is complicated by the fact that without question an imitation Oriental rug is aesthetically better than a jazzy rug from Kilmarnock, and the neo-Georgian title-page in an English book better than the streamlined American title-page spread with its tiresome lines over two facing pages. The neo-period designs one can live with, the modern ones become unbearable after a time. Yet there is a strong argument in favour of even the nastiest of modernistic abominations. They are at least of our age. That endows them with a vitality which even the choicest of Jan Tschichold's new book designs can never possess. Jazz and streamlining are the outcome of the twentieth century; the forms, the colours, the rhythms of Oriental rugs belong to a world which cannot be ours.

This does not imply a denial of their permanent aesthetic values, and to include a Sumak rug or a Sheraton cupboard or some Georgian silver with modern furnishings is one of the best things that can happen to either. But the pieces must be genuine. Mr Farr has in more than one place stressed what is lost of the finesses of originals in the imitations made under twentieth-century conditions.

So much for imitation period furnishings; Mr Farr has other valid things to say about them. He speaks of their snob-appeal, and there is much truth in that. And it is not only, as Mr Farr says, that people wish to pretend to be their betters, it is also that they are in need of an artificially built-up security around themselves, and Hepplewhite chairs and chintzy curtains give them that. In a world which collapses left and right, in a country going through a painful transition—no longer the ruler of an empire and not quite safely established as the cultural centre of a commonwealth of essentially independent partners—some pretence of established prosperity seems especially desirable. Hence the particularly wide appeal of the past in design in Britain, and for a similar reason in France. Nations which have accepted back seats in world politics such as Sweden, Denmark, Holland and Austria are more likely to develop modern design and apply sensitivity to it. Nations which still fight for front seats like Germany and the United States are equally likely to develop modern design and more likely to launch out into adventures. The case of Russia would need a book to itself.

It was design in the United States I wanted to reach at this juncture, because in the United States modern design is showing a particularly robust vitality, and one welcome to neither Mr Farr nor myself. So now that the four questions about good design and modern design are answered separately, we must turn to their combination and try to distinguish between good and bad within modern design. Mr Farr is doing that right through his book, and his result is that curious mark-sheet on p. 252, which seems a twentieth-century revival of Roger de Piles's notorious

Balance des Peintres. How can Mr Farr distribute his prizes and his reprimands so unhesitatingly? And how could I find the courage in 1935 to do the same almost as unhesitatingly?

The most illuminating way to tackle the problem of aesthetic standards in modern design is perhaps to start from the United States and take three equally objectionable modes of expression, streamlining, the mouth-organ radiator fronts of motor-cars, and multicoloured printed ties.

Streamlining, as Mr Farr says, has a perfectly legitimate origin. It is in many ways what he wants design to be. It is decoration enhancing form, it has intensity, it makes a vehicle which is not streamlined appear less relevant, and so on. That is, as far as it is applied to fast-moving vehicles. But already in 1935 I could point to streamlining as a menace to the decoration of stationary things. The streamlined ashtray is, I said, nonsense. Yet there is a justification for it. Associational or evocative appeal is legitimate in art and has never been neglected in the past. The pointed arch in Gothic architecture points upward. It leads away from earthly bonds towards heaven. It is ubiquitous in building and design in the later medieval centuries. Similarly, streamlining surely expresses the spirit of this age, its conquest of space, its faith in speed, its chase for records. So it has a right to appear everywhere. The argument against it is one of qualification. It is incongruous to apply symbols of speed to an easy chair or a perambulator. It contradicts the mood in which we make use of them. The motor-car, on the other hand, is entitled to it, quite independent of the question whether streamlining really helps a car's speed—which almost invariably it does not. One's objection to the way cars used to display streamlining was merely that it was overdone—but overdoing is a common sin of the young and the naive, and the Americans are both.

However, motor-cars are no longer streamlined. The mouth-organ front of the modern American car and its feeble imitation, the British car, does not express speed, it is reminiscent rather of the slow movements of the hippopotamus. Mr Farr has shown how slab-side pressing is influencing form and must continue to influence form. But must the result be that broad and ugly grin of a chromium-plated denture? Its justification is, of course, once again evocative. It symbolizes riches and power, and what is wrong with it is that it does it so grossly. That sort of noisy show comes off in the United States where it is at least in accordance with people. It is extremely un-British, indeed un-European. To condemn it is not to condemn slab-side pressing, but the lack of sensitive designers. As far as I can see there is no reason why so elegant a body as that of the Cisitalia should be impossible under American production methods.

The same is true of the multicoloured tie. If the United States have a place for *Louisiana Story* in the sea of the bad Hollywood films, there might also be a market for good design in printed ties, if they are being produced. I have not come across any. The ties which one does see are perfect counterparts to the modern American middle-class car—jolly, boisterous and self-confident. To accept them in England is to promote the personal qualities that go with them—and is that really what its makers or importers want? It is my opinion that consciously they would deny it; yet they go on promoting as hard as they can. One of the most obvious cases is the imitation Micky Mouse style in the comic strips in the children's papers. Does its rollicking vitality justify it in England? Even admitting that the fag-end of the Kate Greenaway and Rackham style can be thoroughly sickening, one remains in doubt.

In all of these three cases of American decoration—streamlining, the mouth-organ car front and the printed tie—our objection is that they are in contradiction to qualities of the British character. To be gross and flamboyant has never been typical of British art and design. We can count on the appreciation of something subtler and more restrained even among the masses of a nation with so much more tradition behind it.

Scarves correspond in Britain to the ties of the Americans. It may be very difficult to find a printed scarf which can be accepted, for an exhibition, as really well designed—the danger in England being again not crudity, but sentimentality and snobbery, the pink coats and hounds and riding-whips and stirrups. But—and this takes us one step further—one should in any case not argue too seriously about subject-matter and indeed perfection of treatment in scarves. Anything that drapes and amuses has its rights in the fancy-world of the scarf, from maps and written devices to buildings and ballet-dancers and abstract forms and even—though with some misgivings—Henry Moore figures by Henry Moore. Only we don't want fancy in Europe to be quite so elephantine as she appears in much modern American design.

The recognition of fancy in workaday design is indeed what to my mind distinguishes the attitude of 1950 most significantly from that of 1935. Mr Farr, not looking back to 1935 as part of his own experience of design, does not sufficiently insist on this. In 1935—I have had to say that before—good design meant to those who were able to criticize almost universally undecorated design. The various forms of modern ornament which I have tried to analyse belonged to territories out of bounds, and were simply not mentioned. Thereby, that is by disregarding a creation of their own day, critics, architects and designers deprived themselves of response to their work amongst all but the most sensitive class. It is untrue that the common man was hostile to modern design. He went modern as lustily as Le Corbusier, but he went jazzy. The reason why what most people call modern architecture and design never became popular is its uncompromising purism and puritanism more than anything else. The appreciation of proportion, line and surface is confined to few. It remained the privilege of a small group within the professional classes until during the war it found first official recognition in exhibitions and the design and furnishing of hostels. It has now become universal for schools—and it is eminently successful there, also, as far as one can see, in the view of the children. But the attempt to do the same for domestic design, for instance in the utility furniture scheme, failed because the many want something livelier and have a right to it.

I agree with Mr Farr that response to unornamented design would have been wider, if manufacturers and retailers had helped more. But the style of the 1930's could not be the panacea. I was wrong in 1935 to think that a conspiracy of manufacturers was responsible for the garden gate with slats in the form of the rays of the rising sun, or for the slim dancer gracefully lifting her long skirt of opalescent green and raising with her other hand a lamp, perhaps with a shade of what Mr Farr quotes as 'natural imitation vellum', or for the Savoyard boy of earthernware standing in the bay-window, not facing the room but facing the street, because it is more important that passers-by should take him in than that he should smile day in, day out, at his mistress dusting the parlour. All these objects of fancy have their *raison d'être*, and the problem is not to get rid of them, but to raise their aesthetic quality.

Architects and designers between 1935 and 1950 have become aware of this problem. A comparison of the lower-price furniture by Gordon Russell's in 1935 and of furniture now shows that we have grown less stern and exacting. The so-called Victorian Revival belongs very much in this context. It is not another form of historicism, as it is in the United States, but, especially in the circles connected with or influenced by *The Architectural Review*, an antidote. It is never overall, as neo-Georgian tries to be, it is no more than spice added to a modern setting. Its effect is contrast and surprise, in the sense in which Uvedale Price asked for it. Overdo it, and it defeats its object. All the same, it remains *Ersatz*. It is an admission that the twentieth century cannot invent its own fancy, or not enough of it. That is where the exteriors and interiors on the South Bank in 1951 were so revealing, and indeed epoch-making. For here for the first time it was demonstrated how much of fancy can be done within the modern style and with only very occasional and limited Victorian borrowings and how much strictly modern fancy, without any Americanisms will be enjoyed by all and sundry. Such pavilions as The Lion and Unicorn, such interiors as that of the Country pavilion are the best proof of what can be done, and one of the most hopeful things that have happened to design in recent years.

The Lion and Unicorn pavilion was designed by R. D. Russell and R. Y. Goodden, two architects, the interior of the Country Pavilion by F. H. K. Henrion, a professional designer. In 1937 I pleaded for the architect as the best potential designer. Mr Farr now puts the case for the professional designer, or consultant designer, side by side with those for (and against) the architect, the so-called commercial artist, and the staff designer. Considering the architect, he says it would be 'pointless to accept him as the most promising designer'; considering the consultant he calls him 'the best type' in certain cases, which in reality he finds very rare; and so it is the staff designer for whom in the end he votes.

I do not there entirely agree with him. It is true that the staff designer has the best facilities to know about production methods and sales conditions, and that he would therefore be the ideal solution, if he received the right training and achieved the right status in his firm. But how rarely does that happen, probably more rarely, I would say, than ideal conditions for consultant designers. As regards status, the consultant and the architect are almost bound to be treated as superior by the manufacturer than an employee, and as for training, there is to the best of my knowledge no course as yet to emulate that by which a good school of architecture trains architects. Only in schools of architecture exists that combination of aesthetic, technical and commercial points of view which the designer needs. So I should still, as I said in 1935, like to see architectural schools draw up courses for the training of designers open to students after they have reached intermediate standard. The teaching of design for printed textiles, wallpapers, and porcelain decoration may belong to the art schools, but the right introduction to the shaping of everyday things and the development of forms suitable for industrial production is the school of architecture. The Bauhaus, it is true, was unique. But after all, Walter Gropius called it the Bauhaus—that is, the House of Building.

INDEX

Addresses, Lectures and Papers
'Art and Craft of the Machine, The' (Wright, 1903), xxxvii n, 133 n
'Design and the Retailer' (Holmes, 1946), 153 n, 160n, 161 n
'Design in Daily Life' (B.B.C. talks, 1933), 247
'Development of Co-operation between Museums and Industry' (Gordon Russell, 1950), 283 n
'Families, their Needs and Preferences in the Home' (Chapman, 1949), 154 n
'Looking at Things' (B.B.C. talks, 1949, 1950), 247
'Training of the Industrial Designer, The' (Darwin, 1949), 179 n
Aeroplanes, 4, 132
Allard motor-car, Pl. LXVII
Aluminium ware, etc., 8, 42–5, 140, 144, 158, 217, 252, 254, 261, 276, 277. Pl. XVIII
American market, 92, 93, 94, 104, 106 n, 109, 111, 286. *See also* United States of America
Anderson, Green & Co., Ltd., 209
Anderson, Sir Colin, 209, 308
Anthropometry, 144
Architects, 28 *et seq.*, 186, 200, 261, 264–5, 276, 320
Behrens, Peter, 264
Bill, Max, 184
Fry, Maxwell, 233, 234, 247
Gaudí, Antonio, 315
Gropius, Walter, 50, 145 n, 183 and n, 184, 185, 241–2, 264, 320; quoted, 242–3 and n
Hill, Oliver, 234
Holden, Charles, 234
Horta, Baron Victor, 315
Jones, Owen, xxxvii n
Larsson, Fru Lena, 161
'Le Corbusier' (Charles Edouard Jeanneret), 318
Loos, Adolf, quoted 314, 315
Maufe, Sir E., 234
Milne, O. P., 234
Paul, Bruno, 264
Perret, Auguste, 315
Pugin, Augustus Welby, xxxvii n
Rendel, H. S. Goodhart, 234
Riemerschmid, Richard, 264
Robertson, Sir Howard, 201
Rohe, Ludwig Miës van der, xx, 145 and n, 185, 264
Williams-Ellis, Clough, 194, 201
Wornum, G. Grey, 234
Ardil fabrics, 143 n
Armstrong Siddeley motor-car, 142
Art galleries, 198, 282, 296, 301
City Art Gallery, Birmingham, 195, 231
City Art Gallery, Manchester, 295 n
National Gallery, 91 n, 207
National Gallery of Canada, Ottawa, 297 n
Ronald Copeland, 104
Art Nouveau, 30, 53, 315
Articles
'Borax or the Chromium Plated Calf' (Kaufmann, 1948), xxxvii n
'Contemporary Decoration for Furniture' (Tomrley, 1952), 13 n
'Design and the Distributor' (Hardern, 1949), 55 n
'Design Appreciation in London Schools' (Foott, 1953), 221 n
'Design Policy for New Techniques' (Russell, 1953), 144 n
'Experiments with Latex Foam' (Young, 1953), 144 n
'First Design Centre, The' (Belle and Tomlinson, 1949), 223 n
'Gas Cooker' (Stafford, 1954), 52n
'How new moquettes are designed for British Railways' (Colquhoun, 1953), 82n
'Industrial Art and Commercial Design in Glassware' (Riley, 1946), 109 n
'Modern Transfers for Pottery' (Farr, 1954), 100 n
'Operation Fluorescent' (Bell, 1949), 67 n
'Pattern out of Texture' (Farr, 1954), 86 n
'Selling Design in Sweden' (Stewart, 1949), 161 n
'Sitting in Comfort' (O'Donovan, 1954), 144 n
'Street Furniture' (Williams, 1954), 281 n
'Artificial obsolescence', 284
Arts and Crafts Exhibition Society, 191 n
Arts and Crafts Movement, 30, 53, 167, 168, 189
Ashridge House lecture courses, 196; (1949), 230
Ashton, Sir Leigh, 209, 295 n
Association of Canadian Industrial Designers, 230 n
Aston Martin motor-car, Pl. LXVII
Austin motor-cars, 130, 276; Pls. LXVIII, LXX

INDEX

Australia, 22, 23, 111
Austria, 48, 163, 194, 316. *See also* Vienna
Authors
  Akerblom, Bengt, 144 n
  Bayer, H., 183 n
  Bell, Peter, xiv, 67 n
  Belle, J. Cleveland, 223 and n
  Beresford-Evans, J., 309
  Bertram, Anthony, 196, 244–5, 247
  Best, R. D., 62 n
  Brinton, R. S. and J. F. C., 18 n; quoted, 23 and n
  Burridge, F. V., quoted, 279 and n
  Carrington, Noel, 243, 247
  Colquhoun, Ian, xv, 82 n
  Cullen, Gordon, 281 n
  Darwin, Robin, quoted, 179 n; 181, 209, 295 n
  Farr, Michael, 86 n, 100 n
  Foott, Sydney, 221 n
  Geddes, Patrick, 282 n
  Giedion, Siegfried, xxx n, 246–7
  Gloag, John, 209, 243–4, 247
  Gropius, Walter, 50, 145 n, 183 and n, 184, 185, 241–2, 264, 320; quoted, 243 and n
  Hardern, Leslie, quoted, 55 and n
  Holme, Geoffrey, 243 and n
  Kaufmann, Edgar, xxxvii n, 296 n
  Koch, A., 315 n
  Loewy, Raymond, xix, xxxvii n, 148–9, 244, 246 and n
  Logie, Gordon, 144 n
  Martin, J. L., 244
  Morris, William, xxxi, 15, 16, 23, 35, 50, 84 n, 193, 243, 246; quoted, 115
  Mumford, Lewis, quoted, 246
  O'Donovan, Brigid, 144 n
  Olsen, Ivan Munk, 41 n
  Pevsner, Nikolaus, xxvii–xxviii, xxx n, xxxiii and n, 50 n, 145 n, 183 n, 189 n, 241 and n, 244, 310–20
  Piles, Roger de, 316–17
  Powell, H. J., quoted, 110 and n
  Price, Sir Uvedale, quoted, 313; 319
  Pugin, Augustus Welby, xxxvii n
  Read, Sir Herbert, 149, 180 n; quoted, 242–3
  Riley, W. G., quoted, 109 n
  Russell, Gordon, 5 n, 15, 16, 17, 144 n, 208, 234, 319; quoted, 239, 283 n
  Speight, S., 244
  Stafford, Jack, 52 n
  Stewart, Jean, 161 n
  Teague, Walter Dorwin, 148, 245
  Tomlinson, D. H., 223 n
  Tomrley, Mrs C. G., 13 n, 195
  Waterer, John W., xviii, xix, 115 n, 116, 120, 122, 153 n, 295 n
  Williams, George, 281 n
  Wright, Frank Lloyd, xxxvii n, 133 n
Axminster carpets, 17, 18, 22, 23. Pls. XI, XII

Baffle-board receiver, Murphy's, 74–5, 145. Pl. XXX
Bakeland, Dr Leo, 123
'Bakelite', 123, 127 n, 143
Baker, R. W., 171
Barlow, Sir Thomas, 86, 208
Barlow & Jones, Ltd., 86, 87
Bauhaus, *see* Schools
B.B.C. *see* British Broadcasting Corporation
Bent wood, 144, 145
Benz Company, 129
Bentley motor-car, Pl. LXVIII
Berry's Electric Company, 58
Best, R. D., 27 n, 62 n
Bilsland, Sir Steven (now Lord Bilsland), 219
Birmingham, 27, 31, 34, 37, 46, 49, 62, 115, 116, 136, 140 n, 162, 166, 167, 173, 178, 185, 194, 195, 197 n, 230, 254, 256, 258, 281, 302
Birmingham Civic Society, 281 and n
Board of Education, 181, 198, 200, 208. *See also* Ministry of Education
Board of Trade, xi, 4, 6 n, 193, 198, 199, 201, 202, 205, 208 and n, 213, 220–1, 224–5 and n, 226, 232, 287, 309. *See also* Committee on Art and Industry *and* Council of Industrial Design
  Working Party Reports, 225 and n; *on Furniture* (1946), 12 n; *on Jewellery and Silverware* (1946), 33 and n–34, 38 and n, 41 and n; *on Carpets* (1947), 19 n, 21 n *et seq.*, 25 and n, 176 n, 227 n; *on Linoleum, etc.* (1947), 27 n
Books, 241–7, 309. *See also* Year Books
  Academies of Art Past and Present (Pevsner, 1940), 189 n
  An Enquiry into Industrial Art in England (Pevsner, 1937), xxxiii and n, 145 n, 244, 310 n; quoted, 312
  Art and Industry (Read, 1934, 1944, 1953), 180 n; quoted 242–3
  Art and Technics (Mumford, 1952), quoted, 246
  Balance des Peintres (Piles), 317
  Bauhaus 1919–1928 (ed. H. Bayer and W. and I. Gropius, 1938, 1952), 183 n
  Brass Chandelier (Best), 62 n
  Carpet Annual, The (1950), 25 n
  Carpets (Brinton, 1947), 18 n, 23 n
  Cities in Evolution (Geddes, 1950), 282 n
  Conquest of Ugliness, The (La Valette, 1935), 235 n
  Culture of Cities, The (Mumford, 1938), 246 and n
  Das neue Kunsthandwerk (Koch, 1923), 315 n
  Design (Bertram, 1938), 244–5

Books (*cont.*)
Design and Civilization (Carrington, 1934, 1947), 243 and n
Design this Day (*The Technique of Order in the Machine Age*) (Teague, 1940, 1947), 245 and n
Designers in Britain, Vols. 1, 2, 3 and 4, 230, 231
Dictionary of Colours for Interior Decoration, 228
Essays on the Picturesque (Price, 1810), quoted, 313
Flat Book, The (Martin and Speight, 1939), 244
Form in Engineering Design (Beresford-Evans, 1954), 309
Furniture from Machines (Logie, 1947), 144 n
Georg Jensen Silversmith (Olsen, 1937), 41 n
Glass Making in England (Powell, 1923), 110 n
Industrial Art Explained (Gloag, 1934, 1946), 243 and n, 244
Industrial Design (*A Practical Guide*) (Van Doren, 1940), xxxvii n, 245 and n
Industrial Design and the Future (Holme, 1934), 243 and n
Kunstgewerbe (1922), 315 n
Leather in Life, Art and Industry (Waterer, 1946), 115 n, 116 n, 120 n, 122 n, 153 n
Mechanization Takes Command (Giedion, 1948), xxx n, 246 and n
Missing Technician in Industrial Production, The (Gloag, 1944), 243–4 and n
Modern Architecture (Wright, 1931), xxxvii n, 133n
Never Leave Well Enough Alone (Loewy, 1951), xxxvii n, 148 n, 246 n
New Architecture and the Bauhaus, The (Gropius, 1935), 183 n, 184 n, 241, 243 and n
Pioneers of Modern Design (Pevsner, 1949), 50 n, 241 n.
Pioneers of the Modern Movement (Pevsner, 1936), 183 n, 241
Plastics and Industrial Design (Gloag, 1945), 244 and n
Self Training for Industrial Designers (Gloag, 1947), 244 and n
Space, Time and Architecture (Giedion, 1941, 1951), 246 and n
Standing and Sitting Posture (Åkerblom, 1948), 144 n
Stones of Venice, The (Ruskin), 109 n
Technics and Civilization (Mumford, 1934), 246 and n
'Things We See' series (Penguin Books), 214, 245
True and False in the Decorative Arts, The (Jones, 1863), xxxvii n
True Principles of Pointed Architecture (Pugin, 1841), xxxvii n

'Borax', xxxvii and n, 12, 32, 50, 56, 76, 134, 142, 315
Braddell, Mrs D'Arcy, 52
Bradford, 78, 165, 230
Bratt Colbran Ltd, xiii, 53
Brewer, C. G., 193
Bristol, 194, 195, 197 n
British Broadcasting Corporation, 69–70, 196, 217, 247–8, 301
 Talks: (1933), 247; (1937, 1938), 244
British Colour Council, 227–8 and n
British Council, 217, 301
British Industrial Plastics, Ltd., 123, 127, 143
British Institute of Industrial Art, 160 n, 198–9, 232
 Reports: *Art Training of the Middleman, The*, (1927), 160 n, 199; *Industrial Art and British Manufacture*, 199; *Public Departments and Industrial Art*, 199; *Training of the Silversmith, The*, 199; *University Teaching of Art in relation to Commerce, The*, 199
British Electricity Authority, 212
British Export Trade Research Organization, 218
British Intelligence Objectives Sub-Committee, 218
British Jewellers' Association, 225 n
British Nylon Spinners, Ltd., 143
British Overseas Airways Corporation, xxi, 212, 280
British Pottery Manufacturers' Federation, 101, 172
British Railways, xv, 82, 280
British Rayon Federation, 226–7
British Rubber Development Board, xx, 143, 144
British Standard Specifications, 28, 55
British Xylonite Company, 123
Broadcasting, 69–70. *See also* British Broadcasting Corporation
Building Centre, London, 195, 203, 223
Burney, Sir Dennistoun, 133
Burslem, 95, 171, 299

Cabinets, *see* Radio and Television cabinets
Californian potters, 94, 286
Canada, 22, 93, 106 n, 109, 111, 220 n
Carlisle, 170, 282
Carpets and rugs, 17–26, 148, 152, 155, 157, 190, 252, 254, 257, 259, 261, 263, 266, 268, 270, 276, 279, 285, 286, 307, 316. Pls. IX–XII
Celluloid, 123
Central Institute of Art and Design, 91 n, 207, 213. *See also* National Register of Industrial Art Designers
Chantrey Bequest, 282
Chapman, Dennis, quoted, 154 and n; 274
Chiltern range furniture, 5. Pl. II
'Chrysaline' lamp shades, 144. Pl. XXVI
Chrysler 'Airflow' motor-car (1934), 130n.
Cisitalia motor-car, 135–6, 317. Pl. LXVI

Citroën et Cie, 134
City Guilds, 204–5
Civic Councils, 281–2
Clocks, 59–60, 252, 253, 264, 277. Pl. XXIV
Code of Professional Conduct (Society of Industrial Artists), 230 and n
Cole, Sir Henry, 190
Cole Ltd., E. K., xiv, xv, 70, 71, 75, 77, 124
Commercial art, 177, 185, 192, 280, 320
Committee on Art and Industry, 201–2, 308–9
  Gorell Report (1932), 195, 201–2, 204, 295
Competitions
  Design and Research Centre for the Gold, Silver and Jewellery Industries, The, 225–6. Winning entry, Pl. LXXXIV
  Furniture (Scotland), Council of Industrial Design, Scottish Committee, 220
  Hopewell's Ltd., Furniture (1950), 156–7
  Industrial Art Bursaries, xxii, 190
  International Low-cost Furniture (New York, 1948), 145, 146. Pl. LXXIII
  Regent Street Association, 271 and n
Conferences
  Bexhill (1937), 196
  Birmingham (1948), 215
  Furniture Design (1949), 154 n
  Industrial Design (1946), 153 n, 200
  International Council of Museums (1950), 283 n
Congresses
  Design (1954), 220
  Festival of Britain, International Design (1951), 213 and n
Consultant Designers' Group, 231
Cookers, gas and electric, 51, 52–4, 56–7, 252
Copeland Company, Ltd., W.T., 96
Copenhagen, *see* Denmark
Corning Glass Works, U.S.A., 111 n, 147
Cotswold range furniture, 5. Pl. II
Cotton Board, 212, 223
Council for Art and Industry, xxxiii, 199, 202–5, 229, 233, 295. *See also* National Register of Industrial Art Designers
  Reports: *Design and the Designer in Industry* (1937), quoted 203; *Design and the Designer in the Light Metal Trades* (1944), 27n, 39n, 203; quoted, 31 and n, 33 and n, 43–4 and n, 44–5 and n, 64 and n, 258 and n, 284 and n; *Design in the Jewellery, Silversmithing and Allied Trades* (1937), 35 and n, 36n, 202–3; *Education for the Consumer* (1935), 202–3; *Proposal for an Industrial Art Centre* (1937), quoted, 204, 295; *Second General Report* (1936), 202; *Working Class Home, The: its Furnishing and Equipment* (1937), 195n, 203
Council for the Encouragement of Music and the Arts, 301
Council of Industrial Design, xx, xxiii, xxxiii, 153 n, 160n, 179–80, 196, 197, 199, 200, 207, 208–22, 225, 226, 230, 232, 237–8, 240, 244, 270, 271 n, 288, 293, 295, 308–9. *See also* National Register of Industrial Art Designers
  Reports: B.I.O.S., 218, Meynell-Hoskin (1944), 208; *Training of the Industrial Designer in England and Wales* (1946, unpublished), 213; Weir (1943), 200 n, 208, 218, 220, 221, 295, 309; (Annual) 1945–6, 209 and n, 210 and n, 214 and n; 1946–7, 236 and n; 1948–9, 209 n; 1950–1, 211 and n
Courtaulds Ltd., 143
Coventry, 136, 140 n, 254
Crafts Centre, Edinburgh, 220
Crafts Centre of Great Britain, 34, 212
Craftsmen, xx, xxi, 16, 34, 35–6, 37, 41, 47, 96, 98–9, 108, 122, 142–7, 172, 175, 176, 181, 259, 260, 264, 267, 278, 294
Cripps, Sir Stafford, 236
Crossley motor-car (1932), 133 n
Cubism, xxxvii, 30 n, 53, 89, 103, 315
Czechoslovakia, 47, 63, 220 n

Daimler Co., Ltd., 129
Dalton, Dr Hugh, 208
De Havilland Aircraft Co., Ltd., 4
Dearden, Frank, 172
Decca Record Company, 75–6
Decoration and Ornament, 11, 13, 37 n, 44, 50, 61, 64, 91–2, 94, 98 and n *et seq.*, 106 *et seq.*, 129, 154, 171, 311, 314, 315, 318. Pls. II, LI, LIII, LXXVI, LXXVII. *See also* Borax
Denmark, xii, xx, 16, 41–2, 66, 94, 144, 160, 194, 296, 297, 316
Department of Scientific and Industrial Research, 143, 225
Design and Industries Association, xxiii, xxxiii, xxxviii n, 163n, 180, 189, 192–8, 199, 214, 232, 233, 235, 244, 270, 280, 293, 296, 311
Design Centres, 222–9
  Colour, Design and Style, 85 and n, 212, 223–4
  Gold, Silver and Jewellery Industries, 34 and n, 35, 48 n, 212, 225–6, 288
  Interior Decoration, 227–9
  Ottawa, 297 n
  Proposed, 301, 308–9
  Rayon Industry, 85, 212, 226–7
Design Research Unit, xviii, 149
Design standards
  changes in, 251–4
  civic, 281–2

Design standards (*cont.*)
dictatorship and, 272, 285–7
factories and, 254–7
manufacturers and, 22–3, 288–9
price and, 275–8
production methods and, 258–62
purchase-tax and, 34, 40–1, 112–13, 285–6
retailers and, 151–62
sequence of, 251, 253
See also Public Taste
Design Weeks, 215, 221, 222, 231, 270
Designer, the, 6, 93, 102, 127–8, 129, 259–60, 261, 262–8, 294, 305–6
architect, 28, 29, 31, 33, 71, 82, 113, 149, 264–5, 266, 320
consultant, xxvii, xxxvii, 9, 10, 13, 14, 18–19, 25, 51, 54, 56, 57, 64, 66 n, 74 *et seq.*, 113, 119, 120, 126, 127, 128, 148–51, 265, 266, 320; fees, 15, 57, 66n, 71, 76, 119, 148, 150, 265, 266
free-lance, xxxvii, 13–14, 18, 19–20, 33, 36, 37 n, 39, 44–5, 48, 49, 74, 80, 83, 85, 101, 102, 107, 113, 115, 119, 121, 127, 144, 149, 171, 181, 224, 263–5, 266, 300, 308; fees, 18, 40, 83, 121, 263
staff, xxxvii, 13, 14, 16, 18, 19, 20–1, 25, 36, 39, 43, 44, 49, 60, 74, 80, 82, 83, 88, 101, 107, 108, 115, 119, 120, 122, 126–7, 140, 149, 192, 263, 320; salaries, 14, 15, 19, 20, 26, 33, 40, 49, 76, 82, 88, 102, 108, 267–8
Designers
Aalto, Alvar, 144 n
Adams, John, Pl. XLIX
Arens, Egmont, 148
Armfield-Passano, Pl. XLVI
Ashmore, Peter, Pl. LXXII
Barker, John, Pl. XLI
Barman, Christian, 59. Pl. XXIV
Barnes, John, Pls. XIV, XXIII
Barnsley, Sidney Howard and William Edward, 15, 166
Bawden, Edward, Pl. XLV
Baxter, G. P., Pl. LXXXIV
Bell, Jennifer, Pl. LXXXIII
Bell, Peter, 67 n. Pls. XXVII, XXVIII
Bernadotte, Count Sigvard, 41. Pl. XVI
Beverley Pick Associates, Ltd., Pl. XXVI
Bill, Max, 184
Black, Misha, 149
Boissevain, Paul, Pl. XXVII
Booth & Ledeboer, Pl. II
Brandt, Edgar, 29
Breuer, Marcel, 145 and n, 233, 264. Pl. LXXIV
Bridson, Paul, Pl. VIII
Broadbent, Margaret, Pl. LXXXI
Brookes, R. E., Pl. LXIV
Brown, W. Bruce, Pl. LXIII
Casson, Sir Hugh, 233
Chermayeff, Serge, 15, 71, 185, 233, 264
Clarke, Arundel, 161
Coates, Wells, 71–2, 77, 247, 264. Pls. XXIX, XXX, XXXII
Cockerell, Douglas, 192
Colquhoun, Ian, 82n. Pl. XXXVIII
Cooper, Susie, 105 and n, 234. Pls. XLVII, XLVIII
Crane, Walter, 23, 90, 180
Cuzner, Bernard, 234
Dandy, L., Pl. LXXXII
Day, Lucienne, Pls. XLI, XLIII
Day, Robin, 145. Pls. III, VII, LXXIII
de Ferranti, Dr., 58 and n. Pl. XXIII
Dreyfuss, Henry, 148
Drummond, John, Pl. XXXIX
Dunn, Geoffrey, 17, 209, 307. Pl. V
Eames, Charles, Pl. LXXV
Elliott, Eric, Pl. LXXVII
Falding, Jean, Pl. LXXXI
Farina, Pinin, 135. Pl. LXVI
Farquharson, Clyne, 111. Pl. LV
Feeley, Frank, Pl. LXVII
Games, Abram, Pl. LVIII
Gardner, James, 216, 236
Gate, Simon, 111
Geddes, Norman Bel, 148
Gill, Eric, 192
Gimson, Ernest, 15, 166
Goodden, R. Y., 264, 320. Pl. LVI
Goodwin, J. F., Pl. XLVIII
Gray, A. E., 201
Gray, Milner, 149. Pl. LVII
Grey, John, 234
Grierson, Ronald, Pl. IX
Groag, Jacqueline, Pls. XLIV, LXXVI
Guild, L. V. A., 184n
Guyatt, Richard, Pl. L
Hald, Edward, 111
Hammond, David, Pl. LIV
Harvey, A. E., 167, 234
Haxby, Colin, Pls. LIII, LXXVI
Heal, Christopher, 233
Heal, Sir Ambrose, 15, 111, 193, 234
Henrion, F. H. K., 320
Heritage, R. C., Pl. VII
Hindley, V. A., Pl. LXXXIII
Hiscock, R. C., Pl. XXVII
Hogan, James, 192, 234
Holmes, Kenneth, 163. Pls. XVII, LXXVII, LXXVIII
Holt, Constance Mary, Pl. LXXXI
Hooper, Rodney, Pl. LXV

Designers (*cont.*)
Howe, Jack, 233, 264. Pl. XXIV
Hunter, Alec, 86, 234
Hvidt, Peter, Pl. LXXV
Jacobs, Carl, Pl. IV
Jennings, F. M., Pl. LXII
Jensen, Georg, 41
Joel, Betty, 234
Kauffer, E. McKnight, 192
Keith, Howard, Pls. V, VI
Koppel, Henning, Pl. XX
Larrson, Fru Lena, 161
Lamble, Kenneth, Pl. LVII
Lampl, F., Pl. XX
Latimer, Clive, 145. Pl. LXXIII
Leischner, Margaret, 79, 143. Pl. LXXIII
Lennon, Dennis, 227, 264
Leonard, J. W., Pl. LXXV
Loewy, Raymond, 148–9, 244, 246 and n. Pl. LXVI
Loosely, Anne, Pl. XXXIX
Lyons, Eric, 264. Pl. IV
McCredie, Agnes, 169
McGrath, Raymond, 71, 233, 234
Mackintosh, Charles Rennie, 169, 311
Mahler, Marion, Pl. XXXIV
Marx, Enid, Pls. XXXIV, LX
Mason, J. H., 192
Matthew, R. H. and L. Martin, Pl. XIV
Maylor, T. C., Pl. LXI
Mellor, David, Pl. LXXXIV
Milne, A. J., Pl. VI
Minton, John, Pl. XLV
Moorcroft, William, 96, 105. Pl. XLIX
Morphy, D. W., Pl. XXIV
Morris, William, xxxi, 15, 16, 23, 35, 50, 84n, 193, 243, 246; quoted 115
Morton, Alastair, 86
Murphy, H. G., 168, 192, 234
Murray, Keith, 98n, 111, 192, 234, 264. Pls. XLVII, XLIX
Nicholson, Christopher, 233
Nicholson, Robert and Roger, 224. Pl. XLVI
Nielsen, O. M. Pl. LXXV
Nielsen, Harald, 41
O'Rorke, E. Brian, 233, 264, 265
Otto, Carl, 148, 149
Padiak, J., Pl. LXXIX
Palmer, John, Pl. X
Pearson, D. M., Pl. LXXX
Penniall, Edna, Pl. LXXXIII
Pepler, Marion, Pl. X
Poole, Hugo, Pl. LXXII
Priestley, Sylvia, Pls. XL, XLII
Purvis, Tom, 192
Race, Ernest, Pls. LXXIII, LXXV, LXXVI
Ramsden, Omar, 35, 234
Ravilious, Eric, 234. Pls. L, LI
Read, A. B., 247. Pls. XXV, XXVIII
Reich, Tibor, 86. Pls. XXXVI, XXXVII
Ridgley, G. A., 61. Pl. XXII
Robinson, S. W. Pl. XV
Rohe, Ludwig Miës van der, 145 and n, 185, 264. Pl. LXXIV
Rumpler, Dr H., Pl. LXXII
Rushton, James, Pl. LXXXIV
Russell, Gordon, 5 and n, 15, 16, 17, 71, 208, 234, 319; quoted, 239, 283n
Russell, R. D., 71, 234, 320; quoted, 263n. Pls. III, XXIX, XXX, XXXII
Russell, W. H., Pl. XXIV
Sarfatti, Gino, 68 n
Schreiber, Gaby, Pls. LXIII, LXIV
Schuster, F., 15
Scott-Ashford Associates Ltd., Pls. XIII, LIX
Sheringham, George, 192
Smith, A. Sayer, Pl. LI
Spence, Basil, 216, 236, 264. Pls. IV, VIII
Spender, Humphrey, Pl. XL
Stabler, Harold, 192, 193, 234
Stevens, Irene, Pls. LV, LXXVI
Stewart, Margaret M., Pl. LXXXI
Story Design Group, Pl. VIII
Straub, Marianne, 86, 87. Pls. XXXIII, XXXV, XXXVI
Sullivan, Louis, 315
Sutherland, Graham, Pls. XLIV, XLVI
Svedberg, Elias, 161
Syllas, Mrs Phoebe, de, 197
Taylor, Fred, 192
Taylor, Victor, Pl. XXXII
Teague, Walter Dorwin, 148, 245
Thwaites, A. F., Pls. XXX, XXXI
Tijou, Jean, 30
Trethowan, H., 201, 234
Tschichold, Jan, 316
Van Doren, Harold, xxxviin, 57n, 148, 245
Voysey, C. F. A., 23, 90, 192, 264
Wagenfeld, Wilhelm, 114 and n, 147, 183n, 265. Pls. LVII, LVIII
Walton, Allan, 234, 263
Waterer, John, 115, 116, 120, 122, 153n, 295n. Pls. LIX, LX, LXI
Wellstead, D., Pl. LXXXII
Wiggin, A. L. Pl. XVII
Wilkie, Alexander, Pl. LXXVIII
Wilkes, E. G. M., Pl. LXXII
Williamson, A. H., Pl. LVI

Designers (*cont.*)
  Williger, Karin, Pls. XXXV, XLIII
  Wilson, W. J., Pl. LIV
  Woodfull, A. H., Pls. LXIII, LXV
  Young, Dennis, 144 and n. Pl. LXXIV
Deutscher Werkbund, 111, 193, 210 n, 232, 241
Diplomas
  A.R.C.A., 181
  Art Teachers' (A.T.D.), 178, 179
  Arts and Crafts Intermediate Certificate, 178
  Des.R.C.A., 181, 308
  National Diploma in Design (N.D.D.), 178, 179, 308
  See also Examinations
Distinctions
  Royal Society of Arts' Bicentenary Medal, 308
  Royal Society of Arts' R.D.I., 192
Dobby looms and fabrics, 79, 81, 168, 182. Pls. XXXIII, LXXXIII
'Dodge' motor-car (1914), 130
Domestic appliances, 51–61, 252, 259, 265, 276, 278. Pls. XXI–XXIV
'Dominion' design, 94, 103
Door furniture, 28, 30, 31 and n, 32, 33, 183 n, 252, 265, 276, 286. Pls. XIII, XIV, LXV, LXXIX
Dunn, Geoffrey, 17, 209, 307. Pl. V
Dunn's of Bromley, 160, 195, 271

Edinburgh, 105, 170, 195, 219, 220, 237, 239
Edinburgh Weavers, xv, xvi, 86, 147
Edison's phonograph, 69
Edwards, Dr R. S., 208
Ekco Radio, *see* Cole Ltd., E. K.
Ekman, Consul Johan, 110
Electric light fittings, 62–8, 154–5, 158, 252, 256, 258, 265, 276. Pls. XXV–XXVIII, LXXIX, LXXX
Electrical Association for Women, 57
Electro-plate (E.P.N.S.), 34, 37, 38–40, 146, 174, 226, 256 and n, 258, 267, 276, 285. Pls. XVII, LXXXIV
Examinations, 165, 178–9, 181, 184. *See also* Diplomas
  rules, 178 n, 213 n
Exhibitions, 12, 198, 204, 232–40, 295, 296, 309, 312
  Amsterdam, International Industrial Design (1949), 217
  Arts and Crafts (1888), 191 n
  Berlin, German Industries Fair (1950), 217
  Birmingham, Midland Industrial Art (1934), 195
  Britain Can Make It (1946), 200, 208, 212, 216–17, 219, 235–7. Pl. LXXXVI
  British Industries Fair, xxix, 18, 43, 60, 65, 224 n, 226, 309; (1936), 200; (1946), 216 and n, 217; (1947), 200, 217, 221; (1950), 287; (1953), 143 n, 217 n; (1954), 221, 309, 312
  British Plastics (1951), 217
  Burlington House, Arts and Crafts (1916), 193; British Art in Industry (1935), 52, 191, 200, 233–5 and n; Design at Work (1948), 192
  Campania (1951), 239
  Canadian National, Toronto (1953, 1954), 217, 309. Pl. LXXXVIII
  Charing Cross Underground Register Your Choice (1953), 197–8, 275. Pl. LXXXV
  Cologne (1914), 193
  Colour, Design and Style Centre, 223, 224
  Design and Workmanship in Printing (1915), 232
  Dorland Hall (1933), 195, 229, 232; (1934), 233
  Edinburgh, Enterprise Scotland (1947), 219
  Edinburgh, Living Traditions of Scotland (1951), 220, 237
  Festival of Britain (1951), xx, 200, 208 n, 210–11, 213 and n, 216, 219 *et seq.*, 237–40; pavilions, xxiii, 182, 192, 238, 319–20. Pl. LXXXVII. Regatta restaurant, 211
  Furnished rooms, 195, 197. Pls. LXXXV–LXXXVII
  Glasgow, Industrial Power (1951), 237
  Goldsmiths' Hall, Designers' (from 1945 annually), 225–6; Enemy Products (1915), 193, 232
  Göteborg (1923), 111
  Great Exhibition (1851), 36, 190, 191, 237, 241, 294
  Ideal Home, *Daily Mail* (1935), 52; (1946), 217; (1949), 215; (1950), 215–16
  Inn Crafts (1948), 207
  Lansbury, Architecture (1951), 216
  Leipzig, International Industrial Art (1927), 194
  Leipzig Fair (Annual), 204
  London County Council, Educational (1937), 202
  Model houses (1933, 1934), 195
  Motor (1948), 129 n, 141
  Munich (1922), 315; (Permanent) Abteilung für Gewerbekunst, 204
  National Gallery, Historical and British Wallpapers (1945), 91 n, 207
  New York, Good Design (1951 *et seq.*), 296 n
  Paris (1925), xxxvii, 62–3 and n, 315; International (1937), 203
  Pavilion of British Industrial Art, suggested permanent, 221
  Post-War Homes (*Daily Herald*, 1945), 217
  Radiolympia (1949), 76, 77
  Royal Institute of British Architects, 'Everyday Things' (1936), 233
  Royal Society of Arts (early), 190; (1935), 191, 233; (1951), 191

Exhibitions (*cont.*)
Scottish Everyday Art (1937), 202
Stuttgart, Die Form (1923), 210 n; Der Stuhl (1928), 210 n
Sweden, Svenska Slöjdforeningen (regular), 204
Swedish (London, 1931), 111
Tea Centre, London, 217
Travelling, 91 and n, 198–9, 201, 204, 214–15, 219, 239
Victoria and Albert Museum (1920), 198; 'Industrial Art for the Slender Purse' (1929), 199
Women's Fair (1938), 196
See also Design Weeks, Shop Exhibitions, Museums, and Art Galleries
Expressionism, xxxvii, 63, 89

Fachschulen (Monotechnics), 171, 173
Factory organization, 253, 256–7, 304
Faculty of Royal Designers for Industry, 191–2
Fashion jewellery, 47–8, 51. Pls. XIX, XX
Federation of British Industries, 199–200, 201. *See also* Industrial Art Committee; Register and Employment Bureau, *under* Records
Report: *Burlington House Exhibition* (1935), 200
Feininger, Lyonel, 183
Fennemore, T. A., 207
Festival Pattern Group, 211
Films, xxxvi, 123, 185, 214, 293, 315, 318
Finland, 144, 275
Fires, gas and electric, 50, 51, 53, 58–9, 154, 158, 252, 277. Pls. XXII, XXIII
Fischer, Dr Åsge, 296
Flavel & Co., Ltd., Sidney, xiii, 52
Fletcher, B. J., 163 and n, 166, 193
Florence, Professor Philip Sargant, xxvii
Fluorescent electric light fittings, 62, 66, 67 and n, 68, 145. Pl. XXVIII
Ford 'Consul' motor-car, 142. Pl. LXIX
Forest Products Research Laboratory, 144
Fothergill & Harvey, Ltd., 143
France, 25 n, 30, 41, 134, 140, 194, 316
'Freedom' furniture, 6 and n, 287
Fry, Roger, 201
Furnishing textiles, 78–87, 143, 149, 155, 157, 165, 167, 190, 252, 255 *et seq.*, 261, 263, 266, 268, 277, 285, 315. Pls. XXXIII–XLIV, LXXXI, LXXXIII
Furniture, 3–17, 143 *et seq.*, 148, 153–4, 156, 158, 159, 166, 167–8, 190, 252, 254 *et seq.*, 261, 264, 265, 266, 268, 277, 278–9, 319. Pls. I–VIII, LXXIII–LXXVI, LXXIX, LXXXII–LXXXIII

Galashiels, 169, 170
Gallup poll, 274
Gane, P. E., 17, 271, 307
Gas Boards, 55–6, 161, 287–8
Gas Light and Coke Company, 55
General Motors, 136 n
Germany, xxxi, 13, 22, 47, 63, 89, 113, 114, 120, 122, 123, 134, 140, 145, 184, 194, 218, 220 n, 242, 265, 283, 316
Gestetner duplicating machine, 148 n
Glasgow, 17, 169, 198, 208, 220, 237, 239
Glass, 63, 105–15, 144, 147, 172, 252 *et seq.*, 259, 261, 266, 285, 286, 294, 295n. Pls. LIV–LVIII, LXXVI, LXXVIII, LXXXIV
Glass jewellery, 46, 48, 49. Pl. XX
'Golden Cut', 242, 245
Goldsmiths, The Worshipful Company of, 225–6, 232
Good Furnishing Group, 17, 159 n, 307
Goodale, Sir Ernest, 199
Gorell Report, *see* Committee on Art and Industry
Gramophones, 69, 75
Greg & Co., Ltd., R., xx, 79, 128n, 143

Halifax, 78, 195
Hambleden Committee on Advanced Education in London (1936), 180
Report (1936), 180, 200
Handbags, 115, 116, 118, 120, 121, 122. Pl. LXXII
Harnoncourt, Rene d', quoted 297 n
Harold Sanderson Bursary Fund, 88 n
Hartland Thomas, Mark, 237
Harvard University, 184, 185
Hatfield New Town Development Corporation, 216
Heal & Son, Ltd., xi, xx, 111, 233, 271
Heat-resisting ware, 112, 114, 147. Pls. LVII, LVIII
Hélion, Jean, xv, 242
Helios, Ltd., xv, 86
Henderson, James, 146
High Wycombe, 3, 5, 13, 14 and n, 175, 256
'Hillman Minx' motor-car (1932), 130
H.M.V. Household Appliances, Ltd., xiii, 58, 59, 69
Holland, 22, 32, 194, 286, 316
Holmes, H. A., quoted 153 and n, 160 and n, 161 n
Hopewells, Ltd., Nottingham, 156, 162, 270, 274
Hornby, C. H. St John, 193, 201
Huddersfield, 78
'Humber' motor-car, Pl. LXIX
Hunt, Antony, 86
Hupp Motor Company, Detroit, 148

Imperial Chemical Industries, Ltd. (I.C.I.), 143 and n

Industrial Art Committee of the Fe ration of British Industries, xxxiii, 199–200 and n, 230. n. *See also* National Register of Industrial Art Designers, *under* Records
Proposals for a Central Design Council and Industrial Design Centre (1944), 200
Report: *Burlington House Exhibition* (1935), 200
Industrial Design Partnership, 149
Industrial Organization and Development Act, 1947, 224
Industrial Revolution, xxx–xxxi, xxxiv, 189, 254
Industries, old and new, 254–6
Institute of Design, Chicago, 167, 182–6
Institute of Technology, Illinois, 185
Institution of Body Engineers, 140 n
Institution of Engineering Designers, 308
Interior design, 149, 167 *et seq. See* Design Centres
International Council of Museums, 283 n
Investment casting, 48, 144
Italy, 41, 66, 89, 131, 134, 135, 136, 140, 141, 267 and n

Jackson, F. E., 193
Jacquard looms and fabrics, 18, 80–1, 165, 168, 182. Pls. XXXIV–XXXVII, LXXXIII
Janitor Boilers Ltd., xiii, 61
Jazz, xxxvii, 13, 30, 60, 63, 313, 315
Jenaer Glaswerk, Schott & Gen, xviii, 114
Jensen (Silversmiths) Ltd., Georg, xii, xiii, 41–2
Jewellers' and Silversmiths' Association, 173
Jewellery, 46–51, 144, 146, 168, 174, 254, 255, 256, 258, 261, 263, 267, 285. Pls. XIX, XX
Jewellery and Silverware Council, 226
Johnson-Carper Furniture Co., 146
Johnstone, William, 168
Jowett R4 motor-car, Pl. LXXI

'Kabineat' gas cooker, Flavel's, 52. Pl. XXI
Kandinsky, Wassily, 183
Kape, W. J., 175
Kidderminster, 17, 19, 21, 173, 254, 280
Kitchenware, 42–5, 59, 114. Pls. XVIII, LXIII
Klee, Paul, 183
Kogei Shidosho, Kawasaki, Japan, 220 n
Kunstgewerbe style, 315

La Valette, John de, 234, 235 n
Lambert, The Hon. Margaret, Report on Scandinavia (1946), 160 and n
Laminated wood, *see* Plywood
Lancashire, 78, 82, 85, 171, 224, 257
Landsforeningen Dansk Kunsthaandvaerk (Arts and Crafts Society), 220 n, 296
Landsforeningen Norsk Brukskunst, Oslo, 220 n
La Rinascente, Milan, 309
Latex Foam (Sponge Rubber), xx, 144
Leather Footwear and Allied Industries Export Corporation, Ltd. (1940–53), 207, 224 n
Leather goods, 115–22, 143, 174, 190, 224n, 252, 254, 255, 256, 258, 276, 285, 286, 294. Pls. LIX–LXII
Lee, Sir Kenneth, 209
Leicester, 163, 164
Leslie, S. C., 208
Lethaby, W. R., 167, 180, 193
Lewis's Ltd., Birmingham, 270
Design for Living department, 161, 271
Home-Makers' exhibitions (annual), 231–2
Light metalwork, *see* Metalwork, light
Linoleum, 26–7, 146, 157, 257, 259, 276
Lithography, transfer printing, etc., 95, 98 *et seq.*, 103, 104, 144, 164, 171, 172, 185, 261. Pls. XLVIII, LI, LIII, LXXVII
Liverpool, 161, 274
London, xxx, xxxi, 3, 4, 7, 17, 21, 24, 46, 49, 51, 62, 78, 105, 115, 116, 122, 167, 197n, 254, 302
London County Council, 167, 175, 202, 221 n, 227
Lorenz, A., 145 and n

Machine carvings, 13
Maclagan, Sir Eric, 201
Maclean, R. A., 219
'Magicoal' fire, Berry's, 58
Manchester, 78, 170, 194, 195, 197 n, 198, 203, 230, 281, 295 n
Marconi Company, 69, 124 n
Market Research, 273–5
Markets
buyers' and sellers', 27, 283–5
export, 21, 22, 32, 42, 48, 92–4, 95, 103, 104, 109, 111, 135, 223, 224 n, 276, 285–6
working-class, 276, 278
Mass-production, xxxviii, 4, 12, 28, 38, 43, 45, 57 and n, 116, 124, 130, 131, 135, 136, 140, 143 *et seq.*, 151, 176, 183, 243, 258–62, 277, 278, 304
Mercedes motor-car (1934), 133 n
Merchandise Mart, Chicago, 296 n–297 n
Metal-frame furniture, 8, 144, 145, 146. Pls. LXXIII, LXXIV, LXXV
Metalwork, Light, 27–33, 252, 256, 258, 265, 267, 294. Pls. XIII, XIV
Mey, Dr, 114
Meynell-Hoskin Report (1944), 208. *See* Council of Industrial Design
Middleman, 269–71, 309. *See also* Retail Trade
Middleton, A. P., 219
Midland Industrial Designers' Association, 231–2, 270

Midland Retailers' Design Discussion Group, 215, 270
Midlands, xxxi, 51, 149, 152, 154, 166, 203, 215, 231, 236
Ministry of Education, 162, 165, 171, 172, 178 and n, 179, 181, 213 and n, 299, 300. *See also* Board of Education
Ministry of Supply, 217
Mintons, Ltd., 104
Modern Movement, xxx, xxxi, 50, 193, 233, 235, 242, 243, 264, 283
Modernism, xxxvii, 22–3, 27, 31, 32, 36, 40 n, 51, 53, 56, 60, 63 *et seq.*, 68, 70, 89, 91, 112, 113, 128, 194
Moholy-Nagy, Lazlo, 185
Mondrian, Piet, 242
Monotechnics, 162, 171, 173
Moorcroft, Ltd., xvii, 96, 105
Moore, Henry, 318
Moquettes, 82, 83, 280. Pls. XXXVIII, LXXXI
Morris, William, xxxi, 15, 16, 23, 35, 50, 84 n, 193, 243, 246; quoted, 115
Morris Movement, 166, 180, 189, 241
Morton, Sir James, 86, 193
Motor-cars, 129–42, 148, 149, 252, 253, 254, 256, 257, 259, 261, 265, 267, 268, 276, 283, 285, 286, 317. Pls. LXVI–LXXII
Mouldings for furniture (wood), 12, 13
Mouldings, plastic, *see* Plastics, moulded
Murphy, Frank, 70–1
Murphy Radio, Ltd., xiv, xv, 16–17, 72, 74, 75, 77
Museums, 282 n–283 n
  Den Permanente, Copenhagen, 296
  Grassi, Leipzig, 204
  Leicester, 164
  Modern Art, New York, xix, xx, 4, 50n, 145, 296 n, 297 n
  Modern Industrial Art (proposed permanent), 294–7, 308
  Ornamental Art, Marlborough House, 294
  Stuttgart, 204
  Victoria and Albert, 34, 84, 198–9, 203, 205, 216, 236, 282 n, 294, 295
  Wedgwood, 104
  See also Art galleries
Museums Association, 204
Museums Conference, International Council of (1950), 283 n

National House-Building Industries' Joint Standing Committee, 216 n
  Report, *Demonstration Permanent Houses* (1946), 216 n
National Industrial Design Committee, Ottawa, 220 n, 297 n
National Register of Industrial Art Designers, (N.R.D.), 199, 205–7, 212, 224 n
  Report, *Design and Development in the Leather Industries* (1945), 207
National Research Development Corporation, 143
neo-Classic Movement, 189
New Bauhaus, Chicago, 185
New Towns, 216, 221, 280
Nicholson, Ben, 242
Nordiska Kompaniet, 161
Norge Company, 148 n
Northampton, 189, 194, 197 n
Nottingham, 156, 270
Nylon fabrics, 143

Ollman, Wallace O., quoted, 297 n
Opel Aktiengesellschaft, 134
Orient Steamship Line, xxi, 209, 265, 308
Orrefors Glasbruk, 110, 111 and n, 147
Ovenware, glass, *see* Heat-resisting ware
Oxford, 140 n

Pamphlets
  Design and Industry: a proposal for the foundation of a Design and Industries Association (1915), quoted 193
  Design Policy in Industry (1952), 213 n
  Rules governing the Award of the Ministry's Diplomas and Certificates in Art (1951), 178 n, 213 n
  Towards Perfection, 57 n
Paris, 46, 47, 48, 120, 194, 226
Parkinson Stove Company, Ltd., xiii, 52 and n
Peach, H. H., 163 n, 193, 194
Penguin Books Ltd., 214, 244 n, 245 n
Pennsylvania Railroad Company, 149
Period styles, *see* Reproduction styles
Periodicals, 240–1. *See also* Year Books
  Architects' Journal, The, 241, 263 n
  Architectural Forum, xxix n
  Architectural Review, The, xxxvii n, 5 n, 144 n, 211 n, 237 n, 239 n, 281 n, 319; special numbers, 241
  Architektur und Wohnform (Germany), 241
  Art and Design, 207
  Art and Industry, 67 n, 240
  Art et Décoration (France), 241
  Art et Industrie (France), 241
  Board of Trade Journal, The, 202 n
  Commercial Art and Industry, 240
  Country Life, 195
  Dansk Kunsthaandvaerk (Denmark), 241

Periodicals (*cont.*)
Design, 13n, 19n, 52n, 55n, 57n, 82n, 86n, 100n, 127n, 144n, 146n, 148n, 152n, 161n, 170n, 198n, 209, 214, 218, 221, 223n, 240–1, 281n, 287n, 308
Design for To-day, 194
Design in Industry, 194
D.I.A. News, 196; quoted, 206
D.I.A. Quarterly Journal, 235n
Domus (Italy), 241
Everyday Art Quarterly (U.S.A.), 241
Form (Sweden), 241
Furniture Forum (U.S.A.), 241
Handbags Illustrated (U.S.A.), 120
House and Garden, 241
Industrial Design (U.S.A.), 241
Interiors (U.S.A.), 241
Journal of the Design and Industries Association, The, 193–4; quoted, 194, 279 and n
Journal of the Royal Society of Arts, 153n, 191
Kontur (Sweden), 241
Pottery Gazette, 109n
Schuh- und Ledermode (Austria), 120
Stile Industria (Italy), 241
Studio, The, 240, 243n, 245n
Trend in Design of Everyday Things, 194
Werk (Switzerland), 241
Persian carpets, 23
Pick, Frank, 35, 193, 202, 247, 280
Pick Council, xxxiii, 205, 207
Plan furniture (Bauhaus), 15, 71, 275, 307
Plastic, moulded, 31n, 71, 72, 73, 77, 116, 120, 122–9, 144, 145, 149, 252, 253, 261, 265, 276, 277, 304. Pls. XXXI, LIX/LXXI, LXIII–LXV, LXXIV, LXXV
Plastics, 117, 121, 144, 148, 149, 168, 170, 174, 217, 254, 257, 259, 266, 267, 278, 301. Pl. XXVI
Plywood, 4, 8, 116, 120
Plywood, moulded, 144, 145, 146, 176. Pls. IV, LXXIII, LXXV, LXXVI, LXXXIII
Polytechnic Institute, London, 140n
'Portcullis' gas fire, 53
Post-War Export Trade Committee of the Department of Overseas Trade, 1943, 208
Pottery, 92–105, 143, 144, 171, 172, 190, 234, 243, 252, 254 *et seq.*, 261, 264, 266, 269, 276, 286. Pls. XLVII–LIII, LXXVI, LXXVII, LXXXIV
Prince Albert, 190
Printing, 163 *et seq.*, 170, 177, 195n, 232, 282
Public Taste, 22, 25, 26, 57, 152, 153, 269, 271–3, 277, 279–81, 303–4, 307
Purchase-tax, 34, 35, 37, 40, 46, 47, 106n, 112–13, 116, 276, 285–6
Purchase-tax relief scheme, 34–5, 226, 288

Pye Ltd., xiv
'Pyrex' glassware, 114 and n. Pl. LVII

Questionnaires, xxxii, 156–7, 195, 197, 243, 274

Radio and Television cabinets, 69–78, 124n, 125, 128, 145, 148, 252, 254, 257, 259, 261, 264, 265, 266, 269. Pls. XXIX–XXXII
Rasmussen, Prof. Steen Eiler, 160
Rayon, 123, 203, 226–7
Records and Services
Army Bureau of Current Affairs, 196
Bureau of Current Affairs, 215
Canadian Design Index, 297n
Design Centres, 222–9
Design Folios, 214
'Design Review', 211, 218, 238, 308–9. *See also* Stock List
Deutsche Warenkunde, 210n
National Register of Industrial Art Designers, 199, 205–7, 212, 224n
Register Advice Committee (Society of Industrial Artists), 230
Register and Employment Bureau for industrial designers, 199
Registers of Designers, 224, 225, 227
Rural Industries Bureau, 87
Schweizer Warenkatalog, 210n
Stock List, 210–12, 216, 217, 221, 238, 239. *See also* 'Design Review'
Refrigerators, 57–8, 148 and n, 252. Pl. XXII
Regent Street Association, 271n
Renault motor-car (1948), 133n
'Renown' gas cookers, 52 and n. Pl. XXI
Reproduction styles, xxxvi, 9, 11, 13, 23 *et seq.*, 30, 36, 37–8, 40, 50, 58, 63, 64, 69, 70, 77, 83, 84, 129, 155, 156, 273, 277 *et seq.*, 315–16, 319
Retail Trade, 151–62, 269–71, 272, 303, 309
Rover motor-cars, 133n, 142. Pl. LXX
Rowley Galleries, 234
Royal Institute of British Architects, 191, 233
Royal Society of Arts, xxii, 179n, 189–91, 192, 201, 233, 308
Rugs, 20n, 316. Pls. X, XII
Rumpler, Dr H., 133
Ruskin, John, quoted, 109 and n
Russia, 194, 316

Scandinavia, 15, 22, 24, 176, 183, 186, 218, 286. *See also* Denmark, Sweden, Finland; Nordiska Kompaniet
Scholarships, 101, 226, 271
Schools
Advisory Committees, 177

Schools (*cont.*)
- Art Schools, 162–86, 189, 190, 298–300. Students' work, Pls. LXXVII–LXXXII
- Bauhaus, Staatliches, xx, 15 and n, 16, 50, 63 *et seq.*, 68, 71, 79, 89, 114, 145, 177, 182–6, 241, 243, 265, 315, 320
- Bauhochschule, Weimar, 114, 183 n
- Beckenham School of Art and Crafts, xxii, 176
- Birmingham College of Art and Crafts, xxi, 166–7, 174, 176, 185
- Bradford College of Art, xxii, 165, 176
- Burslem School of Art, xxi, 101, 105 n, 171–2, 299
- Carlisle Art School, 282
- Central School of Arts and Crafts (L.C.C.), xxi, 35, 167–8, 180, 227, 267, 308
- Cordwainers' Technical College, London, 174, 299
- Edinburgh College of Art, 170
- Glasgow Royal Technical College, 169
- Glasgow School of Art, xxii, 169–70
- High Wycombe School of Art, xxii, 175
- Hochschule für Gestaltung, Ulm, 184
- Illinois Institute of Technology, 185
- Institute of Design, Chicago, 167, 182, 185–6
- Kidderminster School of Science and Art, 173
- Leeds College of Art, xxii
- Leicester College of Art, 163–4, 176, 282
- Leicester Technical College, xxi, 163
- Manchester Municipal College of Technology, 171
- Manchester Regional College of Art, 170–1
- Royal College of Art, 36, 101, 169, 172, 173, 176, 179, 180–2, 184, 186, 200, 222, 265, 267, 298, 301, 308. Students' work, Pls. LXXXIII, LXXXIV
  - Report of the Departmental Committee (1911), quoted, 180 and n
- School for Jewellers and Silversmiths, Birmingham, 166, 173–4
- School of Design, proposed new, 298, 301–3
- Sheffield College of Art, xxiii
- Stoke-on-Trent Art Schools, 171–2, 185
- Stourbridge Art School, xxi, 172–3, 176
- Technical College for the Furnishing Trades (L.C.C.), xxii, 175
- Technical Schools, 167, 177, 179–80
- Vienna Art School, 49
- Walsall School of Arts and Crafts, 174, 299
- Woollen Technical College, Galashiels, 169, 170

Schott, Dr Erich, 114
Scotland, xxxi, 3, 4, 17, 19, 146, 152, 202, 203, 219–20, 286–7
Scottish Furniture Manufacturers Ltd., The, 286
Sears, Roebuck & Co., 148
Selfridges, Ltd., 153
Services, *see* Records and Services
Shipley, William, 189
Ships' furniture, 6, 7–8
Shop exhibitions, 156–7, 161–2, 195, 216, 219, 231–3, 270, 271
Silverware, 33–42, 173–4, 234, 252, 255, 256, 261, 267, 269, 279, 285. Pls. XV, XVI, LXXVIII, LXXXIV
Smith, Catterson, 166
Smith, Hamilton T., quoted, 197
Smith, Sir Hubert Llewellyn, 198, 201
Society of Industrial Artists, 191, 212, 227, 229–31
- Code of Conduct, 230 and n
- Consultant Designers' Group, 231
- Product Design (Light Engineering) Group, 231
- Register Advice Committee, 230
- Report, *Training of Designers for Industry, The* (1944), 229

Society of Industrial Designers (U.S.A.), 231 n
South Africa, 32, 111
Spode, Josiah, 96 n
Stainless steel ware, 37 n, 41, 42, 145. Pls. XVII, LXXVIII, LXXXIV
Standard Motor Company, Ltd., 170 n
Standardization, 7, 12, 15, 28, 146, 258, 261, 307
Steuben Glass Works, 111 n, 147
Stevens and Williams, Ltd., 111
Stoke-on-Trent, 95, 171–2, 195, 229, 254, 256, 280
Stourbridge, 105, 109, 172, 254, 286, 299
Streamlining, xx, 132, 133 and n, 134, 317
Streat, Sir Raymond, 223
Studebaker motor-cars, 149. Pl. LXVI
Svenska Slöjdföreningen, 110, 204, 220 n, 241
Sweden, 10, 16, 25 n, 48, 94, 109 n, 110, 111, 161, 173, 194, 220 n, 315, 316
Swedish furniture, 10
Switzerland, 48, 49, 220 n, 226

Tatra motor-car (1935), 133 n
Tax-free furniture, 6, 12, 156, 238
Taylor, Crombie, 185
Tea Centre, London, 217
Television Service, 248
Tennyson, Sir Charles, 4–5, 199, 205–6, 207, 209
Terylene fabrics, 143 n
Textiles, *see* Furnishing Textiles
The Times, 295
Thonet, Michael, 144 n
Thonet, Gebrüder, 145
Thorn Electrical Industries, Ltd., xiv, 67
Tibor, Ltd., xv, 86
Time and Life Building, 152 n
Tootal Broadhurst Lee Co., Ltd., 209
Trade slumps, 133, 148, 194, 196, 198, 283–4
Training, Industrial Art, 176–80, 298–303, 320

Travel goods, 115, 116–19, 276. Pls. LIX–LXI. *See also* Leather goods
Tristram, Professor E. W., 201
Triumph motor-car, 142
Trojan motor-car (1923), 133 n
Tubular steel furniture, 144, 145, 146. Pl. LXXIV

Unit furniture, 15–16, 145–6, 307. Pls. III, VII
United States of America, 24 n, 25 n, 49, 57, 59, 67, 69, 111 n, 116, 118, 122, 126, 127 n, 130 *et seq.*, 140, 144, 145 *et seq.*, 185, 213, 231 n, 241, 245, 265, 266, 274, 283, 284, 294, 296 n, 316 *et seq.* *See also* American Market
Universities, 86–7, 293, 294
Universities and Industrial Art, 293–4
Utility furniture, 4–5 and n, 6, 12, 66, 153, 287. Pl. II
Utility Schemes, xi, xv, 66, 236, 285–7, 319

Vereinigte Lausitzer Glaswerk, xviii, 114, 147
Versailles Treaty (1919), 283
Victorian Revival, 319
Vienna, 47, 48, 49, 120. *See also* Austria
'Volkswagen' motor-car (1938), 133 n

Wall Paper Manufacturers, Ltd., The, xvi, 87, 91 n, 307
Wallpapers, 87–92, 252, 256, 258, 259, 263, 276, 307–8, 314. Pls. XLV–XLVI, LXXXI
Walsall, xix, 115, 121–2, 174, 256, 258, 299
Walsh, Ltd., John Walsh, xvii, 111
Warboys, W. J., 208
Warner, Sir Frank, 190, 193
  Memorial Medal, 190
Warner & Sons, Ltd., xv, 86
Watch-cases, 49–50, 252, 256. Pl. XIX
Watson House Laboratories, 55–6
Weaver, Sir Laurence, 194
Weaving, carpets 17–18, 23, 173; textiles 78 *et seq.*, 165, 168, 169, 182, 266
Wedgwood, Josiah, 96, 99 n, 102 and n
Wedgwood & Sons Ltd., Josiah, xvi, xvii, 96 *et seq.*, 102, 255
Weir, Sir Cecil, 208
Weir Report (1943), *see* Council of Industrial Design
Welsh woollen industry, 86–7
'Willys-Overland' motor-car (1925), 130
Windows, metal, 27, 28, 33, 252. Pl. XIII
Wireless sets, *see* Radio and Television cabinets
Woolworth, F. W. & Co., Ltd., 277
Worcester, 95
Worcester Royal Porcelain Co., Ltd., 97, 104
Württembergische Metallwarenfabrik, xviii, 265

Yarns, fancy and plastic, 79–81, 85, 143. Pls. XXXIII, XXXV–XXXVII, LXXIII
Year Books
  D.I.A. Year Books, 194, 220 n; quoted 197 and n
  Year Book of Decorative Art, 240
  Year Book of Education, The, 1936, 183 n
  Werkbund Year Books, 194
Yorkshire, 17, 19, 26, 78, 82, 254

Made in the USA
Las Vegas, NV
28 August 2021

29159582R00254